Container Nursery Production and Business Management Manual

Julie P. Newman, *Technical Editor*

University *of* California
Division of Agriculture and Natural Resources

Oakland, California
Publication 3540

To order or obtain ANR publications and other products, visit the ANR Communication Services online catalog at http://anrcatalog.ucdavis.edu or phone 1-800-994-8849. You can also place orders by mail or FAX, or request a printed catalog of our products from

University of California
Agriculture and Natural Resources
Communication Services
1301 S. 46th Street
Building 478 – MC 3580
Richmond, CA 94804-4600

Telephone 1-800-994-8849 510-665-2195
FAX 510-665-3427
E-mail: anrcatalog@ucanr.edu

Publication 3540

ISBN-13: 978-1-60107-842-1

Editing: Linda Ribera. Design and illustrations: Celeste Rusconi, except as noted in the captions. Proofreading and indexing: Hazel White. Photo research: Evett Kilmartin. Photo credits are given in the captions. Front and back cover photos: J. H. Lieth.

Library of Congress Cataloging-in-Publication Data

Container nursery production and business management manual / Julie P. Newman, technical editor. -- First edition.

pages cm. -- (Publication ; 3540)

Includes bibliographical references and index.

ISBN 978-1-60107-842-1

1. Nurseries (Horticulture) 2. Horticultural products industry. I. Newman, Julie P. II. University of California (System). Division of Agriculture and Natural Resources. III. Series: Publication (University of California (System). Division of Agriculture and Natural Resources) ; 3540.

SB118.5.C66 2014

338.4'76359152--dc23

2014008953

UC PEER REVIEWED

This publication has been anonymously peer reviewed for technical accuracy by University of California scientists and other qualified professionals. This review process was managed by ANR Associate Editors Janet Hartin, Mary Louise Flint, Shermain Hardesty, and Bill Stewart.

Printed in Canada on recycled paper.

2m-pr-6/14-LR/CR

Contents

Acknowledgments

The planning and development of the *Container Nursery Production and Business Management Manual* was a principal activity of the University of California Agriculture and Natural Resources Floriculture and Nursery Workgroup, which began a couple of decades ago. J. Heinrich Lieth and Ursula K. Schuch (chair and past chair, respectively, of the Floriculture and Nursery Workgroup) provided leadership in establishing and developing this project; Julie P. Newman (UC Cooperative Extension Floriculture and Nursery Crops Advisor Emeritus, Ventura and Santa Barbara Counties) served as technical editor, ensuring project success by guiding and coordinating efforts throughout the process; Lorence R. Oki (UC Cooperative Extension Associate Specialist in Landscape Horticulture, UC Davis) assisted in the technical review of chapters; and Linda Dodge (Staff Research Associate, Department of Plant Sciences, UC Davis) provided superlative administrative support. More recently, the project developed into a multistate effort involving the participation of Cooperative Extension specialists and faculty from the University of Arizona, the University of Florida, Michigan State University, and Oklahoma State University. This expanded the expertise of project authors to include business management, marketing, and management of human resources.

The technical editor and authors would like to thank Don Hodel (UC Cooperative Extension Environmental Horticulture Advisor, Los Angeles County) for preparing topic outlines for the chapters, which shaped and guided the development of this manual, as well as Heather Costa (formerly UC Cooperative Extension Entomologist, Department of Entomology, UC Riverside) for her contributions to original draft manuscripts. We also thank the following workgroup members who assisted in the planning of this manual: Ann King Filmer (Director of Communications, Dean's Office, College of Agricultural and Environmental Sciences, UC Davis), Deborah Giraud (UC Cooperative Extension Farm and Community/Economic Development Advisor, Humboldt and Del Norte Counties), and Michael Reid (Professor Emeritus and UC Cooperative Extension Postharvest Specialist, Department of Plant Sciences, UC Davis). In addition, we thank industry representatives who provided additional manuscript review and technical guidance: Dana Groot (Florabunda Nursery), George Gutman (G. Gutman Consulting), Doug Shibberu (Santa Ana Regional Water Quality Control Board), and Roger van Klaveren (Generation Growers). Finally, we would like to provide a posthumous acknowledgment to two people who recently passed away but contributed to the development of original draft manuscripts and technical review of material for this publication: Robert Raabe (Professor Emeritus, Department of Environmental Science, Policy, and Management, UC Berkeley) and Gene Arthur (formerly with Do Right's Plant Growers).

Authors

Roger A. Baldwin, UC Cooperative Extension Wildlife Specialist, Department of Wildlife, Fish, and Conservation Biology, UC Davis

J. Ole Becker, UC Cooperative Extension Specialist and Nematologist, Department of Nematology, UC Riverside

Bridget K. Behe, Professor of Horticulture, Department of Horticulture, Michigan State University

James A. Bethke, UC Cooperative Extension Nurseries and Floriculture Farm Advisor, San Diego and Riverside Counties

Gregorio Billikopf, Area Farm Advisor, UC Cooperative Extension, Stanislaus, Merced, and San Joaquin Counties

David W. Burger, Professor (Emeritus), Department of Plant Sciences, UC Davis

Clyde L. Elmore, UC Cooperative Extension Weed Scientist (Emeritus), Weed Science Program, Department of Plant Sciences, UC Davis

Richard Y. Evans, UC Cooperative Extension Environmental Horticulturist, Department of Plant Sciences, UC Davis

Ben Faber, UC Cooperative Extension Farm Advisor, Ventura and Santa Barbara Counties

Jay Gan, Professor of Environmental Chemistry, Department of Environmental Sciences, UC Riverside

Darren L. Haver, Director of South Coast REC and UC Cooperative Extension, Orange County, and UCCE Water Resources/Water Quality Advisor, Orange County

Rose Hayden-Smith, UC Cooperative Extension Strategic Initiative Leader and Sustainable Food Systems and 4-H Youth, Family, and Community Development Advisor, Ventura County

Alan W. Hodges, Extension Scientist, Food and Resource Economics Department, University of Florida

John N. Kabashima, UC Cooperative Extension Environmental Horticulture Farm Advisor, Orange and Los Angeles Counties

Phil Kenkel, Regents Professor and Bill Fitzwater Cooperative Chair, Department of Agricultural Economics, Oklahoma State University

Steven T. Koike, UC Cooperative Extension Plant Pathology Farm Advisor, Monterey and Santa Cruz Counties

J. Heinrich Lieth, UC Cooperative Extension Crop Ecologist and Professor, Department of Plant Sciences, UC Davis

Deborah M. Mathews, UC Cooperative Extension Assistant Specialist and Plant Pathologist - Ornamental Crops, Department of Plant Pathology and Microbiology, UC Riverside

Donald Merhaut, UC Cooperative Extension Specialist for Nursery and Floriculture Crops, Department of Botany and Plant Sciences, UC Riverside

Julie P. Newman, UC Cooperative Extension Floriculture and Nursery Crops Advisor (Emeritus), Ventura and Santa Barbara Counties

Lorence R. Oki, UC Cooperative Extension Associate Specialist, Landscape Horticulture, Department of Plant Sciences, UC Davis

Karen L. Robb, UC Cooperative Extension County Director (Retired), Mariposa County

Ursula K. Schuch, Extension Specialist and Professor, Environmental Horticulture, School of Plant Sciences, University of Arizona, Tucson

Lawrence J. Schwankl, UC Cooperative Extension Irrigation Specialist, Department of Land, Air, and Water Resources, UC Davis

Trent Teegerstrom, Director of Arizona Federally Recognized Tribes Extension Program and Ag Economics Extension Specialist, Department of Agricultural and Resource Economics, University of Arizona, Tucson

Steven A. Tjosvold, UC Cooperative Extension Environmental Horticulture Farm Advisor, Santa Cruz and Monterey Counties

Cheryl A. Wilen, UC Cooperative Extension Area Integrated Pest Management Advisor, Los Angeles, Orange, and San Diego Counties and UC Statewide IPM Program

Photo: J. H. Lieth

Overview and History of the Nursery Production Industry

Julie P. Newman and Rose Hayden-Smith

CHAPTER ONE

Greenhouse and nursery producers market a broad variety of plant materials for commercial and residential landscapes. Most of the crops produced are ornamentals, including woody plants (such as trees and shrubs, ground covers, and vines), herbaceous plants (such as bedding plants, annual and perennial garden plants, and turf), and potted flowering and foliage plants. However, fruit and nut trees, berry plants, vegetable plants, and herbs are also commonly produced.

This manual includes basic and research-based information on all aspects of the production of landscape plants in commercial nurseries. It will serve as a valuable reference for anyone involved in California's nursery production industry, as well as those involved in the nursery industry in other states. Although the focus of this manual is container production, it also incorporates information pertaining to the production of field stock.

Plant nurseries include a range of different types of businesses, such as wholesale production nurseries, retail nurseries and garden centers, and mail-order nurseries. While providing useful information for retail nursery and garden center managers and plant brokers, this manual is written primarily for wholesale nursery growers, propagators, and contract growers, as well as those interested in starting a wholesale nursery production business. The management of retail operations is not directly addressed in this manual, and the reader who is interested in this subject should refer to other books and resources related to this topic, such as the University of California's *Retail Garden Center Manual* (Pittenger 2006). Similarly, this publication is not intended for cut flower growers or producers of food crops, although it contains related topics that may be of interest to some of these types of growers.

This chapter provides a general overview of the wholesale nursery production industry and a brief history of the industry in the United States and in California. It also examines the present state of the industry, including types of production systems and marketing strategies, the size and scope of the industry, and economic impacts. Furthermore, this chapter discusses key issues and challenges that are affecting the industry today and into the future. Finally, it presents an overview of the information needed for starting a wholesale nursery; however, the reader should refer to other chapters in this publication for more details related to this topic, including the sections on nursery site selection, business management, marketing, managing employees, environmental management, production management, and regulations.

Photo: S. Sim.

History of the Nursery Production Industry

The American Nursery Industry

Robert Prince founded America's first commercial nursery—Prince Nursery—in Flushing, New York, in 1737. Prince (and later, his sons) cultivated and sold a wide range of nursery plants, from roses to pecan trees; their enterprise became one of the largest and most important nursery operations in the country, remaining in business for more than a hundred years (Davidson et al. 2000). Early nursery and seed enterprises established practices that have characterized the industry since: conducting botanical research and exploration, as well as educating the public about plants, their cultivation, and their use. In an era with few formal educational opportunities in horticulture, the early nursery operations also provided the vital service of training growers and serving as a locus of innovation and collaboration. Many of the advances in American horticulture during the eighteenth and nineteenth centuries were made by citizen-scientists and self-taught botanists, including founding father Thomas Jefferson, who ranked botany with "the most valuable sciences" (see the Library of Congress website in the References).

Gardens have been inextricably linked with the development of the nursery industry. Some nurseries, including the one founded by John Bartram Jr. and his brother William in Philadelphia in 1777, had their origins in botanical gardens (Davidson et al. 2000).

The link between the nursery industry and the exploration of the American frontier was also strong; the Prince Nursery propagated and distributed specimens collected during the Lewis and Clark expedition, which had as one of its primary goals botanical exploration. But the foundation of the American nursery industry has roots that reach even further back. Prior to European settlement, explorers identified the enormous potential of America's physical landscape and natural resource base to provide economic benefit to European colonial powers (Cronon 1983; Sackman 2005). Many early expeditions included scientists charged with the task of collecting and cataloging plant specimens. Even in the early years of the republic, the infant nursery industry played a vital role in making the American landscape more fruitful.

The growing strength of the American nursery industry was demonstrated in 1875, when the American Association of Nurserymen was founded (Davidson et al. 2000). The founding of this organization reflected larger American social trends that were encouraging professionalization and specialization in a number of fields, including horticulture (Rossiter 1982; Kohlstedt 1978; Rybcsynski 1999). The industry's growth was also spurred by the extraordinary contributions of individuals such as horticulturist Luther Burbank, whose positive impact on California's nursery industry cannot be overstated. (His contributions are too numerous to recount here.) Other major contributors to the nation's nursery industry include botanist Asa Gray and Extension pioneer Liberty Hyde Bailey (Van Laan 1982).

The World War I era also had a significant (and positive) influence on the nation's industry; parcel post, improved transportation, and legislation prohibiting the importation of certain plants were all factors that fostered growth via increased catalog and retail sales (Davidson et al. 2000). Home-front mobilization in both World War I and II (in the form of the iconic victory-garden movement) kept the industry front and center both nationally and in California. Other post–World War II trends, including the development of a national highway system and the explosion of the American suburban model of living (with its attendant increase in the growth of the American middle class and increased rates of home ownership), created a strong demand for nursery stock and fueled growth in the industry.

A comprehensive bibliography of the history of the American seed and nursery industry is available online at the Smithsonian Libraries website (Woodhams 1999). This is an excellent resource for the reader who would like to learn more about such topics as the important leaders who shaped the industry, early nursery businesses, and the development of trade associations and trade catalogs.

The California Nursery Industry

Today California dominates national production in nursery and floral crops. Much of the perceived promise of California lies in its diverse, bountiful, and productive physical landscape, coupled with a climate that enables a wide range of plants to thrive. Throughout its history, its boosters have presented the Golden State as a virtual Garden of Eden, a place of extraordinary potential and perpetual sunshine. Names of some California cities reflect this belief, including citrus-producing Pomona, named by horticulturist Solomon Gates after the Roman goddess of plenty (Sackman 2005). While California boasts a remarkable native landscape, its garden landscape has evolved from a synthesis of native and non-native plants and cultures. It is a state not only where a diverse population of immigrants thrives but also where an extraordinarily diverse population of plants from all over the world is successfully propagated (Sackman 2005).

The eventual success of the nursery industry in California was foreshadowed by the arrival of Spanish missionaries, who founded the first of 21 missions in 1769—Mission San Diego de Alcala. It is likely that the first European-style garden in California was the one cultivated at the San Diego mission. Missions were intended not only as religious settlements but as a means to provide Spain with a military presence in the New World. Food production, then, was a vital role of missions. Mission gardens became important sites for the introduction of non-native plant species into the California landscape (fig. 1.1). These gardens also contributed to the physical and cultural assimilation of native plant species. Spanish missionaries brought new plant varieties to cultivate in the mission system, applying Native American land management practices to Spanish imports such as grapevines and oranges. In the nineteenth and twentieth centuries, oranges in particular came to represent to non-Californians the allure, beauty, and promise of the Golden State (Sackman 2005). Mission gardens also provided stock for some California gardens and orchards that eventually became commercial enterprises (Van Laan 1982).

The discovery of gold in California in 1848 brought thousands of people to the region, the new American frontier (California was admitted to the Union in 1850). By the 1850s, several commercial nurseries were operating in California. The first nursery catalog produced by a California grower appeared in 1853; it featured numerous exotic ornamentals, including birds-of-paradise, poinsettias, fuchsias, jasmines, and callas. During the decades following the Gold Rush, the growth of the state's nursery industry was reflected in California's increasing national importance as an agricultural producer; the nursery industry was vital to California's social and economic development. Challenges to the industry's prosperity did occur (including political instability due to the Civil War and a brief economic downturn), but the nursery industry continued to grow (Van Laan 1982).

FIGURE 1.1. Mission gardens were important historic sites for the introduction and commercial production of non-native plants in California. *Photo:* R. A. Estremo.

General John Bidwell's nursery, founded during this period, built on old stock taken from the now secularized mission system; his was reportedly the largest fruit tree operation in California. Bidwell's Rancho Chico Nursery catalog for 1889 provided 13 pages of tips and educational advice to its customers. Nurseryman John Rock and the California Nursery Company played a significant role in developing the state's nursery industry, but dozens of other growers also played a pivotal role in its success. Fruit growing proved to be an area of enormous strength for the industry from the late 1870s on, and its importance was institutionalized in legislation through the passage of an 1881 act designed to "protect and promote the horticultural interests of this state." This act, amended in 1909, evolved into the formation of the agricultural commissioner offices currently located in California counties. Also in 1881, the first fruit grower's convention was held in the state; the meetings of this organization led to the adoption of best practices by California nurserymen and nurserywomen (Van Laan 1982).

During the Gilded Age and the Progressive Era, there was enormous interest in horticulture, both in Europe and in the United States, which was reflected in exhibitions, conferences, gardening literature, and the formation of civic organizations supporting gardening and horticultural interests (Moskowitz 2009; Higginbotham 1990). This national trend was apparent in California as well. Women, immigrants, and people of color played an important role in the industry during this period of growth. Kate Sessions, a nurserywoman in San Diego, was a key figure in the development of Balboa Park. Theodosia Burr Shepherd operated a nursery on the hillside behind her home in downtown Ventura, and she contributed in important ways to plant hybridization and cultivation, embarking on an expanded and successful business venture selling seeds and enticing a number of East Coast companies to develop seed trials in California (Hollingsworth 1962; Van Laan 1982).

The Southern Pacific Railroad proved crucial to the growth of California and its nursery industry. In addition to being a major landowner and California booster (the railroad published brochures and pamphlets touting the state's desirability), the Southern Pacific Railroad provided the transportation needed to open new markets to California growers. In 1877, the first railroad shipment of California oranges, produced by Los Angeles grower William Wolfskill, was sent to St. Louis (Sackman 2005; Van Laan 1982). Growth was explosive: in 1889, 1.2 million boxes of oranges were shipped from California; in 1899, 6 million boxes; and by 1909, a staggering 14.5 million boxes (Sackman 2005). It would not be an overstatement to say that, by this point, the nursery industry was vital to the economic development and growth of California. Magazines and advertisements promoted California and its nursery products, generating images that attracted new residents and new businesses. The Southern Pacific Railroad began the publication of *Sunset* magazine, which sold an idealized California landscape to both residents and prospective residents alike. It can be argued that the California nursery industry—nearly as much as Hollywood—has influenced the nation's view of California as the land of eternal sunshine and endless promise (Sackman 2005).

The University of California's Division of Agriculture and Natural Resources (UC ANR) played a critical role in modernizing the California nursery industry. Its seminal publication in this field, *The U.C. System for Producing Healthy Container-Grown Plants* (Baker et al. 1957), shifted the paradigm for California growers. Through robust research, the publication of this book, and Extension work demonstrating best practices, UC ANR helped lead California's nursery producers from being a primarily field production industry to a container production industry. The UC system, as the protocol came to be called, emphasized clean soil (and UC-type soil mixes), clean stock, and best practices in sanitation. Chapter 16 (by R. H. Sciaroni and J. W. Huffman) of the book demonstrated "extension" in action by providing case studies and examples of grower adoption of the UC system.

In 1974, *Environmental Plant Production and Marketing* by Tokuji Furuta was published in California. This important work stressed a highly detailed systems approach to plant production, emphasizing planning of both the smallest and largest details as being vital to efficient operations in production nurseries. Furuta also emphasized the balance of labor, machines, and practices, as well as environmental interactions. His work reflected the evolving nature and continued innovation of the California nursery industry.

In an increasingly complex age, the nursery industry faces challenges related to pests and disease; labor; cost management and profitability; water availability and quality; environmental issues; and regulation, to name a few (see the section "Issues and Challenges" later in this chapter). But the industry continues to thrive, not only due to the unique qualities of the physical geography and climate that characterize California, but also due to the innovation—past and present—of its nursery growers.

Production and Marketing Strategies

Numerous production and marketing strategies are used by the nursery industry. While most nursery businesses offer a diverse array of plants, others provide specialized products for niche markets. Examples of specialized products include uncommon plants, drought-resistant plants, native plants, specimen-size trees and shrubs, palm trees, tree fruits and nuts, ground covers and vines, bedding plants, herbaceous perennials, cacti and succulents, roses, and daylilies. Some nurseries specialize in propagation, providing starting materials such as seedlings, cuttings, plugs, liners, bulbs, or grafted plants for other nurseries.

Plants can be grown in greenhouses, or greenhouses can be used primarily for propagating and growing young plants. In many nurseries plants are grown strictly outdoors or in shadehouses, using plugs and liners produced by specialized propagators.

Wholesale producers market plants to other nurseries, landscape contractors, institutions, agricultural producers, and retail outlets such as garden centers, supermarkets, mass merchandisers (e.g., Walmart, Target), and big-box home centers (e.g., Home Depot, Lowe's). Many wholesale nursery businesses have diverse markets that may include retail sales or landscape design and installation services. Moreover, some wholesale producers generate retail sales by also marketing plants locally at farmers' markets or conducting national mail-order sales.

Production Systems

Two types of systems are commonly used to produce nursery plants: field production (inground growing) and container plant production. A hybrid type of production system that is sometimes used is the pot-in-pot system, which combines some of the features of inground growing with container production.

Field Production

Field-grown nursery stock (field stock) is grown in the ground from seeds or transplants, then lifted by hand or machine and marketed as bare-root, balled and burlapped, or containerized plants. In field production, soil productivity is essential. Nurseries prune the roots of field-grown plants to reduce vegetative growth, develop more compact root systems, and aid in the development of dormancy (Davidson 2000). Survival rates after harvesting may directly depend on the amount of root pruning that is done during production (Avent 2003).

Bare-root plants

A bare-root plant is a plant that is sold with the roots exposed rather than in soil. Bare-root crops include roses, small deciduous shrubs and trees, deciduous berries, small ground covers, herbaceous perennials, and ornamental grass divisions. Bare, woody plants are grown in soil and dug in the fall for shipment in late winter or early spring. These plants should only be planted, harvested, and transplanted while dormant. Cool, moist conditions are necessary, as plants are stored and shipped without attached soil or packaging protection for the root system. The primary challenge in working with bare-root plants is keeping the roots from drying out. Due to the potential for desiccation, few evergreen shrubs or trees are harvested bare root, with the exception of small conifer liners for Christmas tree planting and reforestation. A major advantage of bare-root production is the light weight of harvested plants and the relatively low cost (Eaton and Appleton 2009).

Balled and burlapped plants

Balled and burlapped (B&B) plants are evergreen and deciduous woody plants that are dug with the soil mass attached; the root ball and soil is covered with heavy burlap to keep it together during shipment and sale. Hand-harvesting requires trained staff, and mechanical harvesting requires hydraulic tree spades and other equipment. Although B&B plants are usually harvested when dormant, recent improvements in holding and acclimating techniques to keep balls moist and stimulate summer root growth have allowed increased digging during the growing season (Eaton and Appleton 2009). The B&B treatment can be used for much larger plants than are harvested bare root; some landscapers prefer transplanting B&B plants to container-grown plants to avoid some of the soil interface problems that develop when plants are container grown in soilless media (Eaton and Appleton 2009). B&B plants are heavy—larger plants can weigh more than a ton; unless they are transplanted into containers, they are typically marketed only over short distances from the production fields because of the weight and extra shipping costs.

Container Production

Container plants are produced by various methods. These include

- direct seeding into containers
- direct sticking of cuttings into containers
- transplanting rooted cuttings or liners into containers
- digging up and transplanting field stock into containers

Typically a soilless growing medium is used (see chapter 5). Plants grown in small pots are often placed in structures for protection from undesirable environmental conditions (see chapter 3), while plants in larger containers are mostly grown outdoors in full sun. Most nursery plants used to be grown in the field as bare-root or balled and burlapped plants up until the 1960s and 1970s. Today, however, most of the ornamental plants produced in the United States and in California are container grown.

The transition to containers occurred for many reasons. Soil productivity is not as important as in field production; container plants are more convenient for the consumer and easier to harvest from the nursery; and container production provides a method for producing plants that are difficult to transplant. In addition, container production requires less acreage and generates significantly more sales per acre than field-grown nursery stock due to more efficient use of space. Improved crop uniformity in container production, as a result of better control over factors affecting plant growth, facilitates harvesting operations and enables the container block to be moved or consolidated as plants are removed for shipment or die. In contrast, random harvesting and sporadic plant death in field production result in inefficient space allocation because blocks of plants cannot be consolidated. Moreover, container production takes advantage of year-round growing and marketing, compared with the more seasonal harvesting of field-grown nursery plants, and shipping is less expensive than in the case of ball and burlapped plants (Davidson 2000; Diver et al. 2008; Eaton and Appleton 2009).

On the other hand, container production requires more intense management than field production because plants in containers are more subject to moisture loss and temperature extremes, as well as to blow-over during windy conditions. Field production offers growers in colder climates the option of avoiding or minimizing winter protection methods. Also, the labor required for field production is generally less intensive than for container production (except when harvesting), and it is not as constant. Furthermore, growing in containers can lead to circling roots and problems such as tree failure in the landscape. For these reasons, field production continues to be practiced in many states, including California, and it remains the dominant type of production in some areas.

A comparison between field and container production systems, adapted from Eaton and Appleton (2009), is provided in table 1.1.

Pot-in-pot system

The pot-in-pot system is a specialized type of container production that began in the 1990s, using inground growing to reduce some of the problems associated with conventional container production, such as root zone temperature extremes and blow-over of large, top-heavy trees. This system involves burying a holder pot, or moat pot, in the ground and placing a containerized plant inside this pot. Drainage of these inground pots is crucial and is generally maintained by the placement of drainpipes beneath the rows of moat pots. Some form of microirrigation (see chapter 7) is necessary to avoid inefficiency in water delivery. Overhead sprinklers are inefficient because the large moat containers (generally ranging from 5 to 15 gallons) are spaced far apart, resulting in excess water in the areas between containers.

The pot-in-pot system allows predominantly container-based nursery operations to grow larger nursery stock while still maintaining tight controls over irrigation, fertilization, and other cultural practices of container production. The main disadvantage of the pot-in-pot system is the high initial cost of the moat pots. The moat pots are a long-term investment and are intended to last at least 15 years (Diver et al. 2008). Although the capital outlay can be a challenge for many growers, the pot-in-pot system is the least expensive production system when compared with aboveground containers and field production (Adrian et al. 1998). However, the flexibility in moving and spacing plants is not as efficient and convenient as with conventional aboveground containers. Furthermore, access to certain pesticides may be limited since pesticide products may only be labeled for conventional field or container production.

A variation of the pot-in-pot system is the above ground system (AGS), which is a double-container system for use aboveground. A production container nests in a double-walled holder container that serves to insulate the plant's roots from temperature extremes. A ballast can be added to prevent blow-over. An AGS is easier to harvest than the inground pot-in-pot system (Eaton and Appleton 2009).

Types of containers

Many types of containers are available for the commercial production of greenhouse and nursery plants. When selecting containers, factors to consider include cost, durability, availability, shipping capacity, how the container affects the moisture content and temperature of a growing medium, design features that control root growth, and how the container suits the particular needs of the nursery manager and customers. Following is a description of some of the common types of containers:

- Conventional containers. Most containers used today are blow- or injection-molded plastic containers in sizes ranging from small containers and cell packs (for propagation and production of small plants) to containers of 1 gallon to over 200 gallons. Containers for outdoor nursery stock are typically round and black. A common problem with conventional containers is circling root formation. Transplanting plants at the right time, prior to root circling, is therefore important and is a common best management practice (BMP). When using conventional containers, those with vertical ribs help reduce circling roots, while those with horizontal ribs promote circling root formation. White and light-colored containers produce more root growth than black containers; square pots help keep plants from becoming root bound.
- Engineered containers that reduce root circling. To

Table 1.1. Considerations for field versus container nurseries

	Field (inground) nursery	Container nursery	Notes
Site requirements	larger land area requirement; not all sites usable; soil productivity and good internal drainage are important	smaller land area requirement; ability to use sites unsuited to field production, but good drainage off the bed is critical	—
Number of employees	less labor-intensive; 1 employee per 5–10 acres	higher labor requirement; 1 employee per acre	field production has more seasonal employees
Economic return per acre	less, with fewer plants per acre and longer production cycle	greater, with more plants per acre and realized sooner	—
Production cycle	longer (1–6+ years)	shorter (6 months–2+ years)	—
Production costs per acre	fewer costs for site preparation, labor, and maintenance	greater investment in production area construction, containers, water, substrate, and larger labor force	—
Intensity of management	better soil nutrient reserve and pH buffering result in less intensive nutrient management; less irrigation; root pruning may be necessary	daily irrigation may be necessary (irrigation frequency depends on temperature and plant size); monitoring for blow-over (plants knocked down by wind); more intensive nutrient management (nutrients deplete rapidly, salts may require leaching); frequent repotting may be necessary to avoid root-bound plants; protection from temperature extremes; runoff management	—
Seasonality	periodic with limited harvest during periods of active growth	planting and harvesting year-round independent of weather	both types of production have some downtime in winter
Types and sizes of plants	preferred production system for large-caliper trees; limited holding time after harvest	more efficient production of small shrubs, annuals, and perennials; plants can be repotted to larger containers if not sold; some plants easier to grow or harvest in containers	pot-in-pot and other hybrid production systems allow larger plant sizes in containers
Growing area preparation and maintenance	generally less preparation; unique concerns for removal of acreage from production for soil rebuilding and replacement of soil removed by B&B harvesting	more extensive and expensive site preparation for drainage and water recycling; container beds require a gravel, plastic, or fabric covering	—
Summer stress	root temperatures buffered by soil; less concern for heat stress	greater summer heat stress due to high root zone temperatures in containers and heat produced or reflected from container beds; may use shade structures or syringing (short-cycle overhead irrigation)	—
Overwintering	root temperatures buffered by soil; little protection is generally required	frost protection practices or protection in overwintering structures may be necessary	—
Harvesting	smaller percentage of plant's roots harvested (especially trees); restricted to seasons when soil moisture is adequate and plants are dormant; plants are dug by hand or machine, bare root or with soil ball; root protection required during holding (cold storage, heeling in, balling-in-burlap)	less root loss; possible any time by simply loading plants from their production beds	improvements in irrigation and postharvest handling have allowed some summer digging of field-grown plants
Shipping	bare-root, field-grown plants can be bundled or boxed; large B&B plants spaced on trucks or tied to individual pallets	containers can be stacked can-on-can, boxed, or racked; lighter weight reduces shipping costs	long-distance shipping is possible for container-grown plants because of lightweight media, but cost is prohibitive for field-grown B&B plants
Equipment needs	requires more expensive machine-harvesting equipment; may require larger irrigation and spray equipment	production efficiency requires substrate-mixing and container-filling equipment	—

Source: Eaton and Appleton 2009.

Photo: E. Zhukova.

help reduce or eliminate the problem of circling roots that often develop in smooth or slick-walled plastic containers, a variety of containers have been engineered to deflect or redirect roots; other containers kill root tips when roots encounter holes in the container walls (air-root pruning). Growers should test these types of containers in their own production systems, as they may be ineffective in reducing circling roots as well as expensive and difficult to incorporate into a nursery production system.

- Biodegradable fiber containers. Some bare-root and containerized field stock, small garden plants, and plugs are placed in fiber containers that are biodegradable. These "green" plant containers consist of pressed wood fiber, peat moss, or other organic materials, such as rice hulls. No drainage holes are necessary, and the pot disintegrates into the soil. This eliminates the need to remove the plant from the pot when planting, and it increases the success rate of transplanting.
- Molded fiber containers. Molded fiber cellulose pots are commonly used for small plants and hanging baskets. They typically have thicker walls than many plastic containers, and cellulose has good insulation properties. They may offer better protection against heat and cold than standard plastic containers.
- Inground fabric containers (grow bags). Inground fabric containers are made from porous, synthetic fabrics. They permit exchange of moisture between the soil inside the container and the surrounding soil, while preventing some or all of the plant's roots from growing outside the container. Due to root restriction within the fabric container, drip irrigation is essential. Fabric containers are often removed prior to sale and the root balls containerized to ensure that roots will not be restricted when the plant is transplanted.

Crop Scheduling

Scheduling of nursery and greenhouse crops depends on many factors, such as plant type, production system, temperature, light, day length, and the length of the growing season. In outdoor production of nursery stock, the fastest production cycle is for small 1-gallon container plants, which are usually grown from liner to sale in 1 to 2 years. Bedding plants have a much quicker turnover (measured in weeks), and slow-growing plants, such as Japanese maples, dwarf conifers, cycads, and palm trees, can take years. Since most nurseries market many types of nursery plants, production cycles are diverse. Managers have to implement organized schedules so that the appropriate number and types of plants are ready when needed by the customers. A clear scheduling plan will ensure sufficient production space and supplies. Computer software scheduling programs often used by medium and large wholesale nurseries can be a valuable tool in scheduling diverse crop production cycles.

Value, Size, and Economic Impacts of the Industry

United States

The U.S. nursery and floriculture industry exhibited significant economic growth over the decade preceding the recent recession. In 2006, the industry was valued at nearly $17 billion, with close to a million open acres of nursery and sod production and over 26,000 acres of greenhouses. California has continuously led the nation in total annual sales of nursery and floriculture crops (accounting for approximately 22% of the total sales in 2007), and Florida typically ranks second (close to 13% in 2007) (NASS 2007a, 2007b; Carman 2011).

The largest portion of U.S. nursery and floriculture industry sales is from outdoor nursery stock along with bedding and garden plants. Outdoor nursery stock was valued at over $6

Photo: S. Paisley.

billion in 2007, representing nearly 40 percent of the total industry value; the top three producing states were California, Oregon, and Florida (NASS 2007a, 2007b). Bedding and garden plants were valued at nearly $4 billion and made up approximately 23 percent of the total industry value. California, Michigan, North Carolina, Texas, and Ohio were the top five states in this category, accounting for approximately 79 percent of the total sales.

California

California is the center of nursery production in the United States. California nurseries pioneered container production of plants with the development and improvement of soilless media, drip irrigation, and recycling of excess water. California leads the nation in the production of plants in containers and the production and sale of broadleaf evergreens, fruit and nut trees, ground covers, and roses (Davidson et al. 2000).

California's estimated cash receipts from floriculture and nursery crops increased from nearly $2 billion in 1992 to over $4 billion in 2007, although the economic recession resulted in a plunge to approximately $3 billion in 2008. The average nursery in California had sales of over $1 million in 2007, which was over twice as much as the next highest states, Florida and Texas (Carman 2011).

Nursery production is an important segment of California's overall agricultural output and farm income. From 2002 through 2008, nursery crops ranked number three for every year except 2003 (when they ranked number two) and 2008 (when they ranked number four).

Nursery and greenhouse products are produced throughout California, but production tends to be concentrated in the Central Coast and South Coast counties. Overall, 15 counties produced over 85 percent of California's total nursery, flower, and foliage crops in 2008. In contrast with other agricultural commodities, nursery, flower, and foliage crops are most highly produced in counties that are also California's most populated ones.

Seven counties consistently produced over $100 million of nursery crops (which excludes cut flowers) from 2005 through 2008. These counties, arranged from largest to smallest producers, are San Diego (which dominates the industry with nearly $1 billion in value for nursery products), Ventura, Monterey, Riverside, Orange, Los Angeles, and San Mateo. Nursery crops were the number one ranked crop in value of production in San Diego, Orange, Los Angeles, and San Mateo Counties.

California's floral and nursery sector is closely intertwined with other sectors of the state's economy, and changes in flower and nursery production have income ripple effects throughout the state. For example, the estimated total economic impact of the sector (including both direct and secondary effects) increased steadily from an output of approximately $10 billion and 168,867 jobs in 2001 to an output of over $13 billion and 217,557 jobs in 2007. With the recession in 2008, total output and number of jobs both decreased by approximately 12 percent. However, even with reduced employment, more than one out of every hundred jobs in California during 2008 and 2009 could be attributed to the direct and indirect impacts of California nursery production and retailing (Carman 2011).

Issues and Challenges

Over the past decade, numerous political, economic, regulatory, and environmental issues have led to significant changes in the industry as well as dramatic increases in the cost of producing nursery plants. Moreover, these issues will continue to impact future production and affect the way business in nurseries is conducted. Some of the current major issues facing the industry are discussed below.

The Economy

The nursery industry has gone through several cycles of prosperous growth and decline in the last decades. The national and local economy significantly influence the nursery market, and the demand for nursery products is closely tied to construction, the unemployment rate, and population growth. Wages and unemployment regulate the consumer's buying power, while housing and other construction provide important market outlets. During a recession, as consumers, businesses, and homeowners reduce spending (in response to sharply reduced home values, loss of jobs, rising unemployment, and reduced incomes), the sales of landscaping plants go down. Conversely, when the economy is strong, sales for landscape plants increase. For example, while the California floral and nursery sector's ties to the real estate industry contributed to 15 years of uninterrupted sales growth from 1993 to 2007, the rapid decline of housing values and the ensuing financial crisis beginning in late 2007 resulted in a decrease of floral and nursery sales of nearly 20 percent in 1 year (Carman 2011).

The overall trend in the past decade has been an increase in nursery production costs (e.g., fuel, labor, plastic, health benefits), which have risen much greater or faster than the selling price of ornamental plants. Moreover, during a recession, profit margins are further reduced due to reduced sales. Further, the acquisition of working capital is a major challenge for nursery businesses in a downturn economy. Even businesses and industries with good credit may have difficulty obtaining short-term loans to support good cash flow and longer-term loans to buy new equipment or expand facilities during a recession.

Labor

Labor cost and immigration labor reform are key challenges for nursery businesses. In most production systems, labor represents the greatest portion of the total cost of producing nursery crops, and these costs are increasing. Crops require a substantial amount of hand labor, although automation and mechanization can be used to significantly reduce labor costs (see chapter 4). Labor is also required to hire, train, and supervise employees. Furthermore, pressure to provide retail customers with service in addition to product continues to impact labor costs.

Competitive labor costs depend on immigrant labor, and the size of the labor pool is tied to state and federal immigration regulations. Employment, labor, benefits, and immigration laws are constantly changing. This requires consistently staying abreast of these changes and implementing the required adjustments to business practices.

Understanding the personal work preferences of managers and workers, effectively directed to achieving business goals that are clear and well articulated, is essential to hiring, developing, and retaining good employees. Furthermore, cultural diversity influences many aspects of American businesses and is increasingly important in horticulture management. (See chapter 20 for more information on labor management.)

Water

Water availability and cost, water quality, and water regulations are important issues in the production of ornamental plants. Nursery production depends on reliable sources of high-quality irrigation water, as it can require large quantities of water to produce many container nursery crops and daily irrigation may be necessary, especially during hot or windy weather. Many states, including California, have limited and expensive water supplies. Water shortages in urban areas and local landscape-watering ordinances are also impacting consumer use and selection of landscape plants. In response to water shortages and high water prices, many growers have implemented efficient irrigation practices (see chapter 7 for details).

Poor water quality can lead to excessively high salts and pH extremes in the growing media that may have drastic effects on the availability of nutrients. Furthermore, poor water quality causes diseases, phytotoxicity, plugging of irrigation devices, algae growth, objectionable mineral deposits on leaf surfaces, and other problems. (Refer to chapter 6 for information on managing water quality.)

The Clean Water Act (CWA) of 1972 addresses water pollution and water quality of surface waters on a national level. The use of pesticides and fertilizers in greenhouses and nurseries can contribute to nonpoint source pollution. Water bodies are considered impaired when they are too polluted or otherwise degraded to support their designated and existing uses. A total maximum daily load (TMDL) is the maximum amount of a pollutant that a water body can tolerate on a daily basis while meeting established water quality standards. TMDLs affecting agricultural activities have been established and implemented for impaired water bodies in many states. Furthermore, an increasing number of state and local water quality regulations have been implemented and are being enforced. These impact nursery operations in many areas of the country, including California. Consequently, nutrient management, irrigation water recapture and reuse,

Photo: J. P. Newman.

and mitigation of nutrients and pesticides in runoff water are of increasing importance to nursery operations. Although implementation of practices to protect water quality can lead to increased production costs, practices such as more efficient irrigation and water recycling have been shown to reduce the cost of water and fertilizers. Many nursery growers have already implemented effective management practices for reducing and managing runoff and leaching. See chapter 8 for more information.

Exotic Invasive Pests and Diseases

The speed and frequency of international travel today, combined with the sheer volume of imported food, plants, and other agricultural commodities and materials, have increased the introduction of exotic pests and diseases. Furthermore, some introduced plants have become pests, or noxious weeds, themselves.

Exotic species are plants, animals, and microorganisms that are not native to a particular region. Some exotic species are able to rapidly colonize an area and become serious pests, often because they are no longer being controlled by predators or diseases that limited their numbers in their native habitat; these exotic species are often called exotic invasives. Once established, exotic invasive species are extremely difficult to eradicate and can cause not only ecological disruption but economic problems as well. Invasive plants are a distinct group of generally non-native weeds that impact native plant and animal communities by displacing native vegetation and disrupting habitats as they become established and spread over time. Invasive plants are much more problematic in natural environments than garden weeds because of their ability to disperse, establish, and spread without human assistance or disturbance. The annual damage and effort to control invasive species in the United States alone has been estimated to cost more than $138 billion (Pimental et al. 1999). Economic costs of invasions do not consider the equally important impacts to natural ecosystems.

International plant exploration, introduction, and improvement are important to the nursery industry and vital to new plant research and development. Also, foreign sources of propagative material such as dracaena canes and geranium cuttings are increasingly important to domestic nursery and greenhouse growers. On the other hand, imported plant stock that is contaminated has led to the inadvertent introduction of invasive species that can cause serious economic damage to ornamental crops, as in the case of southern bacterial wilt (*Ralstonia solanacearum* race 3 biovar 2) (Lemay et al. 2003). Moreover, since the nursery industry markets nearly all plant types, a significant share of the burden of responsibility for intra- and interstate movement of potentially infested plants lies with this industry. This requires exceptional management practices and, in some cases, eradication and certification. Consequently, the introduction of new arthropod pests, diseases, and weed species has posed major challenges in the form of shipping restrictions, quarantine regulations, and the development and implementation of new pest management programs, costing the nursery industry billions of dollars. Furthermore, concerns and regulations regarding non-native, invasive nursery crops are also mounting and could impact sales of economically important crops.

Many exotic invasive species are of major concern to California nursery growers. For example, the glassy-winged sharpshooter, Diaprepes root weevil (fig. 1.2), light brown apple moth (LBAM), and *Phytophthora ramorum* are exotic invasive species that reproduce on many host plants. Thus, they have a greater impact on the nursery industry than exotic invasives that affect a limited number of host plant species. Moreover, although many invasive species mostly threaten other agricultural crops or U.S. forests (e.g., LBAM, *P. ramorum*), all invasives impact the nursery industry due to the risk of spreading these species through intra- and interstate trade as well as international trade. More information on invasive species can be found in chapter 17.

FIGURE 1.2. Diaprepes root weevil is an example of an exotic invasive species that reproduces on many host plants. *Photo:* J. N. Kabashima.

Energy

When energy costs and resulting food prices increase, overall consumer demand for nursery products declines. Moreover, nursery operations are energy intensive, and energy costs affect fuel, fertilizer, plastic containers and other plastic products, and production transportation costs. In order for nursery businesses to compete in a national and world economy, implementation of energy-efficient practices, such as those provided in Bartok's *Energy Conservation for Greenhouses and Field Nurseries* (2009), are necessary. Moreover, an energy audit can pinpoint inefficient areas where money can be saved, and most power companies offer an audit for reasonable charges or even at no cost.

In addition to implementing technologies and practices to conserve energy, proactive growers are using alternative or renewable energy sources for pumping water, cooling greenhouses, and reducing overall electrical use. Grants, loans, tax rebates, and other financial incentives are available from various sources, such as the USDA's rural energy program, the Internal Revenue Service (IRS), and local utility companies. Furthermore, automation and mechanization can be used in the nursery to reduce energy costs (see chapter 4).

Nurseries that distribute a large portion of their products to markets in other states are especially vulnerable to rising transportation costs. The cost for driving a refrigerated semitrailer filled with plants across the United States has increased significantly in recent years. Some growers are addressing this challenge by using transportation cooperatives, which can help in consolidating orders and negotiating better freight prices. Moreover, local distribution of products reduces transportation costs and the overall carbon footprint of the operation.

Other Issues

Many other issues challenge greenhouse and nursery producers. Examples include the following:

- Compared with crops grown on larger production acreage, greenhouse and nursery crops are generally considered "minor use" by chemical companies. Significantly fewer pesticides are registered for use or renewed for registration in the case of greenhouse and nursery crops because the cost of registering pesticides is not offset by the expected pesticide sales revenues. Moreover, revenue from pesticide sales may not offset the potential liability of the pesticide registrant if a high-value nursery crop is injured by the pesticide product. Furthermore, due to environmental issues, the greenhouse and nursery production industry has lost a number of pesticides and fumigants that were commonly used in the past.
- Nursery crops and facilities can be severely damaged by catastrophic weather events, such as drought, freezes, tornadoes, and hurricanes. Crop insurance (see chapter 18) is recommended, especially in high-risk areas.
- Consumer attitudes, preferences, and behavior are changing at an increasingly fast pace. Timely, useful consumer research is needed to help nursery managers understand current and future customers.
- According to the California Nursery Growers Association, crime is an increasing problem for nurseries in urban America, with organized gangs and petty thieves heisting commodities that include copper wire, metal, palm trees, and gasoline. In 2007, agricultural theft on U.S. farms was estimated at $5 billion; the California Farm Bureau reported $10.7 million lost in nine Central California counties (USAonWatch 2009). As many farmers do not report thefts, it is estimated the actual farm loss is 9 to 10 times greater than what is being reported.
- Increasing costs and regulation requirements are making it more difficult for small businesses to compete, due to the economy of scale. Also, the mass marketing of nursery and floricultural crops has caused an increase in the overall size of growing operations. While many small to midsize nurseries still exist, the trend is toward a decrease in the total number of operations and an increase in larger firms that expand by purchasing and consolidating smaller businesses.

Starting a Nursery Business

Many people have considered starting a nursery business, and interest seems to peak when the economy is strong and the market for nursery products and services is expanding. However, growing nursery crops requires hard work, long hours, and constant responsibility. Plants are perishable commodities that require daily care and immediate attention during inclement weather or emergencies such as a break in a water line or a power outage. Furthermore, starting a nursery business is a high-risk venture requiring a large capital investment to cover costs such as land, equipment, buildings, supplies, labor, and plants. For these reasons, those interested in starting a nursery business are strongly encouraged to learn as much as they can about the industry and to acquire the skills necessary for starting and managing a nursery business before embarking on this endeavor. If you already have experience and a strong knowledge of plant culture and business management, starting a new nursery or purchasing an existing business could be a viable option, depending on the economic climate and your financial situation.

Required Skills

Successful management of a production nursery requires two different types of skills: first, knowledge and experience regarding the technical aspects of growing plants; and, second, business management and leadership skills. The technical aspects of growing plants include all phases of plant production, from propagation to postharvest handling, including management of the environment (chapters 3 and 10), plant nutrition and irrigation (chapters 6 and 7), and pests and diseases (chapters 11 to 17). Furthermore, nursery managers must know the length of time required to produce marketable crops and how to schedule planting so that the proper number of each species is available for sale at the appropriate times.

Nursery operators must manage a market-driven operation, using sound business practices. Success depends on how well managers listen to their customers and use that information in making decisions on what to grow and what services to provide (see chapter 19). Managers must have effective sales and customer relations skills, and they must be able to analyze and organize marketing channels and distribution of products. Effective labor management and leadership of managers can be especially challenging for new nursery businesses, as working with people is often more difficult than working with plants (see chapter 20). Financial planning and management is another critical aspect of a nursery business in today's marketplace. Especially during a recession, the survival of a nursery operation depends on shrewd fiscal management of all aspects of the business, including production, labor, customers, marketing, and distribution. This requires knowledge and access to expertise in accounting and legal matters as well as the ability to analyze financial reports (see chapter 18).

Business Focus

The focus of the business is one of the most important decisions in the development stage of a nursery. A clear purpose with a well-defined focus is critical to the development of a business plan and to obtaining financing, suitable land, and appropriate permits and licenses. To determine the business focus and objectives of the nursery operation, many questions need to be answered. Following is a discussion of some of the most important questions.

What product will the nursery sell? Is there a market demand and sufficient investment return for the product? One of the most important things to consider before starting production is what crops to grow. In making these decisions, it is critical to consider the crop's marketability. This requires determining customer preferences for products and services, market demand, future market demands and trends, and potential niche markets by examining various types of information. Examples of information sources include local and national industry associations, commercial publications, and market research. Furthermore, competitors need to be identified and a feasibility study conducted to determine if the demand is sufficient to support another business in the targeted area (Schuch 2006). Finally, it is important that the crop can be sold at a price that will offset the cost of producing and marketing it. You may be able to produce a crop, market it, and sell it, but you also have to be able to turn a profit.

What types of production systems are required? In addition to crop selection, it is important to determine if greenhouses, shade structures, or other types of protected facilities are required and if plants will be produced in containers or in the field. Although growers may select several different types of production systems and growing facilities, changing production type or plants after developing the nursery is not advisable and can have disastrous effects. Water rights, building permits, zoning restrictions, and loan payback schedules may all be affected if different equipment, supplies, or structures are needed. They may also be affected if the production schedule and time to market are changed (Schuch 2006).

Who are the customers that will buy the product, and what are they looking for? Customers should dictate the type of product that is produced. Retail garden centers usually want small plants in 1- to 3-gallon containers, while landscaping firms and landscapers want larger plants. Mass merchandisers usually want large volumes of a good mix of fast-moving materials. To better serve their customers, wholesale businesses need to also understand who the final consumers are (i.e., the customers' customers). Types of customers within the target market can be determined using various sources, such as census and local Chamber of Commerce data. Constant monitoring of customer characteristics and purchases is crucial. See chapter 19 for more information on this topic.

How will the products be marketed and distributed? Will you use wholesale marketing methods, retail marketing methods, or a combination of both? The marketing method(s) used will impact the strategy, distribution, and type of advertising or promotion (e.g., sales promotions or promotional events, paid advertisements, publicity, word of mouth) that you use. The tendency in the last century has been to increase the geography for the channel of distribution, and the marketing of ornamental products has become international. However, the cost of shipping goods will determine a nursery's ability to compete in distant markets and will thus affect its customer base. Most wholesale nurseries in California still sell the bulk of their product within the state, but in other states (e.g., Oregon) most of the product is exported out of state. Especially as freight costs increase, a trend toward more local production may ensue. See chapter 19 for more information on marketing and distribution of products.

Legal Structure

Determining the legal structure of the business is critical because the legal structure may affect the actual cost of doing business, the day-to-day operation and management, the type and amount of taxes, and the level of exposure to risks and liabilities. In California, most nursery businesses are sole proprietors, but the number of corporations is increasing (Carman 2011). For a full discussion on legal structure, see chapter 18.

Business Plan

A business plan is a set of guidelines for operating a business that can assist in financing the operation as well as tell the story of a business. For example, a business plan for a nursery business describes what the nursery does, who is involved, who is included in its market, and why the nursery business should be successful against the competition. Developing an organized business plan is an essential step in starting a business. See chapter 18 for information on how to prepare a business plan.

Site Selection and Location

Carefully considering the location and site characteristics for a future nursery is critical. Location and site have numerous effects on the nursery and on the production and marketing of plants, including

- water availability, quality, and cost
- the cost of land
- the types of crops that can be grown
- decisions on marketing, employee hiring, and future expansion

Detailed information on this topic is the subject of chapter 2.

Financing

Establishing a nursery requires start-up funds for creating the necessary infrastructure to produce plants (e.g., land, permits, equipment, structures, supplies, utilities, insurance, licenses, inspections, labor) and for developing a customer base (e.g., specialized reports, consulting). Sources of money for the establishment of a nursery are usually personal savings of one or several owners as well as borrowed funds. The amount owners invest in the nursery indicates the commitment of the individual to the business and may determine the amount of credit from financing institutions (Schuch 2006).

A number of options exist for obtaining borrowed funds. Limited partnerships may be developed in which a number of creditors put up the capital needed for the nursery, and they share in the business decisions and the profits. For example, a nursery may engage in contract growing, a procedure by which a nursery agrees to grow a particular crop at a specified price for another firm and in return receives the financing necessary to produce such a crop (Schuch 2006).

Furthermore, growers could potentially borrow money from a variety of sources. Private banks, production credit associations, equipment dealers, suppliers, cooperatives, packers, shippers, grower associations, and insurance companies are some examples of private organizations and institutions that lend money for agricultural operations. In addition, some government agencies and nonprofit organizations can provide financial assistance through various programs, contracts, or grants (e.g., USDA Farm Service Agency, Farm Credit Services, and the Natural Resources Conservation Service). These programs usually lend at lower interest rates and can include direct loans, cost-share agreements, or contracts and loan guarantees through private banks. The amount of a loan and specific terms should be negotiated with several institutions, preferably those experienced in agricultural loans, to find the most favorable financing for the new business (Lobo et al. 2005).

Regardless of the source for funds you plan to use, the competition for funds is tough. To obtain a loan, the proposed nursery must provide detailed financial data audited by a certified public accountant, information on the objectives and management of the proposed nursery business, and information about the business organization. Having a sound, well-written business and financial plan is extremely helpful in selling the business to the prospective lender and in securing funding.

Capital availability until plants reach market size and are sold is required several years and is vital for the survival of a new nursery. One possibility to tide over new business owners during the first years of no income is to grow fast-maturing crops such as annual bedding plants or other herbaceous crops that can be produced in a few weeks or months and provide cash flow until the target crop is ready for sale (Schuch 2006).

Licenses, Permits, and Regulations

Specific license and permit regulations will vary by state and county. To operate as a nursery business, you may need to consult a number of the following agencies: city or county clerk's office, city or county assessor's office, state department of finance and administration, state department of labor, U.S. Citizenship and Immigration Services (USCIS), IRS (source of Federal Identification Number), county agricultural commissioner's office or agriculture department, and county or city planning department. Other agencies and departments may also have license, permit, or regulatory requirements. In California, a good source of information regarding business licenses, permits, and tax requirements is the website for the California Governor's Office of Business and Economic Development (GO-Biz), listed in the References. Some of the important business regulations for California are described below.

California regulatory requirements

An annual license to sell nursery stock is required by the California Department of Food and Agriculture and is obtained from the local county agricultural commissioner's office. This license is also required following ownership change, including incorporation. In addition, an acreage fee is assessed for land in excess of 1 acre used in production.

Furthermore, most cities and counties in California require a license (certificate) to do business in the respective areas to indicate that the holder has paid the current year's tax. The business certificate must be displayed in a conspicuous location at the place of business. Businesses within the incorporated area of a particular city should contact the business licensing section of that city's government for specific rules and regulations. Businesses operating in more than one city or county may be required to have more than one business license. Moreover, if taxable merchandise is sold or a taxable service is provided in California, a seller's permit is required from the local State of California Board of Equalization office. Taxable merchandise includes all ornamental horticulture products.

In addition, companies in California may be subject to local personal property taxes and be required to register with the county assessor's office. Information on specific requirements should be requested from local city or county clerk's offices. Business owners must file a property statement with the county assessor each year, detailing market value of personal property including machinery and equipment.

New agricultural businesses in California must be familiar with the land use regulations that apply to the potential business location. Zoning and land use information and permit requirements can typically be obtained from the county or city planning department.

Various agencies have regulations regarding runoff management. Growers in California may need to contact the regional water board if a waste discharge permit or "conditional ag waiver" is required (see chapter 8). There may also be county requirements. Consult your Cooperative Extension office for pertinent regulatory authorities to contact in your area regarding discharge of irrigation and stormwater runoff.

License requirements for applying pesticides, as well as certification requirements for shipping quarantined pests, may also apply. See chapters 12 and 17 for more information.

Federal regulations

Once a business expands to include employees, it must comply with federal regulations regarding hiring, benefits and wages, migrant worker protection, and so on. Serious consideration must be given to rules administered by the Occupational Safety and Health Administration (OSHA). OSHA standards cover such worker protection issues as exposure to chemicals, protective gear, first aid, training, and notification laws.

In addition to being aware of regulations regarding employees, it is important to be familiar with other types of federal regulations that affect the nursery industry, such as plant patents and pricing regulations. Runyan (2000) and Davidson et al. (2000) provide good summaries of these laws and regulations; a foremost source of up-to-date information on federal laws and regulations is AmericanHort (see the References and the next section).

Resources and Knowledge

The best approach to starting a nursery business is to acquire as much knowledge and information as possible on plant production, business management, and marketing. If your experience is limited, you can gain valuable skills by working for a reputable grower; employment at a successful nursery is often the springboard to a self-managed nursery. Also, horticulture programs exist at some universities and community colleges that offer courses on nursery production.

Professional organizations are a good source of information. They can provide education, updates on legislative and regulatory issues, buyer and supply directories, representation of member interests, and updates on topics of current interest. Some nursery organizations also publish newsletters and trade magazines, and they organize annual trade shows.

For example, the American Nursery and Landscape Association (ANLA) and the Ohio Florists Association (OFA) merged in January 2014 to form AmericanHort, a national organization that caters to the needs of the entire American horticulture industry with an emphasis on lobbying and education. One of the early activities of ANLA was the development of a standardized system of sizing and describing plants to facilitate the trade in nursery stock. The latest version of the standard (2004) is available at no charge on the AmericanHort website (see the References).

In addition to associations that promote the nursery industry, specialized associations promote the interest of a particular group of plants. For example, the International Plant Propagators' Society seeks and shares knowledge on the propagation of plants; the Perennial Plant Association is dedicated to improving the perennial plant industry by providing education to enhance the production, promotion, and utilization of perennial plants. Many other organizations are dedicated to specific plants such as roses, daylilies, iris, orchids, or native plants. Members of specialty organizations have access to information that can be helpful for a nursery grower.

One of the best ways to seek out other growers who are willing to share their experience is to attend state, regional, and local grower association meetings, which usually provide opportunities for social interaction and dialogue. Grower associations may also provide tours of local nurseries, which can help you gain firsthand knowledge of working nursery operations. In California, state organizations include the California Association of Nurseries and Garden Centers and the Nursery Growers Association of California (see the References); moreover, numerous local plant industry associations exist in California, such as the San Diego Flower and Plant Association and the Santa Barbara County Flower and Nursery Growers Association.

Subscribing to trade publications, such as *American Nurseryman* and *Nursery Management* (see the References), and attending trade shows or conferences are good ways to learn about grower issues. Many trade publications, buyer's guides, e-communications, and other services are available at no cost on the Internet.

Your Cooperative Extension service is also a good source of information. This includes publications on starting a nursery (Eaton and Appleton 2009; Latimer and Eaton 2009; Robbins 2010; Schuch 2006; Sellmer and Dana 1994) and other topics on nursery production and management.

Many resources are available through bookstores and libraries that provide tips on starting a nursery (e.g., Avent 2003; Diver et al. 2008) and information about nursery plants (e.g., American Nurseryman Publishing Company 2006; Armitage 2008; Bailey and Bailey 1976; Holcomb 1994; Nau 1996). Many useful publications and resources related to plant production, marketing, and business management techniques are listed in the References at the end of all chapters in this manual.

References

Adrian, J. L., C. C. Montgomery, B. K. Behe, P. A. Duffy, and K. M. Tilt. 1998. Cost comparisons for infield, above ground container and pot-in-pot production systems. Journal of Environmental Horticulture 16:65–68.

AmericanHort website, http://americanhort.org/.

American Nurseryman Publishing Company. 2006. Pronouncing dictionary of plant names. Chicago: American Nurseryman Publishing.

American Nurseryman website, http://www.amerinursery.com/.

ANLA (American Nursery and Landscape Association). 2004. American standard for nursery stock. Washington, DC: ANLA. AmericanHort website, http://americanhort.org/documents/nursery_stock_standards_AmericanHort_2004.pdf.

———. 2004. Industry guide to marketing container plants. Washington, DC: ANLA. Ball Publishing website, http://www.ballpublishing.com/pdf/container.pdf.

Armitage, A. 2008. Herbaceous perennial plants. 3rd ed. Champaign, IL: Stipes Publishing.

Avent, T. 2003. So you want to start a nursery. Portland, OR: Timber Press.

Bailey, L. H., and E. Z. Bailey. Hortorium. 1976. Hortus third: A concise dictionary of plants cultivated in the United States and Canada. New York: Macmillan.

Baker, K. F., R. D. Durbin, J. Ferguson, J. W. Huffman, O. A. Matkin, D. E. Munnecke, C. N. Roistacher, W. R. Schoonover, and R. H. Sciaroni. 1957. The U.C. system for producing healthy container-grown plants. Berkeley: University of California Division of Agricultural Sciences Manual 23.

Bartok, J. W. 2009. Energy conservation for greenhouses and field nurseries. In T. Smith and P. Lopes, eds., The Massachusetts best management practices handbook for nursery production. University of Massachusetts Extension. 111–115. Mass.gov Energy and Environmental Affairs website, http://www.mass.gov/eea/docs/agr/programs/bmp/nursery-bmp.pdf.

California Association of Nurseries and Garden Centers website, http://www.cangc.org.

California Governor's Office of Business and Economic Development. 2012. Setting up business in California: A guide for investors. California Governor's Office of Business and Economic Development (GO-Biz) website, http://www.business.ca.gov/RelocateExpand/ForeignInvestment.aspx.

California Governor's Office of Business and Economic Development (GO-Biz) website, http://www.business.ca.gov/.

Carman, Hoy. 2011. Economic aspects of the California nursery and floral industry, 2001–2009, Giannini Foundation Information Series Report 11-1. Giannini Foundation website, http://giannini.ucop.edu/InfoSeries/111-Nursery.pdf.

Cronon, W. 1983. Changes in the land: Indians, colonists and the ecology of New England. New York: Hill and Wang.

Davidson, H., R. Mecklenburg, and C. Peterson. 2000. Nursery management: Administration and culture. 4th ed. Upper Saddle River, NJ: Prentice-Hall.

Diver, S., L. Greer, and K. L. Adam. 2008. Sustainable small-scale nursery production. National Sustainable Agriculture Information Service (ATTRA) website, https://attra.ncat.org/publication.html.

Eaton, G. K., and B. L. Appleton. 2009. Getting started in the nursery business—Nursery production options. Blacksburg: Virginia Tech University Cooperative Extension Publication 430-050. VTU CE website, http://pubs.ext.vt.edu/430/430-050/430-050.html.

Furuta, T. 1974. Environmental plant production and marketing. Arcadia, CA: Cox Publishing.

Higginbotham, J. S. 1990. Four centuries of planting and progress: A history of the U.S. nursery industry. American Nurseryman 171(12): 36–59.

Holcomb, E. J. 1994. Bedding plants IV: A manual on the culture of bedding plants as a greenhouse crop (No. IV). Batavia, IL: Ball Publishing.

Hollingsworth, B. 1962. Theodosia Burr Shepherd, 1845–1906: Her garden was her delight. New York: Macmillan.

Kohlstedt, S. G. 1978. In from the periphery: American women in science, 1830–1880. Signs 4(1): 81–96.

Latimer, J., and G. K. Eaton. 2009. Resources for greenhouse and nursery operations and operators. Blacksburg: Virginia Tech University Cooperative Extension, Publication 430-104. VTU CE website, http://www.ext.vt.edu/pubs/greenhouse/430-104/430-104.htm.

Lemay, A., S. Redlin, G. Fowler, and M. Dirani. 2003. Pest data sheet—*Ralstonia solanacearum* race 3 biovar 2. Raleigh, NC: USDA/APHIS/PPQ Center for Plant Health Science and Technology Plant Epidemiology and Risk Analysis Laboratory. USDA Animal and Plant Health Inspection Service website, http://www.aphis.usda.gov/plant_health/plant_pest_info/ralstonia/downloads/ralstoniadatasheet_CPHST.pdf.

Library of Congress. The Thomas Jefferson papers. Library of Congress American Memory website, http://memory.loc.gov/ammem/collections/jefferson_papers/.

Lobo, R. E., B. D. Wallace, K. L. Robb, and S. A. Parker. 2005. San Diego County agricultural directory and guidelines for agricultural enterprises. San Diego, CA: UC Cooperative Extension, San Diego County Farm and Home Advisor's Office.

Moskowitz, M. 2009. Qualities adapted to the country: The place of horticulture in the nineteenth century. Paper presented at the 2009 annual meeting of the American Studies Association. Washington, DC: ASA..

NASS (USDA National Agricultural Statistics Service). 2007a. 2007 Census of agriculture: Greenhouse, nursery, and floriculture operation. USDA Census of Agriculture website, http://www.agcensus.usda.gov/Publications/2007/Online_Highlights/Fact_Sheets/Production/nursery.pdf.

———. 2007b. Nursery crops 2006 summary. USDA NASS website, http://usda.mannlib.cornell.edu/usda/current/NursProd/NursProd-09-26-2007.pdf.

Nau, J. 1996. Ball perennial manual: Propagation and production. Batavia, IL: Ball Publishing.

Nursery Growers Association of California website, http://www.nurserygrowers.org/.

Nursery Management website, http://www.nurserymanagementonline.com/.

Pimental, D., L. Lach, R. Zuniga, and D. Morrison. 1999. Environmental and economic costs associated with nonindigenous species in the United States. Ithaca, NY: Cornell University College of Agriculture and Life Sciences.

Pittenger, D. 2006. Retail garden center manual. Oakland: University of California Agriculture and Natural Resources Publication 3492.

Robbins, J. A. 2010. Starting a wholesale nursery, Part I. University of Arkansas Division of Agriculture website, http://www.uaex.edu/publications/pdf/FSA-6055.pdf.

Rossiter, M. W. 1982. Women scientists in America. Baltimore, MD: John Hopkins University Press.

Runyan, J. L. 2000. Summary of federal laws and regulations affecting agricultural employers. USDA Economic Research Service Agricultural Handbook No. AH719. USDA Economic Research Service website, http://www.ers.usda.gov/publications/ah-agricultural-handbook/ah719.aspx.

Rybcsynski, W. A. 1999. Clearing in the distance: Fredrick Law Olmsted and America in the 19th century. New York: Scribner.

Sackman, D. C. 2005. Orange empire: California and the fruits of Eden. Berkeley: University of California Press.

Schuch, U. 2006. Questions to ask when planning to start a wholesale plant nursery. Tucson: University of Arizona Cooperative Extension Publication AZ1393. TUA CE website, http://cals.arizona.edu/pubs/garden/az1393.pdf.

Sellmer, J. C., and M. N. Dana. 1994. Starting in the nursery business. West LaFayette, IN: Purdue University Cooperative Extension Service, Circular HO-212. Purdue University Cooperative Extension website, www.ces.purdue.edu/extmedia/HO/HO-212.pdf.

USAonWatch. 2009. Farm watch: This land is our land. USAonWatch website, http://www.usaonwatch.org/resource/publication.aspx?PublicationId=79.

Van Laan, G. J. 1982. A penny a tree: The history of the nursery industry in California from 1850–1976. Sacramento: California Association of Nurserymen.

Woodhams, M. L. 1999. History of the American seed and nursery industry and their trade catalogs. Smithsonian Institution Libraries, Horticultural Branch. Smithsonian Institution Libraries website, http://www.sil.si.edu/SILPublications/seeds/bibseednur.html.

Photo: J. H. Lieth.

Nursery Site Selection and Development

Ursula K. Schuch

CHAPTER TWO

Photo: J. K. Clark.

The selection of a suitable nursery site, a realistic business plan, and a functional design are critical steps in the development and success of a new nursery enterprise. This chapter focuses on criteria to consider in site selection for a nursery, development of an efficient layout, and design of the production areas, nursery structures, and supporting nonproduction areas. Developing a business plan is discussed in chapter 18.

The two main production systems for nursery plants are in containers or in the field. Container plants are grown in pots during part or the entire time of production. Field-grown plants are generally planted outdoors in the ground when they are small and are harvested when they have grown to a marketable size. The majority of nurseries in California are container nurseries, which take advantage of year-round growing and marketing, compared with the more seasonal harvesting of field-grown (inground) nursery plants. Depending on which type of nursery is planned (container, field, or both), certain aspects of site selection and development will differ. Both types of businesses, however, have many common requirements, such as the following:

- a favorable growing climate
- the availability of good quality water
- protection of plants from weeds, insects, and diseases
- favorable local ordinances and zoning
- low-cost energy
- competent employees
- a marketable product
- reliable suppliers and distributors
- adequate financing
- proficient management

The development of a nursery is a complex task and requires expertise in many trades. It may be prudent to enlist a team of experts who can help with the initial site evaluation and selection and who have the expertise to critique a business plan and nursery design before any soil is moved. An essential member of such a team is a professional with nursery production experience. Thorough planning is more time consuming and expensive in the beginning, but the advantages are well worth the time and investment.

The general economic outlook for nursery businesses and the specific objectives of a new nursery business have a major influence on the selection and development of a nursery site. A feasibility study and a business plan are vital documents guiding the process of site selection and development. Figure 2.1 is a flowchart that provides an overview of the development of a container nursery business, noting the steps involved in site selection, design, and development. Each step in the process is discussed in this chapter.

Site Evaluation and Selection

Sites suitable for nursery production depend on the plant materials to be produced, their environmental requirements, and the targeted markets. Climate and water, however, are without doubt the most important factors for nursery site selection. Matching the climate requirement of crops to a prospective nursery site and ensuring a consistent supply of quality water will facilitate long-term production. Other

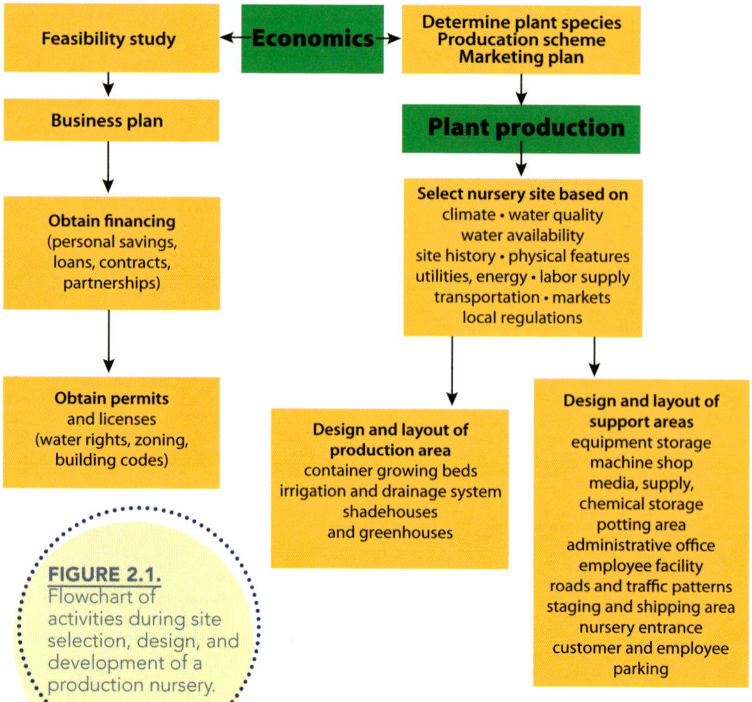

FIGURE 2.1. Flowchart of activities during site selection, design, and development of a production nursery.

considerations include land quality and size, proximity to good infrastructure, availability of natural resources and utilities, a reliable workforce, and accessible service contractors.

The following factors also need careful consideration when evaluating potential new nursery sites. In populated areas, the scarce availability and high cost of land, along with urban encroachment, can cause serious problems for a nursery. Zoning, as well as local and regional regulations, impacts a nursery business. Due to the high cost of hazardous waste cleanup, property should be selected that is free from contamination of toxic chemicals such as petroleum products, heavy metals, and pesticides. Air pollution can also be a problem in the production of plants.

Climate

A nursery's climatic requirements depend on the plants to be grown. A mild climate with high light intensity is most suited for both field and container nursery production. Ideal day temperatures for outdoor production in summer should range from 80° to 95°F, and night temperatures should be below 70°F. Extended temperatures above 100°F impede growth of many plants but are suitable for heat-tolerant and desert-adapted plant material.

Coastal areas of California provide an ideal growing climate for many plant species. Compared with inland areas, coastal areas usually have more foggy days with lower light intensity, but they have less temperature fluctuation. Inland areas have more sunshine and greater temperature fluctuation. In inland areas, nurseries are often located near rivers or large bodies of water to take advantage of the mitigated temperature extremes. To modify the climate for best growing conditions—in a propagation area, for example—a nursery in a coastal location may need supplemental light and heaters while an inland nursery may need heaters and coolers. Building and operating costs of a nursery are greatly reduced on sites where the outdoor temperatures match the optimal growing temperatures of crops and where temperature fluctuations are less frequent.

Knowledge of the optimal temperature requirements of the plants to be produced, as well as historic weather information for different potential sites, can be helpful. Production manuals and other kinds of literature list optimal temperature ranges for some plants. Lists of plants available from wholesale growers producing outdoors in the targeted climate zone are another source of information. California has a system of weather stations throughout the state (California Irrigation Management Information System, or CIMIS) that provides detailed meteorological records to help manage irrigation, including air and soil temperature, relative humidity, and evapotranspiration. Other useful climate information is found in the *New Sunset Western Garden Book* (Brenzel 2012), which assigns detailed climate zones for the western states based on latitude, elevation, ocean and continental air influence, minimum and maximum temperatures, rainfall, humidity, and length of growing season. Many plants growing in California are assigned a range of climate zones where they can thrive. Average minimum winter temperatures are searchable on a plant hardiness map provided by the U.S. Department of Agriculture (USDA), which is now interactive and searchable by zip code (USDA 2012). The plant heat zone map of the American Horticultural Society (AHS 1990) lists the average number of days above 86°F.

Prolonged low temperatures can severely damage container plants and may require the use of overwintering structures or plant covers. These types of structures are described in the next chapter. Overwintering structures are not necessary in southern California, except for tropical or semitropical plants that are sensitive to chilling and frost. Low temperatures often damage container plants more than field-grown plants because the roots are more susceptible to cold than shoots and—especially in smaller containers—they have little insulation. Temporary frost protection through overhead irrigation or other practices may be necessary in areas with occasional frost (see chapter 3). Locations with a high danger of late frost should be avoided because of the potential damage to new, tender shoots in spring. If plants with chilling requirements are to be produced, then the potential nursery site must be selected accordingly.

Windbreaks in areas where sensitive plants are grown may be a good investment. High winds can damage plants through desiccation and physical damage, and windborne soil can blast foliage. In container operations, wind can become a nuisance because it will knock over plants, requiring costly hand labor. Wind will also affect spray patterns of overhead irrigation or chemical applications. Effects of winds, windbreaks, and management practices are described further in chapter 10.

Microclimates are comprised of all environmental factors and structures, which may alter natural patterns of temperature, rain, wind, or radiation. Several microclimates can occur in a nursery, and usually more are found if the topography is heterogeneous. Buildings, hedges, or large trees slow down wind speed and shade the ground. Temperatures in

Photo: U. K. Schuch.

the immediate vicinity of buildings can significantly increase due to heat absorption and radiation. Several structures together can create wind tunnels, intensifying adverse wind effects. Cold air can accumulate in depressions and become frost pockets. Microclimates in a nursery should be identified and used to the advantage of the nursery operation. For example, place plants with chilling requirements in colder areas, but grow frost-sensitive plants in the warmest part of the nursery.

Land Requirements

Land quality and size are important characteristics for nursery site selection. Soil workability, texture, depth, and drainage are vital for a field-grown nursery, while they are of lesser concern for a container production site. Aerial photographs, topographic maps, soil survey maps, and a history of the site's prior use can provide valuable data about a potential site.

Topography

Land suitable for nursery production should be free of physical obstructions that will prevent its use for growing plants. The topography of a future nursery site should be free of valley bottoms or depressions where cold air will settle and create potential frost pockets. In addition, there should be no features that allow standing water on the nursery production site other than planned irrigation ponds. This is because if root balls in containers are partially or entirely submersed in water for extended periods of time, plant health will be compromised and conditions for disease favored. Therefore, level land with a slight slope up to 2 percent is ideal. A slope up to 5 percent is desirable if drainage is slow and the soil has a high clay component. Terracing or grading can bring rough terrain into production but may limit equipment access and mechanization, and it may cause erosion. Mixing topsoil with infertile subsoil during a land-moving operation can negatively affect soil quality for a field-grown nursery.

The elevation and aspect of a nursery site may be of concern depending on the plant material that is grown. Cooler temperatures at higher elevations can shorten the growing season. Southwest-facing sites in lower elevations can be exposed to higher temperatures, and irrigation systems need to be able to meet the demand for high evapotranspiration rates. Large daily temperature fluctuations in winter may decrease the hardiness of plants. Certain elevations or locations may be desirable; for example, a site with a warm wintertime inversion layer without danger of frost is excellent for production of more tender subtropical plants.

Amount of land needed

One step in developing a nursery site is the determination of how much production acreage is needed. It is estimated that the minimum land requirement for a profitable container nursery is about 10 acres compared with 30 acres for a field nursery. Up to 70 percent of the nursery site is usually used for actual production space, including outdoor container growing areas, propagation benches or beds, greenhouses, and shade structures. When calculating how much production area is needed, only the actual production areas where plants are growing are considered. Nonproduction areas such as buildings, roads and parking areas, water reservoirs, and shipping and storage areas are excluded, although they are important in the overall nursery design and usually account for about 30 percent of the land.

The number of plants per unit area depends on the spacing and the size of containers used for plant production, and this will affect the site production potential. Other factors are the percentage of salable plants and the production time of individual plant species or cultivars. With increasing diversity of species and cultivars, determination of the production potential of a nursery will become more complex. Site production potential can be determined on an annual or seasonal basis.

Following is an example of how much space and how many plants it will take to produce 3,000 salable shrubs in #5 containers. (The term #5 refers to the container class according to the American Standard for Nursery Stock [ANLA 2004] and defines a range of acceptable volumes for a particular container class. Referring to a container by the gallon size alone, such as a 5-gallon pot, does not conform to the standards.) We assume that the shrubs have a 2-year production period, and that they will be grown in #1 containers spaced can-tight (6-inch centers) for the first year, then shifted to #5 containers and spaced at 2-foot centers. (This means that the distance from the center of each pot to the center of neighboring pots is 2 feet.) At the end of the second year, we want to have 3,000 salable plants.

Calculations start with the finished product and move backward through the production steps. We assume that 6 percent of the finished #5 plants will not be salable and will be considered culls due to physical damage or insufficient size, and therefore we will need 3,191 plants (3,000 ÷ 0.94) that have been transplanted from #1 to #5 containers. Assuming an estimated 5 percent culls during the #1 container production phase, we need to start with 3,359 plants (3,191 ÷ 0.95) in #1 containers. The #1 containers will require 840 square feet (3,359 × 0.25 ft^2), and the #5 containers will require 12,764 square feet (3,191 × 4 ft^2). The production of 3,000 salable shrubs under these conditions requires a total of 13,604 square feet (840 + 12,764) of container bed space if both sizes are produced at the same time.

Water

Water is the most important factor, and often the limiting one, in nursery production. Container nurseries, more than field nurseries, depend on a steady supply of irrigation water, regardless of rainfall. Water and fertilizer management are generally more intensive in container production because of the limited volume of potting soil or container medium and the limited ability of the organic medium to hold on to nutrients compared with mineral soil in pots. Year-round water quality and quantity need to be evaluated before developing a nursery site. Water rights for current and possible future needs must be obtained, and a backup for emergency situations must be secured.

Irrigation water can be obtained from a variety of sources, including lakes, streams, aqueducts, or wells. Surface water can be contaminated with large debris, vegetation, or weed seeds, but screening filters can effectively exclude water-borne debris or wildlife. Chemical water treatment may be necessary to control pathogens. Well water is an independent, reliable water source for many nursery growers in California. Prior to drilling a well, it may be wise to look at water analyses from existing local wells to determine if groundwater in the area is acceptable for nursery production. Regardless of the water source, reliable water flow and delivery times for a nursery must be negotiated and secured for a nursery to be successful.

Good-quality irrigation water implies that it is low in soluble salts and contains no mineral or other contaminants in quantities that can potentially injure plants. Not all species react uniformly to water quality, and plant age can be an important factor in sensitivity to water quality. If water quality is less than optimal, provisions for purification, such as reverse osmosis, may be necessary to protect sensitive plants, especially in propagation. The pH of irrigation water should be slightly acidic, but it is less critical than soluble salts because it can easily be adjusted. Phosphoric acid or sulfuric acid will lower the pH, while addition of calcium or magnesium to the irrigation water will raise the pH.

The volume of water necessary for plant production varies with the demand of the plant material, its developmental stage, and environmental conditions. The estimated water requirement for an acre of container plants varies from 4 to 10 acre-feet per year. Field nurseries require pumping capacity of up to 10 gallons per minute per acre for peak summer demand. Calculating the yearly amount of water needed for all production areas will be helpful in planning for sufficient sources. Generally, peak demand for water occurs during the hot summer months and possibly in winter for frost protection. The maximum amount of water needed for production should be calculated to negotiate water rights and to estimate water costs.

Photo: D. Rosen.

While the source and supply of irrigation water are of great concern for a nursery, proper management of the runoff has become equally important. California has strict regulations regarding nursery runoff containing fertilizers, pesticides, and other contaminants. In some areas discharge of runoff is entirely prohibited, and runoff needs to be recycled. Water ponds need to be secured for human safety and water quality. Vegetated buffers can also be used to slow down the movement of runoff water and help keep it on-site. For further information on holding ponds, recycling, and vegetated buffers, see chapter 8.

Environmental Contamination

Air pollution is often a problem in densely populated and industrialized areas, and it may damage nursery crops. Pollutants produced by vehicle emissions are photochemically modified into smog. Ozone is the primary toxic component of smog. It enters through the stomata and causes stippling, mottling, and chlorosis, primarily on upper leaf surfaces. Peroxyacetyl nitrate is another smog component, a product of ozone's reaction with incompletely combusted hydrocarbons, that can damage many plants and cause "silver leaf," a bleached whitening or bronzing on lower leaf surfaces that is particularly acute in metropolitan areas with inversion layers. Sulfur dioxide and fluorides are damaging pollutants, particularly in the immediate vicinity of petrochemical manufacture. It is only logical that a nursery site adjacent to or downwind from a major pollution source should be avoided. Nurseries adjacent to a freeway may have problems with particulates deposited on the foliage. Air pollution damage varies for different species—for example, bedding plants are more susceptible to ozone damage than woody plants. Young plants are generally more susceptible than older plants.

Contaminants of soil and water can also be a problem. Agricultural properties where pesticides with long residual life were used in the past are likely to have irrigation and storm runoff that contains concentrations of these pesticides above threshold levels if treatments to prevent movement are not implemented.

Before completing the final purchase of land for a nursery, it may be wise to obtain an environmental audit of the property. Also, sometimes lending institutions require an environmental audit. This would include, for example, assessment of whether soil and water have been exposed to or contain hazardous chemicals or other materials; whether there is a source of net emission of potential air pollutants; or whether environmental permits are required for using the property as a nursery. Environmental consultants provide environmental audit services.

Labor Requirements

Labor requirements for container nurseries are generally around one employee per acre in actual container production (total acreage minus acres in roads, buildings, etc.). This is considerably higher than for field production, which is one employee for 5 to 20 acres. A field nursery tends to have more fluctuating needs for workers, usually with peak demand during planting and harvesting operations.

A nursery needs an adequate, dependable supply of employees with diverse skills. Personnel with knowledge in heavy equipment operation and maintenance, horticulture or agriculture, irrigation management, sales and accounting, and administration are all necessary. In addition to these year-round permanent positions, seasonal trained and untrained labor is necessary to ensure continuity in the production process. Specific tasks that are seasonal, such as harvesting field plants, grading, or weeding, may be contracted out. A thorough analysis of whether in-house or contract labor is cost effective needs to be conducted. The availability of a labor force in one location largely depends on whether nurseries pay competitive wages compared with alternative employment at the same skill level. A more automated and sophisticated nursery operation needs fewer employees, but better-trained ones, than a nursery relying mainly on manual labor.

Competition

Nurseries are often clustered in certain locations because of the favorable growing climate and availability of suitable land. Adequate markets need to exist when several nurseries are located close together. One possible disadvantage is competition among nurseries growing the same type of plants. On the other hand, proximity to other nurseries can have the advantage of drawing more potential buyers to a larger, diverse market. Proximity may facilitate cooperative efforts in buying supplies, marketing, and shipping that may be beneficial to smaller nurseries, and it may encourage exchanging technical information of local importance among growers.

Transportation

A good transportation system is essential for a nursery to accommodate the flow of people, supplies, and finished plants. A nursery needs to have access to major highways as well as the capacity to accommodate large trucks. Access to a nursery should not be restricted during adverse weather conditions, since shipping may be hindered.

Fuel, Utilities, and Other Services

Telephone and Internet service, electric power of different phases and voltages, and a fuel supply are necessary for operating a nursery. The cost of bringing utilities (especially water, electricity, and fuel) to an undeveloped nursery site can be substantial, and it needs to be determined before a land purchase or lease contract is finalized. Prices for commercial power and fuel rates should be compared in different locations.

Proximity to fuel distributors as well as repair and maintenance service contractors for equipment (irrigation pumps, refrigeration or heating systems, plumbing) affects operating costs and is crucial in emergency situations.

Allotment priorities for energy during peak use of electricity are important in order to secure such critical nursery operations as irrigation. Many nurseries lease or rent backup generators to supply electricity during power failures.

Regulations and Land Use Issues

Before purchasing or leasing a site, become acquainted with state and local zoning restrictions, building codes and permits, and business license requirements. Learn about tax rates inside and outside city limits. Consider water rights, irrigation priorities, and regulations governing the disposal of runoff water. Learn about sales licenses, occupational licenses, and material handling licenses. Evaluate the impact of these regulations on nursery operations at a particular site under consideration.

Expansion of nurseries in urban developments is not a realistic expectation, as many nurseries in this situation have to relocate to less populated rural areas. Urban residents may oppose pesticide use, express concerns about groundwater pollution and runoff, or may help enact local zoning laws and ordinances that restrict building greenhouses and hinder business operations.

Evaluation of Potential Sites

Once several potential nursery sites have been identified and inspected by a team of experts, the most suitable one needs to be selected. If obvious deficiencies are found, such as lack of quality water or restrictions in zoning, some sites may be quickly eliminated. However, in most cases there are probably several advantages and disadvantages to a site. The decision-making process can be facilitated by using a decision matrix developed by Kepner and Tregoe (1965) and adapted from Landis et al. (1995), as shown in table 2.1. In the matrix, site selection criteria are defined, sorted by importance, and assigned a weight value on a scale of 1 to

Table 2.1. Evaluation criteria for comparing three potential container nursery sites

Site selection criteria	Weight value*	Site I Rating†	Site I Weighted score‡	Site II Rating	Site II Weighted score	Site III Rating	Site III Weighted score
climate	10	10	100	8	80	9	90
water quality	10	8	80	9	90	7	70
water quantity	10	7	70	10	100	9	90
land availability	8	6	48	8	64	8	64
energy availability	7	9	63	8	56	10	70
zoning, local ordinances	6	4	24	9	54	9	54
microclimate	5	9	45	8	40	10	50
labor supply	5	9	45	7	35	10	50
soil/topography	5	10	50	9	45	9	45
accessibility	4	9	36	6	24	8	32
distance to market	3	9	27	7	21	6	18
potential for expansion	3	5	15	10	30	9	27
Total site score			603		639		660
Site ranking			3		2		1

Sources: Adapted from Landis et al. 1995; based on the Kepner and Tregoe (1965) decision matrix.

Notes:
*1 = lowest, 10 = highest relative importance value.
†Rating is based on the evaluation of the site selection criteria of a particular site.
‡Weighted score is the product of multiplying the weight value by the site rating.

10 depending on the type of production, the species of plants to be produced, environmental requirements, and other factors. For example, if plants will be grown in the field, then soil type, texture, and topography are very important and would receive a high weight value, while energy availability and labor may be less crucial and would receive lower values. On the other hand, for a container nursery, soil type, texture and topography are less relevant and would receive a low weight value, while climate, water, and energy are very important and would be ranked with high weight values. The next step evaluates the suitability of each selection criterion based on the site analysis, again on a scale from 1 to 10 (see table 2.1). The weighted score for each selection criterion at a particular site is then calculated by multiplying the weight value by the rating for each site. Adding up these weighted scores provides an overall total for each site, thus helping to identify and rank the site with the best suitability.

If sites are very close in their final score, then a second evaluation of critical elements may identify further differences in site suitability. The evaluation for Site I can be representative for an urban location, where land availability, zoning and local ordinances, and future expansion may restrict the establishment of a container nursery. The other two sites could be representative for rural areas, where land and water are not limiting but accessibility or distance to market is less than optimal.

The weight value of individual site selection criteria may change over time, as some criteria can become more or less limiting to a successful nursery operation. For example, it appears that local ordinances in urban and coastal areas have become increasingly restrictive to agricultural businesses. Similarly, distance to market may become a greater factor in competition, as fuel prices rise and transportation costs cut into profit margins. Conversely, neighboring growers can save by pooling resources or equipment, thus cutting costs for supplies and capital expenses.

Nursery Layout and Design

Nursery layout and design deals with the general arrangement of production areas, infrastructure, grading, and the size and layout of structures, aiming to ensure an efficient and safe flow of materials and labor. This section concentrates on the factors to consider in the layout and design of container nurseries, although many of the principles apply to a field production nursery as well. The first step before designing the nursery layout on paper is deciding the types and quantities of plants to be grown, which will determine the size of the operation. Next is developing a detailed plan for the production system, including information such as the number of plants and species grown, their size at marketing time (which affects production time), cultural practices, and marketing strategies. These steps result not only in a developmental plan for the nursery, but also in a business plan, which is often required for obtaining a loan or other financing (see chapter 18).

No nursery should be constructed, expanded, or altered without a comprehensive design plan that translates production objectives into the design features of functional areas. Depending on who has the expertise, the design may be rendered by the nursery owner, a manager or employee, an engineer, or a nursery consultant. The design must fit the production, management, and shipping needs.

Nursery design starts with identifying the size and location of the functional areas, including production and support areas. A map of the site, including all topographic features and

FIGURE 2.3. Nursery bed graded for drainage in the middle. Containers are spaced can-tight and irrigated by overhead sprinklers. *Photo:* U. K. Schuch.

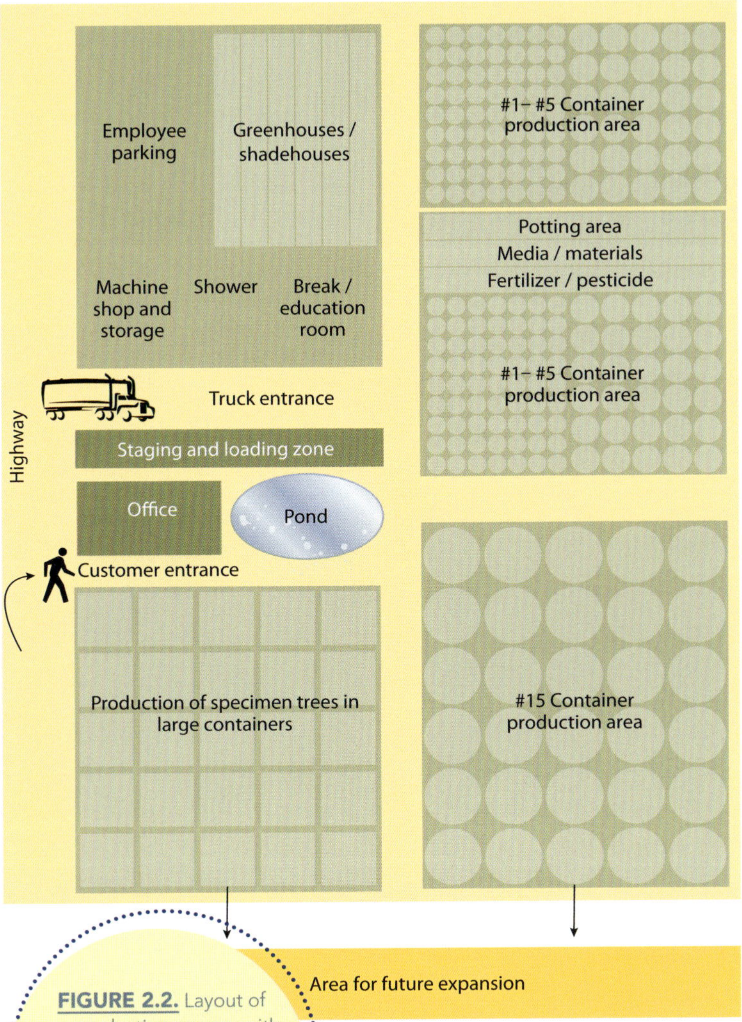

FIGURE 2.2. Layout of a production nursery with central administrative and support buildings surrounded by production grounds (not to scale).

current facilities, should be used in the planning process. Individual areas should be assigned to a nursery location and separate plans should be drawn, featuring size and layout of production areas, buildings, roads, irrigation systems, and underground utilities. The plan should include fencing the nursery perimeter and securing entries for necessary safety precautions.

Specific design goals may include using labor as efficiently as possible; minimizing production time; minimizing the use of water, space, and pesticides; locating similar operations near each other; maximizing the ease of moving plants and supplies; permitting flexibility so that new methods and technology (mechanization) can be incorporated; and allowing for future expansion (fig. 2.2). The design should facilitate several phases of nursery development over time. Keep in mind that later developed areas will increase the volume of plants going through the nursery and that the support facilities should be designed to handle the increased flow of incoming and outgoing plants and materials.

The layout and design of the nursery can start with identifying the size and location of the functional areas. Specific information regarding the development of these areas is provided below.

Production Areas

Container beds

Container production areas are usually arranged in blocks, and their layout and size may be dictated by the slope of the site or the design of the irrigation (fig. 2.3). Container plants with similar water requirements should be grouped together to maximize efficient use of irrigation water. At this point,

FIGURE 2.4. Plants in smaller containers are spaced can-tight and irrigated by overhead spray. The structure can be covered with plastic, as seen here, or with shade cloth from spring to fall. *Photo:* U. K. Schuch.

FIGURE 2.5. The nursery bed is covered with gravel and woven fabric on top to allow for drainage of water and prevent weed growth on the ground. Plants in larger containers are spaced apart to accommodate the growing canopy and are irrigated with spray stakes or drip emitters in each pot. *Photo:* U. K. Schuch.

water delivery, type of irrigation system, and the fate of the runoff water (its collection or drainage) need to be customized (see chapters 7 and 8).

The size of a nursery bed depends on functional use of space. Generally no two nurseries have exactly the same bed size. The goal is to produce the largest number of high-quality plants in the minimum amount of space, and the spacing of containers influences the quality of a finished product. Container spacing in a bed depends on the size of the plant at the beginning of production and the desired finished plant size. For the production of plants with a small canopy and plants in pots up to #1, containers can be spaced tightly; but when larger specimens are grown, containers need to be spaced more widely to allow for sufficient canopy expansion (figs. 2.4 and 2.5). Minimizing labor costs is a concern—and growing plants in their final spacing, even though canopies are initially small, is often the most economical practice.

Small pots are often placed can-tight to maximize the use of space and overhead irrigation efficiency. Containers are mostly grouped by size, and plants in the same size group are usually separated by species. When space is at a premium, two species with different growth habits may be interspaced. For example, tall topiaries or tree forms may be mixed with lower-growing, shade-tolerant species. Growing trees in relatively small containers often requires the building of a support structure such as a wire frame or trellis to keep plants and containers upright (fig. 2.6). Another alternative is the pot-in-pot system, which is generally used for containers ranging from #5 to #15. This is where socket pots are sunk permanently into the ground, into which the snugly fitting containers with plants are lowered (fig. 2.7). The

FIGURE 2.6. Taller plants in containers are attached to a trellis to prevent them from blowing over. *Photo:* U. K. Schuch.

pot-in-pot system is a major initial investment that has been proved especially worthwhile for tree or large shrub production. Plants stay upright when it is windy, their root systems are buffered from temperature extremes, and growth can be faster compared with container plants placed conventionally on top of the bed. With pots spaced at a distance that is suitable for the final plant size, transplanted container plants are placed in their socket pots and remain there until finished. Regularly turning containers with plants by 90 degrees or more, or placing a spacer between the potted plant and the bottom of the socket pot, prevents the roots from growing into the soil through the drainage holes of the socket pot.

Once the size of beds has been determined, construction can begin. Beds are usually graded to accommodate runoff. The amount of grading depends on the rate of water applied, the drainage properties of the soil, and the overall drainage system of the nursery. Beds can slope toward the middle or to one side (see fig. 2.3). Major drainage ditches between nursery beds can be filled with coarse gravel so they are not a hazard for workers.

Container beds are generally covered with ground cover cloth, polyethylene, or gravel to prevent soil erosion, suppress weed growth, and—in the case of gravel—facilitate drainage and reduce disease problems (see figs. 2.4 and 2.5). The choice of cover depends on its major function. If a soil is stable and has the correct slope and good drainage properties, then a woven polypropylene cloth ground cover will provide a level surface for containers. Polyethylene has the potential problem of accumulating water in depressions, and plants standing in pools of water will be at high risk for disease problems. Polyethylene covers are available in different thicknesses and with ultraviolet (UV) resistance. The thicker UV-resistant covers can last several years, while the thinnest sheets without UV resistance may not last even 6 months. Fabric cloth is often used under gravel to prevent it from sinking into the underlying ground and to minimize the addition of new gravel. Gravel may be less expensive, have longer durability, and facilitate drainage compared with polyethylene or a cloth cover. Disadvantages of gravel are that it provides uneven support for containers, offers not as much weed suppression, and allows rooting of plants through container drainage holes into the gravel and soil. Treatment of the ground with preemergence herbicides before or after gravelling the area can keep the container bed free of weeds. Less expensive materials that are suitable as ground covers may be available locally and may be worthwhile as alternatives.

The layout of the nursery bed will affect the design of the remaining areas. Labor efficiency should be a major consideration in arranging the infrastructure, shade and overwintering structures, production and storage areas, office space, and road system.

Growing structures

Some container plants may need shelter from heat, wind, cold temperatures, or full sunlight. Young plants are more susceptible to harsh environments, and propagation areas are usually covered with glass, fiberglass, or rigid or film plastics such as polyethylene. Many greenhouse coverings with various durability and light transmission properties

FIGURE 2.7. Trees are in a pot-in-pot system. The plant in the container is placed in a socket pot that is permanently installed in the ground. Plants do not blow over, and temperatures in the root zone are moderated because of the insulation from the soil. *Photo:* U. K. Schuch.

are on the market. Liners can also be propagated in shadehouses (constructed with a frame covered by lath or shade cloth) or in open protected areas on the ground. In addition to conventional, permanent shadehouses, retractable-roof shadehouses allow growers to manipulate the roof and the sidewalls, creating conditions similar to an open field, a shadehouse, or a greenhouse. See chapter 3 for an in-depth discussion on growing structures.

The propagation area needs to be sheltered from extreme temperatures and wind to prevent damage to cuttings or seedlings and to ensure an even irrigation pattern. In many cases, concrete pads are used for propagation or bottom heat is installed on benches. Overhead misting or fogging systems mostly provide moisture controlled by an electronic timer. However, more sensitive devices are available to adjust the fogging and misting cycles to environmental conditions, saving water and reducing susceptibility to diseases (see chapter 9 for more information).

If water quality is not always high, propagation areas should have a separate water supply from the other production areas. Cuttings or seedlings of some species have a lower tolerance to soluble salts, specific ions, and disease organisms than do well-rooted, established plants. In some cases, expensive systems such as reverse osmosis may be required to ensure successful propagation.

The maximum number of cuttings or seedlings to be produced during one propagation cycle determines the size of the propagation area. This size should include a margin for culls during each step of the propagation cycle. Space within structures is expensive and should be designed for maximum efficiency. For example, if benches are used, the rolling type may be appropriate to maximize greenhouse or shadehouse production space. Also, the entire bench top may be moved in one unit at the end of the propagation cycle to the transplanting area.

Support Areas

Roads

Roads in the nursery are an important aspect of infrastructure and should be designed to minimize the distance required for moving materials and people. Roads heading from the highway into the nursery and shipping area need to be built to accommodate large trucks, while roads within the production area should be designed for smaller vehicles. Good drainage and appropriate road surfaces are of great importance to allow smooth transportation to and within the nursery at all times. Roads between production areas and the container beds should be designed and graded such that they either deflect water or are part of the drainage system. The design should take advantage of the natural drainage patterns of the site as much as possible. Dust from road traffic should be avoided as much as possible to prevent any buildup on the plant foliage.

Support buildings

Support buildings or structures are necessary to accommodate a machine workshop, refrigeration units, and storage for fertilizers, equipment, and tools. Areas for chemical storage need to be separate, and they must comply with state and local ordinances (obtainable from the local agricultural commissioner or fire department). Container media or

FIGURE 2.8. Media and supply storage area are neatly arranged, avoiding clutter. *Photo:* U. K. Schuch.

FIGURE 2.9. Assembling orders for shipping is time consuming and requires a dedicated space in the nursery where trucks can pull up for loading. *Photo:* U. K. Schuch.

their components can be stored indoors or outdoors with facilities nearby for media mixing and pasteurizing (fig. 2.8). Potting or transplanting may be in a central location with finished plants, which are then transported to their growing area.

Staging and shipping area

The shipping area should be close to the entrance of the nursery to minimize expansive truck roads and movement that waste production space and may interfere with the internal nursery traffic. The truck entrance and routes in the nursery should be clearly marked and separated from other traffic. The staging area where orders are assembled before shipping should be close to where trucks are loaded (fig. 2.9). The design should accommodate the maximum number of trucks that need to be loaded at peak time. A water source should be installed to facilitate the final cleanup of plants and containers and also to irrigate plants if necessary.

Entrance, office buildings, and parking areas

Public areas of the nursery, those with high traffic volume and high visibility, should be well maintained and landscaped. They convey the image of the business and could be showcases for the type of plants produced in the nursery. The landscaping can serve as a focal point to present a diverse palette of products or to attract interest in specialty products from the nursery.

Offices with space for administration, accounting, sales, and employees should be close to the entrance and shipping area. A conference/education/break room should be provided, and shower and laundry facilities are needed for workers who apply pesticides. Well-marked customer and visitor parking should be close to the administrative offices. Employee parking may be in the same area if space permits, or it may be located closer to the production facilities.

Movement of Materials

The design of a nursery should strive to minimize the movement of plants, supplies, and people during the production cycle. Labor is generally the most expensive component in the production of container plants, and moving individual containers is a time-consuming and strenuous job. Plants need to be moved from the propagation area to the growing bed, then perhaps to a transplanting area and back again, and finally to a finishing area. Assembling shipping orders is one of the most time-consuming aspects, especially if many small, diverse orders are placed. The work flow, particularly for indoor organization (e.g., propagation), should always be in one direction to prevent blocking and unnecessary delays.

Moving flats or containers as a unit on a pallet with a forklift, or rolling a bench of plants directly onto a trailer, can save considerable time and strenuous activity. Integrated pallet systems have been tested in several commercial nurseries and show great promise. To date they have not been widely accepted, as they require a major shift from current production systems, but they might be considered again when nurseries require further mechanization to stay competitive. One constraint for the current production systems of container plants in implementing integrated pallet systems is the large variability of container size and shape, which makes the development of a uniform handling system more difficult. More details on automation and mechanization are found in chapter 4.

References

AHS (American Horticultural Society). 1990. Plant heat zone map. American Horticultural Society website, http://www.ahs.org/publications/heat_zone_map.htm.

ANLA (American Nursery and Landscape Association). 2004. American standard for nursery stock. Washington, DC: ANLA. AmericanHort website, http://americanhort.org/documents/nursery_stock_standards_AmericanHort_2004.pdf.

Appleton, B. L. 1986. Container nursery design. Chicago: American Nurseryman Publishing Co.

Avent, T. 2003. So you want to start a nursery. Portland, OR: Timber Press.

Brenzel, K. N., ed. 2012. The new sunset western garden book: The ultimate gardening guide. 9th ed. Menlo Park, CA: Sunset.

CIMIS (California Irrigation Management Information System). Department of Water Resources, Office of Water Use Efficiency. CIMIS website, http://wwwcimis.water.ca.gov/cimis/data.jsp.

Davidson, H., R. Mecklenburg, and C. Peterson. 2000. Nursery management: Administration and culture. 4th ed. Upper Saddle River, NJ: Prentice-Hall.

Kepner, C. H., and B. B. Tregoe. 1965. The rational manager. New York: McGraw-Hill.

Landis, T. D., R. W. Tinus, S. E. McDonald, and J. P. Barnett. 1995. Nursery planning, development, and management, vol. 1. In The container tree nursery manual. Agricultural Handbook 674. Washington, DC: USDA Forest Service.

Morby, F. E. 1984. Nursery site selection, layout, and development. In M. L. Duryea and T. D. Landis, eds., Forest nursery manual: Production of bareroot seedlings. The Hague: Martinus Nijhoff/Dr W. Junk Publishers, for Forest Research Laboratory, Oregon State University, Corvallis. 9–15.

USDA (United States Department of Agriculture). 2012. USDA plant hardiness zone map. USDA Agricultural Research Service website, http://planthardiness.ars.usda.gov/PHZMWeb/.

Yeager, T., ed. 2007. Best management practices: Guide for producing nursery crops. 2nd ed. Atlanta, GA: Southern Nursery Association.

Photo: J. P. Newman.

Plant Growing Structures

J. Heinrich Lieth, Ursula K. Schuch, and Julie P. Newman

CHAPTER THREE

Various types of plant growing structures can be used to protect crops from climatic and biotic stresses or to create optimal environments for plant growth and development. Although many nursery crops are produced outdoors, plant growing structures are often used during propagation and the early stages of production. These structures, which range from cold frames to sophisticated computer-controlled greenhouses, vary in size, complexity, and the degree to which environmental conditions inside the structure can be manipulated.

The selection of an appropriate plant growing structure for the nursery is largely dictated by the needs of the crops being grown and by the financial investment one can afford. How closely the growing structure can match conditions required for optimal plant growth depends on the capability of the environmental control system. Thus, if the selected crop requires complete temperature management, then a greenhouse with heating and cooling capability is needed; if it is necessary to time crop development (e.g., for holidays), then the greenhouse must have precise temperature controls. However, if the crop will adequately grow and develop under prevailing outdoor conditions and only requires some shade, then a simple shade structure will suffice.

This chapter provides an overview of the range of available growing structures and equipment that can be used for monitoring and controlling a crop's environment. Cultural practices to manage and manipulate environmental parameters in growing structures are addressed in the chapters that focus on soils, nutrients and fertilizers, and irrigation (chapters 5 to 7) and in the chapter on controlling growth (chapter 10).

FIGURE 3.1. Nursery crops are primarily produced outdoors, but plants may be grown in various structures such as greenhouses and shadehouses, especially during propagation and the early stages of production. Greenhouses are used to protect crops from inclement weather and pests or to provide optimal conditions for plant growth. *Photo:* J. H. Lieth.

Greenhouses

Greenhouses (fig. 3.1) are used to provide optimal environments for plant growth and development or simply to shield crops from pests (exclusion) and the elements (e.g., precipitation, wind, temperature extremes). The functionality of a greenhouse is determined by its structure, glazing material, and environmental control systems. Greenhouse structural materials can be wood, plastic, aluminum, or steel, but they must be matched to the covering (glazing material) and control equipment so as to provide adequate structural support even under the most adverse weather conditions (e.g., ice accumulation, strong winds). Various types of glazing materials are available as greenhouse covers, and they differ in purchase cost, installation and maintenance costs, light transmission characteristics, and heat loss. Control equipment is typically built into the greenhouse for more precise environmental control. For cooling, this may include side or ridge vents, motors to open and close vents, fans to move air out of the greenhouse, and possibly wet pads or fog systems for evaporative cooling. For heating, this may involve mounted unit heaters, duct work, or steam or hot water pipes. Various other pieces of equipment may need to be mounted to the greenhouse structure—for example, high-intensity discharge (HID) lamps for supplemental lighting, shade curtain or black cloth systems, overhead irrigation systems, and horizontal airflow fans. Each additional item requires that an adequate support structure be available beyond what is dictated by the glazing material and climatic forces such as wind, rain, and snow load.

Glazing Material

The National Greenhouse Manufacturers Association (NGMA) lists three categories of materials for glazing commercial greenhouses. Type I glazing materials are thin plastic films. Polyethylene is the most commonly used plastic film, but there are others (e.g., ethylene vinyl acetate, polycarbonate, polyester, polyvinyl chloride, polyvinyl fluoride). Type II glazing materials are rigid plastic panels (e.g., fiberglass-reinforced plastic, acrylic, polycarbonate). Type III glazing materials are glass (e.g., annealed, tempered, laminated). The choice of glazing material is primarily dictated by the local weather conditions and the desired light transmission and insulating characteristics.

Most glazing materials are manufactured to meet industry standards for light transmission according to the Association of Standards and Test Methods (ASTM) guidelines. Glazing materials should have a high transmittance for photosynthetically active radiation (PAR). PAR is the amount of light available for photosynthesis and plant growth in the wavelength range of 400 to 700 nanometers, which approximates the same range as visible light. Light transmittance of the glazing material is the ratio of the measured radiation intensity beneath the covering material to that measured simultaneously above the covering material in the same waveband. It should be noted that other factors besides light transmission properties of the glazing material influence PAR into the greenhouse, such as greenhouse location, season, orientation of the glazing surface to the sun, whether the greenhouse is freestanding or gutter-connected, structural elements (e.g., pipes, light fixtures), and number of glazing layers.

In addition to light transmission, the amount of heat loss from the greenhouse glazing material should be considered, especially in colder climates. The insulating ability of transparent coverings primarily depends on the number of glazing layers. During the winter, multiple-layer covers (e.g., double polyethylene, twin-wall polycarbonate, structured sheets, Thermopane glass) require less energy to maintain the inside air temperature than single-layer covers. However, the trade-off is that with each additional layer of glazing, there is a corresponding reduction in light transmission. Most greenhouses are either single or double glazed. Triple-glazing materials are available, but the loss in light caused by the third layer usually limits their application. Double glazing with glass is generally only economically feasible in areas where winter temperatures routinely get very low.

The energy efficiency of specific glazing materials is rated by the manufacturers with either "U" or "R" values. The U value, or heat transfer coefficient, is the overall rate of heat or energy transfer—the lower the U value, the more resistance to heat transfer. The R value is the reciprocal of the U value (= 1 ÷ U)—the higher the R value, the more resistance to heat transfer.

Heat loss is also affected by air infiltration. Infiltration losses may be significant in older glasshouses, particularly if the joints between glass panes are not sealed. The manner in which the glazing material is attached to the greenhouse also affects air infiltration—gasketed systems are tighter and more energy efficient than nongasketed systems.

The life expectancy of the glazing material is another important consideration. According to the NGMA, the usable life of glazing materials is defined as the number of months or years that a material will function effectively in a commercial growing application. To function effectively, the glazing material must retain some portion of its essential properties of strength and inherent light transmission. Specific ASTM test procedures exist for measuring longevity. Ultimately the life expectancy of a product will be affected by climate and environmental conditions, as well as whether or not the glazing is properly maintained. Average life expectancy for glazing materials reported in the literature is somewhat variable and is best used as a guideline for comparative purposes.

Many other factors are to be considered in the selection of the glazing material, such as maintenance and installation costs, budget, tax laws, and zoning regulations. The Occupational Safety and Health Administration (OSHA) governs the installation of greenhouse coverings. Permits may be required, depending on local zoning and building codes as well as the type of glazing material. Plastic greenhouses may be considered temporary structures with lower taxes and fewer permit requirements than permanent buildings. Check with local agencies that enforce building codes and zoning, as regulatory requirements vary.

A description of some of the more common types of glazing materials follows.

Polyethylene film

The least expensive material to purchase is polyethylene. It can be installed as either a single layer or as two layers (called double poly in the industry). Double polyethylene consists of two layers of film plastic with air inflation between the layers to provide added insulation. The layers are kept about 3 to 6 inches apart by a small fan that pumps air between the layers, sustaining a slight positive pressure. Optimal inflation pressure will depend on weather—if overinflated, the film stretches, making it weaker; if underinflated, the film flaps with the wind and wears prematurely. Overinflated polyethylene can also stress the structure and could result in breakage of trusses and ridges. Monitoring and maintaining proper pressure between the two layers will prevent uneven stress on the material and the greenhouse structure.

The air for inflating the double polyethylene should be taken from outside the greenhouse. Using humid warm air from inside the greenhouse can result in substantial condensation in the space between the layers. Such water is trapped and will accumulate so that a large puddle eventually forms within the space between the plastic film. Algae tend to grow in the condensate, reducing light intensity. Unless this water is removed, a tear will result due to the weight of the accumulated water. Such a tear is difficult to repair and may require replacement of the entire greenhouse cover.

New single-layer polyethylene allows for light transmission of more than 95 percent when untreated and 87 percent when treated with an ultraviolet (UV) radiation stabilizer, although some loss of light transmission occurs with age. There is less light penetration in a double-polyethylene greenhouse; however, the heat retention in the winter is significantly better than for single-layer materials (table 3.1). Plastic glazing generally provides more diffuse light than glass. Diffuse light can potentially reach deeper into the plant canopy than direct light, since diffuse light is scattered more uniformly throughout the greenhouse.

The UV component of sunlight causes polyethylene plastic to significantly degrade and weaken so that it becomes more susceptible to tearing with age. Most polyethylene films used in greenhouses are manufactured with a UV stabilizer to extend the life of the material. These UV films have a specific age rating and may even be guaranteed for 4 years of life. Double polyethylene lasts slightly longer than single because the inflated material provides more rigidity, so that less flapping results in greater longevity. However, the effect of UV radiation is the same, and the life of double polyethylene material is still typically only 3 to 4 years.

Polyethylene that is treated with anticondensation additives (antidrip plastic) is also available. This material reduces condensation damage to plants below it, reduces the incidence of moisture-dependent diseases, and improves light availability to the plants. The surface treatment channels condensation water along the film so that it drains to the side rather than forming droplets that drip onto the plants.

Thermal polyethylene or infrared (IR) film prevents the reradiation of infrared energy. Energy absorbed by objects inside a greenhouse during the day is reradiated at night, leading to rapid cooling inside a structure on cloudless nights. IR film can prevent up to 50 percent of the reradiation loss and may reduce heating costs.

Replacement of polyethylene glazing needs to be scheduled before a tear occurs, since unexpected winds can quickly destroy the entire covering if it is torn. The sudden loss of the cover can lead to crop failures, particularly during the winter months or during storms. Replacement should be based on a schedule that does not expose the plants to harsh, detrimental conditions for the time required to replace torn material. When polyethylene glazing becomes dry and brittle, it is usually a clear sign that replacement is overdue and that the covering could soon fail (tear) during inclement weather.

Fiberglass-reinforced plastic

Fiberglass-reinforced plastic (FRP) has been in use for many years for covering greenhouses. It is a light material that can be installed as A-frame structures or bent into Quonset-style shapes. The material is generally corrugated, providing it with strength so that it can be used where windy and snowy conditions may occur. FRP is the most flammable of the rigid glazing materials.

Light transmitted through FRP glazing is diffused or scattered by the fibers in the panels, resulting in a more uniform light intensity as compared with glass. Light transmission starts out high (88%) but declines with exposure to sunlight. Significant yellowing may occur, particularly if no UV-resistant protective coating is provided. When the plastic that binds the glass fiber becomes abraded by the elements, fibers are exposed and the surface becomes rough. These fibers can catch dust that may build up significantly over time, reducing the amount of light penetrating into the greenhouse. However, UV resistance can be improved by the use of polyvinyl fluoride coatings and acrylic additions to the polyester resin, prolonging the life of the material.

Polycarbonate and acrylic panels

Single- or multiple-layer plastic sheets made from either polycarbonate or acrylic are light and strong. Large sheets can be used to cover relatively few support structures, resulting in high light transmission. Multiple-layered sheets provide excellent insulation characteristics, although the amount of transmitted light decreases with the number of sheets.

Acrylic has excellent outdoor weathering properties. It is UV and shatter resistant with limited impact strength. An impact-modified acrylic is available for locations where hail is a concern, but it is more expensive. Polycarbonate has high impact strength when new and is the most fire resistant of all the plastic glazing materials. Polycarbonate materials usually come with a UV coating to prevent deterioration in sunlight; thus it is extremely important to install the panel so that the UV coating is outside, protecting the plastic from the harmful effects of the sun.

Glass

The oldest greenhouse glazing material is glass. It is still favored by many growers, particularly those who consider

Table 3.1. Common greenhouse glazing materials and their properties

Glazing material	PAR transmission (%)	U values/heat transfer coefficients (Btu/hr/ft^2/°F) = 1/R	Average life span (years)
glass, single layer	88–92	1.13	> 50
polycarbonate panels, twin-wall, 8 mil	79	0.65	10
fiberglass-reinforced plastic	88	0.92	5–20
polyethylene film, 4–6 mil, single layer	87	1.20	4
polyethylene film 4–6 mil, double layer	76	0.70	4

Source: Adapted from Nelson 2012.

FIGURE 3.2. Building codes specify the type of glass permitted in glasshouse roof applications. Tempered and laminated safety glass increases worker safety in case of breakage. *Photo:* J. K. Clark.

greenhouses a long-term investment. Glass is the most expensive material to purchase and install, and it requires more substantial structure than plastic glazing material due to its weight; however, amortization over its long life span may make glass a more cost-effective cover than some others.

As glassmaking technology has improved, it has become possible to use larger panes so that supporting joists and beams can be spaced farther apart. This improves the light levels inside glass greenhouses over those constructed in the past. Clean glass allows for transmission of 88 percent (float glass) to 92 percent (low-iron glass) of the visible light of the sun and has a heat transmission coefficient of 1.13 Btu (see table 3.1). These light and heat metrics are comparable to new single-layer polyethylene. However, the percentage of direct radiation transmitted through glass is the highest of all glazing options. Moreover, glass blocks a major portion of the UV spectrum and, unlike polyethylene, is inherently resistant to UV-radiation degradation.

Tempered and laminated safety glass is not only stronger than conventional annealed glass but also increases the safety of people working underneath it in case of breakage. Often building codes specify the type of glass permitted in roof applications (fig. 3.2).

The main advantage of glass is its relatively long life span. However, heat loss is greater than with double polyethylene (Thermopane glass can be used but is expensive). Maintenance on glass is similar to other rigid materials and includes washing off dust to improve light transmission.

Supplemental Lighting

To increase plant biomass production in areas where sunlight levels are low or days are short, supplemental lighting can be used to provide PAR, either supplementing sunlight or giving extra hours of light to the plant. This is only feasible if light is the only factor limiting productivity. If other variables (e.g., carbon dioxide concentration, water, nutrient status) are insufficient, then supplemental lighting will not achieve the desired effect. Supplemental lighting typically augments existing light by about 10 percent of full sunlight. On days when the ambient light is low, supplemental lighting could double crop photosynthesis, while at high ambient light levels this may boost rates only marginally.

Using lamps to extend the day by lighting at night is feasible for many crops and locations; economic feasibility depends on electricity rates and return on that investment (fig. 3.3). It is important to note that this could interfere with photoperiodic-responsive crops (e.g., prevent flowering of short-day plants), even if these are located some distance from the supplemental lighting (see chapter 10 for information on photoperiodic control).

Supplemental lighting of crops during the day is conducted most economically by only lighting when ambient daytime light levels are so low that productivity is significantly reduced (e.g., during the winter, in the early morning or early evening, and during cloudy days). This can be achieved by using a light sensor and selecting light threshold levels; computer software is available for use in computer-controlled programs for operating the lights. It is best to operate the lights for extended periods, because short cycles of light will greatly reduce the lamp and ballast life. Therefore, when implementing programs based on light threshold levels for operating the lights, the lights should be set so that they are not turned on until the light levels are below the threshold level for a specific period of time (e.g., 30 minutes); alternatively, lights could be set to stay on for a specific time period (e.g., at least 2 hours), regardless of the light levels.

Lamp types

Various types of lamps are available for supplemental lighting, including incandescent, fluorescent, and high-intensity discharge (HID) types (such as metal halide, low-pressure sodium, and high-pressure sodium). These lamps are very

FIGURE 3.3. Supplemental lighting to enhance daily plant biomass production. *Photo:* J. K. Clark.

energy intensive and careful analysis is needed to determine whether their use is economically feasible. HID lamps, particularly high-pressure sodium lamps, are the most commonly used light source for supplemental lighting in commercial greenhouses (fig. 3.4). In the container plant industry, HID lamps are sometimes used in the production of such crops as plugs and flowering plants when the benefits of increased plant growth and yields offset the cost and operation of the lamps. Unfortunately, much of the energy used to power HID lighting is wasted in generating wavelengths of light that plants do not require for growth and in heat generated by the bulb. Adequate cooling must be provided to ensure that the plants do not overheat, which can also affect costs. It should be noted that HID lamps take a few minutes to ramp up to full light-production levels. This needs to be considered when setting up automatic programs for operating the lights.

The evolution of light-emitting diode (LED) technology over the past decade has resulted in recent improvements with potential benefits to the horticulture industry. Most LEDs emit a narrow range of light (e.g., only one color). The newest LED technology for plant growing systems consists of an array of individual LEDs to provide a desirable light spectrum for photosynthesis. Potentially, LED technology could be used in the future to tailor the wavelength spectrum output of the LED array with the light requirements of the specific crop to maximize production. Some other benefits of LED technology compared with most HID systems include a significantly lower energy consumption, a longer operating life, and a smaller and lighter luminaire (lighting unit that includes the light fixture, lamp or LED array, and power supply). LED technology is beginning to be used in commercial greenhouses. Further research is needed to determine the optimal array light spectra for the growth and development of specific crops when LEDs are used as a supplemental light source.

FIGURE 3.4. High-intensity discharge lamps are the most commonly used light source for supplemental lighting in commercial greenhouses. They are used when the benefits of increased plant growth and yields offset the cost and operation of the lamps. *Photo:* J. K. Clark.

Reducing Light Intensity

Sometimes it is necessary to reduce light intensity in the nursery greenhouse, especially during seasons when light levels are very high and temperature control (cooling) is difficult. Two types of materials are available: shading fabrics and greenhouse shading compounds. A shading fabric can be installed over the crop when control over the light intensity is desired (e.g., fabric can be provided only when needed and light can be maximized during low-light conditions; fabric can be applied to just a few benches instead of the entire greenhouse; and fabric can be temporary and easily removed). Spraying a shading compound on the greenhouse is a less expensive method and is used when repeated removal and reapplication is not necessary or practical.

Shading fabrics

Shading fabrics are used in nurseries to reduce the crop solar load and greenhouse temperature. Shading fabrics installed within a greenhouse can also provide additional insulation on cold days and nights by trapping a layer of air between it and the greenhouse roof. Synthetic fabric choices include polypropylene, spunbonded polyester, and saran, which are available in various levels of shading. Shade curtain systems can be automated using light sensors so that the greenhouse is shaded on bright days and light is maximized during low-light conditions.

FIGURE 3.5. Shade curtains mounted within a greenhouse can be pulled to various degrees of closure to provide shade under high-light conditions and help with temperature control on hot days. *Photo:* J. H. Lieth.

Shade curtains may be installed inside or outside of greenhouses. When installed inside greenhouses (as, for example, in the greenhouse shown in figure 3.5), curtains are suspended from the overhead structure and installed in sections. Installation of a shade curtain over a greenhouse is less common but can be accomplished by erecting several tall poles outside, stringing wire or cable between these poles, and using wire, cable, or rods to pull the curtain. While installation inside the greenhouse protects the curtain material and hardware from wind, rain, and UV light, exterior installation also has benefits. Outside shade curtains keep energy out of the greenhouse, which significantly reduces cooling requirements and the solar load on plants. Another reason to install shade curtains outside the structure is the lack of headspace inside some greenhouses. Further, curtains installed inside generally create a certain amount of shade even when completely retracted, whereas in outside installations the curtains can be pulled completely back so that they do not block light when fully retracted.

Shading fabric is also used in propagation, where covering a partial bench of cuttings can reduce the light intensity and increase the humidity while still allowing some air exchange. This is desirable when light intensity and temperature may be too high to prevent water stress even when mist is used. Because shading fabric provides air exchange, there is less likelihood of the cutting environment becoming too hot as compared with polyethylene propagation tents. Note that shading fabric, newspapers, and other materials should never be placed directly on the cuttings, as this practice can favor diseases.

Shading compounds

Commercial shading compounds, whitewash, and materials used as protectants for sunscald on fruits and vegetables have been used for greenhouse shading with good results. This method is considerably cheaper than fabric shade curtains, but it has limitations. These include

- if just a few benches need shading, use of a shading compound is not an option because the entire greenhouse or greenhouse range is typically sprayed
- the sprayed area will be continuously shaded, even during cloudy and low-light days when shading could result in insufficient light in the greenhouse for photosynthesis
- in areas with frequent rain, the shading material is partially lost with each shower, requiring frequent reapplication that increases labor costs
- shading materials can be difficult to apply uniformly

If used with material other than glass, shading compounds that are formulated specifically for nonglass surfaces should be selected to avoid damaging the plastic and to facilitate removal of the shading compound when shading is no longer required. It is important to apply and remove exterior shading compounds in such a way that minimal runoff is produced.

Heating Systems

Several options exist for heating a greenhouse, including hot water, steam, or forced hot air. Forced hot air heaters have self-contained furnaces or may derive heat from hot water or steam generated by a boiler (fig. 3.6). Exhaust gases from forced hot air heaters that are gas fired or use heating oil should be properly vented out of the greenhouse to avoid the presence of harmful pollutants that can affect plants and people working in the greenhouse. Examples of harmful pollutants that can be generated by combustion of forced air heaters include carbon monoxide, sulfur dioxide, and ethylene gas.

Hot water or steam is generated by a boiler located in a nearby building. Insulated pipes and manifolds deliver the steam or water to the zone being heated. Steam can carry much more energy than hot water, but it requires a pressurized system that is more difficult to maintain. Both systems deliver the heat from pipes or tubing that run the length of the benches and along the greenhouse perimeter. The length of pipe needed inside the greenhouse varies according to the heat requirements, the temperature of the water, and whether fins are attached to the pipe to help transfer the heat to the air.

Hot water is also used in bench-top heating systems (fig. 3.7). Here, hot water travels through small black tubes that are laid out at close intervals on the bench. Potted plants may sit directly on the pipes or on some material that may act as a spacer between the tubes and the plants. Care needs to be taken to prevent the plants' root zones from getting too hot. Warmer pots and drier air around the pots also result in greater evaporation from the container medium so that more frequent irrigation may be necessary. This type of heating system can also be installed outdoors to protect the roots of containerized plants from cold temperatures (including frost).

FIGURE 3.6. Forced air heaters should be properly vented out of the greenhouse to avoid harmful pollutants in the greenhouse. *Photo:* J. H. Lieth.

FIGURE 3.7. Bench-top heating. Note that the metalwork above the tubes is there to keep the bottom of the pots from making contact with the hot tubes. Setting pots directly on the tubes results in excessively warm root zones for the plants. This causes poor root growth and excessive drying, requiring considerably more irrigation. *Photo:* J. H. Lieth.

Although not widely used, infrared heating systems can also heat greenhouses, supplying radiant energy and heating plants like the sun (fig. 3.8). Greenhouses heated with infrared heat often feel cold, as the air is not heated. This type of system is used primarily on uniformly low-growing crops such as bedding plants and some potted crops. Crops located near the sides of the greenhouse may not receive enough heat and could suffer cold damage.

Cooling Systems

Passive cooling systems

Most greenhouses are equipped with vents so that outside air can be exchanged for inside air in order to bring the temperature in the greenhouse closer to a selected temperature set point at times when that is feasible. Greenhouse vents typically include roof (ridge) and sidewall vents for natural air movement and cooling. In coastal California, many greenhouses are cooled with these types of passive cooling systems, in which primarily sidewall vents or roll-up walls open to allow natural ventilation from cool ocean breezes. This will work as long as the outside air has a lower temperature or is drier than that in the greenhouse and there is sufficient outside wind velocity (greater than 2.3 miles per hour). Vents can be manually opened, although most systems are automatically raised or lowered with electric motors and integrated with computerized climate control systems.

Exhaust fans and fan-and-pad cooling systems

Fans are frequently used to exhaust hot, humid air from the greenhouse (figs. 3.9 and 3.10). There should be openings in the side or end of the greenhouse wall opposite the fan so that cooler air can enter the greenhouse to replace the hot

FIGURE 3.8. Underbench radiant heaters for greenhouse temperature control. *Photo:* J. K. Clark.

air exhausted by the fans. Fans should be mounted on the side of the greenhouse that is away from prevailing winds. Louvers on the outside of the exhaust fan prevent cold air from blowing into the greenhouse when the fan is off (see fig 3.10).

A wet pad at the opening in the wall opposite the exhaust fans can increase the cooling effect of the fans by evaporating water into the air as it comes into the greenhouse. This is effective when relative humidity is below 90 percent (fig. 3.11).

As the cooled air moves through the greenhouse toward the exhaust fans, it picks up heat from the greenhouse environment. Therefore, such systems experience a temperature gradient between the inlet (pad) and the outlet (fan) sides of the greenhouse. This temperature gradient should be minimal to provide all plants with similar conditions, though a gradient of 7° to 10°F is common.

It is important to select the proper fan size and quantity and to develop an energy-efficient design for fan-and-pad systems. See the following publications (listed in the References at the end of this chapter) for more information: *The Commercial Greenhouse* (Boodley and Newman 2009), *Greenhouse Operation and Management* (Nelson 2012), and *Temperature, Humidity, and Ventilation Interrelationships* (Takakura 1994).

FIGURE 3.9. Outside greenhouse view of exhaust fans with screens for worker protection. *Photo:* J. H. Lieth.

FIGURE 3.10. Outside greenhouse view of exhaust fans with louvered shutters to prevent cold air from blowing into the greenhouse when the fans are turned off. *Photo:* J. H. Lieth.

Fog and mist systems

Fog and mist systems spray water particles into the air around the plants as part of temperature and humidity control. High-pressure fog systems can be used in greenhouse cooling. In these systems, water is forced under high pressure through small nozzles that spray tiny water particles that evaporate and cool the air. Fog systems have water particles that are so small that they can stay suspended in the air and not fall on the plants. Mist systems generally deliver larger water particles that will not evaporate completely before contacting a surface, thus wetting plants, benches, floors, and so on. Fog, therefore, is more desirable for evaporative cooling, whereas mist is usually better for plant propagation. However, the use of a high-pressure fog system that is poorly designed or with imperfect control can also lead to areas of freestanding water. Further, if the generated fog hits greenhouse structure elements, condensation will result in droplets that can drip onto plants, increasing the incidence of moisture-dependent diseases. To ensure that a fogging system works efficiently, high-quality water and filtration systems are extremely important. This is because the small nozzle holes are easily clogged if water used in these systems contains sand, silt, or calcium and magnesium carbonates.

Cooling effectiveness of fog and mist systems depends on ambient air having a relative humidity of less than 90 percent; however, operation of these systems causes air to rise above this level of humidity. Thus these systems must be used in conjunction with careful monitoring of relative humidity and the use of exhaust fans to exchange the humid air with drier air from outside. If drier air from outside is not available, then a fog or mist system will be ineffective as a cooling system.

Mechanical Air Circulation

Air currents due to natural convection cause the temperature to vary at different locations within the greenhouse. The location of heating pipes and fans affects greenhouse airflow, as does wind outside the greenhouse. This can result in a 10° to 15°F difference in nighttime temperature between the coldest and warmest spots in the greenhouse. Moreover, radiant cooling on clear nights can cool plant leaves several degrees below air temperature. When leaf temperatures cool much below the air temperature, the dew point is reached and condensation occurs, favoring foliar pathogens that can harm the plants.

Perforated polyethylene tubing

In the past, greenhouse growers used large fans to move warm air from the greenhouse peak down to plants. Perforated polyethylene tubing (fig. 3.12) was an advancement that resulted in more uniform air distribution in the greenhouse. Air is blown through the tubing and exits through the holes, mixing with the greenhouse air. Air can be blown through the tubes from a heater to improve distribution

FIGURE 3.11. Wet pads on one wall (A) with exhaust fans on the opposite wall (B) are used to cool air by evaporating water into the air passing through the pad. *Photos:* J. H. Lieth.

FIGURE 3.12. Perforated polyethylene tubing used for ventilation and air distribution uniformity. *Photo:* J. H. Lieth.

of heat; for winter ventilation, cold outside air brought in through the tube can be mixed with warm greenhouse air to prevent sudden drops in temperature. A disadvantage of the polyethylene tubing is that it casts shadows over the crop and collects dirt and debris. Many growers have switched to the use of horizontal airflow systems, which are typically the most efficient method for uniform air circulation. However, polyethylene tubing seems to be coming back into use for particular applications, such as in low-profile houses, where the use of tubing in positive-pressure cooling systems can be more efficient than horizontal airflow fans.

Horizontal airflow system

A greenhouse horizontal airflow (HAF) system consists of small overhead horizontal fans that are used to increase temperature uniformity and reduce condensation. When properly installed, the HAF system can overcome air turbulence and friction loss and keep air moving. The moving air dries moisture from the plant canopy, resulting in lower relative humidity. A hot air heating system can be used to replace one or more of the HAF fans, and this could save installation costs; however, operating costs will be significantly higher. The HAF system should be operated 24 hours a day, except when the exhaust fans are on or when the vents are open. Fans should be located below shade curtains so that air movement can be maintained at night. For information on the use of HAF systems, see chapter 10.

Environmental Monitoring Equipment

An important aspect of managing greenhouse crops is to have accurate information on the current and past status of the environment. A graph of environmental data provides a useful tool for diagnosing problems. It is particularly important to track and record temperature and relative humidity data, which can be used to identify many problems. Light and carbon dioxide are other environmental variables that are tracked for monitoring or controlling the environment—accurate monitoring equipment is vital for this.

Environmental monitoring equipment in greenhouses ranges from hand-held sensors (providing instantaneous readings and capable of being used in many locations) to permanently installed sensors that feed into greenhouse control systems. Prior to the advent of electronic monitoring equipment, many growers used a hygrothermograph (an instrument for recording temperature and humidity on a single chart) to monitor the environment. Now many growers have fully automated systems that track temperature, light, relative humidity, wind direction and speed, and rain.

Greenhouse environmental control computers have sensors that monitor the environment in each greenhouse range. Alternatively, a number of individual sensors can be temporarily installed where needed for environmental monitoring (e.g., soil moisture, electrical conductivity). Extreme care should be taken to keep all sensors properly maintained. No sensor should be trusted to provide data for more than 1 year without recalibration. Temperature and humidity sensors need to be shielded from direct and indirect sunlight

FIGURE 3.13. Temperature and humidity sensor unit connected to a computerized environmental monitoring and control system. Air drawn through the aspirated housing protects the sensors from solar radiation. *Photo:* J. K. Clark.

(fig. 3.13), or they will provide inaccurate information. Manufacturers of sensors and computer-controlled systems have the appropriate housing for the different types of sensors to ensure accurate data collection.

Sensors should always be placed in a location that is representative of the crop, which is often near the middle of a greenhouse, away from heaters, drafts, and fan-and-pad systems. If crop canopy temperature is of interest, then the sensing part of the temperature sensor should be located within the crop canopy. Frequently it is tempting to locate such sensors away from the plant so they do not get in the way. This, however, is a bad idea since the temperature in a crop canopy can be substantially different (plus or minus 5°F) from that in the aisle a few feet away. The same is true for light, relative humidity, and carbon dioxide concentration. Sometimes temperature sensors need to be placed in the root zone. For example, monitoring root zone temperatures can be useful for small containers outdoors to alert the grower of conditions that could lead to potential damage from extreme high or low temperatures.

Control Systems

The level of control over environmental variables can vary depending on the heating and cooling equipment available and the complexity of the control equipment. Vent opening (passive cooling) is usually the first cooling stage, followed by stages comprised of activation of various other active cooling systems (e.g., exhaust fans, wet pads, shade curtains, fog). Heating may also be implemented in stages. If many stages are needed, then each system will require its own thermostatic control. While this could be achieved with simple thermostats, most growers find it preferable to use a computerized control system that manages the implementation of each heating and cooling stage. Step controllers are an alternative between single thermostats and integrated computer systems. Step controllers can process information from different sensors and control equipment in a coordinated fashion. Dedicated microprocessors have expanded capabilities over step controllers and can process more complex data and more equipment.

The level of environmental control dictates the monitoring and control equipment used. A simple polyethylene-covered house can suffice to prevent freezing damage to plants in a mild climate. Where outdoor conditions regularly deviate substantially from the desired crop environment, more sophisticated structures and environmental controls are necessary.

In the greenhouse, environmental control computers provide an excellent tool for implementing desired conditions as well as for managing the monitored data. In such systems, the calibration of sensors is potentially even more important since the sensor is used both for monitoring and controlling the environment. For example, if the calibration of the sensor is off by 2°F, then the actual temperature may be 2°F higher than the sensor reading. The grower may get hints of this problem when the plants finish too fast or too tall, or when the heating bill is higher (or lower) than what it should be. See chapter 4 for more information on automated systems for heating and cooling.

In addition to controlling temperatures, controllers are used in some greenhouses to decide when to irrigate or mist crops. Sensors record relative humidity, leaf temperature, and air temperature around the crop, and calculations of the vapor pressure deficit along with other production parameters determine when it is time for the next fogging or irrigation cycle.

Retractable-Roof Structures

Woven UV-stabilized polyethylene or polyolefin provides a fabriclike covering that is used in greenhouses and shadehouses with retractable roofs. In such structures, a motor operates the covering, opening it in sections when outdoor conditions are favorable for plant growth and closing it when protection from the elements is required (fig. 3.14).

When the roof is open, the increased light intensity can help control flowering, crop timing, and growth habit. Crops can be conditioned to outdoor conditions before shipping, without the necessity of moving them to different environments, which saves labor and material handling. Opening the roof increases natural ventilation, which may eliminate the need for cooling fans, thereby reducing energy costs. Fabric sidewall coverings can also be used to completely open the structure and enhance ventilation.

When the roof is closed, crops are protected from excess high radiation, low temperatures, and wet or windy conditions. Frost protection can also be provided, although temperature control without heaters is limited to about 4° to

FIGURE 3.14. In this retractable-roof greenhouse, the glazing that covers the roof is opened when outdoor conditions are favorable for plant growth and closed when protection from the elements is required. *Photo:* J. H. Lieth.

8°F compared with the outside temperature. Heating loss is greater than in conventional greenhouses, due to the single layer of plastic fabric and greater air infiltration through gaps and cracks in the seals. However, internal energy curtains can be used to reduce heat loss.

Retractable-roof greenhouses can be programmed to respond to ambient environmental conditions by controlling the opening and closing of the roof and sidewalls. Control that is based on environmental conditions is superior to operating solely based on time of day or general weather conditions because it optimizes the crop environment. Control can be automated using computerized systems with sensors that track PAR and temperatures inside and outside the structure.

However, frequent daily opening and closing of the roof in response to changes in air temperatures and sunlight results in wear and tear on the fabric, particularly at the crease lines, where hooks connect the curtains to the wire framework or where the fabric slides on wires or supports. Thus, frequent use can limit the life of the cover. Moreover, long-term mechanical wear on the drive system is a concern even with proper and timely system maintenance. These structures, therefore, work most efficiently when the cover is left open for full outdoor growing and is only closed to protect the crop from inclement weather. Computerized systems can be programmed to minimize roof and sidewall curtain movement when appropriate threshold brackets are set.

Retractable-roof structures can be used year-round in areas with moderate climate and rare snowfall or freezing temperatures. In California, such structures have been adopted by a few growers over the last two decades for growing a variety of crops, primarily those in smaller container sizes and susceptible to harsh environmental conditions. These structures can also be useful when nurseries are located away from the mild coastal climate—for example, in desert areas, where high daily and seasonal temperature fluctuations prevail. In desert climates, summer protection from intense sunlight and high temperatures is the major advantage. Retractable-roof structures are used in many climates to extend the growing season of crops that can be produced rapidly or to finish crops early in the season.

Light transmission of the covering depends on fabric thickness. Various shading levels are available—and when covers are selected to provide shade and the opening and closing of the roof and side vents is the only environmental control system, the structure is essentially a retractable-roof shadehouse. The covering material for retractable-roof structures can be watertight or can allow water to drip through. Covering materials that allow water to drip through are primarily used for retractable-roof shadehouses with flat roofs. This works well for plants in larger containers (4 to 6 inches or larger) but not for plants in plugs or container sizes smaller

FIGURE 3.15. Open-roof greenhouses hinged at the gutter line so that the entire roof can be opened. Natural ventilation is considerably more effective than in a fixed-roof greenhouse. *Photo:* J. H. Lieth.

than 2 inches. A-frame retractable roof structures built for areas with ample rain and snow usually use watertight covering for the roof. The expected life of the covering material is typically about 5 to 7 years, although some of the newer products may last 10 years depending on use and climate.

Note that a retractable-roof greenhouse is not the same as an open-roof greenhouse. An open-roof greenhouse has come to be defined in the industry as a rigid roof panel that is typically hinged at the gutter line so that the entire roof can be opened (fig. 3.15). Covering materials used in open-roof greenhouses include glass, polycarbonate, and double polyethylene but not fabric. An open-roof greenhouse is similar to a fixed-roof greenhouse, but natural ventilation is considerably more effective with the open roof.

Shadehouses

Shadehouses are usually provided for crops that look better or grow better if not exposed to full or direct sunlight—plant species that generally have their origin in nature in the understory of forests. Shade structures provide an important tool for outdoor nurseries by creating a cooler, darker microenvironment. Lath houses consist of structures covered with slats of wood, metal, or other material, spaced to allow some sunlight to penetrate between the laths. Shadehouses that use screening materials (fabrics) are generally less expensive to construct; thus they are more common. Modern shade structures are usually made from plastic shading material suspended over the plants from steel poles.

The decision to build a simple shade structure should be weighed carefully against the selection of a more expensive, permanent structure. Without environmental control, for example, the use of the structure is very limited and possibly labor-intensive. Upgrading such a structure later to a greenhouse is difficult, expensive, and subject to permit and code requirements applicable for greenhouses.

The frame for a shade structure can be built from poles or pipes; the structure usually has a flat top with open or closed sides. These frames need to be sturdy enough to withstand strong winds and snow loads. Good ventilation is important, especially in lower structures, to prevent heat buildup. Movable sidewalls of plastic-covered houses are desirable to improve air circulation or protect plants from wind or heat loss.

Shadehouses are covered with woven or knitted cloth, available in a range of densities to provide from 30 to 80 percent shade. Shade cloth also comes in various colors. Common are fabrics made of black-pigmented polyester. The pigment is used as an agent to resist UV degradation, but because of the color, it absorbs light and reradiates thermal energy that can be problematic in some installations. Replacing the traditional black shade cloth with lighter colors, such as beige or light green, reduces temperatures inside the shade structure when outside temperatures and radiation are high. Other fabrics are usually made of polyolefin combined with aluminum (to reflect light and reduce energy absorption) and interwoven with polyester threads.

Colored shade netting affects the spectral quality of light below the net and thus the vegetative or reproductive development of the plants. This type of shade works well when large areas of the same crop are grown and the crop is known to respond favorably to the particular color and density of netting. Further details are provided in chapter 10.

Shade structures also afford protection from wind. This can be an important consideration, not only in keeping plants upright in windy areas but also in improving irrigation uniformity with overhead irrigation. When overhead irrigation is used in shade structures, the sprinkler heads should be installed under the shade rather than over it. This will ensure that the shade frames and covers do not obstruct the flow of water and reduce irrigation uniformity. Refer to chapter 7 for more information about irrigation systems and irrigation uniformity.

Screenhouses

A screenhouse is similar to a shadehouse but is designed specifically for the purpose of creating an enclosed environment for plants that excludes various insect pests. Screen hole size is based on the size of the insect to be controlled. Screening for small insects such as thrips requires very tight screening, which reduces airflow. Thus the interior environment of these screen structures can be very hot during the summer if measures are not taken to cool the structure. Insect screens can also be added over greenhouse vent openings to exclude insects. This is discussed in chapter 12.

Hoophouses

Hoophouses are constructed from pipe bent into hoops and draped with polyethylene film (fig. 3.16). Hoophouses are also known as high tunnels because they have sufficient height to allow people to work in the interior space with the plants, compared with low tunnels and cold frames.

Hoophouses are ventilated through roll-up sidewalls and large open doors at both ends. Some hoophouses are fitted with end walls that frequently can be opened for ventilation. These structures may or may not be equipped with environmental control systems. Hoophouses are normally defined as unheated greenhouses; when a grower installs a permanent heating system and other environmental controls, they transform their hoophouse into a structure more akin to a greenhouse (Blomgren and Frisch 2007). Because hoophouses generally lack foundations and permanent environmental control systems, regulations and permits in California are typically less rigorous for these types of structures than for greenhouses. However, check with your local authorities, as regulation requirements vary.

Hoophouses are a useful, relatively inexpensive tool in mild-climate areas, where they are often used during the cold weather months. In such cases the polyethylene film is fully or partially removed in the spring and reinstalled in the fall or winter. In cold and hot weather, however, crops could be subjected to suboptimal, if not destructive, conditions if the hoophouses are not equipped with environmental control systems.

FIGURE 3.16. A hoophouse is constructed from pipe bent into hoops. These structures are useful in mild-climate areas, where they are frequently used to protect crops during cold weather. *Photo:* J. H. Lieth.

Cold Frames

Cold frames are structures built close to the ground or bench that create a microenvironment more conducive for plant growth or survival than outdoors. Generally these structures are used to protect plants from frost, but they could also be used to control the impact of precipitation. Cold frames are usually simple structures that are put into place and manipulated manually. Consequently, careful attention must be paid to environmental control within the enclosed space. Under full sun, the temperature can rise very rapidly if air exchange does not occur. Like hoophouses, cold frames are essentially small primitive greenhouses.

Germination or Growing Rooms

Special rooms or chambers are used to achieve uniform seed germination or to reduce the production time of a crop. Such rooms are equipped with precise temperature and humidity controls, and they frequently include fog systems to increase humidity. Once the seeds have germinated or desired growth is reached, the crop is moved to the greenhouse and transplanted. With the advent of plug production, specialized growers handle this part of the production cycle so that most nurseries can operate without germination facilities as long as they are willing to purchase the plugs.

References

Bartok, J. W. 2001. Energy conservation for commercial greenhouses. Ithaca, NY: Natural Resource, Agriculture, and Engineering Service, Cooperative Extension. NRAES-3. http://palspublishing.cals.cornell.edu/nra_order.taf?_function=detail&pr_booknum=nraes-3.

———. 2005. Retractable roof greenhouses and shadehouses. USDA Forest Service Proceedings RMRS-P-35: 73–75.

Bartok, J. W., and V. Grubinger. 2011. Horizontal air flow is best for greenhouse air circulation. eXtension website, http://www.extension.org/pages/28043/horizontal-air-flow-is-best-for-greenhouse-air-circulation.

Beytes, C., ed. 2003. Ball redbook, vol. 1: Greenhouses and equipment. 17th ed. Greenville, OH: Ball Publishing.

———. 2010. Open-minded. GrowerTalks 73(11): 44–46.

Blomgren, T., and T. Frisch. 2007. High tunnels. Burlington: University of Vermont Center for Sustainable Agriculture. Burlington, Vermont. UVM Center for Sustainable Agriculture website, http://www.uvm.edu/~susagctr/Documents/HighTunnels.pdf.

Boodley, J., and S. E. Newman. 2009. The commercial greenhouse. 3rd ed. Clifton Park, NY: Delmar.

Both, A. J. 2000. Some thoughts on supplemental lighting for greenhouse crop production. New Brunswick, NJ: Center for Controlled Environment Agriculture, Cook College. Rutgers School of Environmental and Biological Sciences website, http://aesop.rutgers.edu/~horteng/ppt/papers/SuppLightingPaper.pdf.

———. 2003. Introduction into greenhouse environment control. Horticultural Engineering 18(6): 1–5. Rutgers School of Environmental and Biological Sciences website, http://aesop.rutgers.edu/~horteng/newsletter/2003/vol18-6Dec2003.pdf.

Buck, J. 2011. Possibilities of LED lighting. Greenhouse Product News Magazine 21(3): 25–28. Greenhouse Product News website, http://www.gpnmag.com/possibilities-led-lighting.

Fisher, P. R., C. S. Donnelly, and J. Faust. 2001. Evaluating supplemental light for your greenhouse. Ohio Florist's Association Bulletin No. 858. University of New Hampshire Cooperative Extension website, http://www.greenhouse.cornell.edu/crops/factsheets/SuppLight.pdf.

Giacomelli, G. 2002. Introduction to greenhouse glazing. Tucson: University of Arizona Controlled Environment Agricultural Center. UA CEAC website, http://ag.arizona.edu/ceac/sites/ag.arizona.edu.ceac/files/Glazing%20Ball%20Red%20Book%20Chris%20Beytes.pdf.

———. 2003. Evaporative cooling system: Mist and fog. Tucson: University of Arizona Controlled Environment Agricultural Center. UA CEAC website, http://ag.arizona.edu/ceac/sites/ag.arizona.edu.ceac/files/Evaporative%20Cooling%20system%20%20Mist%20%26%20Fog%20MS.pdf.

Nelson, P. V. 2012. Greenhouse operation and management. 7th ed. Upper Saddle River, NJ: Prentice-Hall.

NGMA (National Greenhouse Manufacturers Association). 2004. Glazing standard guidelines and voluntary standard test methods. NGMA website, http://www.ngma.com/standardpdf/Glazing2010.pdf.

———. Helpful hints: Glazing. NGMA website, http://www.ngma.com/industry/Glazing.pdf.

Runkle, E. 2009. LEDs in floriculture. Greenhouse Product News Magazine 19(6): 54. Greenhouse Product News website, http://www.gpnmag.com/leds-floriculture-0.

Takakura, T. 1994. Temperature, humidity, and ventilation interrelationships. In Greenhouse systems: Automation, culture, and environment. Proceedings of the 1994 Greenhouse Systems International Conference: 153–167.

Photo: J. H. Lieth.

Photo: J. H. Lieth.

Mechanization and Automation

Lorence R. Oki and J. Heinrich Lieth

CHAPTER FOUR

Photo: J. H. Lieth.

Nursery growers in California and other areas of the United States produce crops that are labor-intensive and may involve high energy and transportation costs. These factors combined make up a significant part of the total production cost of the product. With increased trade opportunities and competition from offshore markets such as Europe, South and Central America, and Asia, incorporating mechanization and automation into plant production systems is necessary to survive in the global market.

Mechanization is particularly effective in processes requiring an action that is repeated many times and can be easily performed with machinery. When used properly, mechanization can also reduce the negative characteristics of a production task, make the job less demanding, and produce better results consistently. If an operation performed manually is considered repetitive, strenuous, monotonous, or undesirable, it may be difficult for the person doing it to maintain a high level of consistent quality over an extended period of time. Mechanization is also effective in situations where the work is difficult or dangerous without the aid of some sort of equipment. Tractors, for example, can be used to process soils and media much more efficiently than can be done by hand with shovels.

Automation, either alone or in conjunction with mechanization, can increase the level of control over a condition or process. Automation involves the sensing of a condition, assessing the condition, and controlling a process relative to the condition to achieve a desired outcome. A familiar example of automation is the use of a thermostat to control a heating system in a greenhouse. (The thermostat senses the temperature in the greenhouse, assesses [measures] the temperature, and turns on the heating system as the temperature drops below a set point. As the temperature in the greenhouse increases above another set point, the thermostat turns off the heating system.) The incorporation of computers in control processes has increased the ability of machines to perform more difficult tasks such as transplanting and grading.

This chapter describes how automation and mechanization can be used in the nursery to reduce the costs of labor, energy, and other production processes, allowing for increased company and industry sustainability. Facility design and the process of developing a production plan to integrate mechanization and automation are also discussed. In addition, details concerning various types of equipment, machinery, and mechanical aids that can be used for automation are presented, including maintenance aspects.

Cost Considerations

Labor Costs

In most production systems, labor represents the greatest portion of the total cost of producing nursery crops. Plants require a substantial amount of hand labor to produce a high-quality crop that is properly timed. In addition, the actual total cost of labor is not only the direct labor but also includes the cost of hiring employees and training them to the desired skill level. Repeating the hiring and training process seasonally as workloads change can substantially increase these costs. One way to reduce seasonal labor costs is through mechanization and automation.

Utilizing mechanization and automation requires that some investment be made toward the purchase, operation, and maintenance of the equipment that may offset direct labor costs. In some cases, there may be additional costs resulting from the modification of the production facilities in order to accommodate and use the equipment. The decision to invest in automation or mechanization, or both, is made by comparing the cost savings (due to reductions in labor and other resources along with improvements in efficiency and quality) with the cost of purchasing (including financing), installing, operating, and maintaining the equipment.

However, when a production operation uses mechanization or automation, the work done is usually completed in less time and with greater consistency. Time saving is usually the primary factor in deciding to mechanize, but there may be other considerations such as worker health and safety.

The production of greenhouse and nursery crops involves many units: hundreds, thousands, and even millions of plants that may be very small (e.g., plugs or seedlings) or of substantial size (e.g., boxed trees). And because plants need to be exposed to light, large acreages of land are required. This introduces transport and materials handling within the nursery as a production cost factor. Mechanization can reduce the amount of time needed to perform an operation on a single plant, or it can improve the movement of material to and from a workstation. The machine may perform some or all of the work, and in some cases manual labor is reduced to supplying materials to and from the machine or moving the machine to the crop. As the degree of mechanized involvement in a production step increases, the quality of work performed at that step increases in consistency. Completing properly timed tasks at higher quality and greater levels of consistency may translate into higher crop quality, which may, in turn, result in higher product value. This effect is difficult to quantify, but it should not be discounted when considering mechanization. In some cases, it may be the primary purpose for mechanization.

The cost effectiveness of mechanization increases with the number of units that the machine processes. At some smaller quantity, however, it is more efficient to avoid using equipment to do a job. For example, it is not an efficient use of equipment and the supporting people to use a pot-filling machine to fill only twenty 6-inch pots. Setting up the machine, filling the pots, and cleaning up the machine are not worth the time and labor involved.

The seasonal or periodic use of the equipment needs to be considered as well. If equipment may be used only intermittently, the availability of labor during that period should be considered. If labor is not available during that period, then the equipment could mitigate that shortage. However, if labor is readily available, it may not be cost effective to purchase equipment for short periods of use.

A significant and often overlooked benefit of mechanization is an improved employee attitude. Consider the soil-mixing example again. Imagine how much more willing the people assigned to that task would be if a tractor rather than a shovel could be used. Not only could the people do more work in less time, but they could go home less fatigued, more capable to perform the additional duties of home life, and more willing to do the job again the next day.

Energy Costs

Mechanization and automation also have an effect on energy use. Although there may be a reduction in labor, there is generally additional energy used to power the equipment. This increased energy cost is generally lower than the labor costs saved. With the increasing costs of energy and labor, however, these costs must be determined and compared for a proper analysis.

Energy use can also be reduced through automation. Environmental control automation (covered in chapter 3) has been found to significantly reduce energy use for heating and cooling. Where automated irrigation reduces water use (chapter 7), the energy used in pumping irrigation water, as well as the amount of fertilizers used in a liquid feed system, can be significantly reduced.

Unit Costs

One of the most important applications of nursery cost analysis is to estimate the cost of producing a specific crop. This information is the basis for setting prices and selecting a profitable mix of crops. Detailed information on estimating unit costs is presented in chapter 18, but a brief discussion on allocating the cost of mechanization to each plant is provided here. Allocating the cost of mechanization to each plant can be difficult when the equipment is shared with many different crops. One approach is to allocate the equipment costs on a time-per-area (rental) basis. This would involve determining the amount of time a crop spent on a bench and the space that was used. The cost of the equipment can be divided into time (for example, per week) and space (for example, per bench square foot) units. The cost per square foot per week rate is then added to the direct production and other costs of the crop.

Facility Design

Most nurseries do not have the luxury of starting from a clean slate when planning a production facility that fully integrates mechanization and automation into a production system. The more common scenario is the purchase of a piece of machinery for a specific use (e.g., filling cans or pots, material handling, or media mixing) to update an existing production system. A decision that will need to be made is whether the equipment will be used in a fixed location or moved to different sites around the nursery. For example, a pot-filling machine that is used in a fixed location will require space allocated for it. It also means that materials, including pots and media, must be brought to it and filled pots taken away from it. A pot-filling machine that is mobile could be brought into the greenhouse or field for use, but the solutions to the problem of providing supplies and removing filled pots are different.

Production Plan

Integrating mechanization and automation into a nursery or greenhouse operation should be carefully planned and implemented only after completing a thorough assessment. Specific steps in mechanization are

- establishing the specific objectives and purposes for the mechanization or automation
- determining how the use of the equipment integrates with the existing or proposed production system
- developing a plan to use the mechanization or automation
- implementing the plan

A final step is evaluating the implementation of the plan and making necessary adjustments to maximize the effectiveness of the mechanization.

Establishing Objectives

The objectives of implementing the mechanization or automation should be established. Ultimately the purpose is to increase profitability, but how the mechanization will accomplish this should be specific and clearly defined. Some reasons for mechanization or automation have been discussed. The level of mechanization depends on the funds available for capital costs. In a product costing analysis, the costs of capital (depreciation on the equipment and interest costs) and operation should be compared with the labor and other costs that the equipment is intended to displace (return on the investment).

Developing a Plan

The development of a mechanization plan is a stepwise process. Each step of the production process should be listed or diagrammed in the greatest detail possible. Instead of using abbreviated phrases ("fill pots with soil," for example), the list should name all of the supplies required, where they are located, how they will get to the location where the pots are filled, the location where the filled pots will be used, when that space is available, the crop that will be planted in the pots, when the filled pots are needed, and the number of people it will take to accomplish the entire pot-filling process. Important sources of information are the production schedules created for the crops, since they should provide the timing, supplies, space, labor, equipment, environmental needs, and other requirements necessary for developing the mechanization plan.

Each of the steps being considered for implementing mechanization is then determined. This determination can be based on knowledge of existing equipment available for the identified tasks. How the use of the equipment will be integrated into the overall production system needs to be analyzed. Impacts on related processes or conditions need to be determined. Using the pot-filling example again, questions such as these need to be asked: "Will mechanization affect the date when pots are received?" and "When should soil be mixed?" Specific procedures using the equipment should be developed. The procedures should be studied to make sure they are reasonable, effective, efficient, and safe. Procedures for nonproduction tasks, such as moving the equipment or cleaning up after work activities, should be determined as well. Minimum production quantities required to use the equipment, a break-even point, should be determined. This requires determining the total time needed to set up the machine for use, produce a specified quantity, and clean up after use. Then this must be compared with the time it would take to produce a similar quantity using a nonmechanized method.

Visiting other nurseries and even companies of other industries is necessary to compare their production systems as well as their use and integration of mechanization into those systems. Visiting trade shows—not only those that are horticulturally oriented but also those directed toward farm, industrial, food, and plant (manufacturing and factory) maintenance enterprises, as well as product and material handling—may reveal equipment to fill the specified requirements. Comparisons of various pieces of equipment, combinations of pieces, or entire systems are made and the proposed system is refined as necessary. Then the economic decision is made whether to include the equipment in the production system.

Implementation

When the decision to mechanize has been made and the equipment has been purchased, it is relatively easy to integrate it into production since the procedures should have been established as part of the planning process. However, there may be situations that create unexpected results. Attempts to eliminate these conditions should be part of the analysis process. Simple problems may render the equipment partially or completely unusable, although the machinery may be entirely capable of satisfying production requirements. For example, a mobile pot-filling machine that does not physically fit into the greenhouse where it is intended to be used is the result of an oversight that could easily be avoided.

Training employees to use the equipment properly and safely is of the greatest importance. Proper training on the use of the equipment will minimize the learning curve and quickly maximize benefits from the new machine. Training should be provided not only to the people using the equipment but also to those who will be providing support and maintenance to keep the equipment in good condition.

Even as the equipment is used, changes in procedures will occur as the "personality" of the equipment develops. Nursery equipment can be highly customized—and the nuances of the different options on a single piece of equipment, along with the effects of interactions with other equipment and processes, will be revealed over time. As these traits are discovered, slight changes in procedures and even additional modifications to the equipment may result in significant increases in productivity.

All of these steps in evaluating, planning, and implementing the equipment should also include consultation with the manufacturers. They know their equipment and also have experience in introducing it to new users. Their experience will be important in these processes.

Mechanical Aids, Equipment, and Machinery

Irrigation Control and Metering Equipment

Irrigation automation

Hand-watering allows a grower to inspect and custom irrigate each plant on a routine basis. This, however, is a labor-intensive and expensive procedure, and it can result in a very nonuniform application of water that may or may not be intentional. Also, since each individual root zone cannot be seen or inspected to make sure the entire soil volume is wetted, the only way to achieve uniformity in the application is by overwatering each pot. Thus, there is no inherent advantage to hand-watering if it can be done less expensively with an automated system. Highly skilled irrigators can perform this task with sometimes dramatic results, but these individuals are rare and the skill is difficult to teach. In addition, there is not a way to irrigate by hand without generating runoff, although the use of an on/off mechanism to prevent watering the areas between pots will certainly help.

Various automated irrigation systems are described in chapter 7. Any nursery that has not automated its irrigation, at least partially, is wasting money and lowering profits. Consistent uniformity in the applications of water (both control and distribution) will increase consistency in plant quality. In addition, increases in water use efficiency will reduce the generation of runoff and the costs associated with handling that issue.

Fertilizer injectors

Although fertilizer can be applied directly to the root zone of each plant by broadcasting or placing a dry fertilizer product there, this can be labor-intensive and inaccurate. The resulting lack of uniformity will translate into plants of variable quality. In many situations it is feasible to dissolve nutrients into the irrigation water and deliver them with each irrigation. Fertilizer injectors are used to add fertilizer into the irrigation water at a determined rate (fig. 4.1). The feed program can be altered during the production cycle based on crop needs, such as reducing nitrogen to a finishing poinsettia crop.

Injectors vary in complexity, from the simplest hose-end siphons to sophisticated types using variable-speed motors and piston or diaphragm pumps along with meters to measure the flow of water and control systems to regulate the rate of injection. Detailed information concerning fertilizer injectors and their use, including stock solutions, can be found in chapter 6.

Equipment also includes a mixing tank downstream from the injectors to blend the mixture and remove surges or pulses of fertilizer in the irrigation lines. After the mixing, pH and electrical conductivity (EC) meters are used to verify that the irrigation water is of the desired formulation. Data recorders, either digital or analog ones (such as strip charts), can produce a record of the irrigation water's condition over an extended period and may be used to reveal potential problems with the system.

Research is currently being conducted to develop sensors that reliably measure the concentration of each ion in the irrigation solution. These have recently become available and may be able to assist with injecting exactly the right amounts of each nutrient. However, until the use of these sensors is proven to be successful, pH and EC sensors can provide some feedback as to whether the resulting irrigation water is formulated as desired. These sensors can also be tied into alarm systems so that if either factor goes out of a desired range, a warning is activated and the system can be shut down until the problem is corrected. Monitoring the level (tank volume) of each stock solution can also be automated easily to indicate when refilling is necessary.

Fog and mist propagation

Although many types of fog and mist systems exist, their purpose is the same: to introduce water into the atmosphere within a greenhouse. This might be done for two reasons: to raise the relative humidity of the environment or to lower the air temperature inside the greenhouse. Water introduced through mist, fog, or fan-and-pad systems must evaporate in order to lower the air temperature. In the case of mist systems, the water droplets are large enough so that they will wet any exposed surfaces (e.g., floors, bench tops, leaves). Evaporation occurs as the droplets fall from the emitters to these surfaces and also directly from the wetted surfaces. The surfaces, as well as the air, are cooled as a result. When high-pressure fog systems are used, the droplets are small enough (approximately 10 microns) that they dissipate and evaporate as they fall from emitters before any surfaces are contacted and wetted. For cooling to occur in either system, the pre-existing relative humidity must be unsaturated and low enough to allow additional water evaporation.

The controllers used for mist systems may be simple time clocks that turn on the mist for a set duration at a time interval during a chosen time window. Some controllers may operate by simulating the evaporation of water from a leaf to determine when misting should occur. These types of controllers (based on a "mechanical leaf") represent a total integration of all of the environmental conditions that affect the drying of plant surfaces: light, temperature, relative humidity, and wind. Details on the use of time clocks and electronic leaf controllers are provided in chapter 9.

Other controllers may measure environmental conditions and initiate misting based on predetermined criteria entered into the controller. For example, a controller based on solar gain (light intensity and duration) accumulates light units and initiates misting when a set number of units are accumulated. Light units under bright conditions—full sun, for example—accumulate more quickly than when the light is less intense, as might be the case during the early morning or late evening or on overcast days. This controller would start misting more frequently when light is more intense.

FIGURE 4.1. Three injector heads (center) dispense fertilizers at a set rate from stock barrels (right) into the irrigation water. The water then passes through a blend tank (left) to mix the fertilizers. *Photos:* L. R. Oki.

FIGURE 4.2. Moving booms used for irrigation can also be used to apply pesticides or other materials. This one is designed primarily for irrigation in a greenhouse but could also be used for chemigation. *Photo:* J. H. Lieth.

Computers may control misting by using sensors to measure various environmental conditions in the greenhouse and software to analyze the information. The information gathered is used to calculate an estimated (as opposed to measured) rate of evaporation that might be occurring as a result of those conditions in the greenhouse, and mist is started when a set level of evaporation has occurred. Controlling mist successfully using this type of system highly depends on the accuracy of the sensors and the software used.

Spray application equipment

Sprayers are used to apply a number of different materials, most commonly pesticides (e.g., insecticides, herbicides, and fungicides) but also fertilizers, plant growth regulators, and other materials such as leaf polish and antitranspirants. Various factors determine the type of sprayer used, including the purpose of the application, the quantity of material applied, where the application is to be made, and to what the material is applied. The two main categories of sprayers are high-volume hydraulic and low- or ultra-low volume (LV or ULV).

The simplest types of high-volume hydraulic sprayers are the compression sprayers, which may be in the form of a tank carried by hand or mounted on a small wheeled cart or a backpack. Typically 1 to 5 gallons in capacity, they provide a quick way to make applications to small areas. Compression sprayers use a hand pump to compress air within the tank; this pushes the liquid through a wand and nozzle to create a spray. To spray large areas, the substantial volumes of solution that might need to be applied would require many refills, using up a lot of time. But these sprayers are convenient for spot applications—when applying an herbicide, for example.

Motorized hydraulic spray equipment generally includes a larger storage tank and pump that delivers the solution under pressure through a wand or gun, which modifies the spray stream and directs the material to the plants. The tank and pump may be mobile or, in a greenhouse setting, they may be stationary. The size and configuration of the tank, pump, and motor or engine depend on the intended purpose of the spray unit.

A mobile unit may be as small as a 5-gallon tank with an electric pump, but more typically it ranges from 25 to 200 gallons and is matched with a gasoline engine, electric motor, or PTO-driven pump (i.e., driven by power takeoff). The smaller units are usually mounted on a hand cart, but the larger units are commonly mounted on a trailer, motorized cart, tractor, or truck. These types of sprayers are versatile and can be used to make small or large applications.

Other mobile units used to make large applications may consist of a 500-gallon or larger tank, with an air blast sprayer for applying material to large trees, as in an orchard, or those equipped with booms to place a series of spray nozzles over a bed of plants (fig. 4.2). Boom sprayers are capable of applying materials over a large area quickly and with excellent uniformity. A 40-foot boom attached to a tractor equipped with a tank and pump traveling at 3 miles per hour can cover a 300-foot bed (12,000 square feet total) in a little over 10 minutes. Air blast sprayers use a fan that moves large volumes of air at high velocity to carry spray the long distances (50 to 100 feet) required to cover larger, taller trees or younger trees in beds where access is limited. Be aware, however, of issues associated with all types of sprayers and their ability to atomize spray materials, which may be carried to off-target locations.

Stationary systems place a tank and pump at a permanent location. This type of system requires additional plumbing throughout the greenhouse to carry under pressure the liquid that is to be sprayed. At various connection points in the greenhouse, hoses can be attached and the liquid applied through a wand or gun. Chemical materials can be applied through irrigation systems (chemigation). This is feasible only if large areas are to be treated with the same material and the system is capable of applying the chemical with good uniformity and high efficiency. Be sure to understand issues with runoff when using these types of application systems.

Low- and ultra-low-volume sprayers are typically used in greenhouse production and deliver an active ingredient in a small volume of water or other diluent. A high-volume hydraulic application might require 8 ounces of a pesticide per 100 gallons of spray to be applied onto the plants in a greenhouse. For a ULV application, in comparison, the 8 ounces of material may be diluted in a much smaller volume of liquid and applied to the same crop. The LV or ULV application equipment (including thermal foggers, cold foggers, air atomizing, and electrostatic sprayers) uses high pressure or compressed air or both to atomize the spray solution. The small droplets of chemical solution are carried to the targeted plants either via the draft created by the atomizing process or via air moved by fans (similar to air blast) that may be part of the application equipment. The advantage of these systems is that the extremely small droplets may cover the plants more uniformly than hydraulic sprays. This is especially true of the electrostatic sprayers. These are essentially ULV sprayers that impart a small electric charge to the spray droplets that are attracted to plant surfaces. Some of these systems can be set up to start automatically in the evening, thus minimizing exposure to the applicators and other people who might need to access the greenhouse during the day.

Equipment to apply granular materials uniformly over large areas includes centrifugal spreaders and blower applicators. Blower applicators, such as modified backpack leaf blowers, use moving air to carry the granular material. Larger blower types (e.g., Gandy) may be tractor or trailer mounted. They may blow air through tubes to carry and distribute granules, and they may use a boom to cover a large area. Centrifugal spreaders are commonly used and can be smaller, hand-held versions or larger, tractor-mounted ones. These spreaders use a rapidly spinning disk, fitted with vanes to fling the granules over a distance. Although they are simple to use and can rapidly cover large areas, they apply material less uniformly than boom-mounted blower applicators.

In all instances of applying chemical materials, be sure to read and comply with the instructions on the labels, in addition to the current local laws and regulations.

Mixing Equipment Used to Prepare Potting Mixes

Preparing container mixes, even on a small scale, is laborious; consequently it is one of the first processes to be mechanized. Nurseries that produce their own mix do so by combining the components in the desired proportions and blending them until the mixture is homogenous. The blending could be done with a tractor equipped with a front-mounted bucket. In this case, the materials are spread evenly in layers in the proper proportions onto a smooth, hard surface (preferably concrete), and the bucket is used to turn the soil over several times to blend. When the media is sufficiently mixed, it is stacked, ready for use. This method can be used for either small or large volumes of potting mixes. The media components may also be placed, in proper proportions, into batch mixers. These might be cement mixers or equipment that is specifically designed for this purpose. The volume of media mixed per batch, of course, is limited by the size of the mixer. Producing a large volume of media would require the mixing of numerous batches. There is a danger in overmixing, where the media components are physically altered and thus result in a blend that does not have the intended characteristics.

Potting mix preparation can be highly mechanized and automated. An automated system may consist of several feeder bins. These are large, 1- to 4-cubic-yard bins that dispense their contents, each holding one of the components of the mix (e.g., peat moss, perlite, or vermiculite). Smaller bins are used for fertilizers and other amendments. Each bin is adjusted, either manually or automatically, to dispense its contents at a specific rate onto a conveyor belt. This belt deposits the media components into one end of a rotating drum to blend the mixture. The blended potting mix is dispensed from the opposite end of the drum onto another conveyor to be used directly or stacked for later use. These systems are able to prepare large volumes of media as needed, which eliminates premixing and storage. Automation enables changes in the composition as needed. These types of systems can be integrated with can-, pot-, or flat-filling machines so that the media is dispensed directly into the filling machines. The filling machines may be linked to the media mixing system so that more media is mixed and dispensed as it is needed.

Soil and media sterilization and pasteurization equipment

Treating soil or potting mix using heat requires a source of energy to raise the temperature of the media and then maintain the desired temperature for a set time. A common energy source is steam that is generated in either gas-fired or electric boilers. Many nursery operations use steam for greenhouse temperature control, so availability for media sterilization is not problematic. In addition to the steam source, a special container is required to distribute the steam into the soil. A simple version can be a small bin, constructed of wood or metal, with a manifold and perforated pipes at the bottom to distribute steam into the media. A common modification of this type of bin is to put it onto a trailer so that the pasteurized potting mix can be taken to where it is needed. Other containers may be very large (several cubic yards) with perforated false bottoms to distribute steam. They may be mechanized with hydraulic rams to tilt the bins and dispense the pasteurized media onto conveyor belts for delivery directly into pot-filling machines. Vaults may also be used, into which pallets of media-filled containers or other items are placed. The steam is injected into the vaults, which confine the heat. Temperatures of the media must be monitored to ensure that proper levels are attained for pasteurization.

Since the control of temperature is important in order to avoid complete sterilization, air-steam or aerated steam systems can be used to regulate the temperature of the steam. Blowers move air that is heated by steam into the bins containing media. Thermostats are used to control the rate that steam is released into the air stream so that the extremely high temperatures that cause sterilization are avoided. Another advantage of this system is that once pasteurization is achieved, cool air can be blown into the media to quickly lower its temperature and make it available for use.

Equipment can be tractor mounted for field applications to inject chemical fumigants into the soil. The implements used for injection may be a rake or chisel that reaches under the soil surface. Tubes behind the teeth or chisels carry fumigant into the soil. The surface is then covered with polyethylene film to confine the fumigant. For smaller applications or where a tractor is not possible, a winch-pulled rake may be used.

Fumigation equipment used in nursery applications for treating potting mix in batches consists of the following: a tank that stores the fumigant, a device to measure and dispense (Shore dispenser), valves and tubing or hoses to distribute, polyethylene sheeting to confine, and fans to distribute the fumigant under the sheeting. Additional equipment needed for safety include two self-contained breathing apparatus (SCBA) units and monitoring equipment to measure residual gases. Be aware of the materials that can be used in this manner. Be sure to read and comply with instructions on the labels of the materials to be applied, as well as the current relevant local laws and regulations.

The use of chemical fumigants such as methyl bromide is being phased out. In response to the loss of the use of these products, other methods of pasteurizing are being investigated. These experimental methods include ohmic heating and microwave radiation. Ohmic heating uses electrical current and the resistive characteristics of substrates to generate heat. Temperatures high enough to actually melt soils can be achieved. Microwaves can be adjusted to target specific organic substances of pathogens or insects in soils to kill these organisms.

For more details on methods to eliminate pests and diseases from soil and container mixes, see chapters 5 and 15.

Machinery and Equipment for Handling Materials

The moving of plants from one place in the nursery to another can be very labor-intensive due to the large quantities and the distances traveled across a nursery. Each handling of plants costs money and time, and because of the large number of plants involved, the total cost can be very high. Thus, this is generally minimized to avoid the expense. Where feasible, nurseries place the containerized plants in the final spacing at the start of production. The exception to this is where space is at a premium—in other words, where there is not enough acreage to accommodate all the crops fully spaced simultaneously. In these instances, which are common, the youngest plants are placed closest together and the older, larger plants farther apart. However, as the younger plants mature, they will eventually require additional space. This means they will have to be moved.

Tractors, trailers, and other vehicles

Mechanization of the moving process in nurseries is difficult due to the large distances involved, and so a tractor-and-trailer solution is generally used. More options may exist in greenhouses, however, to successfully mechanize plant movement. This is due not only to the shorter distances but also to the higher cost of time and space associated with greenhouse production. Although many more methods are possible for moving plants in a greenhouse, the tractor-and-trailer method is a widely used one. Farm tractors are commonly preferred as the motorized unit because they are widely available, but many other types of vehicles are used to pull trailers. These include golf carts, garden tractors, trucks, and specialized yard maintenance vehicles. Trailers and carts are usually constructed or modified to carry flats or trays in tiers or on shelves. In some cases, these carts and trailers are loaded directly into delivery vehicles and unloaded at the customer's business.

Conveyors, rollers, and racks

Conveyor belts are commonly used to move plants. Specialized versions are available in modular units and can be easily moved. These can be used with other mobile equipment, such as pot-filling machines, to transport filled pots down aisles to be placed onto the benches or on the ground. Conveyors might be adapted with wheels to ease relocating them for use on loading docks, for example, where they move plants into trailers for shipping. Permanently installed belts are usually associated with other permanently placed equipment, such as can-filling or transplanting machines. Other variants of conveyor belts are roller and skate-wheel conveyors that can be portable and motorized. Because of the wheels or rollers, they are useful only for cartons, trays, or plant containers with smooth undersurfaces.

Another method of moving plants in greenhouses is on trolleys that hang from and travel along rails mounted overhead (fig. 4.3). Typically the trolleys are located so that placing plants on them from a bench is a short hand motion. The trolleys are then moved manually to the destination.

FIGURE 4.3. Trolley carts hanging from an overhead rail system are used to carry materials through this greenhouse. *Photo:* L. R. Oki.

FIGURE 4.4. A forklift is fitted with tines to carry 96 soil-filled containers from a conveyor belt (A) onto nursery trailers (B) that will transport them to the field. It only takes two loads for the forklift to fill a trailer with containers. *Photos:* L. R. Oki.

Forklifts

Forklifts, lift trucks, or "powered industrial trucks" (as referred to by OSHA) are efficient in moving material, but their main purpose is for lifting and stacking (fig. 4.4). Ubiquitous in warehousing operations, forklifts are commonly used in nurseries and greenhouses, as warehousing or similar operations often occur here to some degree to receive, store, and distribute the supplies needed for plant production, harvesting, and transport.

Most nurseries use some degree of palletization. Many smaller units (cartons or pots, for example) are placed onto wood or plastic pallets or other containers that accommodate the forks on the lift truck; this is done to minimize handling. The machinery used to move the pallets might be a pallet jack, a hand-operated device to move pallets short distances, or a forklift. Other vehicles—such as farm tractors, with forks installed either in place of a bucket or on a three-point hitch at the back, and skid-steer tractors fitted with forks—can operate as lifts.

Front-end loaders

Front-end, or bucket, loaders come in three basic types: skid-steer, wheel, and track. The skid-steer loaders tend to be smaller but are used to take advantage of their size. The skid-steer method of turning enables the vehicle to turn within its own length, an advantage in tight spaces. They can also be used for other tasks, since many other implements can replace the bucket. These include augers, sweepers, pallet forks, hammers, and many others.

Wheeled loaders are the most commonly used type due to their ability to lift large loads and their ease of operation. Farm tractors fitted with buckets can serve multiple purposes as a loader and an implement carrier. Wheeled loaders typically are front-end steered, as most wheeled vehicles are. Articulated-body loaders, however, are capable of a shorter turning radius and greater lifting capacity than either front-wheel or skid-steered tractors. When compared with skid-steer loaders, articulated-body loaders have less tire wear and disturb the soil surface less. Although loaders are found in nearly any nursery operation, tracked loaders are not commonly used except in nurseries producing field-grown crops. Tracked vehicles distribute their weight over a larger area than wheeled ones, so they are able to move better on softer surfaces. Skid-steer loaders may also be fitted with tracks in place of wheels and tires.

Moving tray systems

Moving tray systems are widely used in Europe, and they are also becoming more common in the United States. In such systems the potted plants are set on a large tray, typically 4 to 6 feet wide and 8 to14 feet long, instead of on a bench. The trays are constructed of aluminum sides with plastic, wire mesh, or expanded metal bottoms. Rails and rollers are installed throughout the production greenhouse, allowing the trays either with or without plants to be moved as necessary. Greenhouses with this system centralize work operations, moving trays to the appropriate workstation to perform necessary tasks on the plants they contain and then returning the trays back to the greenhouse for further growth. In conjunction with fixed equipment for pot spacing, pruning, pesticide application, boom irrigation and so on, these systems can reduce labor costs. In such greenhouses, less room is needed for aisles and so more of the space is used for production, which increases the energy use efficiency of temperature control. These systems, however, demand a large capital investment, so they are only feasible for use in high-value crops and in areas where labor or other costs are very high.

Machinery and Equipment for Crop Production

FIGURE 4.5. This seeding machine uses a drum with holes to hold seed (colored yellow) onto it when a vacuum is applied. A plug tray is moved under the drum to receive the seed. The hole spacing matches that of the cells of the plug tray. *Photo:* J. H. Lieth.

Seeding machines

Sowing millions of tiny seeds into individual cells of plug trays can be tedious and laborious. Errors are frequently made that result in misses or multiple plants per cell. Seeding machines (fig. 4.5) provide a way to accomplish this rapidly, consistently, and with a greater degree of precision. Seeding machines vary in complexity, and their speed of operation depends on the method by which seeds are picked up and delivered. The different types of seeding machines can be categorized into vacuum seeders that use plates, manifolds, or drums and optical seeders.

Vacuum seeders use a vacuum drawn through an orifice that is sized to hold a single seed. There may be many orifices on a template (plate) arranged in a pattern that corresponds to the cells of a plug tray. When a seed is located and held over each hole by the vacuum, the template is inverted over the plug tray and aligned properly so that each seed is located over a plug cell. The vacuum is released (positive pressure may be applied to aid in the release), and the seed falls to the soil surface.

Manifold-vacuum seeders use a manifold to hold needles or nozzles with an orifice drilled into their tips. The number of needles or nozzles on the manifold corresponds to the number of cells in a row in the plug tray used. A vacuum is drawn through the manifold and is then moved so that the tips of the needles or nozzles are positioned into a small tray of seed. Usually this tray is subject to vibration to keep the seed evenly distributed and to increase the occurrence of seed getting picked up by the vacuum. The manifold is moved again so that the tips can eject the seed into tubes that guide the seed as they fall to the soil surface. A conveyor or other type of system advances the plug trays to accept another row of seed.

Drum seeders use a small drum that is located over the plug tray, which is sitting on a conveyor. The drum has holes drilled into it in a pattern that matches the spacing of the cells in the plug trays. Air is sucked into the drum through these holes. As the drum rotates, the holes pass by seed held next to the drum in a small tray. One seed will stick to each hole, held in place by suction. The drum rotation and conveyor speed are coordinated so that a row of holes on the drum is aligned over a row of cells in the tray. As the drum rotates, the backside of the hole is covered. This way the suction is removed (positive pressure may also be applied) at precisely the moment when the seed are over the row of cells where they are to be deposited, and the seed fall or are propelled into the cells. Some drum seeders actually use water jets to eject and drill the seed into the soil.

Optical seeders use light and photo detectors to sense when a single seed passes through a movable tube. These seeders use a vibrating pan to control the rate the seed fall past the photo detector. As a seed is detected, the movable tube is advanced over the open end of another tube that guides the seed as it falls to the soil surface. This is repeated until a row of the plug tray has been sown. The tray is advanced by a conveyor belt to the next row.

Each of these types of seeders has its own advantages and disadvantages. The vacuum seeders tend to operate most efficiently when seeds are round and of uniform size. Of the vacuum seeders, the drum seeders are very fast and the plate seeders (being the least automated) are slower. The optical seeder is not hindered by seed shape or uniformity and is very precise; however, it is much slower than either the drum or the manifold seeders.

The seed itself is part of a mechanized seedling production system. Pelletizing reduces variability in seed size, and it reshapes and enlarges the seed to aid in compatibility with the seeding machines, increasing the sowing rate. Priming the seed increases germination uniformity and may reduce the time for germination. For more information on seed handling, see chapter 9.

Transplanters

A large amount of engineering and design is currently going into the development of machines that extract the individual plants from plug trays and transplant them into pots of various sizes. Transplanting is a laborious and difficult task, so transplanting by machine can save much money. However, the fully automated transplanting machines are very expensive. Acquiring such equipment is only economically feasible if many worker-hours are normally spent on transplanting.

Where purchasing machines is not feasible, transplanting by hand can be done at a conveyor belt (fig. 4.6). Plugs extracted from trays are provided to workers, who plant these into larger containers (ranging from six packs to 6-inch or larger containers) filled with potting media as they move past on the conveyor.

Labeling

Labeling machines are usually associated with transplanters. Because of product labeling requirements, each salable container must have a tag or label with the appropriate botani-

FIGURE 4.6. Conveyor belts are designed to move materials from place to place. Nurseries have outfitted belts to facilitate various tasks such as transplanting. (A) shows a simple but effective system that moves flats to workers who transplant plugs into the flats. Racks hold flats of plugs conveniently for the transplanters. Finished flats are moved on the belt to a trailer for transport to the greenhouse. (B) through (E) show a more sophisticated belt system. (B) Flats are moved by a series of conveyors and are (1) filled with soil by a flat-filling machine (out of frame to the left), (2) watered-in, (3) dibbled, and then transported to workers by the upper belt. (C and D) Workers take a flat from the upper belt to their workstation for transplanting. Completed flats are moved to a lower belt in front of the workers. (E) The belt moves the completed flats past a scanner that reads barcode symbols and automatically makes entries into the computerized inventory system. Flats are then loaded onto racks on trailers for transport to the greenhouse. *Photos:* L. R. Oki.

cal, production, and sales information. Tagging machines insert preprinted plastic labels into the media of a pot, can, or pack, usually after the planting operation. An alternative to tags is placing the necessary information directly onto the plant containers. This might be accomplished by printing the information directly onto the container or by printing onto an adhesive label that is transferred onto the container. A major advantage of printing on demand is not having to keep the preprinted tags on hand. Depending on the number of different plant varieties grown, container sizes needed, and retail pricing codes involved, inventorying the assortment of preprinted tags and labels can be a large investment in direct costs and storage space.

Canning and potting

Pot-filling equipment starts with a supply of containers, and then it unstacks the containers and fills them with a medium (fig. 4.7). The machines may form a depression in each pot, into which a transplant can be placed. More versatile machines are capable of filling a range of pot sizes from 4 to 10 inches. Typically media preparation and pot filling happen concurrently, and the mixed media is deposited directly into the filling machines, eliminating the need for media storage space. This requires that the mixing and filling equipment be located near each other and that the media mixing be done at the same rate as the pot filling. Some production systems bring the filling equipment into the greenhouse or field at the site where the pots will be planted. In this setup, the pot-filling machines are located in a wide, central aisle, and conveyors are used to bring the filled pots to the bench or bed. Media is brought to the machines by tractor, soil feeder bin, bulk bags, or other methods. Depending on the specific equipment and setup, one person could operate this type of system, but maximum rates (6,000 pots per hour on some machines) require three or four people.

Quality Improvement Machinery

Maximizing plant quality sometimes requires a cultural practice such as pruning, pinching, spacing, or grading. Performing a task such as pruning on thousands or even millions of individual plants can be overwhelming when it is done by hand. When the process is mechanized, the task is completed not only in a more timely fashion but also with much greater consistency and perhaps even higher quality.

Mechanized pruning

A simple mechanized method of pruning plants in cans, for example, may use a modified rotary lawn mower that is elevated over the plants. The mower is suspended from rails, which guide the wheels that are raised a selected distance over a block of plants. The mower is passed back and forth across the top of the plants by two operators, one on each side of the block. An extreme version of this concept is a large, self-propelled device that moves over a block of plants. The operator drives the machine over the plants and

FIGURE 4.7. Stacks of empty containers are supplied to a pot-filling machine (A), which fills each container with a growing medium (B). The filled containers are lifted from a conveyor belt onto the bench (C). *Photos:* J. H. Lieth.

FIGURE 4.8. A custom-designed and built machine (A) prunes rows of plants (B). The pruner is self-propelled and operated by one person who drives and controls the height of the pruning (C). *Photos:* L. R. Oki.

controls the mowing height (fig.4.8). In a greenhouse using a moving tray system, an entire tray may be relocated to a pruning machine, where it is passed under a sickle-bar mower, for example, to cut the tops of the plants. The tray of pruned plants is then returned to the greenhouse. Where pruning the sides of plants is necessary, plants in containers can be placed on a machine that rotates the plant across motorized shears. Multiple shears are used, placed at different angles and distances with respect to the plant, to achieve the desired shape. Conveyors may be used with this type of pruning machine to facilitate the movement of plants to and from the machine.

Mechanized spacing

Plants that require additional space during crop production may be spaced mechanically. This process requires that some or all of the plants be moved to another location where additional space is available. In a nursery, plants must be picked up from the ground, placed onto transportation equipment, moved, and finally returned to the ground in a pattern that provides the additional space. In a greenhouse of moving trays, a tray of plants may be relocated to a workstation where the spacing process occurs in an entirely automated procedure.

In greenhouses where containerized plants are grown on slab floors (e.g., concrete), a specialized lift truck can be used that picks up blocks of plants with a rakelike set of long tines that are inserted between the pots. Computerized controls move the truck into position and lower the plants onto the floor. The truck is reversed at a set distance—the diameter of one pot, for example—and the pots are picked up again, leaving those in the row farthest from the lift still on the ground. The process is repeated until all the pots remain on the ground. Another version of the lift has rakes with tines that can move to increase the side-to-side spacing distance.

By modifying greenhouse automation equipment, nursery operators are trying to emulate the level of mechanization currently used in spacing greenhouse plants for outdoor applications. Conditions make spacing plants in the field a challenging task. For example, surfaces such as crushed gravel (on which containers are placed) are usually not smooth enough to allow this process to be easily mechanized. However, container spacing is used in other nursery operations where smooth surfaces exist, such as in canning. Here, at the canning workstation, the planted containers may be mechanically spaced onto pallets or nursery trailers, which are transported to the field.

Automatic grading machines

Crop quality can also be maximized by grading and selecting only those plants that meet established quality standards. Seedling transplant producers, for example, have developed optical methods to scan each cell in a plug tray to detect not only the presence or absence of a plant but also its size and quality. Using this information, substandard seedlings can be removed and replaced to completely fill the tray with seedlings of uniform quality.

Environmental Control Equipment

Greenhouse environment

Many different kinds of equipment are used in greenhouses for environmental control. Fans and wetted pads can be used in conjunction with motorized vents for cooling. Forced hot air units and water heaters or boilers are used in heating. By using simple thermostats, electronic controllers, or computer-based environmental control systems, these functions can be automated. Any level of automation will reduce labor. But some effort will be required to monitor the environmental variables to be sure that the desired conditions are achieved.

Heating and cooling can be automated with simple thermostatic control. However, if several methods of control are used, or if control can be done at various set points, then a system of stages or steps of heating and cooling should be set up. Staging can increase energy use efficiency, but managing this requires more sophisticated controllers. These range from step controllers to computer-based environment control systems. The latter provide the greatest degree of flexibility. More information on environmental control and monitoring systems for greenhouses is provided in chapter 3.

Shading systems

In both greenhouse and outdoor nursery settings, some crops require reduced light levels. Permanently installed shade cloth and whitewashing of glazing or plastic film are ways to subject these crops to lower light levels continuously. Automated shading systems, on the other hand, can be pulled over the crop only when needed as light intensities increase. This is more effective when coordinated with automated control equipment that allows various degrees of shading based on measured light levels or temperature. For example, automated systems can use full sun in the morning and evenings when the light intensity is low, and they can pull the shading over the plants during the middle of the day when light intensity and temperatures are higher. Where photoperiodic control is needed (e.g., for mums, poinsettias), the black cloth can be controlled using a similar system. Materials used for painting greenhouse glazing and shading fabrics are described in chapter 3.

Retractable-roof structures

Retractable-roof greenhouses and shadehouses allow growers to use favorable outdoor conditions when they exist but also to pull a cover over the plants whenever the prevailing natural conditions are suboptimal—during rain or when temperatures are low, for example. Clearly it is essential to automate such systems based on environmental conditions. Detailed information on retractable-roof greenhouses is provided in chapter 3.

Recycling Equipment

Equipment to reduce the volume of compostable waste materials consists of chippers, grinders, and screens. The processed material may be composted for reuse as a soil amendment in a landscape, for example. Chippers are typically used by tree maintenance companies to break down tree limbs into wood chips to reduce the volume and make the transport of the material away from the jobsite easier. Chippers use knives, blades, or hammers on a rapidly rotating disk to cut up the limbs into smaller chips. The chips are usually of a size that is larger than ideal for composting, however. Drum grinders, tub grinders, and hammer mills break up compostable materials into smaller particles using rapidly rotating hammers. These types of grinders have a large rotating drum that contains the material to be processed. The rotation continuously exposes the material to the hammers. Ground material falls out of the drum onto a belt to be transported away from the grinder. Ground material may be passed through trommel screens (rotating drums of steel screen) to separate larger-size particles from smaller ones. The larger particles can be reprocessed and the smaller ones moved to a composting area.

Maintenance Program

Wherever mechanical equipment is used, care must be taken to keep the equipment in good working condition. Usually a maintenance program is recommended by the manufacturer. In any case, one should be established. This is particularly important where safety is a consideration. Vehicles that leave the nursery, for example, must conform to some level of safety that is determined by the Department of Motor Vehicles and is enforced, in California, by the Highway Patrol or other law enforcement agencies. Even minor infractions can cause trucks hauling product to be forced off the road until repairs are made. This can be very damaging to plants that may be subjected to extreme temperatures during a delay and may arrive at their destination in poor condition.

Many pieces of machinery have powerful motors, and improperly maintained equipment can cause injury. Poorly performed maintenance can result in equipment failure and cause severe injury or death to unsuspecting workers. Thus, competent maintenance personnel are needed to carry out a rigorous maintenance program to examine and repair each piece of equipment on a routine schedule. All equipment use should include a training program for proper and safe operation.

Safety should always be of highest priority, but economics is also an issue with regard to maintenance. The cost of an investment in equipment is amortized over many years, so the more product that can be processed with the equipment, the better. Thus it is best to keep equipment well maintained to get the maximum return out of the investment.

Poorly maintained equipment is also much more likely to break down at times that will most likely have detrimental effects on the production schedule. Efficient production schedules usually leave little spare time, so a breakdown of vital equipment can cause serious delays in any facet of production.

References

Posadas, B. C., P. R. Knight, R. Y. Coker, C. H. Coker, S. A. Langlois, and G. Fain. 2008. Production and marketing reports: Socioeconomic impact of automation on horticulture production firms in the northern Gulf of Mexico region. HortTechnology 18:697–704. HortTechnology website, http://horttech.ashspublications.org/content/18/4/697.full.pdf+html.

Thomasson, J. A., F. P. Lee, C. Ji, and R. Sui. 2005. Blooming possibilities: Automating and mechanizing container nurseries in Mississippi. Resource, Engineering and Technology for a Sustainable World 12(6): 11–12.

Soils and Container Media

Richard Y. Evans

CHAPTER FIVE

Growing and marketing plants in containers has a number of advantages over field production, such as convenience for the consumer and easier shipping from the nursery. However, container production requires more intense management than field production because plants in containers are subject to more extreme stresses that develop more rapidly than in field soils (table 5.1). The challenge for horticulturists and nursery growers working with container-grown plants is to develop potting media and management practices that minimize these stresses.

The purposes of a container medium are to provide physical support for the plant and to supply adequate oxygen, water, and nutrients for plant growth and development. Plants can grow in many different types of media, but some require more maintenance than others, adding to the cost of production. Selecting an optimal container medium minimizes the amount of management required for plant production.

This chapter provides a basis for making decisions regarding media selection for container production. First, soil properties (such as soil-moisture relations) will be introduced. Soil moisture concepts are important not only for plant selection but also for understanding how to supply water and nutrients in an optimal way. Fertilization and irrigation are further addressed in chapters 6 and 7, but this chapter lays the foundation. Next, the types of inorganic and organic amendments and chemical amendments commonly used in container media are presented. This includes a discussion on the importance of chemical amendments for establishing favorable pH and fertility, but these topics are more fully addressed in chapter 6. Finally, methods for mixing media, treatments to eliminate soilborne pests and pathogens, and media storage are discussed.

Photo: J. K. Clark.

Table 5.1. Comparison of properties in the root environment of container and field soils

	Container		Field	
	Condition	Time	Condition	Time
Water	near saturation to wilting	1–3 days	field capacity to wilting	1–3 weeks
Aeration	low to high	1 day	adequate to high	season
Nutrition (mainly nitrogen)	high to low (if leached)	1 week or less	high to low	season
pH	± 1–2 units	weeks or month	fairly constant	season
Salinity	chronic low to high	weeks or month	low to high	season
Temperature	40° to 50° change	1 day	fairly constant	season
Root disease	more susceptible	—	less susceptible	—

Physical Properties

A soil's physical properties describe its overall character and behavior in terms of irrigation water infiltration rate, the drainage of water in the soil profile, the amount of water held by the soil that is available to the plant, and the amount of oxygen and other gases present in the root zone.

Particle Size and Distribution of Amendments

Soil texture refers to the size of the various soil particles. The classes of texture, arranged from the largest particle size to the smallest, are gravel, sand, silt, and clay. The physical properties of a potting medium are largely determined by the particle size distribution of the mix. Sometimes growers need to add organic amendments to existing potting mixes to improve texture-related functions, such as drainage and moisture retention. The general recommendation for particle size of organic amendments used in container mixes is that 60 to 100 percent of the particles should be 0.5 to 2.0 millimeters (mm) in diameter, with the smallest percent possible measuring less than 0.25 millimeter in diameter. In mixes that incorporate sand, it is important to avoid sands with a high percentage of "fines," particles in the range of very fine sand (VFS) or smaller (table 5.2), which can plug up pore spaces. In general, the amount of VFS plus silt and clay should not exceed 10 percent by weight. For container plants, the sand mixture should have a narrow particle-size range and should contain at least 60 percent coarse sand (CS) plus medium sand (MS) by weight. A broad particle-size distribution tends to pack well, as small particles slip into pore spaces surrounding larger particles. The result is a dense mix with low permeability, low water content, and high mechanical resistance to root growth. Such sands are better suited to highway construction.

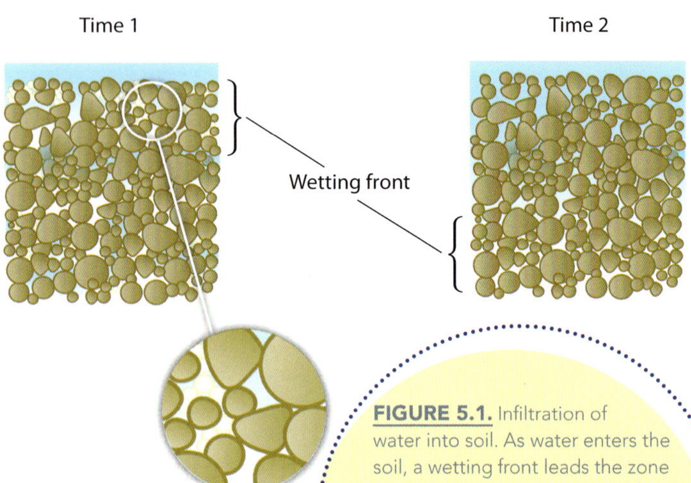

FIGURE 5.1. Infiltration of water into soil. As water enters the soil, a wetting front leads the zone in which all pores are saturated with water. In the wetting front, shown in the detail, all pores near the top are saturated. In the middle, only fine and medium pores are saturated, and at the bottom only fine pores are filled.

Water Movement in Soils and Container Mixes

In a field, as long as water is being applied to the soil surface, the water will move downward through the soil in response to two forces: the downward push of gravitational pressure (the weight of the column of water) and capillary action, which exerts suction at the bottom of the wetting front (fig. 5.1). In a container mix, the downward infiltration of water is almost entirely due to gravitational pressure. Capillary action is relatively insignificant. In both the field and container, all pores from the surface to some point below the surface will become saturated with water as long as irrigation continues.

When irrigation stops in a field, water continues to be drawn downward by fine pores at the wetting front. Coarse pores nearer the surface begin to drain when the water in them is replaced with air. When the downward movement of water becomes so slow that it cannot be detected at the surface (usually by using a tensiometer), the soil is considered to be at field capacity.

Somewhat different forces are at work in a container during drainage. When irrigation of containers ceases, gravitational pressure continues to cause water to move downward. Water empties out of coarse pores near the surface, and it drips out the bottom of the pot. Below the bottom of the pot, of course, there is no dry soil with fine pores to keep pulling water downward, as there is in a field. In fact, there is some resistance to downward movement of water: the fine pores in the potting mix tend to hold onto some water rather than letting it drain downward. This "capillary rise"

Table 5.2. Particle-size classification of sands

Particle size (mm)	Description
2.00–1.00	very coarse sand (VCS)
1.00–0.50	coarse sand (CS)
0.50–0.25	medium sand (MS)
0.25–0.10	fine sand (FS)
0.10–0.05	very fine sand (VFS)
< 0.05	silt and clay

Table 5.3. Height of water column for various pore sizes in container mix

Pore radius (cm)	Equivalent particle diameter (mm)	Height of water column (cm)	Comments
1	20	0.147	
0.1	2	1.47	upper range of sand
0.01	0.2	14.7	fine sand
0.001	0.05	147	very fine sand

Note: The height of a column of water held against gravity depends on the diameter of pores in the container mix.

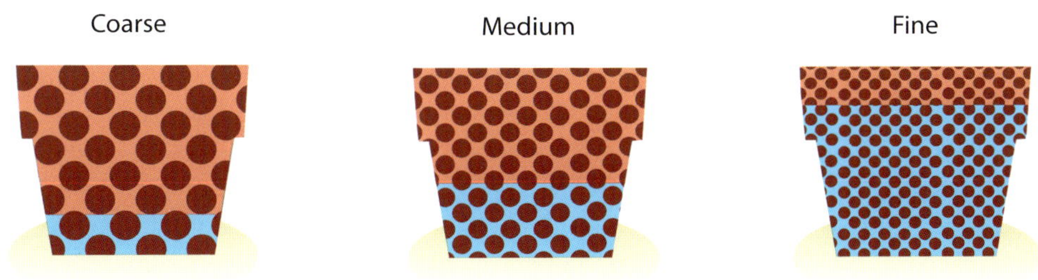

FIGURE 5.2. The height of the saturated zone in a container (blue area) depends on the particle size of the potting mix.

FIGURE 5.3. The depth of the mix in a container does not affect the height of the saturated zone, but it does affect the height of the unsaturated zone.

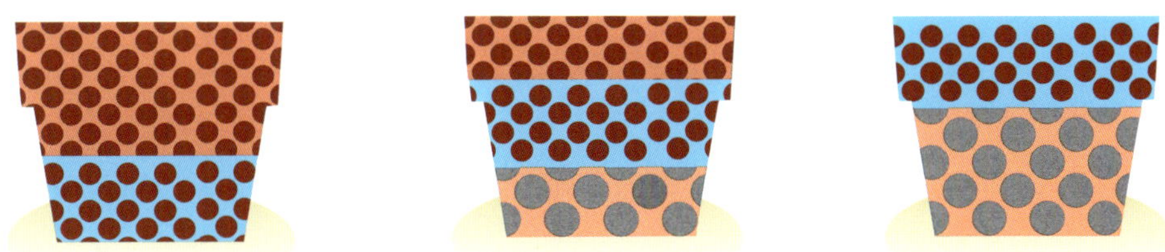

FIGURE 5.4. Gravel (large gray balls) at the bottom of a container does not decrease the size of the saturated zone (blue area), but it does shift the saturated zone up higher in the container.

can be expressed as the height of a column of water that can be held against gravity. Pore size strictly governs the height of this water column (table 5.3). As long as the height of the column of applied water is greater than the height of water that the capillary rise in the potting mix can support, water will continue to drain from the container. Water movement in the container stops when the gravitational force pushing water downward is balanced by the capillary rise holding water in the container. At this point, water stops dripping from the bottom of the container and the potting mix is at container capacity. This is the maximum amount of water the mix can hold in that container.

At container capacity, there is a zone at the bottom of the container in which all pores are filled with water. The height of this zone depends on the particle size of the potting mix (see table 5.3). In a coarse mix, the pores will contain both air and water except in a relatively shallow zone at the bottom of the pot, in which the pores will be filled with only water. A fine mix may be saturated with water from the bottom nearly to the surface (fig. 5.2). The depth of the mix in the container has no effect on the height of the saturated zone (fig. 5.3). The depth of the mix does, however, affect the extent to which the surface of the mix is drained. Thus, a mix containing fine particles may be excellent for tall containers but too wet for shallow containers or flats.

Roots require oxygen as well as water, so a saturated zone that extends upward close to the soil surface will inhibit growth and even cause root death. It is commonly believed that drainage from a potting mix can be improved by putting coarse material, such as gravel, at the bottom of the pot. In fact, because the extent of the saturated layer depends only on the pore size of the mix, a layer of coarse material will not affect drainage. Instead, it will decrease the effective depth of the container (fig. 5.4).

Six Physical Properties of Container Media

The six physical properties listed below describe the characteristics of a potting mix. The values of these physical properties for some typical potting media are shown in table 5.4. A moisture-release curve for a typical medium is shown in figure 5.5.

- air-filled porosity: the percentage of a given volume of a potting mix that air occupies after irrigation and drainage
- water-holding capacity: the percentage of a given volume of a potting mix that water occupies after irrigation and drainage
- available water: the percentage of a given volume of a potting mix that is occupied by water that the plant can extract
- total porosity: the percentage of a given volume of a potting mix that pores occupy (i.e., the space not occupied by solids)
- moisture-release curve: the relationship between the volumetric moisture content and the tension at which water is held in the medium
- bulk density: the dry weight per unit volume of moist soil, expressed in grams per cubic centimeter ($g \cdot cm^{-3}$)

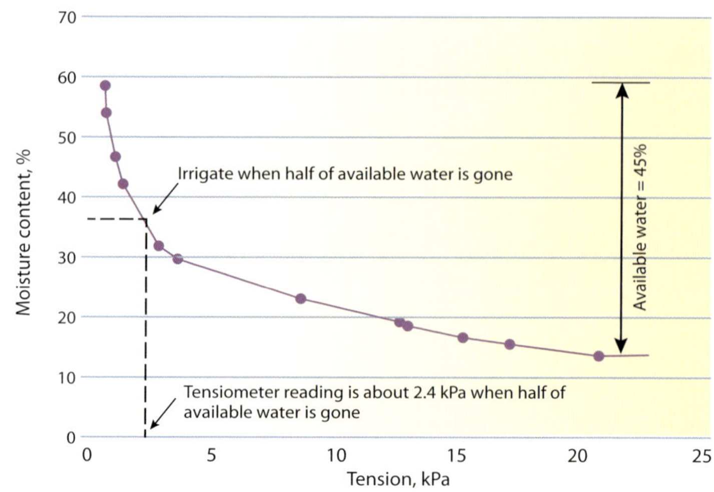

FIGURE 5.5. Moisture-release curve for a mix of coarse sand and redwood sawdust (1:2 by volume). Much of the water is not held tightly by the particles of the mix and is easily available to the plant (steep part of curve, low tension values). As the easily available water is removed, the remaining water is held more and more tightly by the mix (shallow part of curve, high tension values). Available water is the difference between the amount of water at container capacity and the amount present when the plant wilts.

Total porosity and air-filled porosity

All pores in a mix are occupied by either air or water, so total porosity is the sum of air-filled porosity and water-holding capacity. Adequate air-filled porosity is critical for satisfactory plant growth. Typical symptoms of moderate stress caused by poor aeration include less vigorous growth, unusually dark green leaves, and abnormally small leaves. If air-filled porosity is extremely low, plants may wilt even in moist soil (because roots function poorly or die) and leaves may turn yellow and drop. Poor aeration can also cause undesirable chemical changes in a potting mix. When oxygen content is low, some plant nutrients may be converted to toxic forms. Also, poor aeration can render plants more susceptible to soilborne diseases. In general, air-filled porosity should be at least 10 percent.

Water-holding capacity

The water-holding capacity of a mix expresses the size of the water reservoir in the container. Therefore, it is an important indicator of the required irrigation frequency. If water-holding capacity is low, then water must be added frequently. In general, the water-holding capacity should be at least 40 percent.

Available water

A useful property related to the water-holding capacity is the available water, which describes the percentage of the volume of a potting mix that is occupied by water that plants can take up. Recall that water in pots is held against gravity by the capillary rise imparted by the average pore size of the particles in the container medium. This property of the medium is expressed as a water potential and is usually measured as a moisture tension using a tensiometer. As the plant removes water from the container by transpiration, the larger pores are drained, leaving water mostly in the smaller pores, which hold water tightly (fig. 5.6). If water is not replenished by irrigation, then the plant continues to extract water from the largest water-filled pores until eventually the only water remaining is held at very high tensions by the finest pores and by interactions

Table 5.4. Physical properties of selected potting media

Mix	Depth (cm)	Total porosity	Moisture content	Air-filled porosity	Available water	Bulk density ($g \cdot cm^{-3}$)
		(% by volume)				
sphagnum peat	12	94	84	10	—	0.10
peat and perlite (1:1)	12	93	51	42	48	0.14
peat and vermiculite (1:1)	12	94	81	13	60	0.23
UC mix*	12	73	52	21	44	0.67

Note: *UC mix is a 1:1:1 mix of sand, redwood sawdust, and peat.

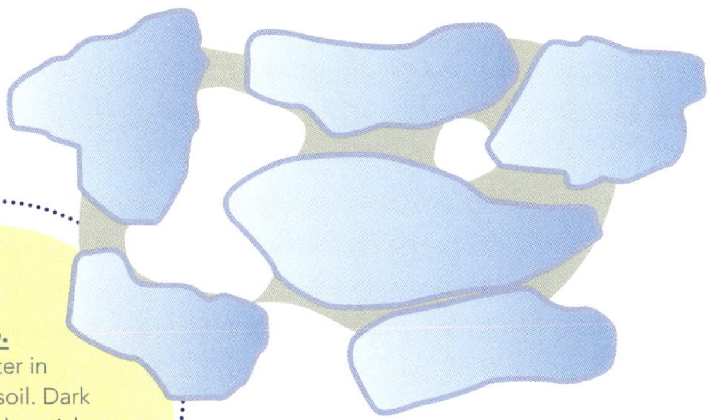

FIGURE 5.6. Available water in unsaturated soil. Dark bands around particles represent "bound" water.

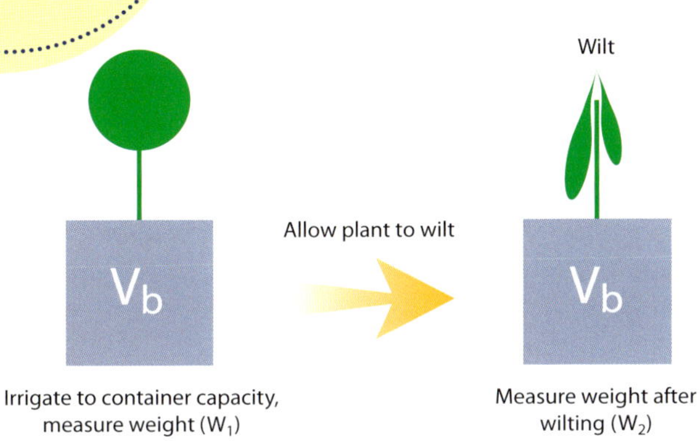

$$\theta_a = \frac{W_1 - W_2}{V_b} \times 100\%$$

FIGURE 5.7. The available water in a container medium (Θ_a) can be calculated from the difference in weight between container capacity and plant wilting, if the substrate volume (V_b) is known.

between water and particle surfaces. The higher the tension at which water is held in the soil, the greater the amount of energy the plant must exert to extract it. When plants stop taking up water and they wilt, there is still water present in the potting mix. Obviously, this remaining water is of no horticultural value, so available water is a useful indicator of the reservoir of water in a container that is usable by plants. The desirable minimum value for available water in container media is 30 percent. A lower value for available water means that containers must be irrigated frequently. The available water in a particular substrate and container can be determined quite easily if the substrate volume is known (fig. 5.7).

Moisture-release curve

A moisture-release curve, in the practical form of a calibration curve, is a physical characteristic of a container medium that provides a useful way to determine the appropriate irrigation interval for a particular container medium and pot size. It expresses the relationship between the volumetric moisture content and the tension at which water is held in the medium. This relationship can be determined for a container medium with a tensiometer and a balance. The basic idea is to irrigate a plant in a container until it reaches container capacity, then weigh it as described in figure 5.7 for determining available water, except that weights will be taken regularly prior to wilting, and the corresponding soil moisture tension will be measured with a tensiometer. Generally a curve of the form shown in figure 5.5 is obtained.

These curves are useful because they identify the available water and the tensions between container capacity and the wilting point. By using a tensiometer and monitoring the tension in a representative pot in a crop, it is possible to determine the appropriate time for irrigation, discussed further in chapter 7. Much of the water in the medium described in figure 5.5 is removed at very low tensions—half the available water (the point at which irrigation is recommended) is gone at a tension of just under 2.5 kilopascals. (A kilopascal [kPa] is a unit of pressure, equivalent to 1 centibar or about 7 pounds per square inch.) This value represents the water present in the coarser pores. As the remaining available water is removed from finer pores, the tension begins to rise rapidly. In this illustration, wilting occurs at a tension of about 21 kPa (21 centibars). Note that there is still water in the medium (roughly 13% by volume) but that it is unavailable to the plant.

Bulk density

Bulk density allows for comparison of the dry weight of mixes. Most container mixes have a bulk density from 0.4 to 0.8 $g{\cdot}cm^{-3}$. Since most of the handling and shipping of mixes done by nurseries involves pots with plants present, the wet weight of mixes is a useful property to know. Wet weights tend to be similar among mixes, even when bulk densities vary greatly. For example, the bulk density of a 1:1 peat: vermiculite mix is 0.23 $g{\cdot}cm^{-3}$ and that of a UC mix (1:1:1 mix of sand, sawdust, and peat) is 0.67 $g{\cdot}cm^{-3}$, but the wet weights are 993 and 1,137 g per liter (62 and 71 pounds per cubic foot), respectively.

Potting Mix Amendments

Amendments are added to potting mixes to increase total porosity, water-holding capacity, and air-filled porosity as well as to decrease bulk density. The factors to consider when choosing amendments of potting mixes are availability, uniformity, cost, ease of addition, stability, ability to hold nutrients and water, low nitrogen draft, and freedom from toxicity, salts, pests, and diseases. Also, the material should have an appropriate particle-size range (see the section "Particle Size and Distribution of Amendments" earlier in this chapter).

Amendments will improve mixes significantly only if they are added in sufficient quantities. For greenhouse crops, the ratio of organic matter to mineral should be about 2:1. For woody nursery crops, the ratio can be as high as 1:1. In most cases, amendments will not change the physical properties of a mix substantially unless they represent at least 30 percent of the mix by volume.

Organic Amendments

Peat

The peats used for container mixes are generally of the high moor type, coming from raised bogs that develop in cold areas with high rainfall. Large deposits of such peats exist in Canada and northern Europe. Canada is the main source of peat for U.S. horticulture.

Under acidic, waterlogged conditions, partial decomposition of sphagnum moss, other mosses, and sedges occurs. Sphagnum peat comes from the lower layers of a bog. After its partial decomposition, it is extremely stable and resistant to further decomposition. It is valued for its fibrous, spongy texture, high porosity, and high water-holding capacity. The water-holding characteristics of peat depend on the degree of decomposition. Less decomposed peats may hold approximately 15 to 20 times their weight in water, whereas well-decomposed peats may hold only 4 to 8 times their weight.

Sedge peats occur in mineral soils. They are usually more decomposed than sphagnum peats but are acceptable as potting mix amendments. Peat humus and sedimentary peat, on the other hand, have a very fine texture, may contain high amounts of salts, and have a relatively high pH. They are not recommended for potting mixes.

Although sphagnum peat is favored for use in container media, it does have some drawbacks. It is harvested from peat bogs, and rates of peat harvest may greatly exceed the rate of peat formation. As a result, its use in ornamental horticulture contributes to loss of peat bogs, which provide habitat for numerous rare species of plants and animals. A more immediate concern for nurseries is the tendency of peat (and some other organic amendments) to become water repellent when it dries out. In many cases, other amendments of the container medium ameliorate this problem. This phenomenon can be tested by placing a sample of the dry mix in a small bowl, adding several drops of water, and measuring the length of time required for absorption of the water and rewetting of the medium. If the water has not been absorbed within a minute or so, it is likely that the medium will become water repellent in containers, at least when its surface dries out. To overcome this, a surfactant (wetting agent) can be mixed into the container medium before planting or added as a drench after planting. The surfactant decreases the surface tension of water and facilitates wetting by enabling the water to penetrate and spread through the mix more completely. It may also cause a slight increase in water-holding capacity and a corresponding decrease in air-filled porosity. See the section "Surfactants" later in this chapter for more information.

Coir

Within the past several years, coir dust has come into use as an amendment for container media. Coir is the mesocarp of the coconut fruit, so it is a by-product of coconut processing. The long fibers of the mesocarp are used in the manufacture of brushes, mattress stuffing, twine, and similar products, but the short fibers, known as coir dust, had been considered a waste product until its value in container media was widely recognized. As long ago as 1949, it was reported to be suitable as a growing medium and was called cocopeat, but coir was not widely used in horticulture until the 1990s. It has excellent physical properties, similar to those of peat. The pH range is 5 to 6.8, so little or no adjustment of pH is needed if coir is the only organic matter in a mix. The quality of commercial coir products varies with source and age of the material. In some instances, coir dust has high concentrations of sodium and potassium salts, which must be leached prior to use.

Wood products

Wood products, including some softwood barks and sawdust, redwood sawdust and bark, and some hardwood sawdusts and bark, are commonly used in commercial soilless media for containers. Redwood and fir bark are sometimes used alone as a medium for growing epiphytic plants, such as many of the orchids. Fresh wood products may contain toxic phenolic compounds and tannins. These problems are overcome by leaching and aging piles of the material.

Some wood products, such as fresh sawdust, are subject to rapid decomposition by microbes, which can result in competition between those microbes and the plant for soluble nitrogen (nitrogen immobilization, also called nitrogen draft). Wood products differ in their rate of decomposition, hence the nitrogen draft differs. Nitrogen drafts in soil, following addition of some barks and sawdust, are listed in table 5.5. Softwood barks are more commonly used as amendments because they decompose more slowly. Hardwoods, including eucalyptus, can serve well as organic amendments if they are composted and if the resulting particle-size range is suitable. Microbes may have other effects in container media besides causing nitrogen immobilization. Some research has shown

Table 5.5. Relative rates of soil nitrogen depletion (draft) in organic amendments

Large nitrogen draft	Intermediate nitrogen draft	Small nitrogen draft
ponderosa pine bark	pine shavings	peat
Jeffrey pine bark	Douglas-fir bark	redwood bark
hardwoods	red fir bark	sugar pine bark
	redwood sawdust	incense cedar bark
		redwood sawdust

Photo: R. Y. Evans.

that properly formulated bark mixes can suppress some root rot diseases and may exert control of *Rhizoctonia* in damping-off, Fusarium wilt, and some nematodes.

Other organic amendments

Many other organic materials are available, but not all are suitable as potting mix amendments. All organic materials should be composted by adding nitrogen fertilizer, moisture, and air prior to use. The suitability of various organic materials is listed below.

- Leaf mold is usually unsatisfactory because it may contain high levels of salts and it breaks down rapidly.
- Composted green waste may be satisfactory, but variation in composition and rate of decomposition makes it an unreliable amendment.
- Sludges come from municipal and industrial wastes, and generally they are unacceptable. They tend to be fine textured and may contain heavy metals (such as cadmium, mercury, nickel, zinc, and lead). Odor may also be objectionable, especially when the nursery is in an urban area.
- Household garbage is variable, may contain toxic heavy metals, high levels of boron, and often has unsuitable physical properties.
- Miscellaneous organic materials, including straw, nutshells, ground palm trunks, and rice hulls, may be satisfactory amendments if prepared properly. For example, fresh rice hulls tend to be too coarse, have a high nitrogen draft, and may contain weed seeds, but after nitrogen addition and aging they make a good amendment.

Inorganic Materials

Vermiculite

A mineral that resembles mica, vermiculite is made up of sheets of alumina held together by calcium and magnesium. The mined mineral is heated briefly at a very high temperature. Water trapped between the sheets of alumina expands in the form of steam, causing particles to expand like an accordion to about 15 to 20 times their original volume. The average particle density is only 0.08 $g \cdot cm^{-3}$. Most sources of vermiculite contain significant amounts of available potassium and magnesium. Vermiculite compresses easily, so it should not be used in mixes that have a high bulk density.

Perlite

Perlite is a volcanic aluminosilicate that is prepared for horticultural use by crushing and rapid heating to very high temperatures, which causes expansion. Perlite is light—the particle density of perlite is about 0.13 $g \cdot cm^{-3}$—and it is mechanically weak. Unlike vermiculite, perlite has a closed structure and does not retain much water. It has no significant chemical properties, but it is useful for reducing the weight of a potting mix.

Pumice

Pumice is a porous volcanic rock that can be used in its natural state as an amendment. Unlike perlite and vermiculite, it is mechanically tough. Depending on its source, it can have excellent physical properties as a potting mix amendment. It is abrasive to mixing equipment.

Calcined clay

Firing clay at high temperatures can produce stable aggregates that are suitable as an amendment. The main purpose for its use is to modify physical properties of a potting mix. Calcined clay holds water in fine pores within each aggregate, so amending a mix with it tends to improve the water-holding capacity. Available water, however, may be relatively low because the water in many calcined clays is unavailable to plants.

Calcined diatomite

Diatomite, also known as diatomaceous earth, forms a stable aggregate after it is heat treated. This material is satisfactory if it is available at a competitive price.

Rockwool

Rockwool is a fibrous mineral material similar to fiberglass. Granular forms have been used to some extent as amendments in potting mixes because of their excellent water-holding properties. Incorporation of granular rockwool into potting mixes can be difficult because of its tendency to clump.

Foam plastics

Polystyrene and urea-formaldehyde foam plastics are sometimes combined with peat to make light potting mixes. Polystyrene is chemically inert and physically stable in normal use. The internal pores are closed cells, so water does not penetrate. In effect, the foam pellets behave like lightweight sand. Large foam particles tend to rise to the surface of the mix after irrigation, and problems sometimes arise during mixing because of the electrostatic properties of polystyrene.

Urea-formaldehyde foam has an open cellular structure that allows water to penetrate. It has physical properties somewhat similar to those of peat, but it tends to degrade in acid conditions.

Use of Hydrogels and Wetting Agents to Increase Water-Holding Capacity

Synthetic hydrogels

The most common hydrogels are polymers of anionic polyacrylamide. Their ability to absorb large amounts of water has created interest in their incorporation into potting mixes. The presence of normal levels of nutrients in potting mixes reduces their swelling markedly, however, and at recommended rates of addition there is no significant increase in available water. Controlled experiments have shown that hydrogels do not increase the time between irrigation and wilting in standard potting mixes.

Surfactants

Surfactants, or wetting agents, are sometimes added to increase the water-holding capacity of container mixes. These chemicals reduce the surface tension of water, thereby enabling the water to penetrate and spread through the mix more completely, especially in mixes that have a high organic matter content (see the section "Peat" earlier in this chapter). Surfactants are useful for some mixes that have excessive air-filled porosity, or for mixes that do not "wet up" readily. In some instances, they also improve the air-filled porosity by reducing surface tension (and therefore the capillary rise of water). Also, growers should take care to use only surfactants that are known to be nontoxic to plants. In general, nonionic surfactants are less likely to be toxic than ionic ones.

Chemical Properties

A soil's chemical properties describe its overall character and behavior in terms of relative acidity and alkalinity (pH) and nutrient-holding capacity (cation exchange capacity, or CEC).

Relative Acidity and Alkalinity (pH)

Chemical amendments must be added to soilless potting mixes to establish a favorable pH. Most organic amendments are acidic (table 5.6). Roots are sensitive to low (acidic) pH, and many nutrients are not absorbed by plants growing in a low-pH medium. Therefore, the pH must be raised to achieve satisfactory growth. For most nursery crops, the container mix should have a pH value from about 5.5 to 6.5. Raising pH is most easily accomplished by adding lime (calcium carbonate) or dolomite (calcium magnesium carbonate). In most cases, dolomite is the preferred amendment because it is also a slow-release source of two essential plant nutrients, calcium and magnesium. In most cases, 3 to 5 pounds of dolomite must be added to each cubic yard of organic amendment. The proper amount should be determined by a horticultural soils testing laboratory.

Table 5.6. Chemical properties of some organic amendments

Amendment	pH	Cation exchange capacity (milliequivalents per 100 cm^3)	Cation exchange capacity (equivalents per yd^3)
sphagnum peat moss	3.5–4.2	15	50
redwood bark	3.2	6	19
ponderosa pine bark	3.6	8	24
hypnum peat moss	5.1	32	96
white fir bark	4.4	16	50
fir sawdust	4.0	3	10
redwood sawdust	3.8	5	16
pine sawdust	4.5	4	13

Table 5.7. Typical thermal death points of plant pathogens and pests

Plant pathogens and pests	Temperature (°F)	Exposure time (min)
most pathogenic fungi	140	30
Pythium and *Phytophthora*	120	30
most pathogenic bacteria	140	30
all pathogenic bacteria	160	30
nematodes	120	10
insects and slugs	150	30
most weed seeds	170	15
resistant weed seeds	212	30
most viruses	212	15

Note: Some organisms in each of these categories may have higher thermal death points.

Cation exchange capacity (CEC)

This property is a measure of the ability of a material to hold on to positively charged ions. The higher the CEC, the better a material is at holding some nutrients (mainly calcium, magnesium, potassium, and ammonium nitrogen) against leaching. CEC is usually expressed as the amount of electrical charge held per unit weight of the material tested. Some potting amendments, such as peat and vermiculite, have high CEC values, while others, like sand and perlite, have low CEC values.

Chemical Fertilizer Amendments

Nearly all organic and inorganic amendments lack sufficient nutrients to meet plant requirements over a period of weeks to months. Nutrients are often added to the container mix before planting. Procedures for preplant fertilizing are described in chapter 6.

Mixing Amendments

Media should be mixed in a location sheltered from wind and away from drainage channels to avoid potential contamination of water bodies with media particles containing nutrients and pesticides. Large batches can be mixed with a front-loading tractor on a smooth, paved surface. Spread out the amendments, one on top of another, then overlay them with chemical amendments. Mixing should be done methodically by lifting bucketfuls from the base of the pile and dropping them onto a new pile, with each bucketful dropped at the top of the new pile. The process should be repeated at least three times to ensure thorough blending of ingredients. Alternatively, mixing can be accomplished with a cement mixer, a continuous belt, or machines that have agitators, rotors, or screws that blend the amendments of the medium. Details concerning the type of mechanized machinery used for mixing are described in chapter 4. Powdered fertilizers should not be added to wet ingredients. If it is necessary to add water, it should be done after fertilizers have been added and mixed in.

The process of mixing can alter the physical properties of the mix by decreasing average particle size. Mechanical mixing can also damage incorporated controlled-release fertilizers and change their nutrient release characteristics, so check the manufacturer's product information for precautionary statements. Hand-mixing is least destructive, and mixing by agitation or other more vigorous methods tends to be most destructive. Among organic amendments, peat is particularly susceptible to damage during mixing or mechanized container filling. Mechanized filling of containers with peat can result in a substantial decrease in air-filled porosity.

Eliminating Pests and Diseases from Media

Heat Pasteurization

Many of the materials used to make soilless potting mixes are free of plant pathogens and do not need to be sterilized prior to their use. In cases where pathogens are suspected, heat pasteurization can be used to eliminate them. Heat pasteurization is effective because most plant pathogens and pests are killed at rather low temperatures (table 5.7). In general, organisms with high moisture content are most susceptible to heat treatment. Thus, it is advantageous to treat potting mixes while they are moist. Nearly all pests and pathogens are killed by exposure to high temperature (about 160°F) for 30 minutes. Weed seeds and viruses may tolerate somewhat higher temperatures. To provide a margin of safety, the recommended treatment is to heat the potting mix to about 180°F for 30 minutes. Since heat pasteurization can alter incorporated controlled-release fertilizers, affecting their nutrient release characteristics or modifying them chemically, check the manufacturer's product information for precautionary statements.

The most effective way to pasteurize a potting mix is by using steam. A plastic tarp should be used to cover the mix during steaming to reduce heat loss. Further details concerning the treatment of media with steam are presented in chapter 4.

Solarization

Another nonchemical control method of pasteurizing soil or media is solarization, that is, the trapping of solar energy beneath clear plastic sheets spread over a soil surface. In container nurseries, a method that has been successfully used is to cover shallow layers of uniformly moist potting mix (about 6 inches deep) with a clear, continuous polyethylene film or polyvinyl chloride (PVC) film that stays in place for 4 to 6 weeks. Furthermore, containers already filled with potting media can be treated using a double layer of plastic.

There is relatively little information on the reliability of solarization for eradication of most soilborne pathogens and pests. Effectiveness usually depends on heating the potting mix to moderate temperatures (about 110°F) for several hours a day over a period of at least 1 month. Although solarization may increase potting mix temperature to levels that kill weeds (see chapter 15) and many pathogens, some heat-tolerant fungi and bacteria are difficult to control by solarization.

Solarization should be conducted during the hottest weeks of the year in an open area to take full advantage of solar radiation. In California, the best time for solarization is usually from June to early September, depending on location, and it is typically most effective in warm, sunny inland areas.

Chemical Treatments

Chemical treatments can also be used to eliminate pests and diseases from soil and media, although in general they are not as effective as steam pasteurization. Fumigants require enclosed treatment and proper temperature. For many years, methyl bromide was the preferred soil fumigant for broad-spectrum activity, but its use today is severely restricted due to global environmental concerns. The use of other chemical fumigants is currently limited. For more information on chemical fumigation, see chapter 15.

Container Media Storage

Good storage practices are important for bulk container media. Media should be stored in a location sheltered from wind and away from drainage channels to avoid potential contamination of water bodies with media particles containing nutrients and pesticides. The amendments of the mix and the bulk medium should be stored on a concrete pad at an elevation higher than that of growing areas to minimize the introduction of pathogens (fig. 5.8). Peat moss and other bagged amendments should be stacked on pallets and covered with tarps to prevent breakdown of the bags by sunlight. Piles of substrate and bulk raw materials should be covered loosely with tarps to retain moisture and minimize wind dispersal. Piles of the substrate and bulk materials should not be more than 10 feet high (but local regulations governing dimensions of piles may specify lower heights for organic materials). Keeping pile heights low will minimize the need to drive heavy equipment onto piles. Heavy equipment driven onto piles of the container medium causes compaction and mechanical damage to the substrate, changes its physical properties, and reduces its quality. Low pile heights also lower the risk of combustion of organic matter in the substrate. Biological activity in the medium or bulk organic amendments can cause them to behave like a compost pile. The interior temperature of the pile should be monitored regularly. The pile should be turned if the temperature exceeds 150°F.

FIGURE 5.8. A concrete storage pad and retaining wall are used to store and mix media and fertilizers. *Photo:* D. Zurawski and J. P. Newman.

The finished container medium should not be stored for long periods. During storage the medium will begin to decompose and its physical and chemical properties will change. Salinity levels in the stored medium will rise, especially if slow-release or controlled-release fertilizers have been incorporated during mixing. Piles that are stored for long periods and are not turned regularly may undergo anaerobic decomposition, which can generate chemicals that are toxic to plants. Use mixes with incorporated organic or coated, controlled-release fertilizers within 1 week of preparation. Mixes with inorganic, noncoated, non-controlled-release fertilizers can be stored indefinitely.

REFERENCES

Baker, K. F., R. D. Durbin, J. Ferguson, J. W. Huffman, O. A. Matkin, D. E. Munnecke, C. N. Roistacher, W. R. Schoonover, and R. H. Sciaroni. 1957. The U.C. system for producing healthy container-grown plants. Berkeley: University of California Division of Agricultural Sciences Manual 23.

Bunt, A. C. 1988. Media and mixes for container-grown plants. London, UK: Unwin Hyman.

Handreck, K., and N. Black. 2002. Growing media for ornamental plants and turf. Sydney: University of South Wales Press.

Nelson, P. V. 2003. Greenhouse operation and management. 2003. 6th ed. Upper Saddle River, NJ: Prentice-Hall.

Pier, J., and D. Barlow. 2012. Western fertilizer handbook: Third horticulture edition. Sacramento: Western Plant Health Association.

Stapleton, J. J., C. A. Wilen, and R. H. Molinar. 2008. Pest notes: Soil solarization for gardens and landscapes. Oakland: University of California Agriculture and Natural Resources Publication 74145. ANR CS website, http://anrcatalog.ucdavis.edu/Items/74145.aspx.

Nutrition and Fertilization

Richard Y. Evans

CHAPTER SIX

Proper use of fertilizers is an essential part of a profitable nursery business. Maximum plant growth rates can be achieved only if nutrients are present in amounts that are neither too small nor too large. A big challenge for growers is supplying crops with the nutrients they require without excessively fertilizing and contributing to nitrogen runoff.

Plants grow poorly if provided with insufficient amounts of fertilizers, especially fertilizers that supply nitrogen. Typical media, such as mixes of sand and organic matter, lack sufficient plant nutrients to meet plant requirements over extended periods of time. The problem of insufficient nutrients is often made worse by the application of excessive amounts of irrigation water, which tends to leach out nutrients, especially nitrogen and potassium. Furthermore, the organic matter used in mixes usually imparts a low pH, which can adversely affect nutrient uptake. To counter these problems, the chemical properties of container media must be altered by adding chemical amendments that establish a favorable pH and good fertility. Additional fertilizer must be supplied subsequently, during crop production, to sustain adequate fertility for continued plant growth.

Excessive applications of some fertilizers may also cause poor growth or direct injury to plants. In addition, excessive fertilizer use can have serious harmful effects on the quality of water and on the health of plants and animals that depend on that water. In fact, nitrogen runoff management has become one of the most critical issues facing California agriculture today.

This chapter provides a basis for the proper use of fertilizers. One can fertilize to supply plant nutrients without knowing why those nutrients are needed, but knowledge of how plants use nutrients can be helpful, particularly when trying to diagnose problems. Diagnosis of nutritional disorders is treated in chapter 11, but a brief introduction to the functions of plant nutrients is presented here. Methods for the application of preplant and postplant fertilization are also discussed, including the use of controlled-release fertilizers and fertigation. Also presented in this chapter is the proper use of fertilizer proportioners and fertilizer storage.

An important topic addressed in this chapter is fertilizer runoff management. This includes leaching management practices for salinity control and the use of container medium and plant tissue tests for basing fertilizer applications. Nitrogen and phosphorus are important nutrients for runoff management, so a detailed discussion is provided that includes fertilizer recommendations and instructions on calculating how much nitrogen your crops actually require during a production cycle. Additional management practices to prevent runoff are described in chapter 8.

Photo: D. Zurawski and J. P. Newman.

Functions of the Essential Elements in Plants

Plant growth requires adequate levels of water, air, and mineral nutrients in the growing medium. The leaf and stem tissue of a typical nursery plant consists of approximately 80 to 85 percent water and 15 to 20 percent dry matter. The dry matter generally is divided among numerous chemical elements, of which 16 are considered essential (table 6.1). Of these, carbon, hydrogen, and oxygen are supplied mostly from the atmosphere or from water (in the case of hydrogen and oxygen). The remaining elements are taken up from the growing medium.

Table 6.1. Essential plant nutrients, expressed as a percentage of leaf dry weight

Element	Chemical symbol	Nutrient classification	Uptake	Leaf content* (% of dry wt)
carbon	C	nonfertilizer	—	
hydrogen	H	nonfertilizer	—	90.0
oxygen	O	nonfertilizer	—	
nitrogen	N	macronutrient	NO_3^-, NH_4^+ in soil solution	3.0
potassium	K	macronutrient	K^+ in soil solution	4.0
calcium	Ca	macronutrient	Ca^{2+} in soil solution	1.0
phosphorus	P	macronutrient	phosphates in soil solution	0.5
magnesium	Mg	macronutrient	Mg^{2+} in soil solution	0.5
sulfur	S	macronutrient	SO_4^{2-} in soil solution	0.5
chlorine	Cl	micronutrient	Cl^- in soil solution	0.1
iron	Fe	micronutrient	ion or chelate in soil solution	0.02
manganese	Mn	micronutrient	ion in soil solution	0.02
boron	B	micronutrient	boric acid or borate	0.006
zinc	Zn	micronutrient	ion or chelate in soil solution	0.003
copper	Cu	micronutrient	ion or chelate in soil solution	0.001
molybdenum	Mo	micronutrient	ion or chelate in soil solution	0.0002

Note: *Carbon, hydrogen, and oxygen together comprise about 90% of leaf dry weight.

Many elements serve multiple roles in plants, but some general classifications are possible. For example, six elements are involved in almost every facet of plant life:

- carbon
- hydrogen
- oxygen
- nitrogen
- sulfur
- phosphorus

Carbon, hydrogen, and oxygen are basic building blocks of all organic molecules, such as those in amino acids, proteins, carbohydrates, fats, and cell walls. Nitrogen is a constituent of DNA, the genetic material of all organisms, and RNA, the molecules needed to translate the genetic code into the proteins that carry out all of the functions of plant cells. Nitrogen is also found in all amino acids and proteins, which is why it is needed in such large quantities. Sulfur is a constituent of some amino acids and proteins, as well as vitamins such as biotin and thiamine, and it is needed to initiate the synthesis of all proteins. Phosphorus has a particularly important role in the acquisition, storage, and utilization of energy, such as the energy captured in photosynthesis.

Plant cell walls would not exist without the six elements named above, but they also depend on other elements. Calcium and boron contribute to the strength of cell walls by helping to stabilize cell wall components. Silicon, an element that most plant scientists do not consider to be essential, also may contribute to the strength and rigidity of cell walls.

Several elements are particularly important for their role in metabolism, the chemical reactions involved in the construction and maintenance of plants. Magnesium is a key component of chlorophyll, the green pigment whose light absorption is critical for photosynthesis. Iron is particularly important in photosynthesis, respiration, and the metabolism of nitrogen and sulfur. Manganese plays a role in photosynthesis and the breakdown of some potentially toxic chemicals. Zinc activates some enzymes that are important in protein synthesis. Zinc is also involved in the transport of carbon dioxide through cells, as well as the detoxification of some potentially harmful compounds. Copper is important in photosynthesis, respiration, and detoxification. Molybdenum is involved in the assimilation of nitrogen.

Many enzymes (the proteins that carry out chemical reactions in plants) are activated by plant nutrients. Potassium and magnesium activate enzymes involved in the transfer of energy through plant cells. Chlorine activates the enzyme that splits water during photosynthesis. Calcium activates calmodulin, which in turn regulates many enzymes. Manganese activates many enzymes and is particularly important in the reactions of the citric acid cycle, in which cells use oxygen as part of respiration. Iron, zinc, and copper are nonspecific enzyme activators.

Some less specific, but still important, functions of nutrients also exist. Many of the soluble chemicals in plants, such as sugars and amino acids, exist in an electrically charged form, and they must be accompanied by a counterion that balances the electrical charge. Calcium, magnesium, and potassium serve as positively charged counterions, and chlorine, nitrogen, and sulfur contribute to negatively charged counterions. In addition, chlorine, nitrogen, and potassium

Photo: J. K. Clark.

are important dissolved solutes in plant cells, useful for maintaining cell turgor.

Preplanting Fertilization

Many plant nutrients can be added to the container medium prior to planting. Most important among these are calcium, magnesium, phosphorus, and the micronutrients iron, zinc, manganese, boron, copper, molybdenum, and chloride. In some cases it is beneficial to add some nitrogen and potassium as preplanting fertilizers. Some of these nutrients may already be present in the components of the container medium or in the irrigation water, so medium and water testing should be done to ensure that fertilizer additions are appropriate.

Calcium and Magnesium

Calcium (Ca) and magnesium (Mg) are often added when the acidity of the container medium is adjusted before planting. For most crops, the pH of the container medium should be 5.5 to 6.5, the range in which most nutrients are readily available (fig. 6.1). The organic amendments in container media usually produce a pH well below this range, so it must be adjusted by adding a chemical amendment that neutralizes the acid. Minerals that contain carbonate, such as lime (calcium carbonate) or dolomite (calcium magnesium carbonate) raise the pH. The amount to add depends on the chemistry of the medium and the pH requirement of the crop, but typical rates are 5 to 10 pounds per cubic yard. The proper amount should be determined by a horticultural soils testing laboratory. Lime also supplies calcium, an important plant nutrient. Dolomite, which supplies both calcium and magnesium, is more widely used for correcting the acidity of container mixes, but lime is preferred in media used for rooting cuttings. Cuttings require calcium, but not magnesium, during rooting, and magnesium can actually inhibit rooting by interfering with calcium uptake.

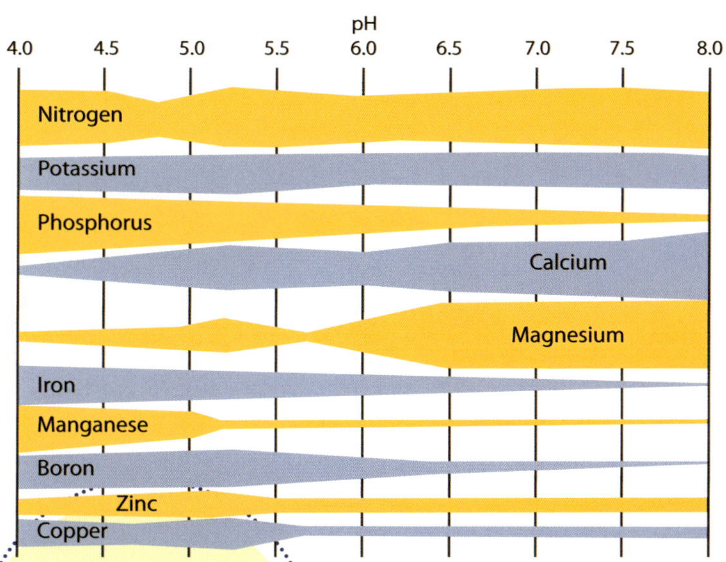

FIGURE 6.1. Effect of soilless media pH on the availability of some essential plant nutrients.

Dolomite is sparingly soluble, which tends to make leaching losses of calcium and magnesium insignificant. In addition, the reaction of calcium and magnesium with sulfates, which are not very soluble, causes these elements to be held tightly in the mix. As a result, it is unusual for plants to require postplanting additions of calcium or magnesium.

Alternative sources of calcium and magnesium for container media exist that do not require pH adjustment. Gypsum (calcium sulfate) can be added as a calcium source, and Epsom salt (magnesium sulfate) is a magnesium source. Neither should be added to a strongly acid medium because they will cause the pH to drop even more.

Phosphorus

Phosphorus (P) may be supplied in a liquid feed or in controlled-release fertilizers, but it is often important to add it as a chemical amendment before planting. Adequate uptake of phosphorus by young plants requires rapid replenishment of the phosphorus in the zone around roots. Young plants that receive phosphorus only in a liquid feed tend to have a slower initial growth rate than those that receive phosphorus as a chemical amendment to the container medium.

Preplanting addition of phosphorus is usually provided as superphosphate, which is made from rock phosphate that is reacted with sulfuric acid, or as triple superphosphate, which results from the reaction of rock phosphate with phosphoric acid (and therefore does not provide sulfur). When added to a moist container medium, both superphosphate and triple superphosphate form sparingly soluble compounds that act as slow-release forms of phosphorus.

In a typical mineral soil, phosphorus solubility is highest at a neutral pH. Phosphate solubility is limited at high pH values by the formation of tricalcium phosphate, which has very low solubility. At low pH values, phosphorus solubility is limited by the formation of aluminum and iron phosphate. In a container medium, in contrast, the normal amounts of calcium, aluminum, and iron are not sufficient to immobilize much of the phosphorus at low pH. Instead, it is subject to leaching and can become a serious pollutant in runoff water unless carefully managed.

Micronutrients

Micronutrients are commonly provided by adding a mixture of micronutrient salts. Most of these are sparingly soluble sulfate salts that are retained in the medium and meet plant needs throughout a crop cycle. Growers sometimes add iron separately. The typical form used is ferrous sulfate ($FeSO_4 \cdot 7H_2O$). Ferric sulfate, a different chemical, should not be used because it reacts with water to form ferric hydroxide, which is insoluble and drastically acidifies the medium.

Potassium

Potassium (K) is sometimes added before planting, especially when the liquid feed program will not begin until 1 or 2 weeks after planting. Potassium fertilizers can damage plants if added in large amounts, so the standard practice is to add either potassium nitrate or potassium sulfate at a rate of 1 pound potassium per cubic yard.

Nitrogen

Along with potassium, nitrogen (N) is the element that must be supplied to plants in the largest amount, both because of high plant demand and because it is easily leached. As with potassium, nitrogen may be added to the container medium before planting if the liquid feed program will not start immediately. Nitrogen is normally supplied as a soluble salt, such as potassium nitrate, calcium nitrate, or ammonium nitrate. All three forms are readily available to plants, but they can be leached easily. Also, these soluble salts would be toxic to plants if supplied in large quantities prior to planting. Therefore, only about 1 pound of nitrogen per cubic yard of container medium should be added before planting.

Fertilization after Planting

Most crop requirements for all of the nutrients except nitrogen, potassium, and sometimes phosphorus are met by the preplanting additions of fertilizers and the addition of micronutrients to the potting mix (see chapter 5). These other three nutrients are required in large amounts and are not held in sufficient amounts by the container medium, so they must be provided by the application of controlled-release fertilizers or the injection of fertilizers into the irrigation water.

Controlled-release fertilizers have a spherical polymeric coating that encapsulates soluble fertilizers. When added to a moist container medium, water vapor diffuses from the container medium into the capsule (referred to as a prill). Once inside the capsule, water vapor condenses to liquid water and dissolves the fertilizer salts, creating a highly concentrated salt solution. This process takes place over approximately 24 hours, during which the capsule swells and becomes translucent. Then the dissolved fertilizer moves through pores in the capsule wall into the container medium. This is a slow process whose rate is controlled by the thickness of the capsule wall, the size of the pores in the wall, and temperature of the container medium. High temperatures increase the rate at which the fertilizer is released. For fast-growing crops, controlled-release fertilizers with thin capsule walls are recommended. These normally release all of their fertilizer contents over 3 to 4 months. For woody plants, a formulation with a thicker coating is usually used, one that releases the fertilizer over 7 to 14 months (depending on the formulation). Controlled-release fertilizers may be incorporated into the container medium prior to planting, but prolonged storage under moist conditions will result in release of the fertilizer. The accumulation of soluble fertilizer in the medium may increase the soluble salt concentration sufficiently to injure plants. Similarly, container media should not be steam pasteurized after addition of controlled-release fertilizer because this will result in rapid release of injurious levels of fertilizer. Furthermore, if controlled-release fertilizers are blended into the medium, steps should be taken to minimize physical damage to prill coatings, which may cause rapid release of fertilizer from prills.

Controlled-release fertilizers can greatly reduce nitrogen losses if they are applied correctly. Nutrient release rates are controlled by the properties of the capsule walls and by temperature and moisture, not by plant needs. This restricts nitrogen leaching losses from overirrigation to the small amount released since the previous irrigation. However, application of nutrients cannot easily be varied according to crop needs. For example, an amount of controlled-release fertilizer that releases enough nitrogen to feed a rapidly growing plant would be far more than the amount needed for a young plant or one that is no longer taking up much nitrogen. The excess nitrogen that is released can be lost to leaching if plants are overirrigated.

The likelihood of nitrogen leaching losses from controlled-release fertilizers is greatest in the first few weeks after planting, when plant root systems are limited, nutrient demand is low, and plants are consuming relatively small amounts of water. It is best to apply controlled-release fertilizers just below the plant roots at the time of planting (sometimes called dibbling) or to broadcast it on the surface of the container medium.

Liquid feeding (fertigation) is widely used by nurseries and greenhouses. It is less expensive than controlled-release fertilizers, after initial capital costs for injectors, and it is well suited to the production of large areas of uniform crops because fertilizer concentrations can be varied according to crop needs (e.g., decreasing the nitrogen supply late in a crop). The appropriate concentration of nitrogen in nutrient solutions depends on the crop being produced. Some suggested concentrations of liquid feeds for many nursery crops are listed in table 6.2. Although exceptions exist, most herbaceous crops require nitrogen and potassium concentrations of 100 to 150 mg/L in continuous liquid feeds. For most woody nursery crops, 50 mg/L is sufficient. The major disadvantages of liquid feeding are its inefficiency in putting nutrients into the root zone and its susceptibility to leaching losses of nitrogen from excessive irrigation.

Forms of Nitrogen for Liquid Feeds

The nitrogen-containing compounds used in liquid feeds must, of course, be highly soluble in water. The most commonly used chemicals are calcium nitrate, potassium nitrate, and ammonium nitrate. Nitrate is the predominant form of fertilizer nitrogen used by nurseries and greenhouses. Although ammonium is readily taken up by plants, it accounts for only a small percentage of crop nitrogen uptake. The microbial process called nitrification rapidly converts ammonium to nitrate in warm, moist container media. The majority of nitrogen taken up by plants is typically in the form of nitrate.

The choice of nitrogen form depends partly on the desired composition of the fertilizer solution (as when both potassium and nitrogen are to be added) and partly on management of pH in the container medium. For example, most liquid feeds rely on potassium nitrate as the source for all of the crop's potassium. But it may be beneficial to include ammonium nitrogen in the liquid feed when the pH of the

Table 6.2. Growth characteristics of representative nursery crops grown in a nursery or greenhouse at the University of California, Davis

		Shoot nutrient concentrations				Recommended liquid feed concentrations		
Species	Dry wt (g)	N (%)	P (%)	K (%)	Water use (L)	N (mg/L)	P (mg/L)	K (mg/L)
Acer palmatum	5.93	2.50	0.67	2.66	4.7	25	8	23
Acorus 'Ogon'	25.10	2.87	0.39	3.73	13.0	49	—	—
Angelonia angustifolia	13.10	3.29	0.49	2.70	3.4	116	17	92
Aptenia cordifolia	26.76	3.25	1.10	7.75	11.3	68	23	160
Argyranthemum frutescens 'Bright Carmine'	21.35	3.20	0.91	5.90	5.6	107	31	199
Armeria maritime	11.36	2.19	0.43	4.38	8.6	29	6	55
Berberis thunbergii 'Crimson Pygmy'	16.04	1.70	0.25	1.82	7.8	38	5	36
Buxus microphylla 'Winter Gem'	18.02	2.28	0.20	1.74	12.1	32	3	23
Calibrachoa 'Superbells Red'	8.60	3.46	0.42	3.35	2.7	97	13	83
Capsicum annuum 'Sweet California Wonder'	21.70	3.97	0.49	2.69	2.9	223	29	232
Caryopteris × *clandonensis* 'Longwood Blue'	15.30	3.98	0.67	4.30	3.2	184	23	127
Coleus 'Tilt A Whirl'	11.20	3.65	0.57	4.34	2.1	179	29	186
Coreopsis auriculata nana	19.68	2.98	0.47	4.84	5.6	88	15	125
Coreopsis grandiflora 'Early Sunrise'	47.41	2.91	0.37	4.99	11.2	89	22	150
Cosmos 'Sonata Pink'	28.10	1.98	0.50	1.91	3.5	253	37	295
Cuphea hyssopifolia 'Lavender Lace'	5.50	4.44	0.56	4.36	2.9	45	7	25
Dianthus deltoides	20.63	2.98	0.98	3.41	7.8	64	24	80
Dianthus 'Wink'	5.30	2.19	0.62	2.15	3.0	25	7	23
Diascia integerrima	17.50	3.46	0.55	4.00	8.2	68	11	79
Echinacea purpurea	14.90	3.03	0.39	2.73	4.3	112	16	123
Euonymus fortunei 'Gracilis'	12.65	2.80	0.61	2.15	5.9	45	11	40
Euonymus fortunei 'Green and Gold'	16.99	2.49	0.30	1.99	12.6	29	4	22
Euryops pectinatus 'Viridis'	25.47	3.62	0.56	4.98	9.3	83	13	123
Gazania 'Majestic Yellow'	19.77	3.62	0.57	6.32	10.1	66	11	114
Hedera helix 'Thorndale'	6.60	3.28	0.36	3.63	2.3	48	8	55
Hemerocallis 'Stella de Oro'	53.67	3.05	0.47	3.94	8.9	140	24	149
Heuchera sanguinea	46.40	2.78	0.40	2.19	10.8	108	16	88
Hydrangea macrophylla 'Nikko Blue'	24.00	2.20	0.23	2.12	11.1	38	—	—
Hypericum calycinum	36.61	1.87	0.34	2.11	16.6	42	8	45
Hypoestes phyllostachya 'Red Splash Select'	19.96	3.75	0.54	4.22	6.7	94	15	119
Iberis sempervirens 'Snowflake'	8.30	2.96	0.34	4.69	3.3	70	7	66

(continued on the following page)

Table 6.2. Growth characteristics of representative nursery crops grown in a nursery or greenhouse at the University of California, Davis (continued)

Species		Shoot nutrient concentrations				Recommended liquid feed concentrations		
	Dry wt. (g)	N (%)	P (%)	K (%)	Water use (L)	N (mg/L)	P (mg/L)	K (mg/L)
Ilex aquifolium 'San Gabriel'	25.06	1.73	0.21	2.18	14.1	26	4	33
Impatiens hawkeri 'Bonfire Orange'	6.70	2.81	0.60	3.38	1.6	93	16	79
Impatiens walleriana 'Double Ole Rose'	8.10	4.28	0.63	4.41	3.9	142	33	167
Imperata cylindrica 'Red Baron'	20.62	1.20	0.34	2.26	9.0	28	8	52
Juniperus scopulorum 'Moonglow'	11.60	2.12	0.15	1.45	11.8	13	—	—
Kniphofia 'Border Ballet'	15.20	3.21	0.42	4.41	3.3	119	9	143
Lagerstroemia indica 'Petite Orchid'	28.37	3.08	0.70	2.84	11.3	65	15	59
Lantana 'Pink Caprice'	29.70	2.04	0.17	1.89	18.6	34	—	—
Lavandula dentata	32.40	2.05	0.19	1.92	13.3	56	—	—
Lavendula angustifolia 'Munstead'	6.70	3.97	0.45	3.75	2.1	87	11	107
Ligustrum japonicum 'Texanum'	25.69	1.81	0.19	1.57	7.4	51	7	48
Miscanthus sinensis 'Purpurescens'	20.84	1.73	0.27	1.72	9.9	30	5	31
Nandina domestica	26.40	1.65	0.15	1.00	10.6	47	—	—
Nepeta × *faassenii* 'Dropmore'	21.40	3.42	0.62	3.15	3.6	224	25	208
Ophiopogon japonicus	4.20	2.55	0.47	2.01	3.6	22	4	17
Osteospermum 'Sunny Mary'	32.30	4.13	0.56	5.10	9.8	112	22	152
Pachysandra terminalis	29.13	2.13	0.25	2.41	7.7	62	11	76
Parthenocissus quinquefolia	23.95	2.69	0.98	2.14	8.0	65	25	59
Pelargonium × *hortorum* 'Flamingo'	24.00	2.93	0.50	3.03	1.7	158	21	165
Penstemon × *mexicali* 'Red Rocks'	9.40	3.20	0.45	4.25	2.7	93	18	95
Perovskia atriplicifolia	13.40	3.61	0.59	4.09	3.4	130	21	151
Phalaris arundinacea 'Strawberries and Cream'	29.79	4.03	0.82	5.14	15.9	62	13	81
Prunus laurocerasus 'Zabeliana'	23.60	2.68	0.44	1.82	15.0	40	8	28
Pyracantha coccinea 'Low Boy'	19.56	2.77	0.32	1.77	8.0	60	5	42
Pyracantha koidzumii 'Walderi Prostrata'	37.14	2.05	0.33	1.51	15.2	45	4	33
Rhaphiolepis indica 'Pink Lady'	29.02	1.66	0.29	1.83	13.0	31	4	36
Rhododendron 'Remembrance'	11.93	2.03	0.27	2.09	5.5	32	5	31
Rosa 'The Fairy'	24.00	2.33	0.42	1.74	8.6	58	5	44
Rosmarinus officinalis 'Prostratus'	31.45	2.58	0.52	3.64	15.9	49	10	69
Rudbeckia fulgida 'Goldsturm'	20.00	4.08	0.41	5.89	4.6	156	24	223
Salvia greggii 'Wild Thing'	23.49	3.07	0.39	3.71	7.6	83	11	105
Salvia leucantha 'Santa Barbara'	14.65	3.43	0.60	4.52	6.1	73	13	99
Sedum spurium 'Fuldaglut'	5.70	2.93	0.46	4.01	2.1	136	11	99
Spiraea japonica 'Neon Flash'	15.82	2.57	0.48	2.34	6.3	55	11	49
Stachys byzantina 'Silver Carpet'	32.20	3.80	0.65	5.21	6.4	199	29	227
Thuja 'Green Giant'	20.88	1.61	0.28	1.64	8.1	32	7	37
Trachelospermum asiaticum	14.80	2.21	0.28	2.24	5.7	44	6	51
Verbena peruviana 'Red Devil'	16.56	3.83	0.57	3.39	5.4	102	16	97
Viburnum suspensum	20.56	2.39	0.53	2.92	6.6	62	15	82
Vinca minor 'Bowles'	13.60	2.87	0.41	2.87	4.6	63	11	69
Weigela florida 'Variegata Nana'	14.60	2.53	0.30	1.70	15.0	18	—	—

Note: Dry weight represents net gain in weight of shoots and roots during about 3 months of growth. Shoot concentrations of nitrogen (N), phosphorus (P), and potassium (K) are expressed as a percentage of dry weight. Water use represents total plant transpiration during about 3 months of growth. The recommended liquid feed concentrations are based on total nutrient uptake divided by total transpired water. Plant growth and water use vary with weather, season, and climate.

container medium tends to rise excessively. The reaction of ammonium with plants and microorganisms causes the pH to decrease (become more acidic), whereas use of nitrate fertilizers will cause the pH to increase.

Fertilizer Management Practices to Minimize Pollution

Fertilizer application is essential to commercial nursery production, but it also can be an environmental issue. Inefficient fertilizer use is a source of two nutrients, nitrogen and phosphorus, that can harm water quality. It is important for growers to follow practices that minimize the loading of these nutrients into surface water and groundwater. Successful implementation of these management practices depends on growers' knowledge of the chemical properties of their irrigation water and growing medium, nutritional requirements of their crops, and proper application and storage of fertilizers.

Salinity Management

Most irrigation water, especially from wells, contains dissolved salts that it picks up from the geologic materials it comes in contact with. Plants exclude some of these salts as they take up water, resulting in salt accumulation in the root zone. Over time, the accumulated salts reduce the amount of water available to plants, and a high amount of dissolved salts reduces yield and crop quality.

Growers who use irrigation water of marginal quality must take steps to avoid negative effects on crops. One possibility is to treat water (usually by deionization or reverse osmosis) to improve its quality, but this approach is expensive. A practical alternative for many growers is to develop an effective leaching program to prevent excessive accumulation of salts in the root zone. Leaching is best achieved on ornamental crops by applying the proper leaching fraction. This is the ratio of the volume of water leached (e.g., the water that runs out of the bottom of a pot) to the volume of water applied (e.g., the total amount of water applied to a pot). The proper leaching fraction depends on the salinity of the irrigation water (including fertilizer in a liquid feed program) and the sensitivity of the crop to salinity. In table 6.3, the term "EC applied" is the salinity of the irrigation water, and "EC leached" is the desired value in the water that has passed through the root zone. The target EC leached should be about 3 deciSiemens per meter (dS/m) for salt-sensitive crops. For other ornamental crops, EC leached should be from 6 to 9 dS/m.

Applying the selected leaching fraction requires using the information known about crop water use and irrigation distribution uniformity. Table 6.4 gives the volume of irrigation water needed for several amounts of plant water use and several leaching fractions. The values for leaching fraction and water use can be rounded off to conform to the table values. For example, a plant that used 8.1 ounces of water should be irrigated with about 10.9 ounces of water if the EC applied is 2 dS/m and the desired leachate EC is 9 dS/m (the leaching fraction is about 0.22).

Container Medium and Plant Tissue Testing

Container media tests give the grower an idea of what plants will take up from the substrate in the near future. It is generally recommended that the chemical properties of container media be tested whenever a new formulation of a mix is used, as well as at regular intervals during crop production. The interval varies, but a good practice is to have tests conducted approximately once per month. It is important to obtain representative media samples in order to make valid decisions after testing. Most laboratories can provide instructions and containers in which to place samples. In general, samples should be taken midway between fertilizer applications. Approximately 8 to 10 samples should

Table 6.3. Recommended leaching fractions

	EC leached (dS/m)			
EC applied (dS/m)	3	6	9	12
0.50	0.17	0.08	0.06	0.04
0.75	0.26	0.12	0.09	0.06
1.00	0.33	0.17	0.11	0.08
1.25	0.43	0.20	0.15	0.10
1.50	0.50	0.25	0.17	0.12
1.75	0.60	0.28	0.21	0.14
2.00	0.67	0.33	0.22	0.17
2.25	—	0.36	0.27	0.18
2.50	—	0.42	0.28	0.21
3.00	—	0.50	0.33	0.25
5.00	—	—	0.56	0.42

Note: Salt-sensitive species require low values of EC leached. Most ornamental crops tolerate EC leached from 6 to 9 dS/m. Salt-tolerant species tolerate higher values of EC leached.

Table 6.4. Volume of water applied to achieve desired leaching fraction

	Plant water use (mL)										
Leaching fraction	50	75	100	125	150	175	200	250	300	350	400
0.075	54	81	108	135	162	189	216	270	324	378	432
0.100	56	83	111	139	167	194	222	278	333	389	444
0.125	57	86	114	143	171	200	229	286	343	400	457
0.150	59	88	118	147	176	206	235	294	353	412	471
0.175	61	91	121	152	182	212	242	303	364	424	485
0.200	63	94	125	156	188	219	250	313	375	438	500
0.225	65	97	129	161	194	226	258	323	387	452	516
0.250	67	100	133	167	200	233	267	333	400	467	533
0.275	69	103	138	172	207	241	276	345	414	483	552
0.300	71	107	143	179	214	250	286	357	429	500	571
0.400	83	125	167	208	250	292	333	417	500	583	667
0.500	100	150	200	250	300	350	400	500	600	700	800

be taken from each greenhouse, or more if the container media or crops within a greenhouse differ. Sampling should be done by inserting a soil probe and withdrawing a sample that represents the profile from the surface to the bottom of the container. Samples should be marked clearly with labels. Fertilizer recommendations from results of analyses depend greatly on the analytical methods used by the soil testing laboratory.

Salinity

Container medium salinity can be expressed as the electrical conductivity of a saturated paste extract (EC_e). Potential effects of EC_e values (in dS/m) are

- < 2: no salinity problem (but values below 0.5 dS/m may indicate insufficient fertility)
- 2 to 4: restricted growth of salt-sensitive crops
- 4 to 8: restricted growth of most crops
- 8 to 12: large yield reductions in most crops
- > 12: only salt-tolerant plants survive

Elemental analysis

The concentrations of chemical elements in container media are usually presented in relation to total media dry weight or in a saturated paste extract solution. These values are intended to represent the amounts that are available to plants, and laboratories use various extraction solutions to assess how much of the total concentration of elements will actually be usable by plants. Since the values depend on the method of extraction, it is usually best to follow the guidelines of the laboratory performing elemental analyses.

Foliar Analysis

Plant tissue testing of ornamental crops is less common than soil testing, but it is a useful tool for diagnosing nutrient deficiencies and monitoring fertilizer practices. Foliar analysis can guide the grower's assessment of how past practices and environmental conditions have affected the nutrient status of crops. The mineral composition of a leaf blade depends on many factors, such as its stage of development, climatic conditions, availability of mineral elements in the container medium, root distribution and activity, irrigation, water status, and so on. The plant integrates all of these factors, and the composition of the blade reflects this integration.

Plants of the same species that receive the same cultural practices can be sampled as a single group: one to three composite samples of 20 to 30 recently matured leaves each is sufficient under these conditions, even for large production areas. The most recently matured leaves should be taken at random from plants within the growing area. The samples can be placed in paper bags and sent to a laboratory testing service for analysis. Some laboratories may advise to wash and dry leaves before sending the samples.

The nutritional requirements of most nursery crops are not well defined, but most analytical labs will provide guidelines. The book by Mills and Jones, listed in the References at the end of this chapter, provides nutrient guidelines for many nursery crops. A grower who keeps records of crop performance and container medium and tissue test results will be able to see how differences in plant nutrition relate to crop growth.

Nitrogen Application

Nitrogen is often applied in excessive amounts to ornamental crops. Although certain crops, particularly herbaceous ones, take up some excess nitrogen, most of the unneeded nitrogen is wasted. Nitrogen that is not taken up by plants has six possible fates:

- Leached below the root zone. Nitrate moves readily with water that percolates through the root zone. Much of the excess applied nitrogen leaches out of the root zone and becomes a potential contaminant of groundwater.

- Soilborne erosion losses. Nitrogen present in soil and container media can be moved by water or wind. Nitrogen present as either ammonium or nitrate will move with sediments.
- Denitrification. Microbes in soil and container media can convert ammonium or nitrate to nitrogen gas, which is lost to the atmosphere. This denitrification occurs to some extent in all media when oxygen levels are low, for example, after irrigation or rainfall when the medium is saturated and has no air. In heavy clay soils with poor drainage, or in soils with restrictive layers that prevent drainage, or in container media with low air-filled porosity, nitrogen losses through denitrification may be 15 to 50 percent of the amount applied in fertilizer. In general, only a small percentage of applied nitrogen is lost through denitrification.
- Immobilization and mineralization. Applied nitrogen may be tied up (immobilized) in organic matter or in the biomass of soil microbes as they work to decompose crop residues. Large amounts of applied nitrogen can be temporarily immobilized into organic nitrogen by soil microbes when plant material containing a high carbon to nitrogen ratio (such as sawdust, which typically has a carbon to nitrogen ratio of 500 or higher) is incorporated into the soil or container medium. This is the phenomenon commonly known as nitrogen draft. Organic nitrogen is continuously converted to plant-available nitrogen at a slow rate by the process of mineralization.
- Residual nitrogen. Nitrogen may remain in soil or container media, available for subsequent season uptake. This residual nitrogen usually accumulates over time, as long as irrigation is controlled to minimize leaching loss. However, residual levels are affected by climate. For example, winter rainfall in most of coastal California leaches most of the residual nitrate from the root zone.
- Ammonium volatilization. When animal manure, urea, or ammonium-containing fertilizers are left on the surface of the soil or container medium, nitrogen can be lost to the air as gaseous ammonia. This loss can be significant when the pH of soils or container media reaches values of 8 or higher. If manure or fertilizers are incorporated within a few hours after application, this loss is negligible.

Plant requirements for nitrogen vary according to growth rate and stage of development. For example, roses and some other woody species exhibit a cyclical pattern of nitrogen uptake that is related to episodes of shoot growth and either pruning or dormancy. In these plants, the rate of nitrogen uptake increases during rest periods between flushes of growth, and then decreases sharply during the growth flushes. Depending on the method of applying nitrogen, it may be possible for growers to adapt the fertilizer program to match changes in plant nitrogen demand.

The total amount of nitrogen (or other nutrients) required by a crop over a production cycle can be estimated easily if the typical dry weight and elemental analysis of the plant at harvest are known. For elements whose concentration is expressed as a percentage of total dry weight, the equation for calculating the total amount of the element needed by the plant is

$$\text{concentration of element (\%)} \times \text{plant dry weight (g)} \times 10 = \text{mg of element needed}$$

For elements whose concentration is expressed in parts per million (ppm, or its equivalent, mg/kg), the equation is

$$\text{concentration of element (ppm)} \times \text{plant dry weight (g)} \div 1{,}000 = \text{mg of element needed}$$

Crop dry weights may vary widely, depending on factors such as cultivar, container size, and cropping time. For example, the dry weight of 1-gallon nursery stock may range from 25 to 60 grams, including roots. The best practice for a grower would be to uproot some plants that are at their commercially salable size, wash off soil or media clinging to roots, dry the plants in an oven for a few days at no more than 160°F, or dry them in direct sun, and then weigh them on an accurate scale.

Inserting the appropriate values for the crop into the above equations, you can establish a value for the plant demand for nutrients. In general, this value will be a high estimate of plant nutrient demand because leaves tend to have higher concentrations of nutrients than roots, stems, or flowers. For example, consider an English laurel *(Prunus laurocerasus)* plant that weighs 24 grams and has a leaf nitrogen concentration of 2.7 percent.

$$2.7\% \text{ nitrogen} \times 24 \times 10 = 650 \text{ mg}$$

The calculations presented here provide an estimate of plant nutrient requirements, but actual fertilizer additions will have to be adjusted to account for fertilizer lost to leaching or volatilization. Methods for minimizing leaching losses are described above, in the section "Salinity Management," and in chapters 7 and 8.

If the plant demand for water is also known, then the concentration of nutrients needed in a liquid feed can be calculated. The concentration of fertilizer in the solution is simply the weight of fertilizer divided by the volume of water. For instance, the English laurel plant considered above probably transpires about 3.7 gallons of water during its production time (see table 6.2; for species not listed in the table, refer to chapter 7 for estimation of plant water use). Therefore, the liquid feed concentration should be

$$\text{mg nitrogen needed} \div \text{L water transpired} = 650 \text{ mg} \div 15 \text{ L} = 43.33 \text{ mg/L}$$

Growers who wish to develop an even more efficient method for managing crop nutrient needs can take into consideration changes in rate of nutrient uptake due to environmental conditions (e.g., seasonal variation in light) and crop developmental stage. Slower growth accompanies the lower light intensities and shorter days of winter, and this is translated into a lower rate of nutrient uptake. In most situations, this does not require a change in liquid feeding concentrations because there is a comparable change in the rate of water use by plants. However, it may be useful for growers to measure finished dry weights of crops at different times of year.

Phosphorus Application

Phosphorus is present in mineral soil in a number of chemical forms. There is a very low concentration of soluble inorganic phosphorus in the soil water. Larger amounts of phosphorus are usually present in less soluble forms, in chemical precipitates, as a constituent of organic matter, or adsorbed onto soil particles. These different phosphorus sources often establish an equilibrium in the soil, so that as plants remove soluble phosphorus, the other forms replenish the soluble phosphorus supply. In mineral soils, phosphorus solubility is low, and most phosphorus is present in chemical precipitates that are relatively immobile. These forms of phosphorus are not usually transported off-site unless rainfall or irrigation results in movement of sediments containing phosphorus.

In contrast, the solubility of phosphorus in soilless container media is relatively high, especially in acidic media. This soluble phosphorus can be leached out of containers by rainfall or excessive irrigation. Leaching losses of 30 to 60 percent of fertilizer phosphorus are common, even when it has been added as triple superphosphate. These leaching losses can be reduced significantly, with little or no decreases in crop yields, by incorporating ferrous sulfate into the medium at a rate of 4.8 pounds per cubic yard. In addition, phosphorus bound to particles of soilless media that have been spilled or washed out of containers can move off-site in irrigation or rainwater. The movement of phosphorus into streams and lakes can lead to algal blooms and subsequent fish kills.

Common laboratory soil test procedures provide an estimate of the amount of phosphorus in the soil or container medium that is available to plants. Unlike nitrate testing, which measures the actual amount of nitrate present, testing for phosphorus, as carried out by most laboratories, gives an index value or ranking of the relative phosphorus supply. Researchers over many years have calibrated these soil test procedures in greenhouse and field trials so that the results can be used to predict whether a crop is likely to respond to additional phosphorus fertilization. For most ornamental crops, plant demand for phosphorus is about 5 to 15 milligrams per day. For field-grown woody nursery plants, the typical recommendation is to supplement existing soil phosphorus with fertilizer to provide a total of 44 pounds of actual phosphorus per acre (100 pounds per acre when expressed as P_2O_5). For container-grown crops, phosphorus is usually added to the growing medium as triple superphosphate (1 pound per cubic yard) before planting. Alternatively, phosphorus can be added as a component of a complete controlled-release fertilizer.

It is not normally necessary to supplement the preplanting addition with a liquid feed source of phosphorus unless the crop will be grown for more than 5 to 6 months. In that case, injection of phosphorus into the irrigation water at a rate of 15 to 20 mg/L is effective. However, liquid feeding with phosphorus is subject to the same leaching and runoff problems that are associated with liquid feeding with nitrogen.

The best way to prevent phosphorus from moving off-site is to reduce runoff. Evaluate your property so that you know the surface drainage patterns and take action to prevent runoff from reaching surface water bodies. Filter strips, grassed waterways, sediment basins, nutrient management, or other Natural Resources Conservation Service (NRCS) practices should be used in conjunction with soil or media testing to be effective at addressing the phosphorus runoff issue. See chapter 8 for further information on runoff management.

Fertilizer Injection

Fertilizer proportioners are used in liquid feeding systems to eliminate the need for immense stock solution tanks. They allow for the measured injection of highly concentrated fertilizer solutions. Several types of proportioners are available, and the choice of a proportioner depends on factors such as the size of the production area, need for mobility, need for multiple injector heads, and desired stock solution concentration. Understanding the operating principles of propor-

Photo: R. Y. Evans.

tioners, as well as their limitations and maintenance needs, is important.

Size of Production Area

Proportioners that have a low flow rate or a low injection ratio are not well suited for use in large nurseries. The low flow rate limits the area that can be irrigated at a given time, and the low injection ratio requires a relatively large stock tank in relation to the volume of irrigation water to be applied.

Injection Ratio

The choice of the dilution ratio between the fertilizer stock solution and the final concentration to be applied to plants depends on the volume of water to be applied and the composition of the fertilizer stock solution. A highly concentrated stock solution permits using a relatively small stock tank. However, some fertilizers are incompatible at high concentrations. Typically a ratio of 1:100 is convenient.

Multiple Injector Heads

Multiple injector heads allow for greater flexibility in the delivery of fertilizers and also help to overcome the problem of incompatibility of fertilizers in highly concentrated stock solutions by relegating incompatible fertilizers to separate stock tanks.

Proportioner Types

The venturi type of proportioner is the simplest. The principle of operation is the same as that for the carburetor on a gasoline engine. The venturi proportioner employs a venturi tube, which places a constriction in the flow of irrigation water. To maintain the volume of flow through the constriction, the water moves faster. This causes the water pressure at the end of the constriction to be low. An inlet tube inserted at that point allows a fertilizer stock solution to flow in because the pressure in the stock tank is higher than in the venturi. In essence, there is a suction in the venturi that draws in the stock solution. The basic models of this type of proportioner are inexpensive and mobile, but the maximum flow rate is low and the injection ratio is low. Thus the stock solution cannot be highly concentrated. It is suitable for a small greenhouse operation. Venturi proportioners are also subject to variation in dilution ratio when water pressure in the mainlines fluctuates. Some of the drawbacks of the venturi proportioner are overcome by introducing a pressurized stock tank, which allows for higher injection ratios.

Positive displacement pump proportioners are more expensive, but they can accommodate a wide range of flow rates and operating pressures. They have chambers of a specific size, usually at a fixed ratio set by the manufacturer, for both the incoming irrigation water and the stock solution. After each of the chambers is filled, pistons driven by either a hydraulic pump or an electric pump force the water and stock solution into the outlet line in a fixed ratio.

More recently, manufacturers have developed proportioners that base the injection of fertilizer on the electrical conductivity and the pH of the output solution. This is effective for almost all fertilizers except urea, which is not a salt and does not conduct electricity. These proportioners are particularly well suited to operations that recirculate the drainage after irrigation.

Management of Injection Systems

A reliable proportioner makes liquid feeding of crops a simple procedure, but it is important to monitor the output from the proportioner and to ensure that the chemicals in the stock solution are compatible and mixed properly.

The ingredients in the fertilizer stock solution must be readily soluble in water to be used effectively with a proportioner. Technical grade chemicals are preferable to fertilizer grade ones, despite their higher cost, because they have fewer impurities that can decrease solubility. Some chemicals are not compatible in highly concentrated stock solutions because they form a solid precipitate. The most common problems occur when fertilizers containing calcium are mixed with others containing sulfate or phosphate, or if iron compounds are mixed with others containing phosphate, especially if the proportioner operates at an injection ratio of 1:200 or higher. If you formulate a new fertilizer stock solution, you should first test for incompatibility by scaling down the recipe to fit in a small volume, then mix the chemical ingredients in a glass jar. Wait for a couple of hours and then check for cloudiness or a fine film on the bottom of the jar, either of which indicates the presence of a precipitate. If precipitation occurs, the formulation must be changed.

In some cases, precipitation can be avoided by heating the stock solution during mixing. Fertilizers usually dissolve more readily in hot water (water at approximately 150°F works well). However, it is wise to follow the edict "When in doubt, leave it out." The alternative is to use multiple injector heads, which use separate tanks to hold incompatible stock solutions.

It is very important to monitor injector function. Just as carburetors in nursery gas-powered equipment can develop problems, so can proportioners. A simple and useful practice is to incorporate a water-soluble dye into the stock solution as a visual indicator that the injector is working. However, it is also important to check the proportioner at least once a month to ensure that it is operating accurately. One method is to measure the volume of stock solution uptake in relation to that of the delivered irrigation solution. The ratio of these volumes should equal the intended injection ratio. An easier method for monitoring the proportioner is to measure the electrical conductivity of the output solution, or to install an EC meter in the irrigation line beyond the proportioner. The EC should equal the sum of the EC of raw irrigation water plus that imparted by the fertilizers.

Backflow Prevention

Backflow prevention devices are required when chemicals are applied through irrigation water. Many types are available, including air gaps, antisiphon valves, and reduced pressure backflow preventers. These devices must be acceptable to both the local health department and the water purveyor. Some local agencies allow users to choose among several authorized types of backflow preventers; in other locations, only one or two specific devices are authorized. Always check with the local water provider and health department before installing a fertilizer injection system to be sure that the backflow prevention device is appropriate. Backflow prevention devices should be tested annually, and the date and results of the tests should be recorded and saved.

Fertilizer Transport and Storage

Fertilizers should be transported and stored in a manner that prevents their direct entry into groundwater or surface water storage. They should be placed in covered and secure locations to prevent direct contact with water from rainfall or irrigation and should be kept as far as possible from wells or surface water. The storage area should provide secondary containment to prevent movement of fertilizers in the event of a spill. The secondary containment should include an impermeable floor, as well as waterproof curbs around the storage area. If liquid fertilizer is stored, the secondary containment should be able to contain at least 110 percent of the volume stored (see the U.S. Environmental Protection Agency website in the References at the end of this chapter). Local regulations may also apply.

In the event of a spill, sweep up dry fertilizer and return it to the storage container or place it in a properly labeled, sealable container. Liquid spills can be recovered by pumping the solution into labeled tanks for reuse.

All fertilizer operations should be performed at least 100 feet downslope of a well. Ensure that fertigation equipment is calibrated and tanks are free of leaks. Be careful not to spill fertilizers when transporting or transferring them. When applying fertilizers in the field from a tractor or rig, shut off fertilizer applicators during turns and use check valves when possible.

References

Bunt, A. C. 1988. Media and mixes for container-grown plants. London, UK: Unwin Hyman.

Handreck, K., and N. Black. 2002. Growing media for ornamental plants and turf. Sydney: University of South Wales Press.

Mills, H. A., and J. B. Jones Jr. 1996. Plant analysis handbook II. Athens, GA: MicroMacro Publishing.

U.S. Environmental Protection Agency website, http://www.epa.gov/.

Irrigation Management Practices

Lawrence J. Schwankl, Richard Y. Evans, and Ben Faber

CHAPTER SEVEN

Nursery crops require a steady supply of good-quality irrigation water for growth and development into healthy, salable plants. Water must be applied in the correct amounts exactly where it is needed to optimize plant growth, while at the same time minimizing runoff and conserving water and fertilizers. In chapter 2, sources of water, volume of required water for irrigation, and water quality are discussed as critical factors in the selection of the nursery site. This chapter focuses specifically on irrigation systems and methods. The topics include an overview of the various types of irrigation systems used in nursery operations, the components of these systems, and maintenance requirements. We describe methods for evaluating irrigation system performance and determining how irrigation efficiency can be improved by using the proper irrigation frequency and duration. We also discuss examples of simple practices that can be implemented in the nursery to improve irrigation efficiency. Additional irrigation practices related to runoff management are provided in the next chapter.

Photo: J. K. Clark.

Overview of Irrigation Systems

Many irrigation systems are available for nursery applications. These range from hand-watering to complex sprinkler systems and from irrigating many plants at a time with a single sprinkler to using drip systems with a single dripper per plant. More sophisticated drip systems provide a higher level of irrigation control at the price of increasingly complex maintenance and operation.

Sprinkler Irrigation Systems

Various kinds of overhead sprinkler systems apply water over the top of plants, covering a wide area and irrigating many plants at a time. These sprinklers are most commonly impact sprinklers, but other application methods are also used. The impact sprinkler head discharge rate is in the gallons per minute range. Other sprinkler types, especially those used in enclosed nurseries, may have lower discharge rates, with each head covering a smaller area. As the discharge rate of sprinklers decreases, their discharge rate is measured in gallons per hour, and they begin to be categorized as a microirrigation system emission device. In this chapter, the discussion of sprinkler systems emphasizes impact sprinklers due to their common use.

FIGURE 7.1. Container-grown plants can be watered with overhead (A) or drip irrigation (B). Typically drip irrigation is more efficient. *Photos:* D. Zurawski and J. P. Newman.

Microirrigation Systems

Microirrigation systems using emitters are often referred to as drip irrigation (figs. 7.1, 7.2, and 7.3). These emitters may be built into drip tubing (in-line emitter), a spray stake, an individual emitter with a weight or stake holding it in place, or some other modification. A low discharge rate is a common trait among these drip emission devices, ranging from as low as 0.5 gallon per hour (gph) for an individual drip emitter to nearly 20 gallons per hour for spray stake emitters designed for large potted plants.

Certain microirrigation devices (microsprinklers or misters) spray water over a wider area. Their discharge rate may be as high as 20 to 30 gallons per hour, and they often irrigate multiple plants at a time.

Microirrigation system emission devices are characterized by small flow passageways that allow controlled low discharge rates but increase the danger of emitter clogging. Drip system maintenance requirements are high but are necessary to maintain a high level of performance.

Some microirrigation systems use capillary action (see chapter 5) to pull water into containers from below. Capillary mats and ebb-and-flow troughs are examples of such subirrigation systems. They can apply water uniformly and efficiently, but they are not advisable for saline irrigation water (EC > 0.75 dS/m) because salts will accumulate in the root zone. In addition, effective capillary mat irrigation requires good contact between the mat and the container medium, so it is incompatible with containers whose drainage holes are on the sides rather than the bottom.

FIGURE 7.2. Some emitter stakes can be turned off when nursery beds are not fully occupied by plants. *Photo:* D. Zurawski and J. P. Newman.

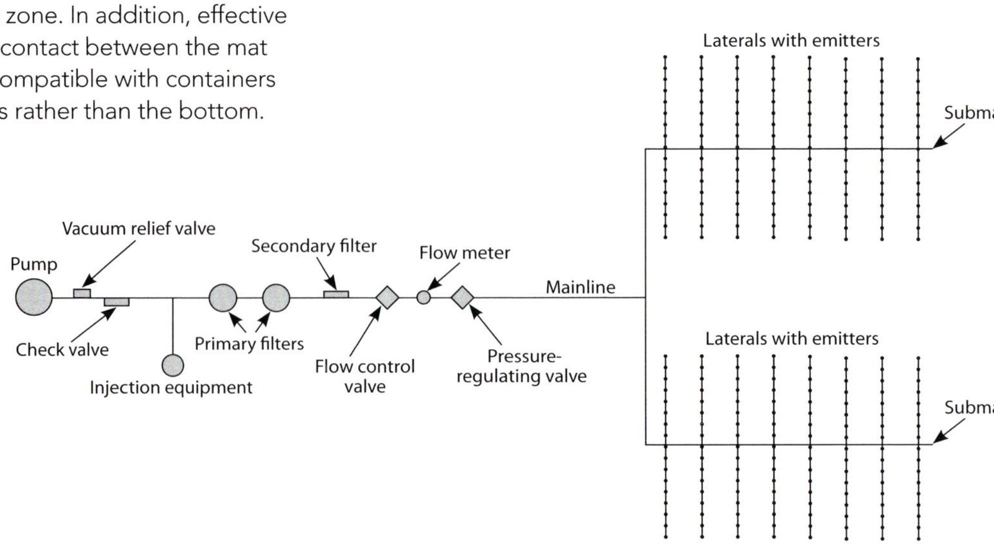

FIGURE 7.3. A typical microirrigation system. *Source:* Adapted from Newman 2008, p. 41.

FIGURE 7.4. Distribution uniformity of sprinklers can be measured with catch cans. *Photo:* A. Storm and J. P. Newman.

Irrigation System Performance

Good irrigation requires applying the correct amount of water to only the locations desired. Two irrigation system performance measures are used to characterize irrigation: irrigation efficiency and irrigation uniformity.

Irrigation Efficiency

Irrigation efficiency is a measure of how much of the applied irrigation water is "beneficially" used. Satisfying plant water needs is the primary beneficial use. In formula form, irrigation efficiency is

$$\text{irrigation efficiency (\%)} = (\text{beneficially used water} \div \text{total water applied}) \times 100$$

Note that two important factors must be known in order to determine irrigation efficiency. First, the amount of water required by the plant must be known. This is usually determined using irrigation scheduling techniques, such as weighing pots or using evapotranspiration values. Second, the irrigation application amount must be known. This requires measuring the applied irrigation water. Using a flow meter installed in the irrigation pipeline is the easiest and most accurate way to measure the application amount. Multiple flow meters should be used when multiple, widely spaced sections of an irrigation system are operated.

Irrigation Uniformity

A sprinkler or microirrigation system should apply water uniformly in an irrigation zone. In other words, all the sprinklers or emission devices in the zone should have the same discharge rate. For sprinklers, uniform irrigation also requires that the sprinkler spacing be the same within the zone.

Irrigation uniformity in a zone is quantified by comparing the minimum amount of water applied to the average amount. Sprinkler system uniformity is measured by placing collection containers (catch cans) in the irrigated area, operating the system, and measuring the water collected in the containers (fig. 7.4). The results are quantified using the measure distribution uniformity (DU), where

$$\text{distribution uniformity (\%)} = (\text{average of the low 25\% of catch cans} \div \text{average of all catch cans}) \times 100$$

Measure the water collected in a catch can either as a volume or as a depth. Both work in the above equation.

Microirrigation system irrigation uniformity is quantified using emission uniformity, which is similar to DU, where

$$\text{emission uniformity (\%)} = (\text{average of the low 25\% of emitter discharges} \div \text{average of all emitter discharges}) \times 100$$

To use the above equation, collect the discharge from a sampling of emission devices in an irrigation zone. Commonly, 30 or 60 seconds of discharge from multiple emission devices is measured in milliliters (ml).

Ideally, the discharge from similar sprinklers or emission devices should be the same. This is seldom the case, since pressure differences and manufacturing or maintenance issues often yield different discharge rates. The causes of irrigation nonuniformity are discussed in greater detail later in this chapter.

FIGURE 7.5. Even a drip system with high uniformity can produce runoff if proper irrigation scheduling is not followed. *Photo:* D. Zurawski and J. P. Newman.

Irrigation Scheduling

Growers can improve irrigation water use efficiency and reduce runoff by proper irrigation frequency and duration (fig. 7.5). All that is needed is knowledge of daily crop water use (evapotranspiration), the water-holding capacity of the soil or container mix in which the crop is growing, and the rate of irrigation water delivery.

Crop water use

Plant water use in container nurseries can be measured directly by weighing plant containers. The weight change in grams over a 24-hour period, with no intervening irrigation, represents the milliliters of water lost from the container during that time (1 milliliter of water weighs 1 gram). As with the other methods described here, it is important to select representative plants when measuring water use. Large plants tend to use more water than small ones, and those on the borders of fields, benches, or nursery blocks tend to use more than those in the interior. Since accurate scales can be purchased for less than $100, this method provides a fast, accurate, inexpensive way to monitor plant water use.

Estimates of evapotranspiration for field-grown crops can be made using evaporation pans, atmometers, or automated weather stations (fig. 7.6), all of which can provide estimates of the amount of water loss in a given surface area. Evapotranspiration values are expressed in units of inches per day or centimeters per day. This value reflects the change in the depth of a pool of water that would occur during a day due to evaporation and transpiration. To calculate the equivalent volume of evapotranspiration, multiply this value by the area affected. Because an evaporation pan provides a measure of water loss from standing water, it can give only a rough approximation of plant water use. Atmometers measure the amount of water that evaporates from a porous surface designed to represent a leaf surface. Weather stations estimate evapotranspiration from calculations that use measurements of sunlight, wind speed, relative humidity, and temperature. (For more information on using weather-based irrigation scheduling, see the California Department of Water Resources CIMIS website in the References at the end of this chapter.) Regardless of which indirect method is used to estimate evapotranspiration, the values obtained must be corrected for the size and type of crop being grown. In addition, since the results are only estimates of evapotranspiration, growers should check soil moisture in the field to verify the accuracy of water use estimations.

FIGURE 7.6. A weather station (left) or evaporation pan (right) can be used to estimate evapotranspiration. *Photo:* L. Schwankl.

Direct measurement of soil moisture in the root zone of a crop can be made using moisture-sensing instruments such as tensiometers, resistance blocks, capacitance probes, and time domain reflectance sensors. Capacitance probes and time domain reflectance sensors measure actual soil water content; tensiometers and resistance blocks measure soil suction, which indicates the relative ease with which plants can remove soil moisture. These instruments can be used to measure moisture in field soils or in container media, but resistance blocks do not perform well in coarse-textured me-

Table 7.1. Approximate available water content of some typical soil textures

Texture	Available water (volume fraction)
sand	less than 0.10
loamy sand, sandy loam, loam, clay loam	0.10–0.15
silty clay, clay	0.10–0.20
silt, silt loam, silty clay loam	0.15–0.25

dia. The sensors should be placed in the root zone of plants selected to represent all of those in the irrigated area. Factors to consider when selecting representative plants include plant size and location. These moisture-sensing instruments can be connected to computers or data loggers so that data can be collected continuously and irrigation valves can be set to operate based on the measured soil moisture values.

Available water

Water in the root zone that a plant can easily extract is known as the available water. The amount of available water can be easily determined for potted crops. Irrigate some plants thoroughly and wait for drainage to stop for up to 1 hour. Weigh the pots, then withhold irrigation until the plants show the first signs of wilting. Weigh the pots again. Subtract this weight from the weight just after irrigation. The difference (in grams) is equal to the volume of water (in milliliters) that is available to plants. You can think of this as measuring the plant's "gas tank," the amount it can hold from one fill-up to the next. In field soils, available water is defined as the difference between the amount of water at field capacity and the amount at the wilting point. It is usually expressed as the volume fraction, percentage, or inches of water in a given amount of soil (table 7.1). For example, a soil with an available water fraction of 0.2 has 2.4 inches of available water per foot of soil depth.

Irrigation frequency

In general, irrigation should take place when about half of the available water has been lost from a pot. For example, if a pot holds 500 milliliters of available water, irrigation should be scheduled when about 250 milliliters has been lost from the pot. The same rule applies to irrigation of field soils: a soil that has 2.4 inches of available water per foot should be irrigated when about 1.2 inches of evapotranspiration has occurred.

Irrigation duration

The length of an irrigation event determines the amount of irrigation water applied. Although decisions about irrigation frequency depend only on replacing the available water that the crop has used, the amount of irrigation water required is more complicated. An irrigation event must replenish the soil water removed by the crop, leach out accumulated salts, and compensate for any lack of irrigation system uniformity.

Growers who use irrigation water of marginal quality must treat the water (usually by deionization or reverse osmosis) or develop an effective leaching program. Leaching is described in chapter 6 in the section "Salinity Management."

See table 6.4 in chapter 6 for the volume of irrigation water needed for several amounts of plant water use and several leaching fractions. The values in table 6.4, however, do not take into consideration the distribution uniformity of the irrigation system. To correct for that, the values in the table must be divided by the distribution uniformity value. Table 7.2 is a worksheet that can be used to determine the proper irrigation duration.

Irrigation Audits

Irrigation system performance can be evaluated by the nursery operator, but it is often easier to have an irrigation audit done by professionals. While professional auditors may provide some irrigation efficiency information, their efforts will primarily be quantifying irrigation system uniformity. Discharge from sprinkler heads or microirrigation devices will be collected to quantify irrigation uniformity and rate. This can also help diagnose problems with clogging or maintenance. Pressure measurements will be taken at key locations in pipelines and tubings to characterize the system pressure distribution. Measuring pressure is important since it controls the discharge rate of most sprinklers and many microirrigation emission devices. Operational conditions of the irrigation components, such as valves, gauges, and filters, will also be noted.

Irrigation audits should be done on a yearly basis or whenever any major changes are made in the system. Audits are particularly important in microirrigation systems, since they may detect clogging problems at early stages when they can be more easily solved. Audits also provide information on changes that occur over time. For example, a decrease in drip emitter discharge from one year to the next may require a change in irrigation times, but it may also indicate a drop in system pressure or an emitter clogging problem. Additional irrigation audit information should be gathered to explain the emitter discharge decrease and to point to steps to correct the problem.

Improving Nursery Irrigation Management Practices

The following irrigation management improvements can be made in any nursery, whether hand-watering, sprinkler irrigation, or drip irrigation is used.

Pulse Irrigations

Applying water in short, "pulsed" irrigation durations can significantly reduce runoff and improve irrigation efficiency. Pots hold only a certain volume of water, and any irrigation in excess of that amount simply runs through or off the pots. Multiple pulsed irrigations during the day can prevent runoff while satisfying plant needs. Determining how often and how long to irrigate is best done by observing the irrigation and noting the operating time at which runoff begins. Setting irrigation durations to this amount of time or less minimizes

Table 7.2. Irrigation calculations worksheet

1. Measure distribution uniformity of irrigation system.

A. Measure weight of empty can. Record weight in grams. ____________ g

B. Space catch cans regularly throughout irrigation block. Run irrigation system for known period of time (record time here: ____________ min). Record weights of cans. ____________ g

C. Calculate amount of water in cans. Subtract **1.A** from **1.B** for each can. ____________ g

D. Calculate average amount in each can (1 g = 1 mL). ____________ mL

E. Calculate average amount in lowest one-quarter of the cans. ____________ mL

F. Calculate distribution uniformity. Divide **1.E** by **1.D**. ____________

2. Calculate flow rate of irrigation system.

A. Record length of irrigation in step **1.B** (in minutes). ____________ minutes

B. Calculate flow rate in mL per minute. Divide **1.D** by **2.A**. ____________ mL/minute

C. For overhead irrigation systems, you must correct the value in **2.B** for pot size.

Use the following calculations or the table that follows this worksheet.

i. Record the radius of the catch cans by measuring the diameter (in centimeters) and dividing it by 2. ____________ cm

ii. Take the square of **2.C.i** (multiply the value times itself). ____________ cm^2

iii. Repeat steps **2.C.i** and **2.C.ii** for finding the square of the radius of the plant container at the rim. ____________ cm^2

iv. Calculate the correction factor. Divide **2.C.iii** by **2.C.ii**. Enter the value in the space to the right. ____________

v. Apply the correction factor. Multiply **2.B** by **2.C**. ____________

vi. The result is the rate at which the overhead irrigation system applies water to the plant containers. ____________ mL/minute

3. Measure electrical conductivity of irrigation water.

A. Use meter to measure electrical conductivity (EC) of irrigation water.

Record value in deciSiemens per meter (dS/m). ____________ dS/m

4. Select desired leaching fraction.

A. Choose desired leachate EC (lower value for more salt-sensitive crops). ____________ dS/m

B. Use table of recommended leaching fractions to select suitable value. (See table 6.3 in chapter 6.) ____________

5. Determine plant water use.

A. Weigh container with plant after irrigation and drainage. Record weight in grams. ____________ g

B. Weigh container with plant 24 hours after **5.A**, with no intervening irrigation. Record weight in grams. ____________ g

C. Calculate daily plant water use. Subtract weight in **5.B** from weight in **5.A**.

Record value as the volume in mL (1 g = 1 mL). ____________ mL

6. Calculate appropriate irrigation volume.

A. Enter amount needed to replace plant water use (use the value in **5.C**). ____________ mL

B. Correct for leaching fraction. Use table 6.3.

Record the value in mL. ____________ mL

C. Correct for distribution uniformity to calculate the total volume of water to apply to each pot.

Divide **6.B** by value in **1.F**. Record the value in mL. ____________ mL

7. Calculate appropriate length of irrigation.

A. Divide irrigation volume (value in **6.C**) by irrigation flow rate value in **2.B** (or in **2.C.v** for overhead irrigation).

Record value in minutes. ____________ minutes

8. Irrigate crop and monitor results.

A. Run irrigation system for appropriate time and catch leachate.

B. Measure leachate EC. Record value in dS/m. ____________ dS/m

C. Measure volume of leachate. Record value in mL. ____________ mL

D. Calculate actual leaching fraction. Divide leachate volume (**8.C**) by volume of water applied (**6.C**). ____________

Record value and compare to desired leaching fraction (**4.B**). ____________

Overhead irrigation correction factor for different catch can and pot sizes (multiply the measured flow rate in step 2.B by this factor to correct for the different diameters of catch cans and pots)

Catch can diameter		Pot size		
cm	inches	4 in	6 in	1 gal
7.8	3 1/16	1.6	3.7	4.0
10.2	4	0.9	2.2	2.4
12.7	5	0.6	1.4	1.5
15.2	6	0.4	1.0	1.0
20.3	8	0.2	0.5	0.6
25.4	10	0.1	0.3	0.4
30.5	12	0.1	0.2	0.3

irrigation runoff. This pulsed irrigation time varies with irrigation systems, plant types, and pot sizes.

A consideration with pulsed irrigations is the filling and draining of the irrigation system during each irrigation event. The lowest elevation point in an irrigation system receives the water draining out of the system after shutdown. This draining water can lead to runoff and sometimes to plant damage. Low-pressure check valves can be installed to minimize this drainage.

Irrigation Zones

Establishing zones in which irrigation is constant and can be controlled improves irrigation water management. It allows a nursery manager to operate only those zones needing irrigation for only the time needed.

Irrigation zones can be controlled using either a manual or an electronically controlled valve. Having more irrigation zones increases operational flexibility. It is often possible to retrofit an existing irrigation system to create irrigation zones. New or expansion irrigation systems should include as many irrigation zones as will be needed in the future.

Group Plants with Similar Water Needs

Placing plants with similar irrigation needs together in an irrigation zone allows efficient irrigation water management. This prevents a situation in which the plants with the greatest water needs dictate the watering schedule, which can cause other plants to be overwatered. Overwatering can lead to runoff or groundwater contamination, as well as plant disease.

Avoid Irrigating Where No Plants Are Present

Avoiding applying water where there are no plants seems like common sense, but it can be difficult to accomplish. It is made easier where increased irrigation control is provided by numerous irrigation zones, each of which can be controlled separately.

As plants are sold, group remaining plants that have similar water needs together to improve irrigation efficiency and reduce runoff. Moving plants is labor-intensive, but it can reduce water applications and minimize impacts on water quality.

Automation

Irrigation automation can improve water management, but it should be used in conjunction with good human management. In automated irrigation, a controller operates solenoid-equipped valves at specified times and durations. While the controller may be scheduled to turn the irrigation system on and off, it takes human management to determine whether there is too much runoff or too few or no plants in an irrigation zone being irrigated. Automation is a tool to help the operator improve water management, not a substitute for human judgment.

Sprinkler Irrigation Systems

System Components and Operation

Sprinkler heads consist of a small orifice, round being the most common shape, through which pressurized water jets. To improve the water distribution, the water stream is broken up by mechanical means. In an impact sprinkler, a spring-mounted sprinkler spoon repeatedly breaks up the water stream. Some water drops thus travel the maximum distance while others fall at various distances from the sprinkler head.

A single sprinkler head does not distribute water uniformly over the area it waters. Sprinkler heads are therefore spaced to provide overlap of the wetting patterns, increasing the irrigation uniformity. Closer sprinkler spacing increases irrigation uniformity, but it also increases the water application rate. An increased application rate can lead to runoff in some soil situations, but in container plants, with their highly permeable soil media, this is less of a concern. It remains important to know the application rate in order to apply the desired amount of irrigation water.

The sprinkler application rate, usually measured in inches per hour (in/hr), is determined from the sprinkler discharge rate and the spacing of the sprinklers. The discharge rate of a sprinkler, usually measured in gallons per minute (gpm), is controlled by the sprinkler nozzle size and the sprinkler operating pressure.

Table 7.3 lists the sprinkler discharge rate (in gpm) for various nozzle sizes and operating pressures. The greater the nozzle size operated at a given pressure, the greater the sprinkler discharge rate. The nozzle size is frequently engraved on the side of the nozzle, but it can also be determined by testing the nozzle opening using drill bits of a known size. The operating pressure can be determined by placing a pressure gauge fitted with a pitot tube, available at most irrigation supply stores, into the water stream just outside the nozzle opening.

The sprinkler discharge rate can also be determined by using a short length of garden hose (approximately 4 feet), a 5-gallon bucket, and a stopwatch. Determine the discharge rate from the sprinkler by placing the hose over the sprinkler nozzle, directing the water into the bucket, and timing how long it takes water to fill the bucket. Use the following formula to calculate the discharge rate (the value of 300 in the equation is a unit conversion constant):

$$\text{sprinkler discharge rate (gpm)} = 300 \div \text{time to fill a 5-gallon bucket (sec)}$$

Measurements should be taken at various locations within the nursery since sprinkler discharge rates vary with the pressure differences.

The sprinkler application rate (in/hr) can be determined, once the discharge rate of the sprinkler (gpm) and the sprinkler spacing (ft) are known, using the following equation:

$$\text{application rate (in/hr)} = [96.3 \times \text{sprinkler discharge rate (gpm)}] \div [\text{sprinkler spacing along row (ft)} \times \text{sprinkler spacing across row (ft)}]$$

Table 7.4 can also be used to determine the sprinkler system application rate. The sprinkler area in square feet (ft^2) is determined by multiplying the sprinkler spacing along the row (ft) by the sprinkler spacing between rows (ft).

Pipeline Flow and Uniformity

Pipelines supply the sprinklers with pressurized water. Ideally, the pressure at each sprinkler head should be the same, facilitating uniform application. This seldom occurs, since the

Table 7.3. Sprinkler discharge (gpm) for various nozzle sizes (in) and pressures (psi)

	Nozzle size (in)										
psi	**3/32**	**7/64**	**1/8**	**9/64**	**5/32**	**11/64**	**3/16**	**13/64**	**7/32**	**15/64**	**1/4**
20	1.17	1.60	2.09	2.65	3.26	3.92	4.69	5.51	6.37	7.32	8.34
25	1.31	1.78	2.34	2.96	3.64	4.38	5.25	6.16	7.13	8.19	9.32
30	1.44	1.95	2.56	3.26	4.01	4.83	5.75	6.80	7.86	8.97	10.21
35	1.55	2.11	2.77	3.50	4.31	5.18	6.21	7.30	8.43	9.69	11.03
40	1.66	2.26	2.96	3.74	4.61	5.54	6.64	7.80	9.02	10.35	11.79
45	1.76	2.39	3.13	3.99	4.91	5.91	7.03	8.30	9.60	10.99	12.50
50	1.85	2.52	3.30	4.18	5.15	6.19	7.41	8.71	10.10	11.58	13.18
55	1.94	2.64	3.46	4.37	5.39	6.48	7.77	9.12	10.50	12.15	13.82
60	2.03	2.76	3.62	4.50	5.65	6.80	8.12	9.56	11.05	12.68	14.44
65	2.11	2.88	3.77	4.76	5.87	7.06	8.45	9.92	11.45	13.21	15.03
70	2.19	2.99	3.91	4.96	6.10	7.34	8.78	10.32	11.95	13.70	15.59
75	2.27	3.09	4.05	5.12	6.30	7.58	9.08	10.66	12.32	14.19	16.14

Table 7.4. Average application rate (in/hr) for various sprinkler discharge rates (gpm) and areas of coverage (ft^2)

	Sprinkler discharge rate (gpm)							
Sprinkler area (ft^2)	**0.75**	**1.00**	**1.25**	**1.50**	**1.75**	**2.00**	**2.25**	**2.50**
500	0.14	0.19	0.24	0.29	0.34	0.39	0.43	0.48
550	0.13	0.18	0.22	0.26	0.31	0.35	0.39	0.44
600	0.12	0.60	0.20	0.24	0.28	0.32	0.36	0.40
650	0.11	0.15	0.19	0.22	0.26	0.30	0.33	0.37
700	0.10	0.40	0.17	0.21	0.24	0.28	0.31	0.34
750	0.10	0.13	0.16	0.19	0.22	0.26	0.29	0.32
800	0.09	0.12	0.15	0.18	0.21	0.24	0.27	0.30
850	0.08	0.11	0.14	0.17	0.20	0.23	0.25	0.28
900	0.08	0.11	0.13	0.16	0.19	0.21	0.24	0.27
950	0.08	0.10	0.13	0.15	0.18	0.20	0.23	0.25
1,000	0.07	0.10	0.12	0.14	0.17	0.19	0.22	0.24
1,050	0.07	0.09	0.11	0.14	0.16	0.18	0.21	0.23
1,100	0.07	0.09	0.11	0.13	0.15	0.18	0.20	0.22
1,150	0.06	0.08	0.10	0.13	0.15	0.17	0.19	0.21
1,200	0.06	0.08	0.10	0.12	0.14	0.16	0.18	0.20
1,250	0.06	0.08	0.10	0.12	0.13	0.15	0.17	0.19
1,300	0.06	0.07	0.09	0.11	0.13	0.15	0.17	0.19
1,350	0.05	0.07	0.09	0.11	0.12	0.14	0.16	0.18
1,400	0.05	0.07	0.09	0.10	0.12	0.14	0.15	0.17
1,450	0.05	0.07	0.08	0.10	0.12	0.13	0.15	0.17
1,500	0.05	0.06	0.08	0.10	0.11	0.13	0.14	0.16
1,550	0.05	0.06	0.08	0.09	0.11	0.12	0.14	0.16
1,600	0.05	0.06	0.08	0.09	0.11	0.12	0.14	0.15

pressure changes within a sprinkler system due to elevation changes and frictional pressure losses. A pressurized line running uphill experiences a decrease in pressure once water has been lifted to the higher elevation. Conversely, a line running downhill gains pressure. Converting elevation changes to system pressure can be done using the following relationship:

1 psi = 2.31 feet of elevation change

Pressure is also lost as water flows through pipelines, fittings, valves, and other devices in a sprinkler system. The amount of pressure loss is determined by the volume of flow and the characteristics of the pipe or line. For pipelines, pressure loss (or friction loss) is determined by the pipeline size and material (plastic, metal, etc.) and by the flow volume. For example, a large volume of water flowing in a small pipe loses more pressure than the same volume flowing in a larger pipe.

The implications of these pressure changes are important when making changes in a sprinkler system.

- Adding sprinklers to an existing system often decreases sprinkler pressure, since more water flows through the pipe and the size of the pipe does not change.
- Using small pipe in a sprinkler system does not increase pressure; rather, it increases frictional losses and decreases pressure.
- The longer the pipeline, the more pressure that is lost. Pressure loss tables usually present pressure losses per 100 feet of pipe.
- Changing nozzles in sprinkler heads affects pressure. Increasing the nozzle size increases the nozzle discharge rate and decreases pressure. Decreasing nozzle size often increases pressure.

Determining pressure losses in sprinkler systems is not difficult. Friction loss tables for pipelines are readily available from irrigation dealers, and friction loss calculators are available on the Web. Pressure losses through devices such as valves or flow meters are provided by their manufacturer.

Improving Sprinkler Irrigation Systems and Management

Sprinklers on Slopes

Sloping ground provides a number of sprinkler irrigation challenges. First, as mentioned previously, pressure varies with elevation change (1 psi = 2.31 feet of elevation change), which in turn affects sprinkler discharge and application uniformity. If the pipeline runs downhill, which increases pressure, flow control nozzles that limit the nozzle discharge rate may be beneficial.

Second, runoff is greater and occurs more quickly on sloping ground. All techniques to minimize runoff described in this chapter are even more important when irrigating on sloping ground.

Third, an irrigation sprinkler system on sloping ground often tends to drain from the lowest point following shutdown. This can lead to runoff, especially when pulsed (short and frequent) irrigations are used. Check valves, which close at low pressure and also prevent reverse flow (backflow), can prevent some of this drainage. The loss of pressure when using a check valve should be considered, however.

Container Spacing

When using overhead sprinkler systems, containers should be spaced as close together as possible while still maintaining adequate light for the plants. This minimizes the empty area where water does not land in a pot. Irrigation water that is not applied to a plant is inefficient water application and is more likely to contribute to runoff.

Overspray

Adjustable sprinkler heads that irrigate only part of a circle should be used on the edges of irrigated areas to prevent water from landing on bare areas adjacent to plants during irrigation. In addition to being an inefficient water use, water applications to these low-permeability areas can lead to significant runoff.

Adjust sprinkler heads to apply water only to the plant area. Adjustable sprinkler heads that turn only a portion of a circle have an increased application rate. For example, a sprinkler irrigating a half-circle has double the application rate of a similar sprinkler head irrigating a full circle. To compensate for this, place smaller nozzles in edge sprinklers. This works well in permanent sprinkler systems, but if portable pipe is used, it can be difficult to keep track of which sprinklers have which nozzle size.

Using Sprinkler Nozzles of Different Sizes

Mixing sprinkler nozzles of different sizes should be done only following the recommendation of an irrigation professional. It is difficult to predict what the consequences will be for the application rate when using nozzles of different sizes. If different sizes are used, an irrigation audit should be done to determine the application rate and distribution uniformity.

Avoid the unintentional mixing of different-size sprinkler nozzles. This can easily happen during maintenance and repairs. Check impact sprinkler nozzles using standard drill bits rather than trying to read the size stamped on the side of the nozzles.

Periodically check all sprinkler heads that are supposed to have the same nozzles to ensure that that is the case.

Vertical Risers

Sprinklers providing overhead irrigation are often located on pipe risers. Ensure that these risers are vertical. The distribution pattern of sprinklers on nonvertical risers is distorted, resulting in lower sprinkler application uniformity. Ensuring that risers are vertical is an attention to detail task that can improve irrigation management and reduce runoff.

Wind

Wind distorts the sprinkler application pattern and leads to lower application uniformity. Some plants may be underirrigated and runoff may increase when irrigating under windy conditions. Avoid irrigating under windy conditions if at all possible.

Routine Maintenance

Detecting leaks and checking sprinkler nozzle wear are two critical sprinkler system maintenance tasks. Leaks are usually obvious and should be fixed immediately, since even a small leak can quickly lead to runoff and erosion.

Water often contains fine particles that can wear a sprinkler nozzle, enlarging it and leaving wear marks. Brass nozzles are particularly susceptible to this. Use the appropriate drill bit to check the size of the nozzle opening for unacceptable wear. Compare the drill bit fit in a used nozzle to that in a new one to get a feel for unacceptable wear.

Sprinkler head bearings also wear with use, and the sprinkler will not turn properly if the bearings need replacement. Sprinkler heads can often be rebuilt for much less than the price of a new head.

Microirrigation Systems

System Components and Understanding How They Work

Microirrigation systems (see fig. 7.1) use drip emitters, spray stakes, misters, or microsprinklers as emission devices. A microirrigation system in a nursery may have a combination of emission devices tailored to fit the mix of plants and containers.

Emission devices are usually connected to polyethylene drip tubing, often in various diameters, customized to fit the plants being irrigated. Barbed and compression fittings allow easy assembly and modification of the system.

The polyethylene tubing is frequently connected to PVC pipelines, often installed belowground. These PVC pipelines, referred to as submains and mainlines, carry clean, pressurized water from a pump and filter station that often can also inject nutrients and chemicals.

Filtration of water used for microirrigation is very important to prevent emitter clogging. Microirrigation emission device passageways are small and can be clogged even by very small particles. Screen, disk, or sand media filters can provide the level of filtration suggested by the emission device manufacturer. The choice of filter is influenced by the water quality and the clogging hazard. Groundwater sources, which tend to be clean, can often be filtered using screen or disk filters. Water supplies from open canals, ponds, or reservoirs can contain organic contaminants and are best filtered with sand media. Screen or disk filters remove organic matter but must be cleaned frequently, a maintenance headache.

The following sections discuss a number of irrigation management issues unique to microirrigation systems.

Emission Device Selection

Choosing a microirrigation emission device is usually based on the desired discharge rate and coverage. For example, a large container plant with a high water demand requires an emission device that wets most of the pot area and has a high discharge rate. In general, the higher-discharge emission devices have larger flow passageways and are less susceptible to clogging. Another consideration in choosing an emission device is ease of inspection for clogging.

If drainage from emitters after system shutdown is a problem, drip emitters are available that have a built-in check valve that closes at low pressure. Drip systems that are pulse irrigated—multiple, short irrigations per day—may benefit from these specialty drippers.

Pressure-Compensating Emission Devices

The discharge rate of standard drip emitters and microsprinklers is related to the operating pressure (see fig. 7.8). As pressure changes, the discharge also changes. Pressure changes due to elevation differences and friction losses are common in a microirrigation system. Pressure changes that result in varying emitter discharge rates reduce the system's emission uniformity. This can lead to overirrigation of some plants and lower irrigation efficiency.

Pressure-compensating (PC) emission devices maintain a constant discharge under a range of pressures (see figs. 7.7 and 7.8). They can be an advantage in maintaining high emission uniformity, especially in systems with variable pressures such as facilities on sloping ground.

Pressure Control and Regulation

Maintaining constant pressure in a microirrigation system facilitates high application uniformity. Adjustable pressure-regulating valves and pressure regulators preset to a pressure (e.g., 35 psi) are available. These valves are actually pressure-reducing valves since they can reduce their inlet pressure only to a constant downstream pressure.

Pressure-regulating valves may be placed at various locations in a microirrigation system. Commonly, one is placed at the head of the system, downstream of the pump and filter station. Other designs place them throughout the system to maintain constant pressure. Using multiple pressure regulators (e.g., one at the head of each drip lateral) is more common in systems on sloping ground, where pressure differences due to elevation changes must be compensated for.

Mixing Drip Emitters

Care should be taken when using drip emitters with different sizes or manufacturers in the same system. Mixing emitters makes it extremely difficult to maintain high application uniformity. Mixing different-size emitters from a single manufacturer may make it difficult to tell what emitters are being used, since discharge rates (such as 0.5 gph, 1 gph, etc.) may look similar. Mixing emitters of the same size from various manufacturers leads to poor irrigation uniformity: sel-

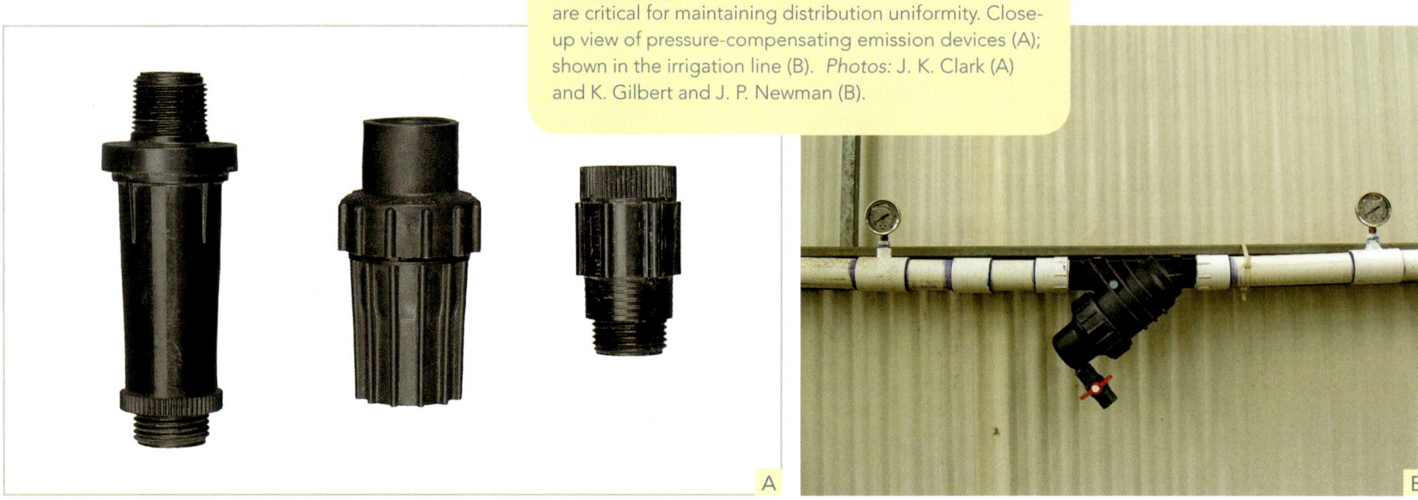

FIGURE 7.7. Proper pressure regulation and filtration are critical for maintaining distribution uniformity. Close-up view of pressure-compensating emission devices (A); shown in the irrigation line (B). *Photos:* J. K. Clark (A) and K. Gilbert and J. P. Newman (B).

dom is the emitter discharge-pressure relationship the same for emitters from various manufacturers. For example, a 1 gph dripper from Manufacturer A discharges 1 gph at 15 psi, while a 1 gph dripper from Manufacturer B discharges 1 gph at 12 psi and 1.1 gph at 15 psi. Avoid mixing emitters from different manufacturers in the same irrigation zone unless there is a good reason to do so.

Monitoring and Audits

Microirrigation systems should be monitored to detect clogging or other maintenance issues and to check the volume of applied water for irrigation scheduling purposes. Pressure gauges and flow meters are excellent monitoring tools. Readings from pressure gauges on the downstream and upstream sides of filters can detect when a filter is dirty and needs to be cleaned. Monitoring and keeping flow meter records can detect clogging in emitters, since the flow rate drops as the emitters clog.

The best health check on a microirrigation system is a system audit. An audit measures the discharge from a sampling of emitters to determine the system's emission uniformity and pressure, and it also gives a general assessment of the system's condition. Microirrigation systems should be audited on a yearly basis, and the audit reports should be kept so that any year-to-year changes can be detected.

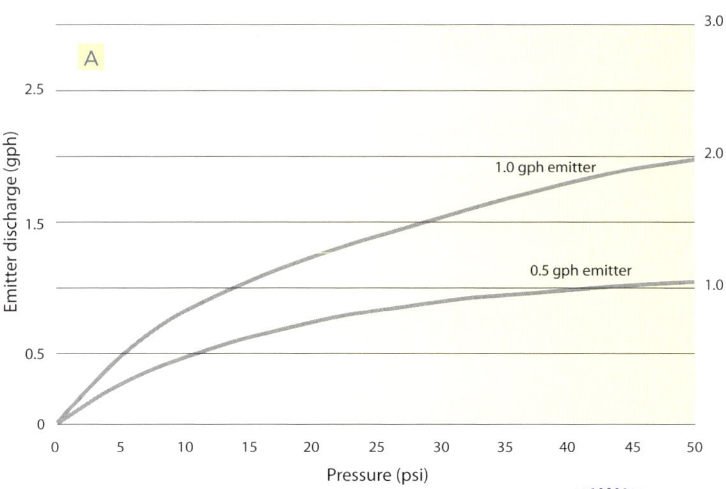

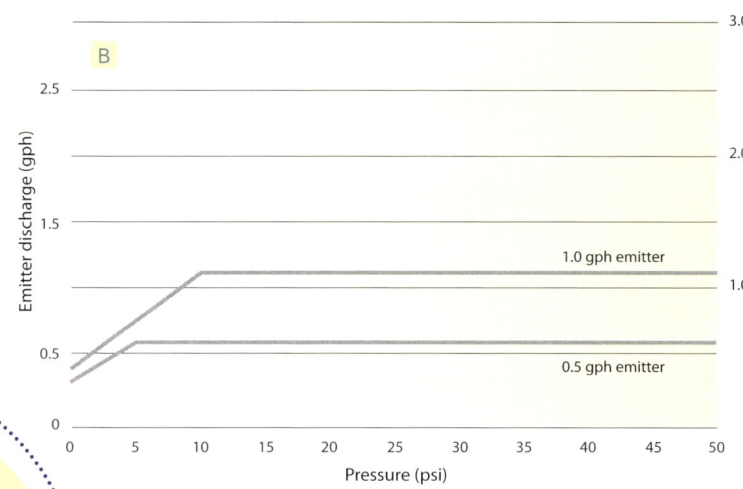

FIGURE 7.8. Pressure compensation helps control line pressures so that emitters can have greater potential for distribution uniformity. The graphs show discharge-pressure relationships for drip emitters that are (A) non–pressure compensating and (B) pressure compensating. *Source:* Adapted from Newman 2008, p. 43.

Emission Device Clogging

Emitter clogging can be caused by particulate matter such as sand, chemical precipitates such as lime or iron, or biological growths such as algae or bacterial slime.

Good filtration is a major factor in solving all clogging problems. Particulate matter clogging can usually be solved simply with good filtration. Preventing biological clogging requires good filtration, usually sand media, along with periodic use of a biocide, such as chlorine, injected into the irrigation water.

Lime (calcium carbonate) precipitate clogging requires the addition of acid materials to lower the water pH and dissolve the lime deposits. Lime clogging can often be handled with periodic pH adjustment or adjustment only for the last portion of each irrigation, but serious lime clogging may require continuous pH adjustment. Iron precipitate clogging is difficult to solve. The most common solution is to place water in a pond or reservoir and allow the iron precipitate to settle out before using the water for irrigation.

Flushing

Microirrigation systems should be flushed on a regular basis, every 2 weeks or at least monthly. Silt and clay particles in water are very small and may pass through filters to settle in the pipelines, particularly in the lateral lines, where they can clog emitters. Flush valves installed on the pipelines should be opened and the water should be allowed to flush clean. The lateral line ends should be opened and allowed to flush clean. If little material is evident at flushing, increase the flushing interval. Very dirty lines should be flushed more frequently.

References

ASABE (American Society of Agricultural and Biological Engineers). 2007. ASABE standards. St. Joseph, MI: ASABE.

Burt, C. M., A. J. Clemmens, R. Bliesner, and J. L. Merriam. 2000. Selection of irrigation methods for agriculture. Reston, VA: American Society of Civil Engineers.

CIMIS (California Irrigation Management Information System). Department of Water Resources, Office of Water Use Efficiency. CIMIS website, http://wwwcimis.water.ca.gov/cimis/data.jsp.

Davidson, H., C. Peterson, and R. Mecklenburg. 2000. Irrigation of nursery crops. In H. Davidson, ed., Nursery management: Administration and culture. 4th ed. Upper Saddle River, NJ: Prentice-Hall.

Hanson, B., L. Schwankl, and A. Fulton. 1999. Irrigation scheduling: When and how much. Oakland: University of California Agriculture and Natural Resources Publication 3396.

Israelsen, 0. W., and V. E. Hansen. 1962. Irrigation principles and practices. New York: Wiley.

Jensen, M. E., ed. 1980. Design and operation of farm irrigation systems. St. Joseph, MI: ASAE Monograph No.3.

Merriam, J. L., and J. Keller. 1978. Farm irrigation system evaluation: A guide for management. Logan: Utah State University.

Newman, J., ed. 2008 Greenhouse and nursery management practices to protect water quality. Oakland: University of California Agriculture and Natural Resources Publication 3508.

Pair, C. H., ed. 1983. Irrigation. Silver Spring, MD: The Irrigation Association.

Schwab, G. 0., R. K. Frevert, T. W. Edminster, and K. K. Barnes. 1981. Soil and water conservation engineering. 3rd ed. New York: Wiley.

Schwankl, L., B. Hanson, and T. Prichard. 2008. Maintaining microirrigation systems. Oakland: University of California Agriculture and Natural Resources Publication 21637.

SCS (USDA Soil Conservation Service). 1982. Florida irrigation guide. Gainesville, FL: SCS.

———. 1987. Farm irrigation rating method. Gainesville, FL: SCS Engineering Technical Note FL-17.

Smajstrla, A. G., B. J. Boman, G. A. Clark, D. Z. Haman, D. S. Harrison, F. T. Izuno, D. J. Pitts, and F. S. Zazueta. 1991. Efficiencies of Florida agricultural irrigation systems. Gainesville, FL: University of Florida, Florida Cooperative Extension Service.

Photo: J. P. Newman.

Controlling Runoff and Recycling Water, Nutrients, and Waste

Julie P. Newman, John N. Kabashima, Donald Merhaut, Darren L. Haver, Jay Gan, and Lorence R. Oki

CHAPTER EIGHT

Nursery runoff may contain fertilizers, pesticides, sediment, trash and debris, and other contaminants in quantities that exceed water quality standards set by federal, state, and local regulatory agencies. These pollutants have been detected in surface waters, where they may disrupt ecosystems and lead to impairment. Moreover, contamination of groundwater by fertilizers and pesticides is of human health concern.

To protect water quality, a number of federal and state laws have been enacted. In California, most growers are participating in agricultural conditional waiver programs or have permits that require reduction or elimination of runoff from their nurseries by regulatory agencies. Furthermore, many states including California have limited and expensive water supplies requiring the implementation of more efficient irrigation systems. Whatever the reason, there are numerous methods for reducing and managing water runoff that have been used successfully to mitigate both water quality and water supply concerns. A summary of practices and technologies to reduce, capture, and treat runoff, as well as a discussion of options for recycling water, are presented in this chapter.

The greenhouse and nursery industry also generates a significant amount of waste that ends up in landfills, where it may contribute to environmental pollution. Growers should adopt management practices to reduce and recycle nursery waste. Recycling plant waste and plastic products, along with other methods for reducing waste, are described in this chapter.

Water Quality Regulations

While many states rely on federal regulations and voluntary compliance to protect water quality, California has also enacted state laws that mandate the ceasing or mitigation of wastewater discharges from nursery operations if discharges exceed established water quality criteria. This section provides a discussion on the federal regulations that affect nursery growers in all states and a summary of some of the most important water quality laws in California that impact agriculture. Jones et al. (2003) published a guide to California water pollution legislation that covers this issue in detail (see the References).

Photo: J. P. Newman.

Federal Regulations and Programs

Clean Water Act

The Clean Water Act (CWA) of 1972 addresses water quality of surface water, including lakes, rivers, streams, estuaries, and wetlands on a national level. Major amendments were enacted in the Clean Water Act of 1977 and the Water Quality Act of 1987. While the CWA is administered federally by the U.S. Environmental Protection Agency (EPA), in California, the State Water Resources Control Board (SWRCB) and the nine regional water quality control boards (regional water boards) administer many of the CWA's provisions. Local governments may also be involved. For example, in 1999, San Diego County and two of its cities lost a lawsuit for failing to enforce the CWA and, as a result, nursery and greenhouse activities in these areas were heavily regulated by city and county governments.

The CWA provides legal definitions for two types of pollution sources: point source pollution and nonpoint source pollution. A point source is defined in Section 502(14) as any "discernible, confined and discrete conveyance" (e.g., a pipe) "from which pollutants are or may be discharged. This term does not include agricultural stormwater discharges and return flows from irrigated agriculture." A nonpoint source is defined as "any source of water pollution that does not meet the legal definition of 'point source' in Section 502(14)." Generally, point source pollution refers to water pollution that comes from a single, easily identifiable, distinct place, such as a water treatment facility, industrial factory, or urban storm conveyance system. In contrast, nonpoint source (NPS) pollution comes from many diffuse sources, with agricultural runoff as the single largest source. NPS pollution is caused by water moving over and through the ground. As the runoff moves, it picks up and carries away pollutants, finally depositing them into lakes, rivers, wetlands, coastal waters, and groundwaters. Most NPS pollutants fall into six major categories: sediment, nutrients, acid and salts, heavy metals, toxic chemicals, and pathogens.

The CWA developed two programs to address both point and nonpoint sources of pollution: the National Pollutant Discharge Elimination System (NPDES) program and the Total Maximum Daily Load (TMDL) program. These are described below.

NPDES program

The NPDES is a national program under Section 402 of the CWA that regulates discharges of pollutants from point sources to water within the United States. Discharges are illegal unless authorized by an NPDES permit. The NPDES program requires point source dischargers to mitigate their effluent to meet specific water quality standards based on the best available control technologies. While agriculture has historically been exempt from the NPDES permit program, there have been exceptions. For example, in 2003, greenhouses in one area of California were declared point pollution sources by the Central Coast Regional Water Quality Control Board, with NPDES permit requirements for any waste discharges.

TMDL program

Section 303(d) of the CWA requires each state to make a list of impaired water bodies and develop total maximum daily loads (TMDLs) for each pollutant causing the impairment. Water bodies are considered impaired when they are too polluted or otherwise degraded to support their designated and existing beneficial uses, such as for wildlife habitat, recreation, source of drinking water, agriculture, and shellfish harvesting. A TMDL is the maximum amount of a pollutant a water body can tolerate on a daily basis while meeting established water quality standards. The process requires that all point and nonpoint sources be identified and assigned a pollutant allocation that may not be exceeded. The 303(d) list of impaired water bodies contains information describing which beneficial uses are impaired and by what pollutants. The list is updated every 2 years. California's 303(d) list is available for viewing on its State Water Resources Control Board website or regional water board websites (regional board links are available at the state board website). TMDLs affecting agricultural activities have been established and implemented for impaired water bodies in various areas throughout the state. More TMDLs can be expected to be implemented and enforced in the future.

California Regulations and Programs

California Porter-Cologne Water Quality Control Act

In 1969, prior to the enactment of the federal 1972 Clean Water Act (CWA), California implemented the Porter-Cologne Act, which is the principal law governing water quality in the state. The Porter-Cologne Act established a comprehensive program to protect water quality and the beneficial uses of water. Unlike the CWA, Porter-Cologne applies to both surface water and groundwater. The Porter-Cologne Act established the State Water Resources Control Board as the statewide water quality planning agency. It also gave regulatory authority to the nine regional water quality control boards that were founded 20 years earlier by the Dickey Water Pollution Act. The Porter-Cologne Act requires each regional water board to establish basin plans containing policies, beneficial uses, water quality objectives, and implementation plans.

Under the Porter-Cologne Act, regional water boards can issue Waste Discharge Requirements (WDRs) to individuals discharging into state waters to ensure that water quality objectives are not violated. If it is determined that a nursery's discharges result in the impairment of a water body, the regional water board may issue a cease and desist order and require a WDR permit for further discharges. The WDRs set water quality requirements that must be met by all discharges that occur during the period of time that the permit is issued. Several regional water boards have issued WDRs to nurseries in areas of the state adjacent to sensitive water bodies or when concerned local citizen or environmental groups expressed concern over the discharge of surface runoff from a facility.

Senate Bill 390

The California Water Code authorizes state and regional water boards to conditionally waive waste discharge requirements (WDRs) if it is in the public interest. Although waivers are always conditional, waivers in the past had few conditions. In general, they required that discharges not result in violations of water quality objectives but did not require water quality monitoring. Senate Bill 390 (SB 390), signed into law in 1999, required the regional water boards to review their existing waivers and to renew them or replace them with WDRs.

To comply with SB 390, comprehensive conditional waivers of waste discharge requirements for discharges from irrigated lands—known informally as conditional waivers or ag waivers—have been adopted or are in development by water boards for the Los Angeles, Central Coast, Central Valley, San Diego, and Santa Ana regions. Moreover, some

Photo: D. Zurawski.

water boards have adopted conditional waiver programs, or are planning to develop them, specifically to implement TMDLs in their regions (e.g., the San Francisco and Lahonton regions and the north coast of California). The conditional waiver allows the discharge to occur, but growers are required to take specific actions. These include adopting best management practices (BMPs) to prevent pollutants from entering water bodies, as well as monitoring downstream waters where agricultural lands drain. Monitoring is accomplished through participation in monitoring coalitions or groups to provide more cost-effective results, although individual growers can also act independently of any group. Conditional waivers are good for 5 years, can be extended, and are conditional. They can be revoked at any time. For example, the Central Valley regional water board is currently working on replacing its conditional waiver with WDRs.

If your nursery operation is in one of the California regions with adopted conditional waivers or WDRs, you must comply with mandatory requirements. Contact your regional water board if you are not familiar with these requirements. Furthermore, other discharge requirements may affect agriculture, such as conditional prohibitions incorporated into the region's basin plan.

Groundwater regulations

Assembly Bill 2021, California's Pesticide Contamination Prevention Act (PCPA) of 1985, was passed to prevent further pollution by agricultural pesticides of groundwater used for drinking water supplies. It required the California Department of Pesticide Regulation (DPR) to establish a program to identify pesticides that have the potential to pollute groundwater, to determine from sampling if those pesticides are present in groundwater, to maintain a database of all wells sampled by all agencies for pesticides, and to conduct a formal review to determine whether the use of the detected pesticides can be modified to protect groundwater. Under the PCPA, pesticides detected in groundwater were expected to be prohibited unless future contamination could be controlled. The regulatory program focused on limited mitigation measures and applied only to the 1-square-mile pesticide management zones (PMZs) around contaminated wells. By 2003, those zones covered about 313,000 acres statewide.

In 2004, DPR adopted regulations designed not only to stop continued contamination but to prevent it in the future. This increased the size of the regulated area to about 2.4 million acres of California farmland identified by the DPR as most vulnerable to pesticide contamination from leaching and runoff. These regulations prescribe actions to prevent pesticides from reaching groundwater in Ground Water Protection Areas (GWPAs) so that contamination can be avoided. Vulnerable areas are classified as either runoff type or leaching type, and management practices are written into regulation for each area type. The GWPAs are based on detections (and include all PMZs) or on soil characteristics and depth to groundwater. To use a pesticide regulated as a groundwater contaminant in a GWPA, the user must obtain a permit from the county agricultural commissioner that specifies the enforceable management practices required in each GWPA. Additional statewide restrictions pertain to pesticides applied in canals, ditches, and artificial recharge basins, as well as those applied by chemigation. Wellhead protection measures were also adopted by the program. DPR samples a network of wells to determine the effectiveness of these regulations. For a current list of regulated pesticides, visit the DPR's website, www.cdpr.ca.gov. (Click on "Programs and Services" button, then on "Ground Water Protection Program.") The DPR groundwater-protection fact sheet is listed in the References at the end of this chapter.

Water Management Plans

It is imperative that growers be fully aware of any negative impacts of their operation on the beneficial uses of local water bodies. Each nursery should develop a plan on how runoff and groundwater leaching will be mitigated. In developing a plan, the following steps should be taken by existing and planned nursery operations in California:

- Determine which regional water board regulates your operation.
- Inventory streams, lakes, estuaries, or other water bodies that may be impacted by discharges from your operation.
- Obtain a copy of the appropriate regional water board basin plan to determine the local water quality objectives.
- Be an active stakeholder in the development of TMDLs for water bodies that your facility impacts.
- Be an active member in your local conditional agriculture waiver group, if one exists.

- Be aware that if you discharge, a WDR permit may be requested by your regional water board.
- Be prepared to work closely with your regional water board, local water agencies, and other regulatory agencies to develop appropriate controls for discharges from your facility.
- Inventory chemicals used in your operation, especially those likely to be present in runoff.
- Monitor runoff flow and basic nutrient concentrations in discharges, whether or not you are operating under a WDR permit or conditional agriculture waiver. It is also recommended that you test for other contaminants according to the products used, such as specific pesticides that you suspect may be present in runoff.
- Compare results of your water analyses against local, state, and federal water quality standards and regulations for potential noncompliance issues.
- Establish and maintain water quality runoff records and use them for planning and evaluating future improvements. Records will help protect you against inaccurate data.
- Know the basic properties of the pesticides used at your operation and how they potentially impact local water.
- Be aware of management practices that can reduce, treat, and contain runoff, and implement those that are appropriate to your operation. To remain compliant with the numerous water quality regulations, it is imperative that you design, implement, and maintain efficient fertilization, irrigation, and pest management programs.
- Keep a record of management practices implemented to protect water resources and the date of implementation. This information can be used to help complete documents required by regulatory agencies, such as conditional waiver and permit documents required by California regional water boards. It can also be used to demonstrate good stewardship. Newman et al. (2008) published a checklist of management goals and practices to mitigate runoff and leaching that can be used as a template.

Practices That Minimize Surface Runoff and Leaching

Perhaps the most effective and obvious method for mitigating surface runoff from nurseries and surrounding areas is to reduce the quantity of runoff produced by your operation wherever possible. The less runoff produced, the fewer resources (e.g., labor, equipment, and land) are needed to mitigate runoff. Many irrigation, fertilizer, and pesticide management strategies can be employed to reduce the quantity of runoff leaving your operation; effectiveness generally depends on the integration of numerous practices.

Irrigation Strategies

Under dry weather conditions, irrigation is the single most important contributor to runoff. Therefore, it is the most important practice to correctly manage to prevent or reduce runoff. Irrigation runoff water carries dissolved nutrients, pesticides, and other contaminants, as well as soil and potting mix particles with adsorbed pollutants. Irrigation efficiency, irrigation uniformity, methods of irrigation, and timing of irrigation events can all contribute to runoff. The following is a summary of some of the irrigation practices that can be implemented to reduce runoff. Further details are provided in chapter 7.

Irrigation System Design

Irrigation systems should be designed or retrofitted for maximal irrigation uniformity and efficiency. Irrigation system performance can be measured by irrigation efficiency (how much of the applied water is used by plants) and uniformity (the capability of an irrigation system to evenly deliver water). Distribution uniformity testing provides valuable information on where adjustments can be made to the system to minimize the misapplication of irrigation or excessive leaching due to the overapplication of water.

Use of pressure regulators to reduce pressure to the optimal level for the chosen emitters is very important. Emitters that minimize pressure differences or pressure-compensating emitters should be selected when using drip systems. For overhead systems, flow control nozzles can be used when pressure is too high or variable. For slope irrigation, systems should be designed to compensate for pressure differences at the top and bottom of the slope. This may include installing additional pressure regulators at key points. Furthermore, to prevent decreased uniformity, emitters with different flow rates should not be combined in the same watering zone. Emitter flow rates must be correlated with plant types and container size. Flow rates that are too high will apply water faster than soil can absorb it, resulting in runoff. Irrigation system design should also employ appropriate and uniform nozzle sizes and sprinkler heads with a high uniformity rating that are properly spaced for optimal distribution uniformity.

Overhead irrigation can result in only 20 to 50 percent of the water reaching the target due to crop spacing and high susceptibility to wind and evaporation losses. If necessary, other irrigation systems should be implemented to more efficiently deliver water, such as microirrigation (e.g., drip, spray stakes, microsprinklers) or subirrigation (e.g., ebb and flow, capillary mats).

Irrigation system maintenance

Regular maintenance of irrigation systems is necessary for continued efficient operation. This includes inspecting and repairing all leaks and replacing worn, outdated, damaged, or inefficient irrigation system components and equipment. It also includes flushing and unclogging lines, emitters, and sprinkler heads, as well as regularly cleaning filters and screens.

Irrigation scheduling and rates

Irrigation scheduling should ideally be based on environmental conditions and plant moisture requirements. Methods to estimate plant water use include weighing plant containers, estimating evapotranspiration, and directly measuring soil moisture in the root zone of a crop with moisture-sensing instruments such as tensiometers. Plant types with similar moisture requirements and in the same container sizes should be grouped and watered together. This avoids overwatering some plants in order to provide adequate moisture to others. To maintain a high distribution uniformity when using outdoor overhead irrigation systems, avoid irrigating under windy conditions. Pulse irrigation can reduce the amount of water applied by irrigating in small increments that can be more effectively used by the plants. Automatic timers can assist in implementing more complicated irrigation schedules such as pulsing, and they can prevent operator errors associated with manual systems. However, timers must be checked for accuracy, and they must be managed to properly schedule irrigation based on changing environmental conditions and plant growth stage. Computer-controlled irrigation using sensors that measure factors such as solar load or soil moisture content can be very efficient, conserving both water and labor.

Efficient irrigation practices

Efficient irrigation requires applying the correct amount of water to the desired location. Spray patterns should be evaluated to ensure that water is being applied only to plants and not running onto walkways or roads. To reduce runoff associated with overhead irrigation, containers should be placed as closely as possible to one another without reducing plant quality. Emitters should be placed within each pot when using drip systems.

When containers are moved, such as during harvesting operations, plants should be consolidated and the irrigation should be shut off in unused areas. Some emitters, such as spray stakes, can be turned off when not in use. Overhead emitters with check valves can prevent line drainage and drip damage. Hand-watering should be performed carefully to avoid overwatering and creating runoff between pots and on walkways; using an on/off mechanism will help accomplish this.

Irrigation practices need to be assessed for their effectiveness in reducing runoff. The best determiner of the effectiveness of an integrated irrigation management system is to compare the volume (gallons) of water applied to the entire nursery to the volume of runoff from the entire nursery over a period of time. The optimal monitoring system for providing irrigation management information operates continuously (24 hours a day, 7 days a week).

Fertilizer Management Strategies

A major challenge for growers is supplying crops with sufficient amounts of required nutrients without overfertilizing and contributing to nitrogen runoff. Chapter 6 describes management practices that growers can implement to reduce nitrogen loads in runoff. Objectives include evaluating irrigation water, soils, growing media, and plant tissue; using this information to schedule fertilizer application rates and timing to optimize plant growth and avoid overfertilization; conducting efficient fertilizer and leaching practices; and avoiding fertilizer spills during all phases of transport, storage, and application.

Pesticide Runoff Management Strategies

Implement an IPM program

The best method for mitigating pesticide runoff is to reduce chemical inputs by establishing an effective integrated pest management (IPM) program, as described in chapter 12. This includes using good sanitation and other preventive control techniques to avoid pest and disease problems; using nonchemical control tactics to reduce overall pesticide use; selecting pesticides that best target the problem pest to optimize natural predators; applying pesticides in a safe manner to reduce pesticide loads and potential runoff; and avoiding pesticide spills and leakage during all phases of transport, storage, and application.

Know the factors that affect pesticide runoff and leaching

Before selecting and applying pesticides, it is important to understand how pesticides move in the environment and how site characteristics can affect movement.

Soil properties. Soil properties, including potting soils used for container production, have a significant effect on the fate of pesticide applications. Soils that are groundwater sensitive or subject to erosion and surface water runoff are particularly problematic. Groundwater-sensitive soils often have high leaching potentials, facilitating rapid movement of pesticides to groundwater. Leaching potential is affected by soil permeability, water table conditions, and soil adsorption potential.

Permeability is the rate at which water moves through soil. Permeability is controlled by the size and continuity of the soil pores, and it is influenced by soil texture, organic matter, and density. Coarse-textured sandy and gravelly soils have the largest pores and the most rapid permeability, whereas fine-textured clayey soils have tiny pores and slow permeability rates. Organic matter is also important because it contributes to the aggregations of grains of sand, silt, and clay. These aggregates have relatively large spaces between them, permitting more rapid water movement. Soils that are dense, compact, or made of cemented layers have very slow rates of permeability. These soils are prone to surface runoff and may contribute to shallow water tables. Plant roots and burrowing insects and animals also affect permeability by creating large voids, or macropores, that are connected to the soil surface. These macropores carry pesticides vertically into the soil during heavy rainfall or irrigation events.

Water table conditions include the height and duration of water tables in the soil. Shallow water tables that persist for long periods increase the risk of groundwater contamination. Well-drained soils rarely have water tables that persist for long periods above a depth of 6 feet. They are much less sensitive than poorly drained soils, which may maintain water tables at or near the surface for several months. Perched water tables occur when a zone of saturation is separated from permanent groundwater by a soil layer of very low permeability. The soil layer that perches above water serves as a barrier, preventing contaminants from moving to the permanent groundwater supply, reducing risk.

Soil adsorption potential refers to the binding of chemicals to particles of organic matter and clay in the soil. Adsorption retains pesticides in the soil, where they can be degraded. A soil with a high adsorption potential has a low risk of groundwater contamination and surface runoff. A soil that has both a high leaching potential and a low adsorption potential has a high risk of groundwater contamination. Adsorption of a given pesticide varies dramatically in different soils, and it generally increases with soil organic matter and clay content. Organic matter provides the greatest number of binding sites because it has a large surface area and is highly chemically reactive. Clay particles also have a large surface area that is chemically reactive. The higher the clay content, the greater the number of binding sites for pesticide retention. Clay content is particularly important in the subsoil, where the organic matter content is generally much lower than in the surface soil.

Water movement. Water movement is the driving force for pesticide leaching and runoff. Rainfall and crop irrigation can lead to environmental contamination when their infiltration rates exceed the plant's uptake rate and the soil's ability to absorb water. Careful management of the amount and timing of irrigation applications can be very effective in reducing the risk of contamination. Maintaining high distribution uniformity coupled with proper irrigation scheduling can greatly preserve surface and groundwater quality. Crops grown outside greenhouses should not have pesticides applied when rain is forecast.

Pesticide properties. The risk of a particular pesticide is largely related to chemical properties that determine its mode of transport in the environment. These include pesticide adsorption and persistence, as well as the effect of these properties on runoff and leaching potential. Pesticide adsorption refers to how well a pesticide adheres to soil particles. Typically, as pesticide solubility in water increases, adsorption in soil decreases because water-soluble pesticides tend to stay in the soil solution. Pesticide adsorption is also a function of soil properties, especially soil organic matter content. Pesticide persistence is a measure of how long the chemical resists degradation and stays in its original form in the soil. Typically pesticides are not stable, and degradation by soil microbial and chemical reactions eventually detoxifies a pesticide. This is a desirable process from the perspective of environmental safety, although fast degradation may render a pesticide ineffective for pest control. Persistence depends on the properties of the pesticide itself. Some pesticides are stable, while others are degraded more rapidly. The use of highly persistent pesticides poses the greatest risk of contamination to groundwater and surface water.

Organophosphates, carbamates, and synthetic pyrethroids have runoff potential with a high associated risk for contaminating surface water and harming aquatic life. These include many commonly used pesticides for ornamental crops, such as chlorpyrifos, diazinon, malathion, carbaryl, bifenthrin, cyfluthrin, fenpropathrin, permethrin, and many others. Pesticides that are weakly adsorbed in the soil (e.g., diazinon, chlorpyrifos, and many herbicides) often move in runoff in the dissolved form. Pesticides that are strongly adsorbed (e.g., synthetic pyrethroids), on the other hand, can move in runoff along with loose soil particles, including potting soil mix. Pesticides that are of risk to groundwater (e.g., imidacloprid, metalaxyl) have high leaching potentials; they do not adsorb to soil readily, and they have a long persistence. Groundwater-risk pesticides include many herbicides that cannot be used in greenhouses but may be used in areas outside of the nursery.

Select reduced-risk pesticides

An important method to prevent contamination of groundwater or surface water with toxic chemicals is to select reduced-risk pesticides. Each pesticide has characteristics that determine its mode of transport in the environment. These include formulation, application timing, runoff potential, and leaching potential. Pesticides should be selected for low migration risk based on site conditions as well as pesticide label and hazard warnings. Whenever possible, pesticides should be selected that will not potentially contaminate surface water or groundwater.

Consider weather and site conditions

Be especially careful when pesticide applications are planned for the rainy season in areas of shallow water tables or where soils are permeable or have low organic matter content. The use of pesticides in DPR's Ground Water Protection Areas was described earlier in this chapter in the section "Groundwater Regulations."

The *UC Integrated Pest Management Guidelines* (found on the UC IPM website) contains information that can be used to determine if a pesticide active ingredient has the potential of water-related risks. Site-specific conditions can be factored in and comparisons made with other pesticides used in nurseries by clicking the "Water Quality Compare Treatments" button at the top of the list of treatments.

Minimize pesticide movement on soil particles

Since many pesticides adsorb onto soil particles, it is important to clean up any potting mix spills and to store media containing pesticides and fertilizers in a covered area sheltered from rain and wind. Minimizing erosion is also important. Practices to reduce erosion and contain sediment are described in the next section.

Treating, Controlling, and Capturing Sediment and Surface Runoff

If preventive practices do not sufficiently reduce runoff, measures should be implemented to contain runoff on-site or to treat it so that it does not contain pollutant levels above water quality objectives. Numerous practices can be used to mitigate runoff. A summary of common methods, including practices for settling soil particles, use of water impoundments to capture water on-site, use of vegetated buffers, erosion control practices, and practices that control water movement, are provided in this section.

These systems should be designed and constructed to comply with all federal, state, and local laws and regulations. Local zoning regulations and other permit requirements may apply. Multiple agencies may be involved in the permitting process, especially if the project is near a stream or wetland, involves constructed wetlands, or affects natural habitat or endangered species. When planning and designing runoff and erosion treatment systems, it is best to consult a licensed engineer or the Natural Resources Conservation Service (NRCS). Local NRCS service centers in each state are listed at the national website (http://www.nrcs.usda.gov).

Methods for Settling Soil Particles

A considerable percentage of the constituents of concern in surface runoff are adsorbed to suspended soil particles and can be removed by settling. The methods listed below have been shown to effectively remove sediment from surface runoff. The specific use of these methods in recycling systems is described later in this chapter. Under most circumstances, the implementation of several methods for settling soil particles will be required to effectively mitigate runoff. For example, a pilot water pollutant mitigation system developed at a 100-acre nursery (fig. 8.1) in Orange County, California, consisted of flocculation (polyacrylamide [PAM] addition), sediment traps, and sediment ponds and basins (Kabashima et al. 2004). A vegetative filter that removes sediment and absorbs dissolved pollutants was also used (see the section "Vegetated Buffers" later in this chapter).

Sediment pits and traps

Large, heavy particles such as sand used in container mixes can be removed by allowing the drainage water to collect in a sediment pit or trap (fig. 8.2). Generally much smaller and shallower than a sediment basin or pond, sediment pits or traps slow the movement of surface runoff and allow for the settling of these large soil particles. Sedimentation pits or traps should be large enough to allow large, heavy particles to settle out, and they should be designed for easy clean out. Lining the sediment pit or trap with concrete will facilitate thorough removal of collected sediment. Sediment pits

FIGURE 8.2. Shallow sediment trap used by a small nursery to remove large, heavy particles such as sand. Sediment trap shown without the cover (A) and with a grate on top so vehicles can drive over it (B). *Photos:* J. N. Kabashima.

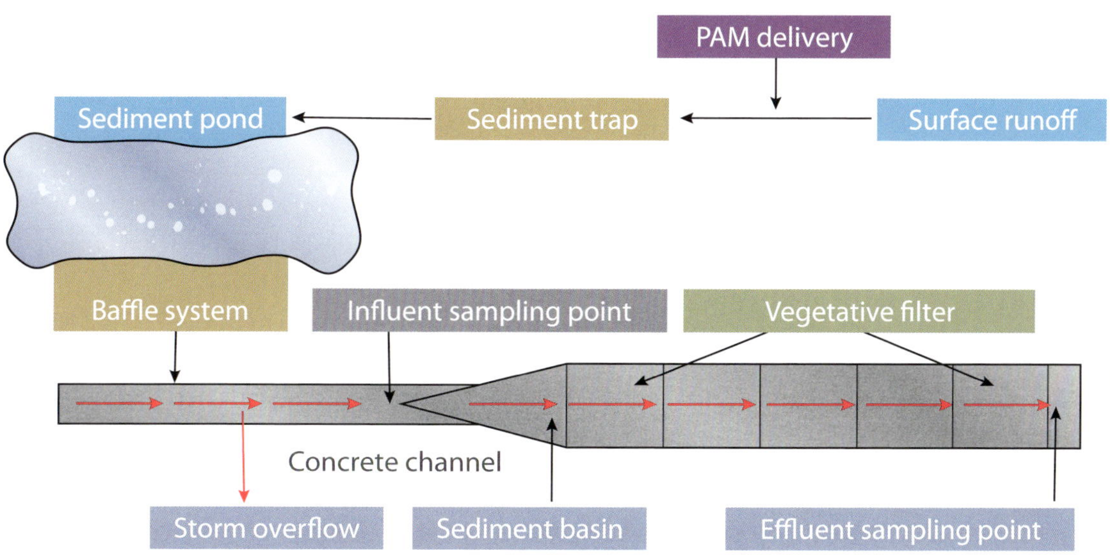

FIGURE 8.1. Arrangement of mitigation methods to improve the quality of runoff from a 100-acre nursery.

or traps should be sloped less than 5 percent to ensure that water is slowed to a velocity that allows sedimentation to occur. Information on design can often be obtained from public agencies or companies specializing in water treatment, in addition to the NRCS.

Sediment basins

For intercepting larger amounts of sediment, a sediment basin (fig. 8.3) can be constructed at the base of a nursery field. Sediment basins temporarily detain sediment-laden runoff under quiescent conditions, allowing sediment to settle out before the runoff is discharged. In order to avoid seepage into groundwater and for easier removal of sediment, the basin should be lined with an impervious material such as concrete, gunnite, or polymer sheeting.

Flocculants

The most difficult suspended solid to remove from the runoff water is the colloidal fraction—usually consisting of clay. An individual particle of clay is less than 2 microns in size (less than 8/100,000 of an inch) and is invisible to the naked eye. However, when present en masse, it will make water appear milky-brown. Such small particles settle out with difficulty and only in quiescent water. If the clay is flocculated (i.e., aggregated electrically and physically into larger particles), it will settle much more easily. Nonflocculated clay takes 1 to 5 days to settle in quiescent water, whereas flocculated clay will settle in as fast as 5 minutes. Polyacrylamides (PAMs), flocculants composed of long-chain polymers, have been shown to be effective in removing the colloidal fraction—and thus a majority of the contaminants—from runoff. Research by the University of California Cooperative Extension has shown that the application of PAM at 10 parts per million directly to runoff reduced suspended solids by greater than 90 percent. The level of bifenthrin and permethrin, pesticides that are strongly adsorbed to sediment, was also reduced by greater than 90 percent (Kabashima et al. 2004).

Water Impoundments

Impoundments (e.g., retention basins, infiltration basins, ponds, and reservoirs) can be constructed to collect large amounts of runoff flow and retain (store) runoff water. In these types of impoundments, water dissipates by infiltrating into the ground (in unlined basins) and by evaporation. Overflow may be prevented by using the collected water in recycling systems (described later in this chapter) or in the irrigation of landscapes and other noncrop areas.

Vegetated Buffers

Vegetated buffers are areas or strips of land maintained in permanent vegetation to prevent erosion and improve water quality by trapping and treating contaminants. They help to slow down runoff, thereby encouraging the settling of soil particles. In addition, plant roots take up water and dissolved pollutants from runoff water. Vegetated buffers can offer many other benefits, including supporting populations of beneficial insects for biological control of crop pests, protecting stream banks, and providing secondary crops (fig. 8.4).

Although different types of vegetated buffer systems exist, most work in a similar manner. Runoff containing soluble nutrients and pesticides, along with sediments with adsorbed pesticides, enters the buffer. Vegetation in the buffer slows surface flow and sediments drop out. Some water infiltrates into the root zone and subsoil, while the remainder becomes lateral subsurface flow. When the roots of buffer plants grow to sufficient depth, they intercept infiltrated water and take up the soluble nutrients and pesticides. Nutrients and pesticides adsorbed to soil particles become trapped in the root zone, and high soil organic matter supports microbial growth

FIGURE 8.3. Two sediment basins used in a field outside greenhouses. Sediment basins temporarily detain sediment-laden runoff under quiescent conditions, allowing sediment to settle out before the runoff is discharged. *Photo:* J. P. Newman.

FIGURE 8.4. The canna lilies in this vegetative bioswale, used in an outdoor container nursery to filter nutrients and pesticides and trap sediments, are marketed as a secondary crop. *Photo:* D. Haver.

that contributes to denitrification and pesticide degradation. In addition to improving water quality, some vegetative buffers are also effective in controlling erosion.

Types of buffers

Many different types of vegetated buffers can be used in ornamental production. Some examples include vegetated bioswales, vegetated filter strips, and constructed wetlands. A vegetated bioswale (see fig. 8.4) is a conveyance system that channels stormwater or irrigation runoff water. This type of buffer system improves water quality by reducing flow velocity, increasing sedimentation, and filtering pollutants. It may also allow infiltration into the underlying soils in unlined conveyance structures. A filter strip is a band of vegetation that can be used between a runoff source (e.g., nursery) and a water body to be protected. The purpose of the filter strip is to slow runoff from the production area and trap sediment, fertilizers, and pesticides before they reach the water body. A constructed wetland is an artificial marsh or swamp for treating wastewater, controlling floodwaters, and reducing erosion. In nursery production, a constructed wetland can be built to further remove pollutants in the effluent from a retention basin.

Site selection and design

A licensed engineer should be consulted in the design, as the overall efficiency of the filter depends on the residence time of the runoff, the volume of runoff, the extensiveness of the root system, and the vigor of the plant material in an aquatic or semiaquatic environment. Many factors must be considered in developing a good design for a vegetative treatment system. Depending on the type of buffer, these factors can include slope percentage, erosion rate, density and height of the vegetation, runoff volume, connecting structures such as pipes, and auxiliary systems such as treatment ponds and detention basins. General design specifications for buffer width and length, minimum capacity, outlets, and other parameters can be found in the Natural Resources Conservation Service practice standard by contacting your regional NRCS office.

Vegetative buffers that serve primarily to trap and filter pollutants must be designed to increase water infiltration and maximize runoff contact with soil and vegetation. Vegetative buffers are most efficient when runoff water spreads out evenly across the buffer at a shallow, uniform depth. This type of flow is called sheet flow, in contrast to concentrated discharges with high flow rates that do not make enough contact with the buffer or bypass it. A proper buffer design can encourage sheet flow by including the use of stiff-grass hedges or the use of engineering innovations such as level spreaders that create a wide, flat, level flow path to limit velocity and reduce flow depth.

Plant selection

Plant species should be chosen based on the need to remove sediment, nutrients, or pesticides, or a combination of contaminants. Some plant species are more effective in removing contaminants than others.

Plant selection decisions should also be based on climatic and microclimatic adaptability, including plant hardiness and heat tolerance. Seasonal rainfall is important, especially if supplemental irrigation is not planned for dry periods. Furthermore, soil and root media conditions need to be considered in plant selection, including soil type and depth to water table. If plants are grown in irrigation channels or constructed wetlands, it is essential that species are selected that can tolerate standing water and low root aeration.

Growth rate and season of growth are also important factors in plant selection. Adequate plant growth during seasonal storms is important in the uptake of pollutants generated during these events. Plants that grow rapidly are easier to establish, and they take up more nutrients from runoff. However, they may have high costs associated with pruning, mowing, and other maintenance associated with rapid growth.

The type of buffer system also needs to be considered. A good choice for reducing gully erosion in a grassed waterway may be a poor choice for taking up nutrients in a filter strip. Dense stands of perennial grass are better than shallow-rooted annual species for maintaining safe flow velocities in grassed waterways. Flexible grass species are required in water channels, whereas stiff-stemmed grasses are better for filter strips.

When vegetated buffers are used in natural areas that may surround the nursery, it is preferable to use native plant species that blend with existing vegetation. When ornamental plants are desired for vegetative buffers, invasive characteristics should be considered. The NRCS is a good source for lists of commercially available native plants adapted to your specific area.

Some nursery growers have selected plant species for vegetated buffers that have served as secondary crops for generating supplemental income. It is important to make sure that there is a market for the plants or cut flowers and foliage harvested from the buffer. The market price ought to offset at least some of the production and maintenance costs. Growers who are interested in developing techniques to produce secondary crops in vegetative treatment systems should contact their state Cooperative Extension and the NRCS for guidance.

Planting and maintenance

Plants used in buffers are typically seeded because seeds are generally the least expensive propagation method available. However, it is important to include the costs in extra labor involved in weed control during establishment of seedlings, which is higher than those related to transplant plugs or larger plant parts that are purchased or produced on-site. Moreover, stands of vegetation produced by transplanting may be more uniform than seeded buffers.

Planting should be timed so buffers are established prior to expected runoff. The use of mulch or straw dikes may be necessary to protect vegetation until it is established. Supplemental irrigation may also be warranted, especially to promote seed germination and vegetation establishment. Continuous protection from vehicular traffic is critical.

Maintenance of vegetative buffer systems is necessary to sustain buffer function and effectiveness. Buffers should be inspected regularly, especially after storm events, and damaged areas immediately repaired. Trapped sediment changes the shape of buffers, and it may cause runoff to flow parallel to buffers rather than across them. Therefore, periodically remove sediment and reshape the buffers. Dredge ditches lined with plants before seasonal storms.

Vegetative buffers require management of pests such as noxious weeds, insects, and vertebrates. A regular maintenance program should be established to control weeds and invasive species. Vigorous grass growth prevents many annual weeds from establishing, but perennial weeds may require mowing or spot treatment with herbicides. Insect pests in buffers that can move into production areas may require treatment. On the other hand, vegetated buffers can also harbor beneficial insects that could be used in IPM programs when management promotes their development. If the buffer has areas of standing water, mosquitoes need to be controlled, as they may spread diseases that affect human and animal health and safety. Protection from wildlife and domestic animals may also be necessary. For example, invading pocket gophers can cause excessive damage to vegetative buffers, and prompt eradication is necessary. Vegetation may also need to be protected from animal grazing.

If chemical treatments are necessary to control pests, such treatments should be selected after considering potential risks to adjacent aquatic ecosystems. Reduced-risk pesticides, such as formulations containing *Bacillus thuringiensis israelensis* for mosquito control, should be selected whenever possible.

Mowing and pruning may be necessary for controlling plant growth and weeds, as well as for aesthetic reasons. Mowing can stimulate growth and biological activity, leading to increases in nutrient uptake and pesticide degradation. Conversely, it can increase runoff flow and cause injury to some species if the mowing height is too short. Mowing must be conducted to avoid peak nesting seasons and reduced winter cover.

Periodic harvest of vegetation used in buffers removes trapped pesticides and nutrients from the system, preventing eventual release to the soil and potential movement to water. Harvested plant material can potentially be marketed or used to produce compost.

Erosion Control

Water erosion

It is important to determine the location, amount, and causes of erosion and runoff following irrigation and storm events in outdoor nursery production areas and nonproduction areas (e.g., roads, hillsides). In addition to irrigation management practices to reduce dry weather erosion and runoff, practices must be implemented to retard water movement so that sediment and runoff are trapped on-site (e.g., the use of terraces, mulches, buffer vegetation, cover crops, and landscaping).

Roads. Nursery roads must be properly designed and constructed to prevent erosion and contaminated runoff. They must also comply with applicable grading regulations and other local and state regulations, including permit requirements. Poorly planned roads can cause excess runoff and erosion during storm events. This leads to sedimentation that can pollute water supplies, increase flooding potential, and trigger landslides.

In the design process, soil type should be considered for erodibility, excessive slopes should be avoided, roads should generally follow natural contours and slopes to minimize disturbance of drainage patterns, and all cuts and fills should be designed to have stable slopes. Locating roads near water bodies should be avoided; use buffers such as vegetated filter strips to protect waterways by absorbing road runoff and trapping toxic sediment.

Adequate road drainage from both the road surface and hill slope is important. The type of drainage structure used will depend on the intended use and runoff conditions. Water breaks (water bars)—that is, dips built at an oblique angle across the road with a berm at the end—are recommended on steep grades where runoff and erosion are anticipated. Diverted water from water breaks should flow only into stable areas, avoiding septic fields or waterways. Water break discharge areas must be vegetated or covered with other erosion-resistant materials. Surface crowning can also help direct road runoff into the side drainage ditches. Alternatively, when crops are grown adjacent to roads, surfaces can be graded to slope slightly toward the center line of the road rather than into ditches where erosion can threaten crops.

Unpaved roads require protection from concentrated flows of water. Thoroughly compact all backfill in irrigation pipeline trenches along unpaved roads. In Mediterranean climates, seed moderately used road surfaces and exposed soils with traffic-resistant vegetation like annual or perennial grasses. Do this prior to winter rains, and mulch during establishment. Alternatively, stabilize roads with gravel.

Inspect culverts, roadside ditches, water breaks, and outlets after each major runoff event, and restore flow capacity as needed. Maintain vegetated areas in adequate cover, reseeding and mowing as needed. Inspect roads with water breaks periodically to ensure that proper dimensions and slope are maintained and that outlets are stable. Avoid excessive road maintenance. Regrade to remove deep ruts or damaged areas caused by storms and to maintain proper road dimensions and slope; remove roadside vegetation that obstructs line of sight or overall road maintenance. Regrade existing roads and construct new roads only during dry weather. Using roads during wet weather may aggravate erosion and drainage problems, particularly when roads are not protected by vegetation or gravel. Contaminant-laden dust from traffic and wind erosion should be avoided by sealing or watering unpaved roads, ensuring that dust control with applied water does not create runoff. Minimize damage to vegetative buffers adjacent to the road when it is necessary to chemically treat the road surface.

Sediment barriers. The simple use of sandbags, erosion control blankets, jute mesh, wattles, and gravel provide a significant degree of erosion control over the absence of any mitigation method. Erosion control blankets, rolls, or mats made from natural or synthetic polymer fibers can provide ground and slope stabilization, facilitating vegetation establishment. Jute materials are available to reduce erosion and increase sedimentation, with the increased benefit of lasting longer than sandbags or straw bales. These materials are also available in assorted sizes and configurations to match all situations.

Ground covers. Bare soil areas in the nursery must be covered with ground covers such as mulch, cover crops, rock protection, and landscaping to prevent erosion from concentrated flows of water. Plant covers reduce the amount and speed of runoff and can trap sediment, thereby reducing soil losses considerably compared with bare soil. Some vegetated buffers (see the section "Vegetated Buffers" earlier in this chapter), such as grassed waterways, can also be used for erosion control. Ground covers include inorganic materials (e.g., plastics and gravel), plant cover (cover crops and landscaped areas), and mulching. Ground covers protect the soil surface from exposure to rain or irrigation droplet impact, improving infiltration.

Terracing. Terracing is an effective treatment to control erosion on steep slopes. Terraces are level earthen steps that follow the contour of a hillside, breaking a long slope into shorter segments and intercepting the flow of water. Terraces serve as small dams on a hillside, intercepting runoff water and guiding it to a safe outlet. Some terraces are designed to collect water and temporarily store it until it can filter into the ground or be released through a stable outlet. Other terraces are designed as a channel to slow runoff and carry it to a stable outlet such as a grass waterway.

Wind erosion

Wind erosion wears away topsoil and has a direct effect on the productivity of the nursery land, in addition to causing air and water pollution. Wind erosion can be controlled by maintaining good soil structure and using plant covers. Furthermore, trees, shrubs, or other vegetation can be planted as windbreaks (shelterbelts) along the upwind boundaries of the production field. This reduces wind erosion and protects sensitive nursery stock from wind burning, sand blasting, and spray drift. Container media should be mixed and stored in a location sheltered from wind and away from drainage channels so that wind does not move media into water bodies. If your property is affected by discharge sediment or runoff from upslope or upstream properties, it is a good practice to contain this sediment or runoff to prevent contaminants from entering your property. You can do this by using engineered diversions, filter strips, sediment basins, and underground outlets (conduits installed beneath the surface of the ground to collect surface water and convey it to a suitable outlet). Coordinate these efforts with neighboring nurseries and farms.

Control of Runoff Water Movement

Practices that improve water-holding capacity and the infiltration of water into soil or container media include the use of mulches and plant covers on bare soil areas in the nursery as well as the incorporation of amendments. Amendments improve soil structure so that the soil can absorb and hold more moisture, thus reducing runoff and related damage. Wetting agents may increase the water absorption of some soil textures, allowing faster wetting and reduced channeling down sides of pots. Wetting agents can also be used in field soils to improve infiltration rates where water repellency is a problem. Wetting agents should not be overused because they can be toxic to plants and a contaminant in runoff.

Conversely, in areas where groundwater contamination is a concern, percolation into the soil profile can be minimized by grading the soil (minimum 1.5% slope) so that water will flow to drainage ditches. Percolation can also be minimized by lining drainage ditches with a water-impermeable material, such as cement, plastic, or gunnite, or by installing drainage tile, which prevents water from percolating into the soil.

Greenhouse roof runoff contains contaminants. When greenhouses are used for nursery production, roof runoff should be directed so that it does not contaminate waterways. Management is needed to prevent roof runoff from flowing across concentrated waste areas and roads, to reduce pollution and erosion, to prevent flooding, and to improve drainage. In some cases, stormwater from greenhouse roofs can be collected for reuse. Check regulations in your area regarding potential restrictions on stormwater recycling.

Water Recycling

Water recycling is another method to mitigate nutrient and pesticide runoff. In addition to mitigation, water recycling conserves water and reduces irrigation and fertilizer costs. However, there are many factors to consider related to the feasibility of installing a water recycling system at your nursery. The runoff must be routed to a collection system and properly handled. The suspended solids and floating debris must be removed, salt and nutrient levels controlled, and pathogens removed or managed. Also, impacts and benefits of herbicide use should be assessed. You will need to consider how by-products from the water treatment will be handled and what the overall plant response to irrigation with recycled water will be. Additionally, costs for necessary capital investment in structures, equipment, maintenance, and operational expenses should be determined. Public relations and safety issues will also need addressing.

Runoff can be collected by gravity flow into a reservoir or pumped from a settling pond into a reservoir. Factors that determine whether a nursery builds a simple or sophisticated recycling system are the quantity and quality of the water collected; the availability of fresh water to blend with recycled water; the potential loss of water before it reaches a collection site; and contaminants in the water such as disease organisms, salts, organic material, and pesticides. Other important factors in determining the type of recycling system to install are the complexity of the operation regarding the type and number of crops grown, production and irrigation methods, the topography of the land, area of land available to build a reservoir, and budget available to build a system.

Collection of Runoff Water

Grading and drainage ditches

As described in the section "Control of Runoff Water Movement," water losses due to evaporation and percolation into the soil profile can be minimized by grading the soil and lining drainage ditches or by installing drainage tile. Drainage ditches and canals should be designed to handle peak runoff flows. Underdesign can lead to choke points or overflow leaving the nursery.

Reservoir capacity

Reservoir capacity should be designed based on recycling objectives while also meeting all local, state, and federal regulations. Every situation is unique and should be judged independently. It is best to consult a licensed engineer or the NRCS to ensure that reservoir capacity is accurately estimated and designed to meet all applicable regulations.

Reservoirs that recapture only a portion of the irrigation water applied have lower holding capacities than reservoirs built to recover all runoff or runoff from rainstorms. If all runoff will be captured, runoff from all potential sources needs to be considered, including production sites and nonproduction areas such as greenhouse roofs, parking lots, and driveways. A rule of thumb might be to design reservoir capacity to recapture 1½ days' maximum irrigation runoff. Or, in the case of rainwater, to consider the maximum rainfall per hour, annual rainfall for the area, percolation potential, and exposed surface area.

In addition, a second reservoir should be available to hold all treated water used for irrigation, taking into consideration that treated water is usually blended by about 50 percent with fresh water before reuse. Where adequate land is available, a large reservoir or body of water is preferable because it is less prone to drastic quality changes.

Location of collection basins

The location of collection basins should be carefully considered. Nursery topography and space limitations will dictate the location of collection basins, as well as the size and number of reservoirs that are feasible and necessary. When different types of production systems are located within the nursery (e.g., propagation, field crops, containerized crops, or specialized crops), there may be a need for more than one collection basin for multiple water recycling systems. If the nursery is situated on a hill, the collection basins can often be placed in the lowest elevation to allow gravity to direct runoff.

Removal of Floating Debris and Suspended Solids

Floating debris can be filtered out by using a baffle system, which physically stops the particles before they enter the sedimentation basin or pumping systems. Large, heavy particles, such as sand, can be removed by allowing the drainage water to collect in a sediment pit (described earlier in this chapter) before it is processed through sand filters or other treatment. Design the sediment pit so that there is a separated pumping pit that collects water flow from the sediment basin. A baffle system (fig. 8.5) can also be used to remove suspended solids. In a closed recycling system, the collection basin or reservoir can act as a sediment basin to hold large deposits of sediment, retaining runoff water only as long as is necessary to settle large particles.

Removal of Suspended Colloidal Material

The most difficult suspended solid to remove from the runoff water is the colloidal fraction, which is usually clay. Such small particles settle out only in quiescent water. When nursery infrastructure and cultural practices are designed so that contamination of water with clay is avoided or minimized, the use of costly filtration processes can be avoided. However, if clay and other suspended particles need to be removed, one of the following options should be used: settling, floccula-

FIGURE 8.5. A baffle system can be used in the recycling system to remove suspended solids. *Photo:* J. N. Kabashima.

tion with filtration, or filtration methods without flocculation. Sedimentation and flocculation were described earlier in this chapter. Details specific to recycling and filtration methods are provided below.

Use of settling in recycling systems

In simple recycling systems, settling is used to remove colloidal material because it is the least costly method; however, the use of sedimentation to remove clay requires up to 5 days, which is too lengthy for most nursery production systems. Nevertheless, some nurseries are willing to sacrifice some water clarity for economy and use a simplified system to recycle water that relies on sedimentation and forgoes filtration. With such a system, one has to be satisfied with water that has some inherent cloudiness. Some of the cloudiness is due to bacterial contamination and can be clarified by chlorination. A method to effectively remove suspended solids should be implemented to prevent excess wear of pumping equipment.

Flocculation

In sophisticated water systems, such as those used in drinking water clarification, the clay and other colloidal fractions are flocculated into larger particles by alum or ferric sulfate. This process, which has been used by a nursery in southern California, settles out the colloidal fractions within several minutes rather than several days. This kind of clarification produces water of excellent quality after the flocculated colloidal fraction is removed by filtration. For sophisticated water clarification systems, alum dosages can range from 20 to 160 parts per million depending on the clarity of the water. The flocculant aids, either cationic or anionic polymers, may range from 0.6 to 20 parts per million. When using higher dosages of alum, there is a resulting reduction of the pH of the water, which needs to be adjusted from 6.9 to 7.2 for proper precipitation of the alum and flocculation of the clay. The use of ferric sulfate for flocculation does not have as much of a pH restriction since it will precipitate above pH 5.5. An analytical laboratory that processes water samples can perform a jar test to determine the proper dosages for your particular water (Skimina 1992; Skimina and Kabashima 1994).

Filtration

Once all floating debris and larger particles have settled out of the water, filtration is necessary. The usual sand filter allows the nonflocculated clay fraction to pass through because the pore sizes are larger than the clay particles. Filters that are fine enough to trap clay will quickly plug, rendering them useless. Most drinking water plants filter their clarified water (after clay flocculation) through anthracite coal over sand on gravel before disinfection (Skimina 1992). The coal permits storage of large volumes of flocculated clay before it plugs the sand.

Six types of filtration systems are available:

- slow sand filtration
- rapid sand filtration
- microfiltration
- ultrafiltration
- nanofiltration
- reverse osmosis

These filtration methods are described below.

Slow sand filtration

Slow sand filtration (SSF) is usually implemented for pathogen control and is discussed in the next section.

Rapid sand filtration

Rapid sand filtration (RSF) is a physical filtration process that separates particulates suspended in the water. Water to be treated is pumped through the sand bed. Particulates are trapped by the sand as water continues to flow through the pores between the sand grains. The size of particulates that can be removed depends on the size of the sand grains in the filter. Generally, these sand grains are larger than 1 millimeter in diameter. The quality of the sand is important: the grains should be round and not sharp, because sharp sand packs tightly and restricts water flow. The sand grains should be of uniform size, since a mixture of sizes can also restrict flow.

Because it is a physical filtration process, rapid sand filters do not remove pathogens, dissolved pollutants, or very fine particles. However, they are capable of treating large volumes of water with capacities of 18 to 180 gallons per minute per square yard of bed surface area. Periodic maintenance is necessary, which consists of backflushing the sand bed, and the frequency depends on the quality of the water being treated. This process can be automated. Several rapid sand filters can be used in series to treat water in stages; subsequent filters should contain sand that is finer than in the previous filter. This increases the interval between backflushing.

Microfiltration

Microfiltration uses membranes with pore sizes of about 100 to 10,000 nanometers (0.0001 to 0.01 mm). This filtration process requires the least amount of energy to pass water through the filters. However, this process does not remove most pathogens, and additional sanitation is required. Microfiltration is occasionally used before reverse osmosis.

Ultrafiltration

Ultrafiltration uses membrane filters with pore sizes of 1 to 20 nanometers. Dissolved salts (fertilizers and nonessential salts) are not removed with these relatively large pore sizes. Clay particles, however, are removed. In addition, some pathogens such as bacteria, nematodes, most fungal spores, and some viruses are removed. As in nanofiltration, small viruses and viroids are not removed, and additional sanitation processes are required for complete control of pathogens.

Nanofiltration

Nanofiltration uses membrane filters with larger pore sizes than those used in reverse osmosis; however, the pores are still small enough to filter out molecules ranging in size from 200 to 1,000 daltons (1 dalton = 1 atomic mass unit [amu]). These pores are usually large enough to allow chelated nutrients to pass through, since most chelates, such as iron-EDTA, have a molecular weight under 500 daltons. In addition, nanofilters may also filter out some charged particles. Complete pathogen control does not occur through nanofiltration, and additional sanitation is required.

Reverse osmosis

Reverse osmosis (RO), also known as hyperfiltration, uses membranes with the smallest pores available. The pores are so small that the filters remove dissolved salts, charged particles, most pathogens, and any compounds with a molecular weight of 200 daltons or greater. Nurseries that use water sources high in dissolved salts usually must use reverse osmosis (see chapter 7). Since this filtration process removes dissolved fertilizers as well as nonessential salts, filtration should occur before fertilizer is added to the irrigation water.

Sanitation and Treatment for Pathogens

The presence of plant pathogens in recycled water presents the possibility of introducing or spreading pathogens such as *Phytophthora* and *Pythium* species, as well as bacteria, throughout the nursery. Prevention of plant diseases in the nursery is the most effective way to reduce pathogens in recycled water. This can be achieved by practicing good sanitation, purchasing or growing clean nursery stock, growing plants so they are not stressed (by excessive heat, salt, light, etc.), and using the latest research-based technology to irrigate most effectively (not over- or underwatering plants). Refer to chapter 13 for more details.

Disease monitoring

Common practices used by nurseries to prevent the potential for a disease outbreak are prophylactic fungicide treatments of nursery stock or chlorine treatments for the water, or both. However, without specific knowledge of which pathogens are present, their population levels, their seasonal fluctuations, or the efficacy of fungicide treatments on populations within the nursery, treatment recommendations are difficult to make with adequate accuracy. An important aspect of disease prevention and mitigation is timely and routine monitoring of nursery crops for the presence of disease so that action can be taken at an early stage to prevent further spread. If disease organisms such as *Phytophthora* or *Pythium* are detected before they become a major problem, the crop can be treated with appropriate cultural and chemical therapies and will likely remain viable and healthy. Root rot fungi in recycled water can be detected using baiting, culture plating, or a combination of these methods.

Although there are over 16 baiting techniques and selective media for detecting and estimating *Phytophthora* spp. populations, the best method should be chosen based on the availability of the bait in your area, the species of *Phytophthora* or *Pythium* to be detected, and the availability of laboratory supplies (selective agar media, petri dishes, etc.) and facilities. When using these techniques to sample for *Phytophthora* in soil, the soil samples should not be allowed to dry out and should be collected during cool, wet periods. One baiting method that will detect several common *Phytophthora* spp. (*P. cactorum, P. cinnamomi, P. megasperma* f. sp. *glycinea, P. nicotianae* var. *parasitica, P. boehmeriae, P. cambivora, P. citricola, P. cryptogea, P. dreschsleri, P. nicotianae* var. *nicotianae, P. richardiae,* and *P. megasperma megasperma)* is the use of pine needles. Young basal segments or whole needles of *Pinus radiata* or *Cedrus deodara* are floated or immersed with basal portion downward in water above a soil sample in an open container for 3 days. After disinfecting the surface, all needles should be removed (even those without symptoms) and then plated on prune-extract agar or corn meal agar amended with various antibiotics or fungicides to enhance specificity (Skimina and Kabashima 1994).

Various selective media have been developed for isolating *Phytophthora* spp. Although these methods are effective, they are also labor-intensive and cumbersome and require knowledge of fungal taxonomy. With the development of rapid assay *Phytophthora* kits, a grower only needs to detect the presence/absence of *Phytophthora* at the genus level, and thus the need for selective media is eliminated. A detection method has been developed at the University of California at Davis to detect *Phytophthora* and *Pythium* species in water samples (Ali-Shtayeh et al. 1991). This method uses commercially available ELISA (enzyme-linked immunosorbent assay) tests marketed by Neogen Corporation to detect the presence of these organisms (Kabashima et al. 1997).

Pathogen control

Treating pathogens in water can be done by several physical and chemical methods: slow sand filtration, chlorination, ultraviolet light, ozonation, copper ionization, or heat treatment. These methods are discussed below.

Slow sand filtration (SSF). Slow sand filtration is an older water treatment technology that is reappearing in horticultural applications in Europe, but it is not yet widespread in the United States. A common misconception is that SSF and rapid sand filtration are the same, aside from different flow rates. Though they use the same type of substrate, they are quite different; SSF is a biological treatment method that can remove pathogens (Wohanka 1995).

Some physical filtration in SSFs occurs when particulates in the water become lodged in the sand surface and decrease flow rates. In order to delay this fouling, pretreatment such as settling and rapid sand filtration may be desirable for water that contains suspended particulates. As water moves through a freshly built sand bed, a biofilm develops on the surface of the sand grains and eventually can become relatively thick at the surface of the sand bed. This thickened layer, also known as the schmutzdecke, is primarily responsible for the treatment. The biofilm is a diverse and dynamic community of microorganisms, and its composition depends on the contents of the water. The composition also changes in response to variations in that content. Most of the biological activity occurs within the top 6 inches of the bed, just below the sand surface. Organisms that have been identified in the biofilm include algae, bacteria, diatoms, and zooplankton (Calvo-Bado et al. 2003; Joubert and Pillay 2008). However, the specific mechanisms of treatment are not fully understood. Inoculation of the sand bed with microorganisms is not necessary.

SSF systems are very simple (fig. 8.6). The sand filter can be constructed in any container that can hold sand and water (e.g., drum, steel water tank, concrete septic tank, or lined earthen reservoir). Sand filters are constructed with an underdrain so that treated water can be collected once it has passed through the sand bed. The sand must have round (not sharp) uniform-size sand grains of about 60 mesh (about 0.3 mm) with a coefficient of uniformity less than 3 (UC < 3). Quartz blasting sand works well. The sand bed should be about 3 feet deep and submerged under 3 feet of water. This water head is used to push the water through the sand bed. It is important to include a method to control the flow rate through the sand. Other components that are required include a reservoir to hold captured runoff to be treated, a reservoir to store treated water, and a method to move the water between these components. Water can flow through the system by gravity if the site is appropriate.

Management of the slow sand filtration system requires frequent monitoring of the flow of water through the filter. If the flow rate is too fast, complete treatment may be compromised and the sand bed may quickly become plugged. A flow rate that is too slow does not generate the desired volume of treated water and may cause degradation of the biofilms, due to lack of sufficient aeration and carbon supply.

If the desired flow rate cannot be achieved, the sand bed requires maintenance. This requires draining the sand filter to expose the bed surface. The top layer of sand (about 0.5 to 1 inch deep) should be removed and the sand bed resubmerged. The water running through the scraped sand bed should be returned to the reservoir, capturing untreated water for at least 1 day before returning the filter to service. It is recommended that two SSF beds be used so that one can remain in operation while maintenance is conducted on the second one. Once half of the sand is removed through maintenance, the sand bed should be reconstructed with fresh sand to the full 3-foot depth.

Slow sand filters are effective in removing pathogens and other pollutants. The advantage of these systems is the low requirement of additional inputs. Chemical (e.g., chlorine, chlorine dioxide, or ozone) or radiation (e.g., UV) treatment requires chemicals or energy to provide the treatment. While slow sand filters require energy to move water, like other treatment systems do, they have no additional chemical or energy requirements.

Because of the slow flow rates, the tradeoff for these systems is space. A flow rate of 4 to 6 inches per hour through the sand bed is recommended. A sand bed with a surface area of 1 square yard operating at a flow rate of 6 inches per hour equates to about 90 gallons per day per square foot of sand bed area. For example, a 4-by-8-foot septic tank (totaling 32 square feet) used for containing the sand would treat about 2,880 gallons per day. To treat 100,000 gallons per day, two sand beds in 30-foot diameter tanks (1,414 ft^2 total surface area) would be needed.

Slow sand filtration does not remove a significant amount of nutrients. If treated runoff is integrated back into the irri-

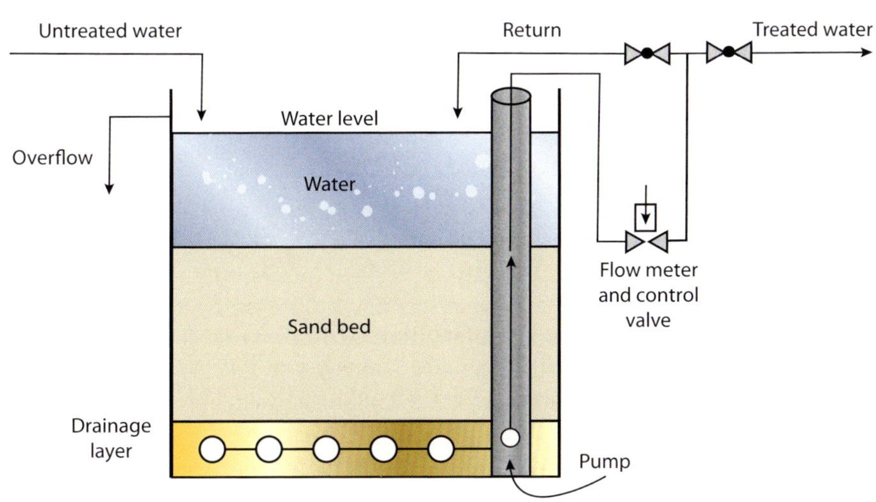

FIGURE 8.6. Slow sand filters consist of a sand bed submerged in water. Incoming untreated water slowly flows through the sand. Treated water is held in a storage tank for later use. Illustration by L. R. Oki.

gation system, the residual nutrients can supplement a plant nutrition program. However, if the treated water is stored for a long period, the nutrients can promote algal growth and become problematic.

Chlorination. Chlorine kills organisms through the oxidation of cell membranes, proteins, and other organic biochemicals. The three methods of incorporating chlorine into irrigation water include sodium hypochlorite, calcium hypochlorite, and chlorine gas. Always check with local agencies regarding the regulations associated with the use and storage of chlorine products.

Effectiveness of treatment depends on several factors, including water cleanliness, concentration of chlorine, pH, and temperature. The dirtier the water, the more chlorine it requires to treat it, because chlorine binds to other mineral and organic particles. As the concentration of chlorine increases, the effectiveness increases along with the rate of disinfection. Most crops will tolerate 100 parts per million chlorine. Certain pathogens require higher chlorine concentrations or longer exposure times for effective treatment.

When using chlorine, you should strive for a residual of 0.5 ppm available chlorine for water not containing ammonia. Water containing ammonium nitrogen will form chloramines, and your total residual combined chlorine should range from 0.5 to 1.0 ppm. Typically the dosage ranges from 4.5 to 8.0 ppm. The residual will always be less than the dosage since much of the chlorine reacts with contaminants in the water. Too much residual chlorine can damage sensitive crops.

Chlorine molecules are most stable, and therefore most effective, at a neutral pH (7.0). Low water temperature (less than 50°F) or high temperature (greater than 68°F) may reduce chlorine effectiveness. The disadvantages of chlorine are that residual chlorine can damage or kill plants, and chlorine does not break down or remove most pesticides or herbicides. Also, brown coloration, due to dissolved organic matter and acids, will not be removed with chlorination. Moreover, the fumes are highly dangerous, leading to a worker/operator exposure concern.

Ultraviolet light (UV). Ultraviolet light, which has a wavelength from 100 to 400 nanometers (nm), may be used to kill pathogens. The mode of action is the production of free radicals that disrupt cell membranes and kill organisms. In order to be effective, the water being treated must be relatively clear and colorless. Nurseries that use hydroponic or soilless systems are candidates for UV treatments. Nurseries using traditional methods of producing plants in soil would need to subject the water to other filtration processes to remove any suspended fractions or discolorations in the water.

Four types of lamps can be used to emit UV: low-pressure mercury lamps, high-pressure mercury lamps, excimer lasers, and xenon flash lamps. Low-pressure mercury lamps emit a wavelength of about 254 nanometers. High-pressure mercury lamps emit a wavelength of 190 nanometers, which also causes the formation of sanitizing ozone in the water. Excimer lasers emit pulses of light at 248 nanometers. Xenon lamps emit light over a larger spectrum, some of which is not UV; therefore, this light source is not as energy efficient.

There are several advantages of using UV for water disinfection. The cost of operation is low if the water source is relatively clean. No chemicals are used in this process, and very few components must be maintained. Finally, even though it kills all organisms in the water, there is no resulting residue that is toxic to plants.

A major disadvantage of using UV is that the water must be fairly clear and clean of debris for effective treatment. If the water source is not clean, a UV exposure time of longer than 30 seconds will be required for complete pathogen kill. Light sources may chemically denature some chelates that are used to keep micronutrients in the soluble form. UV does not remove other chemicals from the water, such as herbicides and pesticides. UV treatment also does not remove discoloration caused by organic acids.

Ozonation. Ozone is an oxidant that, like chlorine, kills organisms by disrupting cell membranes. The ability of any specific chemical to cause oxidation is measured as oxidation reduction potential (ORP). ORP values of 700 millivolts should provide complete disinfection. ORP values less than 300 millivolts are usually considered safe for most aquatic life.

The advantages of ozonation are that no residual chemicals remain after treatment, no chemical storage is required since ozone is manufactured on-site, the system efficiency is inexpensively monitored by measuring ORP values, and ozonation effectively oxidizes most pesticides. A major disadvantage of ozonation is that it requires a fairly clean water source to work properly. Therefore, water with large amounts of organic matter, clay, or other debris requires increasing the ozone exposure time (20 minutes to 1 hour). Other disadvantages of ozonation include an increased cost due to the use of electricity to produce ozone, the risk of some chelated nutrients precipitating out, and an increase in water pH, which may require acidification. Furthermore, ozonation is not completely effective in killing chlamydospores and microsclerotia of some pathogens.

Copper ionization. Copper ionization is the process of adding copper ions to water. Copper electrodes are inserted into the water, and an electric current is passed through the electrode, releasing copper ions into the water. Effective pathogen treatment requires an ion concentration of approximately 50 to 300 parts per billion (ppb). While copper ionization has been used to treat algae in greenhouse coolant pads, there has been limited use for treatment of irrigation water. In hydroponic systems, a copper concentration of approximately 50 parts per billion is recommended, which also contributes to plant nutrition since copper is a necessary plant nutrient. However, a disadvantage is that certain crops are sensitive to the concentration of copper recommended for effective treatment—and in closed recirculating systems, toxic levels may accumulate.

An advantage of copper ionization is the relatively low cost for installation and maintenance of the system. The copper ionization system is also portable, alleviating the need for large holding tanks; there is also no chemical storage necessary with this process. However, like many other disinfectants, the effectiveness of copper ionization is reduced if the water contains excessive amounts of organic matter, clay, or other debris.

Heat disinfection. Heat treatment has also been successfully used by the nursery industry to sterilize soils and growing media. European countries use heat to sterilize water, and this method has also been used in California greenhouses. Viruses are killed at temperatures of 130°F, with an exposure time of 1.5 hours. As the temperature is increased, exposure time decreases. Heat exchangers (for heating water) are located at one point of the water system. Prior to heating, the water pH is reduced to approximately 4.5 to prevent calcium accumulation on the heat columns.

The primary advantage of heat treatment is that the water does not have to be as clean as is required in other chemical treatment processes. The other major benefit is that no chemicals are used, alleviating chemical storage and residue issues.

Some disadvantages of heat treatment include the following: water must be cooled before being applied to plants, heating water adds to the cost, heating duration may be as long as 1 hour, calcium removal from heat exchangers needs periodic maintenance, additional space is required for tanks to hold treated water, and high heat denatures chelates that are used for micronutrients.

Fertilizer Injection

Recycling runoff involves a semiclosed system that captures and reapplies container and field leachate. Recycling, consequently, conserves water and nutrients. If your water costs are low, the value of the recycled nutrients may exceed the value of the water.

In starting a recycling system, it is wise to have a laboratory analyze your source irrigation water and runoff water for all nutrients so that you have an estimate of the concentration of nutrients being recycled. Samples should be taken throughout the year to adequately measure water quality changes over time. Runoff water is commonly higher than irrigation source water in all nutrients, including nitrogen, phosphorus, potassium, calcium, magnesium, iron, manganese, zinc, copper, and boron.

In a constant fertilization system, where the irrigation water is fortified with nutrients, the nitrogen and potassium concentrations may closely approximate the concentration of these elements in the applied water. It is usually better to mix the phosphorus into the growing media rather than adding it to the water, because many phosphorus fertilizers have low solubility. However, supplemental phosphorus in irrigation may be required for long-term crops (greater than 5-month crop cycle [see chapter 6]). Because runoff water contains a higher concentration of trace elements than the applied water, these do not have to be added to the fortified water or to the growing media unless a large percentage of purified water is added.

Salt Management and Blending

Runoff water will inevitably be higher in salts than the applied irrigation water because it contains fertilizers that leach from containers. Additionally, runoff water is constantly dissolving substances as it runs over soil surfaces. This salt concentration varies by season, amount of water applied, and amount of fertilizer injected into the irrigation water. During the summer months, salinity is highest as a result of evaporation and a greater ability of warm water to dissolve various minerals. The salinity can vary from a 7 percent increase to as much as a 24 percent increase in salinity for each irrigation cycle (Skimina and Kabashima 1994). Recaptured runoff should be blended with fresh water before reuse to reduce the salt concentration to levels that will not damage plants.

Anyone contemplating recycling runoff water should invest in an electrical conductivity meter, which will quickly indicate the salinity level of your water. You should not attempt recycling until you have such a meter and understand the influence of higher salt levels on plant growth. Salinity should be adjusted when blending recaptured runoff with fresh water to ensure that the electrical conductance of the water does not exceed levels that can be tolerated by the plant material. Many plants can tolerate soil salinity levels of up to 3.2 $dS \cdot m^{-1}$ if adequate moisture levels are maintained.

As mentioned earlier, reverse osmosis (RO) can be used to remove salts before the water is injected with fertilizer. RO is expensive, and RO-treated water has a tendency to corrode metals. Despite this drawback, RO-treated water can be used for limited back-blending.

Detection and Removal of Herbicides in Recycled Water

The use of preemergence herbicides to control weeds in containers, as well as in the soil underneath and between containers, is a common practice in container nursery production (see chapter 15). Preemergence herbicides that are currently available not only do a very good job of controlling weeds, but they are also much less expensive than the cost of hand-weeding. Postemergence herbicides are also frequently applied. Most herbicides are moderately hydrophobic and are effectively retained (adsorbed) by the planting media, although some amount of herbicide may be expected to leach through containers. In addition, most herbicides are readily degraded in the retention pond through chemical, photochemical, and microbial transformations. These factors, coupled with the fact that herbicide residues entering a retention pond are highly diluted, suggest that it is uncommon for the accumulation of herbicides in retention ponds to reach a level that can injure crops. However, certain practices could lead to levels of herbicides in recycled water that could damage herbicide-sensitive plant species, and these should be avoided.

The herbicide concentrations in the runoff water will depend on its solubility in the water, affinity to be adsorbed on the colloidal fraction, biodegradation rate, chemical degradation rate, volatility, and degree of UV degradation. If there is a concern about herbicide concentrations in recycled water, these concentrations can be measured analytically, which is highly accurate but expensive.

Alternatively, bioassays can be used, which are not as accurate but are simple, versatile, and inexpensive. For example, bentgrass is an especially sensitive bioassay for detection of oxadiazon (Ronstar) and oxyfluorfen (Goal, Rout, OH II), barnyardgrass is very sensitive to oryzalin (Surflan), and sorghum is very sensitive to oxadiazon (Skimina and Kabashima 1994). The effects of the various herbicides on the chosen indicator plants can be assessed in several ways. Sublethal herbicide concentrations frequently produce an inhibition of radicle or shoot elongation in seed germination bioassays. Furthermore, many herbicides produce characteristic symptoms in indicator plants that are dose related, such as malformation of leaves, necrosis, chlorosis, and stem deformations. For current information on the use of bioassays, contact the manufacturer of the particular herbicide you are testing, your state nursery association, Cooperative Extension service, or university weed scientist.

Research conducted at a nursery in southern California demonstrated that a substantial amount of the lower-solubility preemergence herbicides applied to land and in spiked simulated runoff was removed following runoff treatment. Conventional water treatment was used, involving flocculation, settling, and filtration. Bioassays conducted in these experiments showed the most phytotoxicity with processed water from the higher-solubility herbicides. Spiked water tests using glyphosate (Roundup) and oxyfluorfen (Goal) indicate that there could be some suppression of growth in herbicide-sensitive plant species at approximately 100 parts per billion (ppb) glyphosate and 2.5 parts per billion oxyfluorfen (Skimina 2001).

Although the levels of herbicides in runoff water in commercial nurseries would likely be less than those produced in these experimental conditions designed to model worst-case scenarios, it is recommended that growers take the following precautions: treat runoff with flocculants before filtration, or allow 2 to 5 days settling in a quiescent pond before recycling, and blend the runoff with at least 50 percent fresh water to dilute herbicide concentrations. Herbicide use should be limited to those herbicides having a low water solubility (less than 3.5 ppm) and strong adsorption characteristics (table 8.1). Highly soluble herbicides that are strongly adsorbed or fixed by clays, such as glyphosate, can be used safely if limited spot treatment is applied and the suspended clays are allowed to settle before recycling. Applications of oxyfluorfen should only be used on dormant stock showing no budbreak activity (Skimina 1992, 2001).

Other precautions that should be observed to control the amount of herbicides in a water recycling system include

- treating the growing media to kill existing weed seeds before you use it
- rotating classes of herbicides from the same mode of action
- using herbicides only when necessary, such as on newly planted container stock that has a high percentage of exposed soil surface, since weed germination is dramatically suppressed when the plant canopy completely covers the soil surface
- activating the herbicide with only enough water to facilitate herbicide adsorption onto the soil
- spot treating existing weeds or areas, if possible, and staggering the applications so that the entire field is not treated at the same time (Skimina 2001; Skimina and Kabashima 1994)

A side benefit to recycling water that contains herbicides is weed suppression of sensitive weeds species such as *Poa annua*. Another unexpected benefit is that at very low levels some herbicides may stimulate plants or act as a growth regulator on sensitive plants, resulting in an enhanced appearance (Skimina 1992, 2001; Skimina and Kabashima 1994).

Table 8.1. Characteristics of commonly used and experimental herbicides for ornamental plants in container production nurseries

Common name	Product name	Oral LD_{50} mg/kg	Water soluble ppm	Toxicity fish	Soil adsorption	Soil half-life
metolachlor	Pennant	2,780	492	2–15 ppm	weak	15–50 days
napropamide	Devrinol	5,000	74	> 10 ppm	medium	8–12 weeks
oryzalin	Surflan	10,000	2.5	low	medium to strong	60–90 days
oxyfluorfen	Goal	5,000	0.1	yes	strong	30–40 days
oxadiazon	Ronstar	8,000	0.7	> 2 ppm	strong	60 days
pendimethalin	Stomp	3,277	0.5	yes	strong	42–101 days
fluazifop	Fusilade	3,000	2.0	moderate	medium to strong	< 60 days
sethoxydim	Poast	2,676	4.8	low 2,676	weak	4–11 days
glyphosate	Roundup	4,300	6,380	86–120 mg/L	strong	nil
diquat	Diquat	230	677,000	low	strong	persistent
dithiopyr	Dimension	> 5,000	1.3	0.48–0.72 mg/L	strong	17 days
isoxaben	Gallery	> 10,000	1.0	> 1.1 mg/L	medium	90–150 days
cacodylic acid	Montar	700	667,000	low	strong	nil

Sources: California Department of Pesticide Regulation 2004; Ahrens 1994; Meister 1990.

Sludge Waste Handling

Water treatment processes that remove suspended matter result in a sludge that requires disposal. Tests conducted by a southern California nursery indicate that flocculated clay sludge may be recycled into container mixes. Incorporation rates as great as 8 percent pose no phytotoxicity problems. If the sludge has a high clay content, however, avoid adding more than 5 percent of it to the growing medium. The sludge should be mixed with an organic amendment such as compost and aged for at least 3 months under warm conditions to stimulate the biological breakdown of any residual herbicides that may be present. Herbicides tend to adsorb on the clay fraction and settle with the clay (Skimina 1992; Skimina and Kabashima 1994).

Cost Considerations

Capital costs

An initial capital outlay is needed for building the necessary structures to collect, maintain, and treat runoff water. Because the pumping equipment and the structures have a limited useful life, the recovery of these expenses can be depreciated. A study of six nurseries in southern California found median capital costs for recycling systems were $8,000 per acre, with a range of $3,680 to $17,200 (Mangiafico et al. 2008). Because costs are directly related to production size, larger nurseries may benefit from positive economies of scale in the installation of recycling systems.

Operating costs

Maintenance and energy expenses and cost of chemicals need to be considered in the cost analysis. Because a water recycling system may use pumps, equipment for pathogen control, pH meters, conductivity meters, handling pits, and reservoirs, the required maintenance would include servicing, cleaning, waste handling, and control. The chemical costs to process runoff water will be determined by the dosages of chemicals (such as chlorine disinfectant, alum, or polyacrylamide) that are required to do an adequate job of treatment, which will be greater for dirtier water and higher volumes of water. Costs will also be determined by the prices of these chemicals, which vary by locality. To get a rough estimate of costs to treat water, find out what it costs your local sewage treatment plant to process water through secondary treatment. It will probably cost you a little less than that. In Los Angeles County in 1991 to 1992, the average cost to treat sewage water was $435 per million gallons (Skimina 1986, 1992).

Costs saved and payback

In determining the costs of the recycling system, it is important to include the costs saved. The first obvious benefit is water conservation. This depends on the slope of the land and the impermeability of the runoff soil surface, which typically is at least 50 percent (Skimina and Kabashima 1994). A 50 percent recovery of the applied water equals a 50 percent savings in the cost of the water minus the cost to process it. In considering the cost economics of recycling water, be sure to include the value of the recycled nutrients. If a nursery is on a constant feed system of 200 parts per million nitrogen and 100 parts per million potassium, the value of the fertilizer in recycled water can exceed $1,000 per million gallons of water (Skimina 1992).

The percent volume of the applied water that runs off and is recycled will determine the amount of nitrogen and potassium that is saved. If the system uses constant fertilization, in which the majority of nitrogen and potassium are supplied in the fortified irrigation water, total nitrogen concentration of the runoff will be almost equal to that of the applied water. Tests also indicate that only about 2 percent of the potassium is lost in the recycled water. If ammonium nitrate is used as the nitrogen source (50% ammonium nitrogen and 50% nitrate nitrogen), the runoff will have a decreased ammonium concentration and an increased nitrate nitrogen concentration. This is due to the conversion of ammonium nitrogen to nitrate nitrogen by the nitrification bacteria. The pH will generally drop a fraction of a unit due to this conversion (Skimina and Kabashima 1994).

In systems using slow-release fertilizers rather than fortified water for the source of nitrogen and potassium, the concentrations of these elements will be less in the runoff due to dilution of the leachate from the container or the soil with the applied fresh or recycled blended water. Most nutrients increase in concentration at each cycle with the exception of copper, which does not change (Skimina and Kabashima 1994).

In addition to compliance with government regulations, the decision to build water treatment and recycling systems should be dependent on the return on the benefits of recycling. Payback time for the system can be determined based on the capital cost of installing the recycling system, as shown in the following equation (Skimina 1992):

capital cost of installing the recycling system ÷
[(annual savings in nutrients and water costs per million gallons – annual operating costs per million gallons) × (number of million gallons treated per year)]
= payback time in years

As mentioned earlier, the capital costs should include depreciation of equipment. Loan costs should also be factored in to more precisely estimate payback time.

Public Relations and Safety Issues

A water recycling system will result in benefits beyond water and nutrient savings for a nursery. Nurseries located near impaired water bodies or in areas where water is scarce during a drought can demonstrate to water districts, the media, or government officials that they are doing their part to save water and reduce pollution through their recycling efforts.

Negative public relations can result, however, if neighbors view the reservoir as an eyesore or if noise from the pumps and filters is audible. To prevent this from becoming a problem, growers should establish landscape screens to conceal the reservoir and install appropriate sound barriers to prevent or reduce noise pollution.

Reservoirs also pose a safety hazard to workers and animals. Safety measures should include, but not be limited to, the following: a safety fence to exclude unauthorized entry, a small boat, flotation vests that must be worn when working on the reservoir, and a safety line that can be used to exit the reservoir.

Minimizing Nursery Waste: Composting and Waste Recycling

The greenhouse and nursery industry generates a significant amount of waste that must be managed to protect water quality. Trash and debris that is blown off the nursery by wind or carried off-site by irrigation or stormwater may clog drains, litter waterways and beaches, and endanger marine life. In fact, trash TMDLs exist for agriculture that impact nursery growers in some areas of California. Furthermore, agrochemicals, petroleum products, paint, batteries, and other potentially hazardous materials used in nurseries may pollute the environment if not disposed of properly. Sewage and human waste from portable toilets could pose serious human and animal health hazards from fecal coliform bacteria if it enters municipal stormwater or sewer systems.

Organic waste generation from nurseries, which includes yard waste from plant prunings, weeds, and discarded plants, varies with the type of crop grown and the turnover rate. One study found that a container nursery can generate 60 cubic yards of waste per acre per year (Skimina 1992). Much of the organic waste produced by nurseries ends up in landfills. Many states have enacted regulations that require cities to recycle their wastes, especially yard waste, which may comprise as much as 40 percent of the waste volume sent to landfills.

Most nurseries use large amounts of plastic, which include pots, flats, hanging baskets, greenhouse film, drip irrigation components, plant labels, and containers for pesticides and other agrochemicals. The extensive use of plastic has resulted in a significant waste disposal problem. In California, the amount of agricultural plastic disposed of on a yearly basis by the greenhouse and nursery industry was calculated to be over 19,648 tons (Hurley 2008). Trays and flats accounted for the largest percentage of plastic disposed of on a yearly basis.

Growers should adopt management practices to reduce and recycle waste for several reasons. First, resolving the nursery industry's waste problems is an environmentally sound response to the larger societal concern. Second, nursery growers who reduce waste and recycle are perceived to be good stewards of the land, and they help to promote the sustainability of the entire nursery industry. Further, waste disposal costs money. In California, for example, nursery growers spent an annual average of $594 in tipping fees and $559 per ton in hauling costs to transport agricultural plastic to a landfill (Hurley 2008). These disposal costs can be reduced by implementing an effective waste reduction and recycling program.

Waste Management Practices to Protect Water Resources

To prevent nursery waste from moving off-site, where it can contribute to pollution, it is important to have an adequate number of waste containers available where needed and to service them regularly to avoid overflow. Nursery trash containers should be maintained in good condition and kept closed. Waste containers, collection areas, storage areas, and stockpile areas should be located indoors or covered when outdoors. If water bodies are located on-site or near the property, extra care should be taken so that containers and other plastic products do not end up contributing to pollution downstream from the property with debris and litter.

Employees should be trained to recycle nursery waste such as metal, oil, paper, cardboard, and plastic. Nursery employees should also be trained to properly dispose of batteries, paints, agrochemicals, petroleum products, and other potentially hazardous materials used in the nursery.

Toilets, sinks, and floor drains in restrooms should be properly hooked up to the sanitary sewer system. Restrooms and portable toilets should be regularly maintained to prevent sewage and human waste leakage. Portable toilets should be located where wet weather will not wash waste into a municipal stormwater system.

Recycling Plant Waste

Organic waste material can be ground and composted, reducing the volume by 20 to 72 percent depending on the type of waste (Skimina 1992). A national survey showed that 64 percent of greenhouse and nursery growers compost plant waste (Dennis et al. 2010). A nursery that grinds and composts its waste for reuse in growing media saves money by avoiding the costs of disposal and the need to purchase organic matter for media. There may also be tax credit incentives for the cost of equipment such as tub grinders and chippers needed to grind green waste materials. The green waste, once it has been ground, may be recycled in community-based composting programs or an in-house composting program. Allelopathic chemicals in raw plant waste used as a source of organic matter have not been shown to be a problem (Skimina 1992). However, culled plants and used growing media awaiting recycling can be a major source of insect pests and pathogens. Cover green waste, store debris well away from crops, or promptly compost crop residue to avoid this problem.

Recycling Plastic

Opportunities for recycling horticultural plastics have increased in recent years. This is due to high oil prices, which have resulted in increased prices for recycled plastics, as well as to growing interest from consumers and retailers in keeping plastic out of landfills. The number of community curbside programs for plastic collection is increasing, and many programs exist in garden centers for recycling containers. Plastic recycling companies, including some suppliers of recycled horticultural containers, are developing and improving methods for collecting, sorting, cleaning, grinding, and selling horticultural plastics. Some states, such as New Jersey, have developed programs to facilitate the collection of horticultural plastics. Moreover, big-box retailers, garden centers, and progressive growers have initiated programs to collect and recycle plastic containers and trays from gardeners.

One of the most successful recycling programs in the

United States was developed at the Missouri Botanical Garden. Since the inception of its Plastic Pot Recycling program in 1998, the Missouri Botanical Garden has processed over 980,000 pounds of plastic pots, cell packs, and trays. The garden serves as a central collection site, and retail garden centers throughout the area participate as satellite collection sites. Also contributing to the collection are horticultural businesses, which include nursery growers. The collected plastic is transported to a single location for grinding into small chips that are easily transported for recycling. The marketed plastic chips are eventually made into plastic lumber and other recycled products.

Another successful program is the Ag Container Recycling Council (ACRC). This nonprofit organization safely collects and recycles plastic crop-protection product containers, including pesticide, fertilizer, and adjuvant product containers. Thousands of farmers and applicators nationwide participate in this free recycling program.

Nurseries that have in-house collection programs typically do not receive much money for their used plastic, although they do save on disposal fees. Some recyclers do not pay growers for the plastic but do not charge for transportation costs, which are often high. Despite not yielding large profits, the practice of recycling horticultural plastic is increasing among nursery growers and their retail customers. A nationwide survey reported that over 69 percent of the greenhouse and nursery crop growers recycled plastic containers. Although less than one-fourth of surveyed growers used recycled plastic, those that did reported that recycled containers accounted for just under half of all container types used (Dennis et al. 2010). In a California survey, 46 percent of the nursery producers reported that they recycle some portion of their agricultural plastic (Hurley 2008).

Types of plastic

Several different types of plastic are used in the horticultural industry. It is important for nursery growers to know what types of plastics they use, because these need to be sorted carefully and handled separately to maintain the highest quality in the recycled product. Moreover, some recyclers only take specific types of plastics.

Low-density polyethylene (LDPE). This is a relatively inexpensive film plastic used as a covering for greenhouses and hoophouses. It is also used as a mulch material for food and ornamental crops as well as plastic bags.

High-density polyethylene (HDPE). This is a flexible plastic that is usually black. It is commonly used in blow-molded containers for outdoor nursery stock from 1 to 65 gallons. HDPE resists breakage and degrades slowly from UV radiation. It is extensively used for containing pesticides and herbicides because it is thermally and chemically resistant. HDPE is also used in other screw-top containers and some types of irrigation pipes. Outside of agriculture, the most important use for HDPE plastic is for milk containers. HDPE is the most commonly recycled plastic in the United States. Recycled HDPE is typically made into plastic timber and other durable plastic products.

Polypropylene (PP). This is commonly used in the nursery industry in decorative containers, injection-molded flats and plug trays, and hanging baskets. It is generally durable, stiff but flexible, lightweight, and resists breakage. Many PP products are shiny and tear resistant. Spunbonded polypropylene (or woven polypropylene) is used as row covers for frost protection. Virgin PP is white, whereas recycled PP is always darker, or even black, because of contaminants.

High-impact polystyrene (HIPS). This material is inexpensive, extremely lightweight, and more rigid than polypropylene. It is commonly used in containers for seedlings and small plants, and it includes thermoformed bedding-plant packs and flats, transport trays, plug trays, square pot trays, pots, specialty packs, and propagation sheets. Pure solid polystyrene is colorless, but plug trays are generally black because they are made from recycled plastic. HIPS is relatively easy to recycle in many parts of the United States. Some plastic recycling companies grind, melt, and mold HIPS directly back into trays, which eliminates a step in the recycling process and enhances the financial return. HIPS trays and liners are increasingly made from recycled plastic, and some producers are displaying this information directly on the trays. Expanded polystyrene (Styrofoam) has been used for many years in the production of small plant containers.

Barriers to recycling plastic

Plastic types must be sorted and stored separately. Mixed types may be rejected by the plastic buyer or a sorting fee may be charged, because most plastics do not mix when they are melted during recycling. PP is one of the more difficult horticultural plastics to recycle, though some recyclers mix it with HDPE to make landscape timber. Styrofoam is generally difficult to recycle; it has a reputation for accumulating in the environment (Cameron 2009). When used as a greenhouse cover, LDPE films break down from UV radiation, which limits their usefulness when recycled. In fact, a nationwide survey found that only 30 percent of greenhouse and nursery crop growers recycle plastic greenhouse coverings (Dennis et al. 2010).

Transporting horticultural plastic to recycling facilities can pose problems. According to California nursery growers, their greatest difficulty with recycling is that the current recycling facilities are too far from their operation (Hurley 2008). Many recyclers require a minimum volume of plastic before picking it up, which necessitates a large storage area. Plastic containers are bulky, and cost-efficient transport may require baling or grinding equipment to compact used containers. Further, unless LDPE is baled, it is difficult to transport due to its very low bulk density. In some cases, recycling companies will temporarily provide the baling equipment for growers to use while they are removing greenhouse coverings (Cameron 2009).

Pots, flats, LDPE films, and other horticultural plastics are typically contaminated with soil and other organic materials, which reduces the quality of recycled plastic compared with virgin material. Every time plastic is recycled, contaminants and structural degradation reduces the quality of the plastic, making it more difficult to remanufacture the same product. Therefore, some recyclers will not take plastic unless it is washed to remove soil (Cameron 2009).

Tips for recycling plastic

- Know your plastics, and keep them separate. If you do not know them, ask your supplier. Most plastic has a code stamped on it (e.g., #6, keynotes polystyrene; #5, polypropylene; and #2, high-density polyethylene). Each type of plastic should be segregated.
- Identify potential plastic buyers that are close to your operation or pick up in your area, either on-site or at local collection points. Your Cooperative Extension office or local nursery association may be able to help you. Match plastic buyers with the types of plastic that you use. A list of horticultural recycling programs and companies is available in the online *Grower Talks Magazine* at the Ball Publishing website. A searchable national directory of plastic buyers and suppliers is available at the Plasticsmarket.org website. In California, a directory of recycling facilities is available at the CalRecycle website. And a list of contractor service areas, instructions for preparing containers for recycling, and other information concerning recycling plastic crop-protection product containers is available at the ACRC (Ag Container Recycling Council) website.
- Contact the recycler to find out the price you can expect to receive for used plastic (if any). Also ask for the collection and bundling guidelines that are recommended or required for acceptance. Plastic film from greenhouses and mulch may need to be cut, rolled, or folded. For hard plastic containers, remove metal hangers and rings and then nest like containers together. Tie plastic to be recycled onto pallets, compress with baler, or shrink wrap. Plastic crop-protection product containers, once emptied, must be triple rinsed or pressure rinsed to remove all standing residue.
- Plastic needs to be as clean as possible. Excess soil, rocks, and other debris must be removed, and film plastic must not be muddy. Store collected plastic on clean surfaces away from dusty or dirty areas. Keep bales off the floor when moving them so that they do not pick up dirt and contaminants.
- When storing plastic on-site, maintain conditions that are safe and prolong the life of the product. If storage must be outdoors, plastic should be kept dry with minimum exposure to the sun, which can cause UV degradation. Safe stacking height should be based on integrity of the bales, ceiling height, bale dimensions, and fire codes or building regulations. When stacking against a wall, the bottom bale should be about 6 inches out from the wall; otherwise, subsequent bales will lean away from the wall instead of into it.
- Volume makes a recycling program a realistic option for growers, and it is the key to success. Some growers have found that working with their retail customers helps to achieve the critical mass of plastic required: retailers collect empty pots and trays for recycling from customers, and the collected scrap is sent back to the nursery production facility. Smaller nurseries that have trouble storing a large amount of plastic for recycling may be able to combine waste with other growers in the community. When multiple growers consolidate their plastic material at a central location for pickup, it is important that they mark each individual skid with their business contact information to receive credit.

Other Solutions for Reducing Plastic Waste

Reusing plastic containers

Spent plant containers and trays, either generated from the plant production cycle or returned from customers, can be sanitized and reused. However, issues related to sorting, cleaning, and disease control will need to be addressed (Cameron 2009). HDPE and PP containers are the most common types that are reused. HIPS trays and liners are sometimes reused, but more often they are discarded or recycled. Unless they are all of the same design, they do not stack or nest. The trays may be clean, but the inserts are typically dirty.

Debris, soil, and plant material that cling to used containers may contain pathogens. *Thielaviopsis* black root rot spores can persist in containers and infest subsequent crops of susceptible species. Heat treatment is effective in killing plant pathogens that adhere to containers or exist in the debris. Where steam is not available, hot water or solarization can be very effective. Most plastic can be treated with hot water at temperatures that cause minimal softening. The minimum water temperature should be 140°F whenever possible. Treatment time can be as short as 1 minute. Longer treatment times are more reliable, and the containers must reach at least 140°F. For solarization, containers should be moistened, stacked, and placed beneath a double-layer tent. Incubate them for 30 minutes at or above 158°F or 1 hour at or above 140°F. Sodium hypochlorite (the active ingredient in bleach) is effective in killing some types of fungal spores and bacteria—but it is effective only as a surface disinfectant, so containers must be free of soil and plant material and clean before treatment. Quaternary ammonium compounds are excellent bactericides and viricides, and they are effective in killing some kinds of fungal spores. However, they penetrate plant material and soil poorly, so containers must be clean before treatment.

Biodegradable containers

Some container manufacturers in the horticulture industry are working to reduce waste by developing biodegradable pots. Biodegradable plastics use an oxo-degradable film additive that helps the product to degrade. Biocontainers are containers that are not comprised of petroleum-based plastics, so they degrade rapidly when planted into the field or when placed in a composting operation.

Two types of biocontainers are available. One type is left intact with the plant root ball when planted so that the roots grow through the container walls, and the container eventually decomposes. The other type is a compostable biocontainer that is removed from the plants before the planting, and the container is composted. A major concern of growers is the ability of biocontainers to physically withstand production and shipping conditions prior to reaching their final destination. In fact, a national survey of greenhouse and nursery crop growers found that only 14 percent used biodegradable plant containers, and only 12 percent planned to implement them in the next 1 to 3 years (Dennis et al. 2010). In two studies that were conducted to evaluate biocontainers (Taylor et al. 2010), container strength and degradation varied significantly among the different types of biocontainers tested. Growers who decide to use biodegradable containers will need to choose a container that best fits their production techniques and the needs of end users.

Long-lasting plastic products

Another solution to the plastic waste disposal problem is to offer or use horticultural plastics that are designed to last for periods longer than just 1 year. For example, one way to reduce the volume of LDPE plastic waste is to use multiyear UV-stabilized films for greenhouse covers. Containers are also commercially available with removable liners that allow for continued use of the product. One polypropylene hanging basket currently available is guaranteed to last consumers 10 years.

References

ACRC (Ag Container Recycling Council) website, http://www.acrecycle.org.

Ahrens, W. H. 1994. Herbicide handbook. 7th ed. Champaign, IL: Weed Science Society of America.

Ali-Shtayeh, M. S., J. D. MacDonald, and J. Kabashima. 1991. A method for using commercial ELISA tests to detect zoospores of Phytophthora and Pythium species in irrigation water. Plant Disease 75(3): 305–311.

Arent, G. L. 1996. The greenhouse wastestream. HortTechnology 6(4): 365–366.

Ball Publishing. Grower talks magazine. Ball Publishing website, http://www.ballpublishing.com/growertalks/ViewArticle.aspx?articleid=16578.

California Department of Pesticide Regulation (DPR). 2004. About DPR and its programs: A better way to protect ground water. Consumer Fact Sheet. Sacramento: CDPR. CDPR website, http://www.cdpr.ca.gov/docs/emon/grndwtr/factsheet.pdf.

California State Water Resources Control Board website, http://www.swrcb.ca.gov/.

CalRecycle website, http://www.calrecycle.ca.gov/Plastics/Recyclers/RecyclerList.aspx.

Calvo-Bado, L. A., T. R. Pettitt, N. Parsons, G. M. Petch, J. A. W. Morgan, and J. M. Whipps. 2003. Spatial and temporal analysis of the microbial community in slow sand filters used for treating horticultural irrigation water. Applied and Environmental Microbiology 69:2116–2125.

Cameron, A. 2009. Horticultural plastic recycling: The future is brighter. Greenhouse Product News 21(10): 20, 23–24.

Dennis, J. H., R. G. Lopez, B. K. Behe, C. R. Hall, C. Yue, and B. L. Campbell. 2010. Sustainable production practices adopted by greenhouse and nursery plant growers. HortScience 45(8): 1232–1237.

Dhingra, O. D., and J. B. Sinclair. 1985. Basic plant pathology methods. Boca Raton, FL: CRC Press.

Diver, S., L. Greer, and K. L. Adam. 2008. Sustainable small-scale nursery production. National Sustainable Agriculture Information Service (ATTRA) website, https://attra.ncat.org/publication.html.

Gan, J., D. Haver, and J. Newman. 2008. Pesticide use and water quality protection. In J. Newman, ed., Greenhouse and nursery management practices to protect water quality. Oakland: University of California Agriculture and Natural Resources Publication 3508. 73–82.

Hayes, B. 2010. Degradable pots could help resolve disposal issues. Greenhouse Management 30(1): 25–27.

Horowitz, M. 1976. Application of bioassay techniques to herbicide investigations. Weed Research 16:209–215.

Hurley, S. 2008. Postconsumer agricultural plastic report. Publication #IWMB-2008-019. Sacramento: California Integrated Waste Management Board. CalRecycle website, http://www.calrecycle.ca.gov/Publications/Documents/Plastics%5C2008019.pdf.

Jones, A., T. Harter, M. Bianchi, and J. Harper. 2003. Water pollution legislation. Oakland: University of California Agriculture and Natural Resources Publication 8088. ANR CS website, http://anrcatalog.ucdavis.edu/FarmWaterQualityPlanning/8088.aspx.

Joubert, E. D., and B. Pillay. 2008. Visualisation of the microbial colonisation of a slow sand filter using an environmental scanning electron microscope. Electronic Journal of Biotechnology 11(2): 119–125. Electronic Journal of Biotechnology website, http://www.ejbiotechnology.info/index.php/ejbiotechnology/rt/printerFriendly/v11n2-12/49.

Kabashima, J. N. 1993a. Innovative irrigation techniques in nursery production to reduce water usage. HortScience 28(4): 291–293.

———. 1993b. Watering and fertilizing practices to avoid groundwater contamination and surface runoff. Proceedings of the 9th Conference on Insect and Disease Management on Ornamentals: 60–72.

Kabashima, J. N., S. J. Lee, D. L. Haver, K. S. Goh, L. S. Wu, and J. Gan. 2004. Pesticide runoff and mitigation at a commercial nursery site. In J. J. Gan, P. C. Zhu, S. D. Aust, and A. T. Lemley, eds., Pesticide decontamination and detoxification. Washington, DC: American Chemical Society (ACS) Symposium Series 863. 213–230.

Kabashima, J. N., J. MacDonald, S. Dreistadt, and D. Ullman. 1997. Easy on-site tests for fungi and viruses in nurseries and greenhouses. Oakland: University of California Division of Agriculture and Natural Resources Publication 8002. ANR CS website, http://anrcatalog.ucdavis.edu/Floriculture/8002.aspx.

Lu, J., L. Wu, and B. Faber. 2008. Erosion control and runoff management. In J. Newman, ed., Greenhouse and nursery management practices to protect water quality. Oakland: University of California Agriculture and Natural Resources Publication 3508. 83–96.

Mangiafico, S. S., J. Newman, D. J. Merhaut, J. Gan, L. Wu, J. Lu, B. Faber, and R. Evans. 2008. Detention and recycling basins for managing nutrient and pesticide runoff from nurseries. HortScience 43:393–398.

Meister, R. T., ed. 1990. Farm chemicals handbook. Willoughby, OH: Meister Publishing.

Newman, J. 2009. Manage runoff with vegetated buffers. Greenhouse Management and Production 29(10): 32–33.

Newman, J., D. Haver, V. Mellano, and K. Robb. 2008. Management goals and management practices to mitigate runoff and leaching. In J. Newman, ed., Greenhouse and nursery management practices to protect water quality. Oakland: University of California Agriculture and Natural Resources Publication 3508. 11–28.

Newman, J., V. Mellano, K. Robb, and D. Haver. 2008. Conducting an environmental audit. In J. Newman, ed., Greenhouse and nursery management practices to protect water quality. Oakland: University of California Agriculture and Natural Resources Publication 3508. 115–135.

Plasticsmarket.org website, http://www.plasticsmarkets.org.

Skimina, C. A. 1986. Recycling irrigation runoff on container ornamentals. HortScience 21(1): 32–34.

———. 1992. Recycling water, nutrients, and waste in the nursery industry. HortScience 27(9): 968–971.

———. 2001. Herbicides in recycled water. Proceedings of the 2001 California Weed Science Society (53):70.

Skimina, C. A., and J. N. Kabashima. 1994. Water recycling in container crop production. In C. W. Heuser and P. E. Heuser, eds., Recycling and resource conservation practices. Harrisburg: Pennsylvania Nurserymen's Association. 13–19.

Taylor, M., M. Evans, and J. Kuehny. 2010. Comparing strength and biodegradability of biocontainers. Greenhouse Management 30(9): 14, 16–17.

UC IPM (University of California Statewide Integrated Pest Management Project). UC integrated pest management guidelines. UC IPM website, http://www.ipm.ucdavis.edu/index.html.

Wohanka, W. 1995. Disinfestation of recirculating nutrient solutions by slow sand filtration. Acta Horticulturae 382:246–255.

Photo: D. L. Haver.

Plant Propagation

David W. Burger

CHAPTER NINE

Photo: D. W. Burger.

Propagation is the reproduction of new plants from existing plants or plant parts by sexual or asexual means. The two fundamental objectives of plant propagation are to increase the number of a particular plant and to perpetuate the essential character of the plant. Retaining a plant's valuable characteristics and multiplying it efficiently on a large scale are critical to the success of the nursery industry. Accomplishing both objectives requires knowledge of the science of plant propagation as well as practical experience in the art of performing propagation techniques.

When a plant's flower or inflorescence is involved as an essential organ in plant increase and an exchange of genetic material occurs (e.g., sperm, egg), propagation is sexual. Sexual propagation of plants in the nursery industry focuses primarily on reproduction via seeds (seed germination, seedling growth), but it also takes advantage of newer in vitro tissue culture techniques using reproductive structures, such as pollen, ovules, seeds, or spores.

When exchange of genetic material does not occur during plant multiplication, propagation is asexual, or vegetative. Depending on the plant species, asexual propagation techniques may include the following:

- production of bulbs and other modified stems and roots
- production of plants growing on specially selected rootstocks by grafting or budding
- production of plants growing on their own roots, resulting from layering or cuttings
- production of entire plants from vegetative plant cells, tissues, or organs by tissue culture

This chapter covers both sexual and vegetative propagation, and it compares the two types of propagation methods. Basic concepts of seed biology and methods used for handling and sowing seeds will be examined. Vegetative propagation techniques presented will include cuttings; budding and grafting; separation, division, and layering; and tissue culture.

Sexual Propagation

Sexual propagation is the primary method by which most plants naturally reproduce, and it is one of the most widely used propagation methods for agricultural crops. Sexual propagation is often referred to as seed propagation because the sexual union of egg and sperm results in the production of seed. Seed propagation is a common means of propagating self-pollinated plants, and it is often used for those that are cross-pollinated. The initial step in seed propagation is the sowing of the seed, but the seed itself is the result of a long process of growth and development that usually involves pollination, fertilization, zygote formation, and embryo development.

Seed Development

The sexual cycle includes the development of male (pollen) and female (ovule, ovary) reproductive structures. Upon flowering, pollen grains are transferred from mature anthers to the stigma, where they germinate if suitable conditions exist (related plant, receptive stigma, etc.). After germination of the pollen grain, a tube is formed (pollen tube) that penetrates the style and carries the male gametes into the ovule (fig. 9.1A). The generative cell of the pollen grain divides to form two sperm. The sperm move through the pollen tube down the style toward the embryo sac housed in the ovule and join with egg nuclei contained within (fig. 9.1B). Fertilization occurs when one of the generative sperm nuclei (male gametes) fuses with the female egg cell, creating a zygote that combines the genetic material of the male and female parents. The zygote (fig. 9.1C) ultimately develops into the embryo of the seed (embryogenesis). The other generative nucleus fuses with the two polar nuclei in the embryo sac, resulting in a triploid endosperm, which is a tissue used as a nutritive (food) source during germination (fig. 9.1D and 9.1E). The nucellus (a diploid, maternal tissue) is sometimes also associated with the endosperm as a storage tissue. The integuments surrounding the embryo sac become the seed coat (testa). The embryo of the seed consists of a shoot apical meristem (plumule), an embryonic axis, at least one cotyledon (seed leaf) attached to the axis, and a root apical meristem (radicle) (fig. 9.1E).

Seed Anatomy

A seed consists of three basic parts: an embryo, food storage tissues, and seed coat. The embryo and food storage tissues are products of fertilization. The seed coat is usually several cell layers thick and can be very hard due to cell wall sclerification. The seed coat of a mature seed protects the embryo from mechanical damage, provides a means for long-term survival, and facilitates dispersal. It can also, however, present a barrier to germination, depending on its physical or chemical properties.

Seed Dormancy

Seeds may or may not be ready to germinate when they are formed. Many seeds have a protective mechanism (dormancy) to prevent them from germinating during suboptimal times of the year. Seed dormancy may be due to physical properties of the seed coat, chemical properties of the seed itself, physiological properties of the embryo, or some combination of these.

Seed coat dormancy is usually due to a heavily sclerified, impermeable seed coat. Hard seed coats can be softened using hot water, fire, or concentrated acid, or they can be abraded mechanically using a file or sandpaper. The goal of these scarification treatments is to make the seed permeable to water and oxygen or to make it possible for growing embryos to overcome any physical resistance of the seed coat so the germination process can commence.

Hot water scarification is performed by placing seeds in 4 to 5 times their volume of water at 170° to 212°F. The water is allowed to cool as the seeds soak for 12 to 24 hours. Acid scarification is done by exposing dry seeds to concentrated sulfuric acid (H_2SO_4) in a ratio of 1 part seed to 2 parts acid. The treatment should be done in containers that will withstand the corrosive action of concentrated acids (e.g., glass or special plastics). As the seeds soak, the mixture should be carefully stirred to ensure uniform exposure of the seeds to the acid. The treatment time will vary from 10 minutes to several hours, depending on the plant species. Mechanical abrasion is also performed to provide an entry for water so the germination process can begin.

Dormancy caused by chemical inhibitors in the seed can be broken with a moist, cool treatment of the seed called stratification. Stratification treatments mimic the cool/cold winters that seeds normally experience in nature, when they are produced in the summer or fall and germinate the following spring. To stratify seeds, fully imbibed seeds are mixed with a material (e.g., peat moss, sand, vermiculite, perlite) that will stay moist but also provide adequate aeration. The mixture is placed in a plastic bag and stored at temperatures from 32° to 50°F for varying lengths of time (usually 4 to 10 weeks), depending on the species. During this stratification period, seeds may go through an after-ripening stage in which the embryo develops fully or internal chemical changes may take place, allowing germination to occur when the temperature rises.

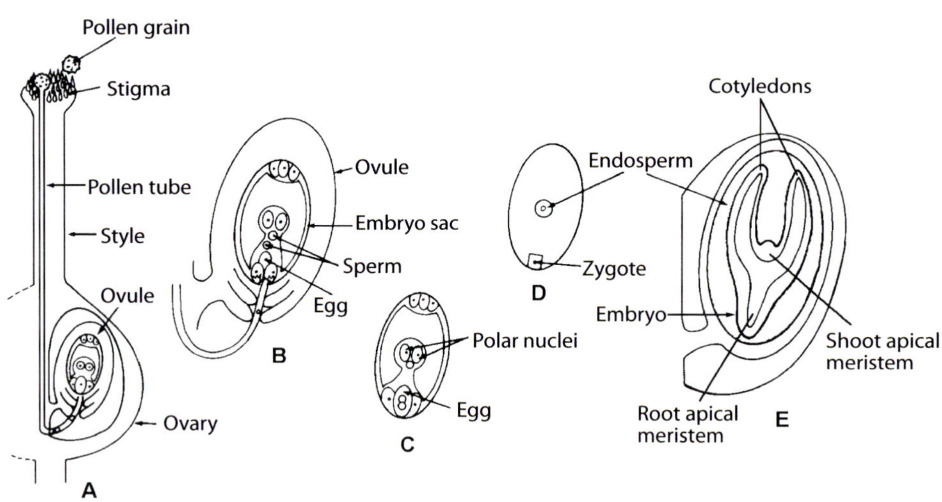

FIGURE 9.1. Pollination (A), fertilization (B), and embryonic development (C, D, E), resulting in the formation of an intact seed.

Photo: Courtesy of Ever-Bloom Inc.

Table 9.1. Stages of seed germination and important characteristics of each stage

Stage 1 - imbibition of water	Stage 2 - digestion and translocation	Stage 3 - seedling growth
The moisture content of the seed increases as water is taken in by diffusion. Seed tissues soften and swell. Tremendous pressures can be generated, usually capable of breaking the seed coat. Water imbibition triggers the rapid synthesis of new enzymatic proteins and reactivates stored enzymes.	Stored complex materials (carbohydrates, proteins, fats) are enzymatically converted to simpler substances and translocated to the embryo.	Cell division and enlargement commences in the shoot and root apical meristems. The radicle emerges, growing downward, and the shoot begins growing upward.

The growth regulator gibberellic acid (GA_3) can be used to substitute for a stratification treatment for many seeds. Gibberellic acid has been shown to overcome internal physiological dormancies and stimulate germination in seeds with dormant embryos when used at concentrations ranging from 200 to 1,000 mg/L (ppm). GA_3 at 500 parts per million is commonly used as either a soak before germination or as a solution to water seeds in after sowing.

Leaching can also overcome dormancy due to chemical inhibitors found in the seed coat or surrounding fruit tissues. Leaching can be done by soaking seed in water that is changed at least every 12 hours, or, preferably, seed can be washed under running water for 1 to 2 days.

All of these dormancy-breaking treatments mimic strategies nature has employed to prevent premature germination in the fall or winter and hasten germination in the spring. Alternate freezing and thawing, soilborne microorganisms, and passage of seeds through the digestive tracts of animals can gradually soften hard seed coats; heavy spring rains can leach germination inhibitors; and changing light intensities and day lengths can speed germination.

Seed Physiology

Plants take varying lengths of time to complete their life cycles from one seed generation to the next. Annual plants complete their life cycle in one growing season and then die. Biennials require two seasons to complete their life cycle. The first season is characterized by vegetative growth and development; during the second season, usually after a cold period, biennial plants flower and form seeds. Perennial plants live for more than two seasons, and some can live for hundreds of years. When they mature, they complete the reproductive seed-to-seed cycle each year; yet, they always maintain the capability to grow vegetatively.

The germination processes for seeds of annuals, biennials, and perennials are all similar. Several conditions must be met before a seed will germinate. First, the seed must be viable. That is, the embryo must be active and capable of germination. Next, conducive environmental conditions must exist around the seed: water must be available, the temperature must be within the range appropriate for the seed, and oxygen must be available. The appropriate light level must be present, which may mean no light for some seeds. Lastly, no internal or external barriers to germination can be present—the seeds must not be dormant. Seed germination proceeds continuously, but it has well-defined stages (table 9.1). Variations in the environmental parameters important to seed germination (water, temperature, oxygen, and light) can positively or negatively affect the rate of germination and the germination percentage.

Seed Handling

Recent advances in seed preparation and coating technologies have improved the uniformity of seed germination for many plant species and have enhanced the capability to handle seeds mechanically. Seeds are now graded routinely using various physical properties such as density, length, size, shape, and weight to separate seeds from other tissues and inert materials. Seeds may be altered (e.g., removing the "tails" of marigold seeds) and then coated with special films or graphite to facilitate easier handling by automated seeding machines, described in chapter 4. Small seeds may be coated with manufactured clay materials, creating pellets that may include nutrients or fungicides to enhance germination (fig. 9.2). Pelleting is performed to increase seed size and improve the uniformity of a seed lot; however, pelleting may provide a barrier to water if it is too hydrophobic (repels water) and a barrier to oxygen if it is too hydrophilic (attracts water). Film coatings are usually applied to seal in fungicides and promote dust-free handling. Coatings come in a variety of colors so that seed lots can be color coded, making them less prone to animal consumption since they are not recognized as food. Graphite coatings create a smooth exterior surface and add weight (see fig. 9.2). These coatings are particularly helpful when seeds are used in mechanical seeders.

Seeds can be germinated before they are sown. Pregerminated seeds with radicles 1 to 2 millimeters in length that are embedded in a viscous liquid to protect from desiccation may be used. This process is called fluid drilling, and it can improve germination percentages and reduce germination times.

Seed quality is an important factor for the effective use of seeds in propagation. Here, quality is defined as the purity of the seed lot and the viability of the seeds. Purity is expressed as a percentage of other materials in the seed lot. Inert materials, seeds from other species, and noxious weed seeds should comprise a very small percentage of the seed lot. High-quality seed packages may provide information, such as botanical and common name, purity, germination percentage (viability), germination times, collection date and location, and planting instructions.

Seed viability is an important measure of seed quality, and it can be determined by cutting open a sample of seeds from a seed lot (the cut test), by direct germination tests, or by indirect viability tests. Cut tests show whether seeds contain intact, fully developed embryos. Germination may be tested directly by sowing a small, known number of seeds from the seed lot on moistened paper towels or filter paper and then counting germinated seeds. Viability may be measured indirectly using tetrazolium chloride (2,3,5-triphenyltetrazolium chloride, TTC, or TZ). Viable seeds soaked in a solution of TTC turn red within 24 hours if they are viable. Samples from seed lots can be treated with TTC and the number of red-colored versus noncolored seeds counted. A very simple test for seed viability is to float the seeds on water. Quite often, seeds that float are empty or have poorly developed or aborted embryos while seeds with complete embryos and dense endosperm sink. This test is not always effective, however, since all small seeds tend to float due to surface tension and seeds having "wings" or "tails" will often not sink even when viable.

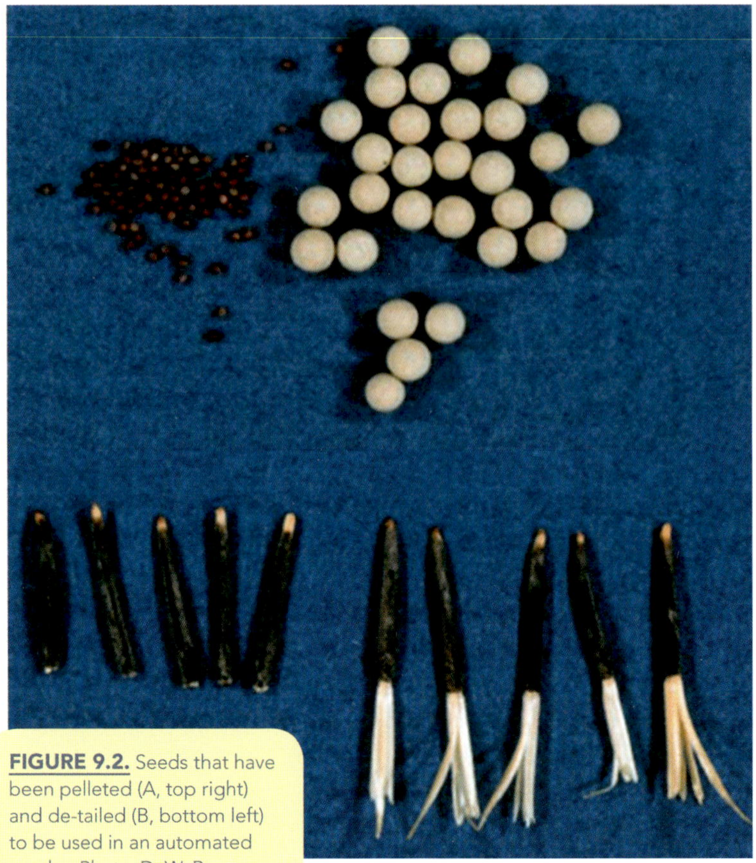

FIGURE 9.2. Seeds that have been pelleted (A, top right) and de-tailed (B, bottom left) to be used in an automated seeder. *Photo:* D. W. Burger.

Methods for Sowing Seed

Seeds may be planted directly outdoors or they may be started in a protective environment (e.g., greenhouse, cold frame) and the resulting seedlings transplanted outdoors. In either case, seedbed preparation must provide a moist environment where seeds are in intimate contact with the soil. Seedbeds also need to supply adequate aeration and nutrition, and they must be free of disease, insects, weeds or weed seeds, and any other soilborne problems; thus seedbeds or media used for seed sowing should be pasteurized before use. Seeds need to be planted at the correct time of year, which is determined by the species. Planting depths also vary with species. Since some seeds require complete darkness to germinate, they should be planted at a depth 2 to 4 times their minimum diameter. Seeds requiring light to germinate can be placed on the seedbed surface and lightly shaken (if done in flats) or raked in. Sowing rates (i.e., seeds per unit area) also depend on the species and can affect the quality of germinated seedlings. Viability percentages should be included in the calculations to determine the desired seedling density.

Diseases Associated with Seed Germination

The environment for seed germination is conducive to the development of pathogenic organisms, especially fungi. Warm temperatures, readily available moisture, and young, succulent plant tissues highly susceptible to pathogen attack can often result in devastating losses due to disease. Damping-off is a term used to describe the rapid death of germinating seeds and young seedlings, and it is characterized by stem necrosis (deterioration or collapse) at the soil surface. Several fungal organisms cause damping-off, including *Pythium ultimum, Rhizoctonia solani, Botrytis cinerea*, and *Phytophthora* spp. These pathogenic fungi are usually soilborne organisms; however, infected plant tissues and seeds may also contribute to the spread of damping-off problems. Control measures for damping-off include the complete elimination of viable pathogenic organisms from the germination environment, heat pasteurization or fumigation of germination media, and careful control of the environment around germinating seeds. Excessive moisture around seeds, along with temperatures in the 68° to 86°F range, increase the probability that damping-off will occur. Plants having optimal germination at temperatures in this range are more susceptible to damping-off, whereas plants with optimal germination temperatures below 68° or above 86°F are less susceptible simply because germination is enhanced and the growth of pathogenic fungi is inhibited. See chapter 13 for more information about damping-off.

Asexual Propagation

Comparison between Sexual and Asexual Propagation

As the term implies, asexual propagation is a means of increasing the numbers of plants without any exchange of genetic material. The resulting plants are theoretically identical to the plant from which they originated, based on the concept of totipotency, in which a single, isolated cell can develop into a whole plant. Therefore, a population of asexually, or vegetatively, propagated plants will be quite uniform. Depending on what plant is being propagated, sexual or asexual propagation methods (or both) may be selected (table 9.2).

Although the terms *sexual propagation* and *seed propagation* are often used interchangeably, seed of some plants can be asexual in origin. Normally, pollination and fertilization must both occur to produce a single embryo in a viable seed—but not in all seeds. In some plants (Valencia orange, *Citrus sinensis*; mango, *Mangifera indica*), the nucellus is stimulated to form somatic (nucellar) embryos during fertilization, resulting in one or many embryos of maternal origin growing alongside the single zygotic embryo, a process known as polyembryony. Apomixis, which is common in Kentucky bluegrass *(Poa pratensis)*, is similar to polyembryony except that embryos form from unfertilized eggs. Embryos in seeds produced from either nucellar embryony or apomixis have not undergone any exchange of genetic material and are, therefore, genetically identical to the maternal parent and asexual in origin.

Table 9.2. Comparison of sexual and asexual propagation methods and benefits

Sexual propagation	Asexual propagation
• variability in seedling populations • less expensive • more convenient • used for plants that cannot be propagated by separation, division, grafting, or cuttings • seeds can be stored for long periods of time • less likely to transmit diseases	• uniformity of clones • requires skill • faster and easier • some plants produce no viable seed • takes advantage of rootstock characteristics

Methods of Asexual Propagation

Cuttings

Asexual propagation by cuttings is the most common and most important type of vegetative propagation. A cutting is a part of the plant that is treated in such a way as to stimulate the regeneration of missing organs (fig. 9.3). Cuttings may be classified based on the part of the plant from which the cutting was taken (stem, leaf, leaf bud, or root) or on the type of wood (herbaceous, softwood, semihardwood, or hardwood). The formation of adventitious organs (organs such as roots and shoots arising from an unusual location) is the basis for cutting propagation, and it involves the concept of totipotency. With stem cuttings, adventitious roots may develop from preformed root initials or they may be the result of cellular dedifferentiation (loss of specialized cell function) and redifferentiation (regaining the ability of specialized cell function) phenomena in the stem. Several types of cuttings exist besides stem; they include leaf (consisting of the leaf blade with or without the petiole), leaf bud (consisting of the leaf and part of the stem containing an axillary bud), and root.

FIGURE 9.3. Rooted stem cutting of walnut (*Juglans hindsii* × *J. regia* 'Paradox'). *Photo:* E. G. Sutter.

For a new plant to develop, stem and leaf bud cuttings must initiate a new root system, root cuttings must initiate a new shoot system, and leaf cuttings must initiate both a new shoot and root system. Adventitious organ formation is not an uncommon natural occurrence. Brace roots on corn and aerial roots on English ivy *(Hedera helix)* are examples of naturally occurring adventitious roots.

Four types of stem cuttings exist: herbaceous, softwood, semihardwood, and hardwood. These categories reflect the kind of wood used to make the cutting. Each type of cutting has distinctive characteristics and rooting responses (table 9.3). Generally, herbaceous and softwood cuttings root easily in a relatively short period of time, whereas semihardwood and hardwood cuttings usually take longer to root and often have reduced capacity to form adventitious roots. This is a broad generalization with many exceptions, since rooting potential is an inherently unique characteristic for each species. Stem cuttings containing the stem tip or apical meristem are called tip or terminal cuttings, while those not containing an apical meristem are called basal or subterminal cuttings. Root formation may differ, depending on whether the cutting is terminal or subterminal.

Asexual propagation with leaf cuttings is possible with some plants, such as African violet *(Saintpaulia ionantha)*, snake plant *(Sansevieria trifasciata)*, and begonia (*Begonia* spp.). Since most leaf cuttings contain no meristem (undifferentiated tissue made up of cells that divide to form specialized tissues and organs), *de novo* (renewed) synthesis of a meristem is required through dedifferentiation and redifferentiation. The new plant will have the genetic makeup of the leaf cells forming the new meristem. In the case of a chimera, where the leaf is not genetically uniform, the unique characteristics of the plant may be lost through leaf cutting propagation. An example of this is a cultivar of *Sansevieria trifasciata* 'Laurentii' that has a yellow leaf margin. The yellow margin is genetically distinct from the rest of the leaf and is a chimera. Leaf cuttings taken from this plant will lose the yellow leaf margin attribute since only leaf cells in the middle area of the leaf undergo *de novo* meristem formation.

Plant growth regulators can be used to enhance the formation of adventitious roots in stem cuttings. Auxins are the type of growth regulator most commonly used for this purpose. Indole-3-acetic acid (IAA), a naturally occurring auxin, can enhance rooting, but the synthetic auxins, indole-3-butanoic acid (IBA, formerly indole-3-butyric acid) and naphthaleneacetic acid (NAA), have been shown to be more effective. These compounds can be used as a powder or talc applied to the base of stem cuttings before they are placed in a rooting medium, or they may be dissolved in a solvent (either potassium hydroxide or rubbing alcohol) and applied as an aqueous liquid (fig. 9.4). NAA or IBA are the active ingredients in most commercially available powders, talcs, or liquids. IBA can also be obtained as a potassium salt that is freely soluble in water. If solutions are used, cuttings can be treated in high concentrations (1,000 to > 10,000 ppm) for short periods of time (1 to 60 seconds, known as a quick dip), or they can be treated in low concentrations (10 to 1,000 ppm) for longer periods of time (1 to 24 hours, known as a slow dip or a soak). Once treated, the cuttings are placed in a flat containing a pasteurized rooting medium usually comprised of some combination of peat moss, vermiculite, perlite, and sand. The flat is then placed on a rooting bench under mist. Bottom heat may be provided, if necessary, to accelerate the rooting process.

Other techniques may be used instead of or in combination with growth regulators to enhance adventitious root formation. Wounding of the cut stem base stimulates root formation in some plants. Stems are wounded by cutting into the stem with a sharp knife, by scraping the outer surface of the stem, by splitting the base of the cutting, or by slicing away a portion of the stem base. Wounding may improve rooting by improving water or growth regulator penetration into the stem or by reducing physical barriers in the stem that may inhibit the development and emergence of newly formed adventitious roots. Shading parts of the plant where cuttings are to be taken with opaque tapes or Velcro strips, an etiolation technique known as banding or blanching, can improve rooting in some plants (e.g., maples, *Acer* spp.; birch, *Betula* spp.; oaks, *Quercus* spp.; and lilac, *Syringa* spp.).

Many factors can influence the formation of adventitious roots from cuttings. The stock plant from which cuttings are obtained must have no nutritional deficiencies, no disease infections or insect infestations, and no environmental stress (e.g., water, temperature, light). Stock plants can be managed to maximize rooting success by careful irrigation, fertilization, and pruning methods. Pruning can be done to increase axillary bud break, thereby increasing the number of cuttings

Table 9.3. Characteristics and rooting responses of various cutting types

	Herbaceous	Softwood	Semihardwood	Hardwood
Description of the cutting	succulent stems from nonwoody plants	new, tender growth on woody species	partially matured wood from current season's growth	dormant leafless stems or evergreen conifers
Time of year	all year	spring to early summer	summer to early fall	late fall to early spring
Cutting length	< 15 cm	5–15 cm	5–15 cm	5–75 cm
Relative humidity	humidity tent, mist or fog	humidity tent, mist or fog	humidity tent, mist or fog	direct field planting; stool beds
Examples	*Begonia, Coleus, Chrysanthemum*, carnation	lilac, *Pyracantha*, crape myrtle	*Citrus, Camellia, Rhododendron*	walnut, grape, fig, willow, poplar
Rooting response	roots form easily and quickly; most susceptible to disease	most root in 2–6 weeks; bottom heat may improve rooting	most root in 2–12 weeks; wounding, auxins enhance rooting	roots form after callus; bottom heat may help; inexpensive

FIGURE 9.4. The application of auxin as a liquid to a group of stem cuttings. Spraying the solution onto the cuttings instead of dipping cuttings into the solution prevents the spread of disease and contamination of the solution. *Photo:* D. W. Burger.

FIGURE 9.5. Propagation benches covered with polyethylene to increase the relative humidity around propagules. *Photo:* D. W. Burger.

FIGURE 9.6. A typical overhead mist system for propagation of seeds and stem cuttings. *Photo:* U. K. Schuch.

available. Plants must be fertilized, but not so that cuttings are rapidly growing when they are taken. A thorough watering can be done before cuttings are taken to ensure that the cuttings are not water stressed. It is usually best to avoid stock plants that are in flower. In some plants, cuttings from flowering shoots do not root as well or as fast as those from nonflowering shoots.

The environment where cuttings are rooted plays an important part in the overall success of cutting propagation. Adequate soil moisture, high relative humidity (close to 100%), suitable light, and favorable temperature must all be provided to root stem cuttings. Adequate moisture is probably the most important factor, especially at the beginning of the rooting process. Water must be available at the cut stem to allow for water uptake. At the same time, oxygen must be available for adventitious root formation. Therefore, the rooting medium must be able to hold water well while providing adequate aeration. Rooting media comprised of ingredients having large particle sizes (e.g., perlite, vermiculite, coarse sand) can supply the required moisture and air for the rooting process.

Important interrelationships exist between light and temperature during the rooting process. Photosynthesis is important for the cuttings to meet the heavy respiratory demand of the rooting process. However, excess light and its associated heat may increase transpiration above what the cutting can withstand and may be detrimental to rooting. Ideally, the ambient temperature should be in the range of 55° to 70°F and the temperature of the rooting medium should be higher than the ambient temperature by about 10°F. This differential soil:air temperature can help optimize rooting, since higher temperatures around the cutting base will enhance adventitious root formation, and low temperatures around the leaves will reduce evaporation and transpiration.

High relative humidity must be maintained around cuttings as they root to reduce the water lost through transpiration. Relative humidity can be increased by placing the cuttings in plastic-covered enclosures (fig. 9.5), by using high-pressure fog systems, or by using low-pressure mist systems (fig. 9.6). Mist systems are most commonly used to increase relative humidity. The mist system may be controlled by time clocks that activate the system at regular intervals. A combination of two clocks is often used; one clock controls the frequency and length of each mist application, and the other clock turns the first clock on at dawn and off at dusk. The time settings must be changed manually by the propagator throughout the year.

A common setting for the mist clock is for mist to be applied for 5 seconds every 5 minutes from dawn to dusk.

Devices that simulate leaf evaporation can also be used to control mist systems. These "electronic leaves" are based on the weight of water (balance type) or the conductivity of water (electrode type). The balance-type mist controller is comprised of a screen attached to the end of a lever that activates a mercury switch (fig. 9.7). As water evaporates from the screen (the simulated leaf), the lever rises until it reaches a point where the mist is turned on. Mist accumulates on the screen, causing it to go down and thereby turning the mist off. The electrode-type mist controller is comprised of a platform containing two electrodes separated from one another. As long as there is a water film on the platform, thus completing the circuit, the water remains off. When the circuit is broken (by water evaporating from the platform surface), the mist system is activated and remains on until the circuit is completed again. The electrode mist controller cannot be used on mist beds that spray deionized water. Both types of electronic leaves are more environmentally sensitive than time clocks and can help conserve water. Mist cycles will occur based on what the cuttings need instead of being based on time. However, electronic leaves require maintenance to prevent excessive algae growth or salt buildup when tap water is used. New mist-controlling devices are incorporating photocells to override the time clock during the night or to control the mist applications fully. The use of computer controls, as well as mist controllers that rely on environmental sensors, are described in chapter 4.

Budding and Grafting

Budding and grafting involve human intervention to join two or more pieces of plant tissue together so they will unite and grow as one plant. Generally, two pieces are used: a scion, which develops into the aboveground portion of the plant, and a rootstock, which becomes the root system. Scions in grafting are usually whole or relatively large sections of a branch containing more than one bud, whereas, in budding, the scion is a single vegetative bud. Budding and grafting techniques have been in practice for more than 3,000 years. The ancient Chinese practiced the art of grafting, and Aristotle wrote quite knowledgeably on the topic. Modern propagators practice budding and grafting for several reasons:

- Budding and grafting may be the only effective choice to propagate some plants asexually.
- Scion characteristics (e.g., flower or fruit quality, branching habit, disease tolerance or resistance) and rootstock characteristics (tolerance to soil conditions such as low aeration, salinity, alkalinity, disease resistance) can be chosen to best suit the environment in which the plant will grow.
- Budding and grafting provide the opportunity to change an established field from one scion cultivar to another without removing the established root systems.
- Damaged plants can be repaired. Rootstocks can influence the quality and characteristics of the scion, such as fruit yield and quality.

Not all rootstocks and scions of a particular genus can be budded or grafted. When either anatomical, physiological, or biochemical barriers exist between a particular scion and rootstock, they are said to be incompatible. In some cases, when a scion and rootstock do not join together successfully due to some kind of incompatibility, an interstock (usually a section of stem) may be used between them. Several factors can affect the success of a particular scion and rootstock union, including incompatibility, temperature, moisture,

FIGURE 9.7. A "mechanical leaf" used to control the frequency of mist applications based on the evaporation of water. *Photo:* D. W. Burger.

season, disease, and oxygen. The many different types of budding methods include T-budding, inverted T-budding, patch budding, and chip budding. Grafting methods include whip grafting, splice grafting, side grafting, cleft grafting, wedge grafting, bark grafting, approach grafting, inarching, and bridge grafting. Detailed drawings and procedures for grafting and budding can be found in *Plant Propagation: Principles and Practices* (Hartmann et al. 2001).

Separation, Division, and Layering

Specialized vegetative structures (geophytes) can be used to propagate particular plants asexually. Naturally detachable structures (e.g., bulbs, pseudobulbs, corms) can be separated from parent plants to begin new plants. Other organs (e.g., rhizomes, tubers, tuberous roots, tubercles) may require cutting or dividing to produce new propagules. In either case, the resulting plants are genetically identical to the plant from which they originated, the parent plant. Separating and dividing techniques have been practiced since the beginning of the agricultural age.

Bulbs are specialized underground organs consisting of a short, fleshy stem axis enclosed in thick, fleshy scalelike leaves. Tulips (*Tulipa* spp.), daffodils (*Narcissus* spp.), Easter lilies *(Lilium longiflorum)*, onions and garlic (*Allium* spp.) are examples of bulbs. Pseudobulbs are specialized organs produced by many orchid species and consist of an enlarged, fleshy section of stem made up of several nodes. Corms (e.g., gladiolus, *Gladiolus hortulanus*; and crocus, *Crocus* spp.) are swollen underground stems enclosed in dry, scalelike leaves, and they are predominantly made up of stem tissue with distinct nodes and internodes. All of these underground organs are useful in propagation by separation (fig. 9.8).

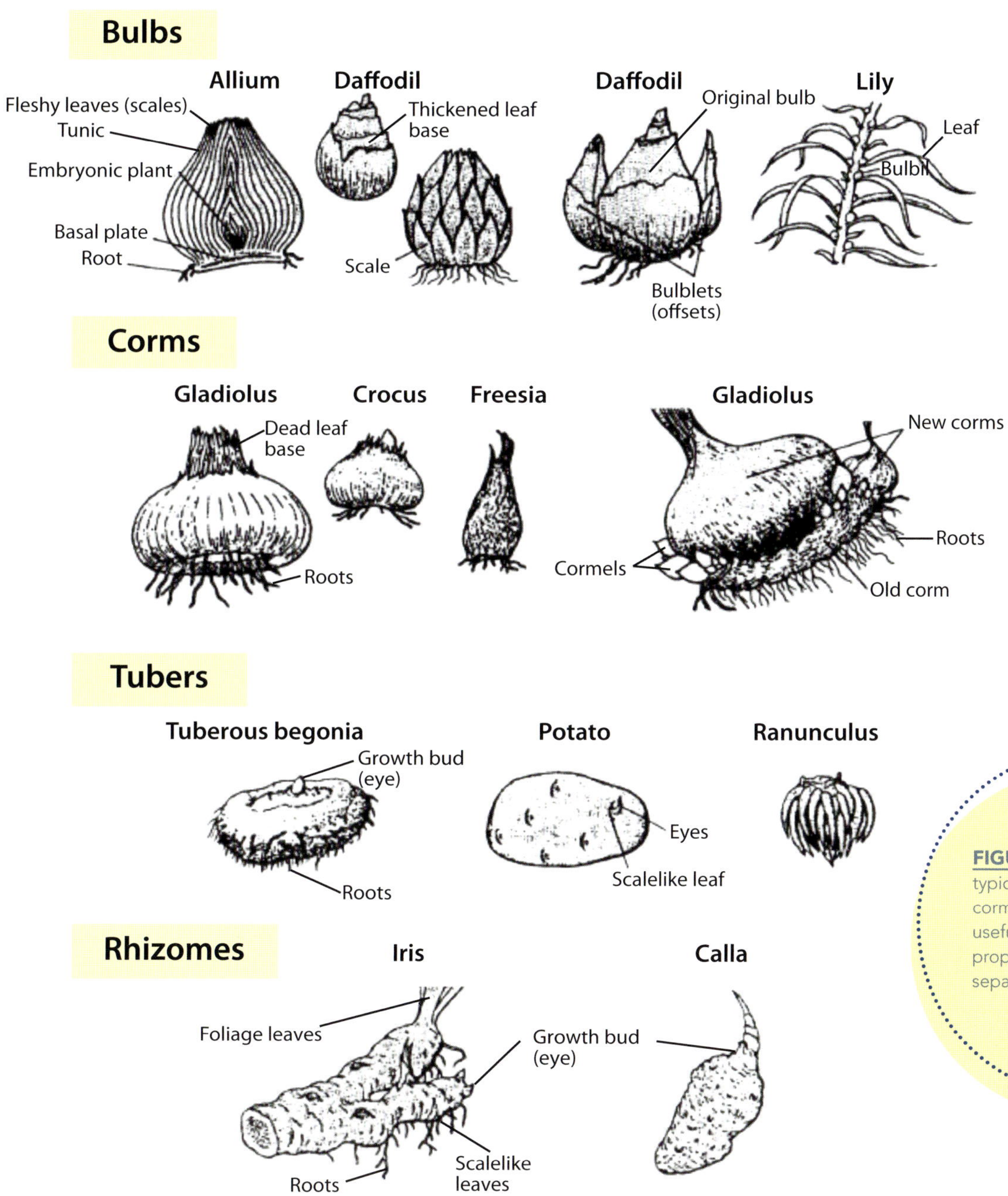

FIGURE 9.8. Examples of typical geophytes (bulbs, corms, tubers, and rhizomes) useful in vegetative propagation by division or separation.

Asexual propagation by division occurs with the use of tubers, rhizomes, tuberous roots, and tubercles. Tubers (e.g., potato, *Solanum* spp.; and ranunculus, *Ranunculus* spp.) are swollen underground stems that function as storage organs. The "eyes" of a tuber are nodes or small buds subtended by leaf scars (see fig. 9.8). Tuberous roots and tubercles are similar to tubers except that tuberous roots are swollen root tissue as opposed to stem tissue, and tubercles occur aboveground as opposed to belowground. Plants with tuberous roots or tubercles are propagated by dividing these structures into smaller sections, each containing a shoot bud. Rhizomes are specialized stems that grow horizontally at or just below ground level. Rhizomatous plants (e.g., Bearded iris, *Iris* spp.; calla, *Zantedeschia*; bamboo, *Bambusa* spp.) are propagated by dividing the many attached rhizomes into single pieces containing a stem axis and associated roots (see fig. 9.8).

Layering is a long-practiced method of asexual propagation in which adventitious roots are induced to form on plant stems while they are still attached to the parent plant. Layering is the natural method of asexual propagation in some plants (e.g., trailing blackberries, *Rubus* spp.). For plants in which layering is a natural occurrence, there may be specialized organs, such as runners, stolons, offsets, or suckers, that make layering possible. Layering is performed for various reasons:

- It may be the natural asexual propagation method worth exploiting to maximum advantage.
- Other asexual methods may not be successful or possible with some plants.
- Relatively large plants can be produced in a short time.
- Layering does not require extensive facilities or equipment; however, it does require considerable labor.

Several different layering techniques exist, including tip layering, simple layering, compound layering, air layering, mound or stool layering, and trench layering. The procedures for each type of layering technique differ, but basic similarities can be found among them. First, all layering methods involve the manipulation of a branch that is still attached to the parent plant. The branch may be bent, wounded, or girdled in some way to help induce the formation of roots. Next, the area where roots are to form is exposed to an environment conducive to root formation (e.g., buried in soil or wrapped in a moist substrate); and last, once roots have formed, the new offshoot is removed from the parent plant and transplanted to new soil to grow on its own.

Tissue Culture of Vegetative Structures

Tissue culture, also called micropropagation, or in vitro (i.e., in glass) culture, is the newest development in asexual propagation. The aseptic culture of plant cells, tissues, and organs has been researched and practiced since the early 1950s. The application of tissue culture to propagation started in the late 1960s and today is an important means of propagating many plants. Many greenhouse and nursery crops are propagated via tissue culture. These include most of the common foliage plants *(Ficus, Philodendron, Syngonium, Aspidistra,* and *Peperomia)*, woody ornamentals *(Rhododendron* and *Nandina)*, and orchids *(Cattleya* and *Cymbidium)*.

Tissue culture has three unique characteristics: the plant is in a sterile environment; the part of the plant (explant) placed in vitro is isolated from other tissues and organs; and the environment (temperature, light, culture media) is under strict control.

Required equipment, materials, and general procedures

Micropropagation requires specialized equipment and highly skilled technicians (fig. 9.9). An autoclave (or pressure cooker) is required to sterilize tools, glassware, and solutions used in the culture of plant tissues and organs. A near-sterile environment in which to work is also needed. Most micropropagation laboratories use a laminar flow hood for this purpose (fig. 9.10). This hood takes in room air, forces it through a series of highly efficient filters to remove almost all particulates, and distributes the filtered air over a work surface. This gives technicians confidence that they are working in a very clean environment and reduces the risk of airborne organisms contaminating culture media. Other standard laboratory materials and necessary equipment include a pH meter, dissecting microscopes, high-quality (deionized or distilled) water, and a microwave oven.

The culture environment is usually equipped with fluorescent or incandescent lights (or both) to provide intensities at 50 micromoles of photons per square meter per second ($\mu mol \cdot m^{-2} \cdot sec^{-1}$). Temperature is also carefully controlled to

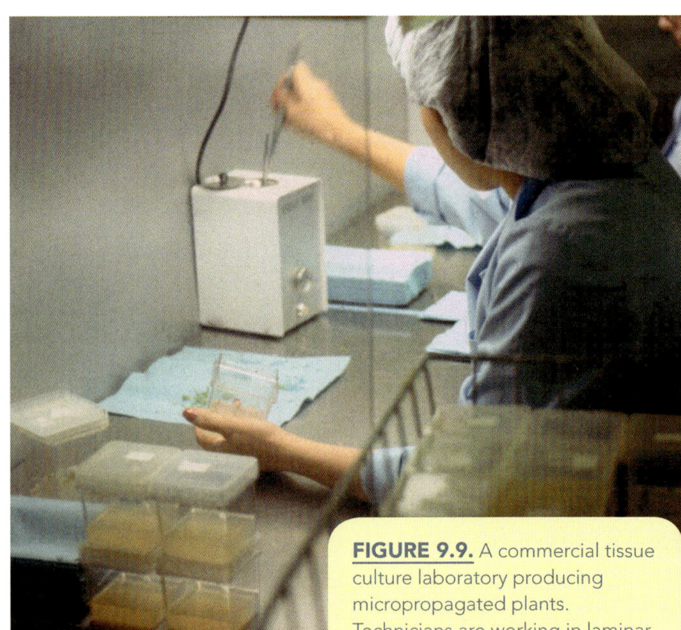

FIGURE 9.9. A commercial tissue culture laboratory producing micropropagated plants. Technicians are working in laminar flow hoods. Various types of culture vessels are shown. *Photo:* D. W. Burger.

FIGURE 9.10. Tissue culture procedures require a nearly sterile environment, such as this laminar flow hood, in which to work. *Photo:* D. W. Burger.

maintain a nearly constant temperature in the 68° to 82°F range.

General procedure

The general procedure for micropropagating plants is as follows:

1. The explant to be placed in vitro is chosen (axillary buds are most often the starting material).
2. The explant is treated with disinfestants, such as sodium dichloroisocyanurate (NaDCC), bleach, or ethanol, for 10 to 30 minutes to kill any surface microorganisms (fungi, bacteria).
3. The disinfestant is rinsed from the explant with sterile water.
4. The explant is placed on the culture medium. (The culture medium is prepared previously and has been distributed into a culture vessel and sterilized.)
5. The explant in the culture vessel is placed in a culture room in the light or dark, depending on the needs of the explant.

The goal is to induce the explant to proliferate many growing points (shoots) so they can be dissected and placed back into culture. The increase in the number of new plants depends on the rate of proliferation. If the rate is high, many thousands of plants can be propagated from a single, original explant in a short period of time.

Tissue culture media

A number of different culture media are used in micropropagation. The presence of macronutrients, micronutrients, vitamins, amino acids, a carbohydrate source, growth regulators, and some kind of support are common to all media. Many micropropagation media used today are modifications of the Murashige-Skoog medium (MS salts) developed for tobacco tissue culture in the early 1960s. Most modified media differ in the types and concentrations of growth regulators (cytokinins, auxins) added and the type of carbohydrate source (sucrose, glucose, fructose). Auxins and cytokinins are added to culture media to induce organ formation. Cytokinins, such as kinetin (6-furfurylaminopurine), BAP (6-benzylaminopurine), 2iP (6-[3-methyl-2-buten-1-ylamino)-purine, zeatin, and thidiazuron (N-phenyl-N-1,2,3-thiadiazol-5-ylurea), are added to stimulate the formation and multiplication of shoots. Auxins, such as IAA (indole-3-acetic acid), IBA (indole-3-butanoic acid or indole-3-butyric acid), NAA (naphthaleneacetic acid), 2,4-D (2,4-dichlorophenoxyacetic acid), and NOA (naphthoxyacetic acid), are added to stimulate the formation of adventitious roots.

Shoot multiplication, organogenesis, and embryogenesis

Micropropagation of plants may proceed via existing meristem multiplication, *de novo* organogenesis, or *de novo* embryogenesis. Depending on the initial explant and its response to culture conditions, existing meristems may be stimulated to grow and proliferate; this is an example of multiplication. When new organs are formed in the explant or from tissues derived from the original explant, it is a result of organogenesis (formation of an organ from unorganized cells). Embryogenesis occurs when explants undergo cellular dedifferentiation and subsequent redifferentiation to form somatic embryos. These embryos are genetically identical to the parent explant, assuming no mutation occurred during the differentiation steps.

From a propagator's standpoint, meristem multiplication is the preferred response, since it generally does not involve a callus stage, a stage in which cells are totally unorganized and more likely to undergo mutation. In organogenesis and embryogenesis, a callus stage is usually involved, and therefore these developmental processes are characterized by higher rates of mutation than multiplication.

References

Dirr, M. A., and C. W. Heuser Jr. 2006. The reference manual of woody plant propagation: From seed to tissue culture. 2nd ed. Athens, GA: Varsity Press.

George, E. F., and P. D. Sherrington. 1984. Plant propagation by tissue culture: Handbook and directory of commercial laboratories. Hants, UK: Exegetics Limited.

Hartmann, H. T., D. E. Kester, and F. T. Davies Jr. 2001. Plant propagation: Principles and practices. 7th ed. London: Prentice-Hall.

Hartmann, H. T., A. M. Kofranek, V. E. Rubatzky, and W. J. Flocker. 1988. Plant science: Growth, development and utilization of cultivated plants. 2nd ed. Englewood Cliffs, NJ: Prentice-Hall.

Janick, J. 1986. Horticultural science. 4th ed. New York: Freeman.

Macdonald, B. 2006. Practical woody plant propagation for nursery growers. Portland, OR: Timber Press.

Young, J. A., and C. G. Young. 1992. Seeds of woody plants of North America. USDA Agriculture Handbook 450. Portland, OR: Dioscorides Press.

Photo: J. K. Clark.

Controlling Plant Growth

Ursula K. Schuch, J. Heinrich Lieth, and Julie P. Newman

CHAPTER TEN

Plant growth characteristics, such as plant height, stem size, and flowering, are regulated by endogenous plant factors and the environment. Ideally, an environment should be maintained that provides optimal conditions required for plant growth. However, most nursery crops are primarily produced outdoors with limited or no control of environmental factors. Nevertheless, some degree of control is important during periods of seasonal temperature extremes. Moreover, during propagation and early stages of production, container nursery plants often grow in a variety of structures, such as greenhouses, shadehouses, or retractable-roof greenhouses, that allow various degrees of environmental control of plant growth. These types of structures are discussed in chapter 3 and should be reviewed prior to reading this chapter.

Photo: U. K. Schuch.

Even when nursery crops spend the entire production cycle inside greenhouses with sophisticated environmental controls, they can be exposed to conditions that result in undesirable growth responses. For example, if the greenhouse is not adequately cooled during a hot spell or if plants are shaded, plants may stretch and grow too tall. Sometimes cultural practices can substitute for a lack of environmental control. In outdoor container nurseries, plant growth is regulated mostly by cultural practices.

An understanding of the natural growth cycle of a plant is important in making decisions that affect the overall efficiency of nursery crop production. Although it is assumed that the reader already has some understanding of how plants grow and develop, the fundamentals of plant growth and physiology are reviewed in this chapter. More information about basic botany and plant physiology can be found in many horticulture textbooks (e.g., Ingels 2001) and in the *California Master Gardener Handbook* (Pittenger 2002). It is also important to understand how plant environments can be manipulated to optimize crop yields and control plant growth characteristics. This chapter discusses the atmospheric conditions that affect the aboveground portion of plants, as well as management of aerial environments in outdoor production and in protected structures.

Moreover, it is critical that the nursery manager is knowledgeable of the methods that can be used to affect plant growth. Cultural practices to control plant growth, such as root media preparation, container selection, irrigation, fertilization, chemical growth regulation, spacing, pruning, and staking, are described in this chapter. However, detailed information on root media preparation, irrigation, and fertilization are discussed in chapters 5 to 7. Pest control also affects plant growth and is discussed in chapters 11 to 17.

Basics of Plant Growth and Physiology

The basic requirements for plant growth are air, water, minerals, and sunlight. Plants have different parts: leaves, stems, roots, buds or meristems, flowers, fruit, and seed, all of which carry out vital functions.

Leaves and other green surfaces are the sites of photosynthesis and produce carbohydrates, which is the energy source for plant growth and reproduction. Photosynthesis is the process where carbon dioxide from the air, in combination with water and energy from sunlight, produces carbohydrates. Leaves are also the site of gas exchange, where carbon dioxide enters through stomata (small openings in the leaves), and oxygen and water vapor are released. Transpiration (the loss of water through stomata) regulates the temperature of the leaves and aids the flow of water throughout the plant.

Plant stems contain the vascular system that consists of xylem and phloem. In the xylem, nutrients and water are transported from roots to leaves; in the phloem, carbohydrates from photosynthesis are moved throughout the plant from leaves to stems and roots. Roots absorb water and minerals from the soil, they anchor the plant in the soil or growing substrate, and they store carbohydrates for later use. Plant growth occurs on buds or in the cambium, a layer between the xylem and phloem that adds annual growth rings in woody perennials.

In addition to the plant parts described above that are vital for vegetative growth and development, flowers, fruit, and seed are necessary for sexual reproduction.

Atmospheric Conditions That Affect Plant Growth

Nurseries contain microclimates, where atmospheric conditions subtly differ from the prevailing climate primarily due to local topography, plant cover, and buildings. Microclimates in outdoor production areas vary considerably from those in protective structures such as greenhouses and shadehouses. In general, light, temperature, soil moisture, and nitrogen are the most important environmental parameters that affect nursery crops. For the aboveground portion of the plant, the most important environmental parameters are light, temperature, carbon dioxide concentration, humidity, air quality, and wind. We will discuss these aerial environmental parameters in this section. For plants growing in containers, many of the belowground parameters (temperature, water availability, nutrient concentrations, oxygen) are directly affected by the aboveground parameters. In fact, while any one environmental factor can, by itself, have a pronounced effect on the crop, factors typically interact, resulting in a complex picture. For more detailed information on belowground parameters, see chapters 5 to 7.

Temperature

Temperature affects all plant processes, including photosynthesis, respiration, development, and morphological development such as stem elongation. The rate of plant growth generally increases steadily from the base temperature to the optimal temperature. The base temperature is the temperature below which no plant growth occurs. However, individual plant processes (e.g., photosynthesis), plant species, and plant organs (e.g., roots, leaves) vary in their base and optimal temperatures. For example, photosynthesis generally has an optimal temperature from 68° to 104°F, depending on the plant. In general, growth increases as temperature rises above 32°F and is greatest in the range of 68° to 95°F, depending on the species and plant process.

If temperature falls within the range between the base and optimal temperature, the rate of plant growth depends on the daily 24-hour average temperature. Temperatures of 32°F or lower can freeze tissue (fig. 10.1) and temperatures of 50°F or less can cause chilling damage in tropical or subtropical species.

Temperatures above the optimal range reduce growth and may result in flowers of smaller diameter in flowering plants. In addition, high temperatures can stress plants because they increase transpiration and the rate of water loss, and they may cause wilting if not enough soil moisture is available to rehydrate the plant. Insufficient water quickly leads to reduced photosynthesis, transpiration, and nutrient uptake. High temperatures can also lead to the closure of stomata in some plants, reduced photosynthesis, and tissue damage. Plants grown permanently well below or above their temperature optimum will be stunted, damaged, and unable to

FIGURE 10.1. Freezing temperatures caused this bleached white and yellow necrotic foliage on agapanthus. *Photo:* J. K. Clark.

Photo: J. H. Lieth.

develop into marketable crops. All plants sustain irreversible damage when grown above certain high temperature levels, resulting in tissue death and possibly plant death.

Base and optimal temperatures are generally different for roots and shoots. For many plants, the optimum for root growth is 59° to 81°F. When plants are grown in containers, the roots are exposed to temperatures similar to the ambient air, and freezing temperatures as well as high temperatures can damage or kill roots. The smaller the container, the greater the potential is for damage from extreme temperatures. Root zone temperatures exceeding 125°F have been recorded inside containers that are exposed to direct sunlight. Measures that maintain root zone temperatures within the optimal growth range, such as using pot-in-pot systems outdoors, will support optimal growth.

Vegetative and reproductive growth are most rapid when temperatures are within an optimal range for a particular species. Maps for average low temperatures and plant hardiness (USDA 2012), heat zones (AHS 1997), and western climate zones (Brenzel 2007) are available, and they will help in determining which plants are suitable for production and growth outdoors in different climates. Plants not suited for a particular climate will require some environmental control of the production facilities.

Light

Solar radiation provides energy that is perceived as sunlight and heat. Light affects plants in several ways:

- It provides energy for photosynthesis.
- It provides energy that is absorbed, heating up the plant.
- It triggers various physiological processes (such as flower induction or onset of dormancy).

For each of these, three characteristics of light are important: quality (color), quantity or intensity (brightness), and duration (day length).

Light quality

The spectral quality of radiant energy refers to the wavelength or color of light. The solar (radiant) energy from the sun is in the form of waves of different lengths with ranging effects: the longer the wavelength, the lower the energy; the shorter the wavelength, the higher the energy. The unit of measure for wavelength is the nanometer (nm), or one billionth of a meter. The human eye perceives only a small portion of the total spectrum of light (visible light) in the range of 390 to 700 nanometers. Plants can use the ultraviolet, visible, and infrared portions of the spectrum. For photosynthesis, plants need light in the 400 to 700 nanometer range, which approximates visible light.

Light intensity

Great variation exists in the light intensity that falls on the surface of the earth. Differences are due to cloud cover and other atmospheric conditions, geographic location, season, and time of day. Typically light consists of direct and indirect radiation. The former is light coming directly from the sun while the latter is light reflected from particles in the atmosphere and surrounding objects. The amount of light available to plants inside a greenhouse or shade structure is less than that outdoors, and it depends on the type of glazing and the amount of structural beams and other obstructions above the canopy.

Scientists measure light intensity in various ways. The most common unit of measure used in the greenhouse and nursery industry is the foot-candle (fc), which is an instantaneous measurement of the visible light that the human eye can see. Full sun conditions measure around 10,000 foot-candles while interiors with bright natural light can range from 400 to 1,000 foot-candles. Light meters that determine irradiance in foot-candles are used by some growers. Other types of light meters that measure light intensity at different wavelengths are also available. For example, quantum sensors measure photosynthetic active radiation, which is the light plants use for photosynthesis. They express measurement in micromoles of photons per square meter per second ($\mu mol \cdot m^{-2} \cdot s^{-1}$). Full sun conditions measured with a quantum meter would equal about 2,000 $\mu mol \cdot m^{-2} \cdot s^{-1}$.

Many woody plants produced outdoors in full sun are more compact, with shorter internodes and more branches, compared with plants grown under 50 percent or more shade. The latter are taller, with fewer branches and a more open canopy, but they have larger leaves and a greater leaf area. Plants growing outdoors under high light conditions can develop into higher-quality plants when leaf tempera-

Table 10.1. Light and temperature requirements for selected herbaceous perennials to flower when grown in 5-inch containers

Species	Cold duration (at 45°F)	Long days required	Weeks to flower (at 68°F)
Achillea 'Moonshine'	no	yes	4
Achillea millefolium 'Fireland,' 'Paprika,' 'Galaxy,' 'Terra Cotta'	for some	yes, for most	6–9
Asclepias tuberosa	no	yes	8
Coreopsis grandiflora 'Early Sunrise'	no	yes	7
Coreopsis grandiflora 'Sunray'	10–15 weeks	yes	7
Echinacea purpurea 'Magnus' and 'Bravado'	no	yes	14
Gaillardia × *grandiflora* 'Goblin' and 'Baby Cole'	10+ weeks	yes	7
Gaura lindheimeri 'Siskyou Pink' and others	no	yes	6
Geranium dalmaticum	6 weeks	yes	7
Gypsophila paniculata 'Double Snowflake'	12–15 weeks	yes	12
Hemerocallis 'Stella De Oro'	15 weeks for bare-root crown	no	8
Lobelia × *speciosa* 'Compliment Scarlet'	6 weeks	yes	9
Penstemon digitalis 'Husker Red'	9 weeks	yes	7
Perovskia atriplicifolia	5 weeks	no	11
Rudbeckia fulgida 'Goldsturm'	10 weeks	yes	13
Salvia nemerosa 'May Night'	9 weeks	no	4

Source: Cameron et al. 2000.

tures remain cooler, compared with plants growing indoors. Adequate moisture is a prerequisite for plants to grow at optimal rates, and it is especially critical under high light levels. As with temperature, different species of plants differ in their range of optimal light levels.

Day length

Day length (photoperiod) refers to the number of hours that a plant receives light during a 24-hour period. Plants that require more than 14 hours of light to initiate flowering are long-day plants, and these include many of the herbaceous perennial species listed in table 10.1 (Cameron et al. 2000). Plants that require less than 12 hours of light are short-day plants and include chrysanthemum, kalanchoe, and cosmos. Day-neutral plants, such as coral bells, delphinium, and roses, are not affected by day length; other environmental conditions can trigger flowering in day-neutral plants.

Humidity

Humidity is an important variable that interacts with other variables such as temperature. Humidity is important because it partly controls the moisture loss from the plant. The humidity inside a plant is close to 100 percent, but when a plant grows in dry conditions it loses moisture because water moves from high to low humidity. When the difference in humidity is large, the loss of moisture from the plant is rapid. Humidity is also important because high humidity favors some foliar diseases.

Relative humidity is expressed as a percentage of the moisture saturation of air. Hot air can hold much more moisture than cold air. This is important when plants are grown in greenhouses because if moist, humid air is trapped in a greenhouse at the end of the day and then cools during the night, the cooled air will hold less water and the relative humidity will increase as the temperature declines. As the relative humidity reaches 100 percent, the air at that temperature will be saturated and will contain as much water as it can hold. If the air cools further, water will condense.

The temperature at which condensation occurs can be calculated using a chart produced from a psychrometer (a device used to measure humidity). Condensation will first form on anything that is colder than the air, such as glazing material and plant foliage. Prolonged periods of wet foliage can result in disease problems.

Typically the optimal range of relative humidity is 40 to 80 percent. Relative humidity values below 40 percent may be suboptimal because the plant will evaporate water rapidly, possibly leading to internal water stress. If temperatures are high, relative humidity above 80 to 85 percent can increase the incidence of disease and reduce plant transpiration.

Carbon Dioxide

Carbon dioxide is a fundamental plant nutrient required for photosynthesis. Ambient carbon dioxide concentration varies over the day from 300 to 600 parts per million with an average of 350 parts per million (higher in urban areas). This level is adequate for plant growth, but more rapid growth rates may be achieved at higher concentrations as long as other environmental factors, such as light, are not limiting.

All plants accumulate carbon through photosynthesis when light is present. If there is sufficient light, then the rate of photosynthesis is limited by such factors as temperature and carbon dioxide concentration. If there is sufficient carbon dioxide, then other factors may limit photosynthesis.

Air Quality

Air pollutants can cause injury to nursery plants, especially in urban areas. Pollutants include sulfur dioxide, fluorides, ozone (fig.10.2), and photochemical smog formed by the interaction of sunlight and pollutants. These pollutants are produced primarily by industry, vehicles, and the generation of electricity.

Furthermore, in closed environments (e.g., greenhouses and nursery storage areas) ethylene gas can adversely affect nursery plants, although this is not usually a problem in outdoor production. Sources of ethylene include cigarette smoke; exposure to damaged, decaying or dying plants; exhaust from engines of cars and forklifts; contact with ripe or decaying fruit; and improperly functioning or unvented greenhouse heaters. When ethylene accumulates in sensitive tissues at concentrations from 0.01 to 0.1 parts per million for periods of up to 24 hours, it can cause distorted growth of stems, curling of leaves, and abortion of buds and flowers (fig. 10.3). Higher exposures for several hours can kill plants.

Greenhouse heaters that are unvented to the outside can also release other contaminant gases. These include carbon monoxide, nitric oxide, and nitrogen dioxide.

Wind

Wind can have an effect on nursery crops in many ways, causing the following:

- physical damage to the plants
- rapid desiccation of plants, thus exposing them to water stress
- modification of the overhead irrigation pattern, resulting in under- or overwatering of some plants
- soil erosion

Physical damage caused by wind in outdoor production is usually due to containers falling over or to other objects (plants, shade cloth, structures) falling onto crop areas. When a pot falls over, the plant will generally receive far less, if any, irrigation water. Plants that are not returned to their upright position after the wind ceases will dry out in a short period of time. Also, the foliage that contacts the ground in the fall may be injured and exposed to pathogens, which could later result in disease. Plants on the border of a block may fall into the path of nursery traffic.

FIGURE 10.2. Ozone air pollution damage can be subtle, as shown by the yellowish and brownish patches on these Japanese maple leaves. *Photo:* J. K. Clark.

FIGURE 10.3. Shattering (petal drop) occurred in the geranium on the right after the plant was exposed to low levels of ethylene. *Photo:* J. K. Clark.

High winds cause plants to transpire considerably more water than they would in still conditions, and edge effects are more pronounced even in light winds. Moreover, overhead sprinkler irrigation is affected by wind, which reduces distribution uniformity (see chapter 7). Where winds are high, the water can be blown so far off course that a higher percentage of water is delivered to noncrop areas than to the roots of the plants. Wind also affects spray patterns of overhead chemical applications.

Frequent strong winds contribute to soil erosion, as well as movement of root media. This can result in increased sediment loads and off-site movement of pesticides in nursery runoff if not properly managed (see chapter 8).

Environmental Management

This section provides some guidelines for managing the specific atmospheric factors that affect the aboveground portion of plants in outdoor production and in protected environments. However, it is important to keep in mind that environmental management involves the entire crop rather than individual plants. Several cropping characteristics directly affect profitability, such as crop production efficiency, and must be computed for entire blocks of plants. Marketing and pricing are also based on groups of plants. Perhaps the most important crop factor that must be controlled as tightly as possible is crop uniformity. Thus, in considering how the various environmental factors influence crop production, it is important to focus on how to control the individual variables at the crop level and how to create a uniform production environment with respect to each variable.

Temperature

Timing of flowering and crop maturation

Temperature can be manipulated to initiate, accelerate, or delay flowering. For example, some plants require exposure to extended cold temperatures to initiate flowering. Research on the flowering requirements of herbaceous perennials shows that many species benefit from treatments of cold temperatures, and some species require long days to initiate flowering and shorten the time to marketing (see table 10.1). Examples in table 10.1 demonstrate the great variability among species and even among cultivars of the same species regarding flowering requirements for production in a timely manner.

Manipulation of temperature to force flowering has been investigated more in floriculture crops than in landscape plants. Nevertheless, temperature manipulation is conducted in commercial nurseries on some crops (such as azaleas) to increase marketability, especially when there is access to a greenhouse.

Temperature effects on the production of roses in the nursery provide an example of how manipulating temperature can be used to hasten flowering, increase flower size, and reduce production time. Dormant bare-root roses are generally transplanted into containers from late winter to early spring. Plants are ready for marketing when they have grown a full new canopy, with flower buds ready to open approximately 6 to 8 weeks after transplanting. Two strategies can be taken to reach this goal: one is to place the plants in a greenhouse or protected structure with night temperatures at 62° to 65°F and day temperatures at 68° to 70°F, and the second is to place the plants outdoors where temperatures will vary (Pemberton et al. 1997). Forcing plants under the optimal temperature regime results in a predictable time to flowering, which will be hastened by higher temperatures and delayed by lower temperatures within the optimal range. Some rose cultivars grown outdoors without protection can tolerate daytime temperatures as high as 90°F, but serious delay in flowering occurs below 55°F. Furthermore, marketing is delayed in locations where temperatures are consistently below the optimal temperature range for specific cultivars. Higher temperatures, on the other hand, not only affect crop maturation but also cause roses to produce flowers of smaller diameter than those on roses grown under lower temperatures.

Regulating plant height

The difference between day and night temperatures (day temperature – night temperature = DIF) can be used to regulate plant height. For many plant species, stem elongation increases with increasing DIF. For example, a day temperature of 65°F and a night temperature of 50°F gives +15 DIF, which typically results in tall, weak shoots. A DIF of 0 (when day and night temperatures are the same at 58°F) results in intermediate growth, and a DIF of -5 (night temperatures of 60°F and day temperatures of 55°F) produces the shortest plants.

In floriculture crops, where plant height is of importance, growers monitor height development on a frequent basis by using graphical tracking to regulate the rate of stem elongation with DIF. Since most of the stem elongation occurs in the early morning hours, a drop in temperature during that time can control stem elongation to a certain degree. DIF is used successfully to control the height of many floriculture crops.

Previously a technique confined to greenhouses, DIF can now be used in retractable-roof greenhouses by trapping cold air early in the morning. This exposure to cooler temperatures in the early morning is effective because most of the stem elongation occurs during that time. However, excessive warm temperatures later in the day can negate some of the benefits of cooler temperatures in the morning.

Reducing temperatures

The use of greenhouses and shade structures to reduce temperatures is described in chapter 3. Cooling outdoors is more difficult and generally involves shading or overhead irrigation, or both. It is possible to have a retractable shade curtain system outdoors to provide shading on demand. A more common method of cooling outdoors is to irrigate with overhead sprinklers to wet the foliage and cool the root zone. The main cooling effect comes from the evaporation of water from the foliage. Consequently, it is important to use high-quality water with few (or no) dissolved compounds, as these generally remain behind as residue on the foliage when the water evaporates. If the relative humidity is high, then effectiveness of this cooling method is reduced. Under such conditions, which typically occur in the late afternoon, the water on the plant may remain for many hours, favoring foliage disease development. The most effective time to irrigate container plants to cool the root zone is usually between noon and 2 p.m.

Root zone temperatures can also be reduced by using containers that are poor conductors of heat or by using light-colored containers, which will reflect some of the solar energy. The use of the pot-in-pot system to stabilize root media temperature is another option (Davidson et al. 2000).

Maintaining temperature uniformity

Normally it is difficult to achieve high crop uniformity in outdoor production because uniformity is only possible when every plant experiences the same environmental conditions. Crop uniformity can be improved when nursery plants are grown in greenhouses using horizontal airflow (HAF) to maintain uniform temperature (fig 10.4).

HAF uses the principle that air moving in a coherent horizontal pattern in a building like a greenhouse needs only enough energy to overcome turbulence and friction loss to keep it moving. Small, 1/10 to 1/15 horsepower, circulating fans are used to provide the required air movement. The general rule is 2 cubic feet per minute of fan capacity for each square foot of floor area. The total fan capacity required for an efficient system can therefore be calculated by multiplying the floor area times two. However, a slightly greater capacity is needed for tall crops or hanging baskets to overcome additional turbulence. The number of fans needed can be determined by dividing the calculated total fan capacity by the fan output. The first fan should be placed

FIGURE 10.4. Crop uniformity can be improved when nursery plants are grown in greenhouses using horizontal airflow (HAF) to maintain uniform temperature. The HAF system, as shown here, is comprised of horizontal circulating fans that operate continuously to create a circular pattern that properly mixes the air from each end of the greenhouse and from ceiling to floor. This also helps to reduce condensation on plants and to control many foliar pathogens. *Photo:* J. K. Clark.

about 10 to 15 feet from one end wall to pick up the air that is coming around the corner from the other side. Subsequent fans should be located 40 to 50 feet apart to keep the air mass moving (Bartok 2005).

Besides improving temperature uniformity, HAF systems can reduce the incidence of foliar diseases. The moving air removes moisture from the plant canopy, resulting in a drier microclimate. Furthermore, HAF reduces the difference between leaf and air temperatures, which results in less condensation.

Frost protection

Many frost-hardy crops exist, so the threshold at which damage occurs varies. Only plants that cannot tolerate the low temperatures at a particular nursery location need protection from cold. It is wise, therefore, to group frost-sensitive plants together in the warmest locations of the nursery. Dormant plants generally don't need to be protected unless a very severe freeze is due to occur.

Shadehouses can often serve as overwintering structures in milder climates by adding insulating covers or by replacing shade cloth. Even with just shade cloth, temperatures warmer than those outdoors are maintained at night. The main purpose of overwintering structures is to prevent desiccation of plants and heat loss from the ground and the atmosphere. The best choice for overwintering is a cover with poor heat-conducting properties such as white polyethylene or thermal blankets. Structures should allow for irrigation and ventilation while covered.

Roots are generally more susceptible to freezing damage than are shoots. Cultural methods in the fall should discourage new growth and promote hardening of plants. In some cases, insulation of the container to protect roots from low temperatures may suffice to overwinter plants.

For sensitive crops, frost protection outdoors in unprotected areas is usually accomplished in two ways. Both are only effective under mild frost conditions, temperatures reaching as low as 28°F; if temperatures fall below that, little can be done with temporary measures. One method of protection is to cover the plants with some sort of material designed for this purpose, such as batting material or frost blankets. The material should be lightweight to avoid causing any damage to the plants, and the sheet size must readily accommodate the area of the crop to be covered. This will provide protection down to 28°F for a few hours and may provide some degree of protection at even lower temperatures. Two of the most significant effects of the covers are to reduce evaporative cooling and wind chill. The covers also trap energy from the ground, which may be significant enough to reduce damage.

Another method of frost protection is to spray water onto the plants to wet the foliage (fig. 10.5). This will prevent damage only for plants that tolerate temperatures down to

FIGURE 10.5. This sprinkler is one of a series linked to an environmental monitoring system. If frost is expected, the outdoor container plants are automatically blanketed by a continuous spray of water to reduce the likelihood that they will suffer freeze damage. *Photo:* J. K. Clark.

32°F. The idea is to constantly spray water onto the plants. As the water freezes, it releases energy. The temperature of the water will be at 32°F even if the air surrounding the water is below 32°F. Only when all the water has turned to ice will the temperature proceed to drop below 32°F, so water needs to be constantly present to prevent the temperature of the ice and water from dropping below 32°F. Thus, water has to be sprayed so that the plants are always covered by liquid water in the process of freezing. It should be noted that clear water must be used for this, since any dissolved materials (e.g., fertilizers) will drop the freezing point to some level below 32°F, which may be damaging.

Clearly, spraying plants with water to protect them from frost has limitations. If significant amounts of ice build up on the plants, the plants may break under the weight of the ice. Overhead spraying should not be combined with cloth covering, since this will result in a large sheet of ice that will crush the plants. If ample water is available and drainage is good, continual spraying can prevent ice buildup, as any ice crystals would be washed away by the water. If drainage is poor (perhaps due to snowfall or other blockages), using large amounts of water for frost protection can be disastrous. If the danger of frost exists, pipes may need insulation to avoid pipe ruptures due to freezing water. Even partial ice buildup inside the plumbing should be avoided since ice crystals could easily clog emitters or spray heads used for irrigation and frost protection.

Light

Light quality

Greenhouse glazing material may alter the light spectrum, but not enough to be of much use as a tool for directing plant growth. However, different colors of shade netting have been developed to alter the spectrum of light, scatter diffuse light, or alter the thermal components of light. Effects of colored nets on plant growth have been documented for a number of ornamental, vegetable, and fruit crops, and they may substitute for chemical growth regulators for some species (Shahak et al. 2004, 2008). The nets cause different physiological responses based on light, and they provide physical shade.

For example, *Pittosporum variegatum* grown under red nets shows increased shoot elongation and develops larger leaves, which is desirable for this plant, which is grown for green decorative branches. Plants growing under blue nets are dwarfed and have good variegation compared with those grown under the commercial black net. Plants under gray nets develop into bushy plants with short branches and little variegation.

Use of these colored shade nets can also affect flowering. For example, *Lupinus luteus* and *Lisianthus*, grown for cut flowers, yield longer and heavier-flowering stems under yellow nets, but flowering is delayed for 2 weeks compared with plants under black nets. Plants under gray nets produce the highest number of flowering stems.

Light intensity

Modifying light intensity can be important in commercial nurseries. It can be done either by increasing light intensity (e.g., artificially lighting) or by shading (see chapter 3). However, when plants are suddenly moved from shade into an environment of high light intensity, they can sustain sunburn on leaves and branches, resulting in permanent damage. Gradual adjustment to higher light intensity can prevent this damage.

Light duration

Photoperiodic control is commonly conducted in commercial floriculture operations but much less frequently in commercial nurseries. However, nursery growers who have greenhouses can control photoperiod with lights and blackout curtains. Long-night plants (or short-day plants) require long, uninterrupted dark periods in order to flower. For example, chrysanthemums and poinsettias require 12 hours of dark to flower; these plants will stay vegetative if the night period is shorter than 7 hours. Blackout curtains that promote flowering can be applied using automated systems for photoperiod control (see chapter 4). Flower initiation can be prevented by interrupting the night with light. For example, chrysanthemums are usually exposed to light for 2 to 5 hours in the middle of the night, depending on latitude and seasonal day length.

Humidity

Reduction of relative humidity outdoors is generally not feasible. Furthermore, relative humidity can only be increased outdoors by applying water as overhead spray or mist.

In greenhouses, high humidity and conditions conducive to the formation of condensation can be managed by sufficient venting or successively heating and venting. For example, relative humidity may be lowered by cycling through the following procedures: allow the greenhouse air to reach 95 percent relative humidity; vent this air to the outside, allowing cooler air to replace it; and then shut the vents and heat the cooler air up to higher temperatures. As water evaporates off the soil and transpires from plants, the relative humidity will again rise, so that at 95 percent relative humidity the cycle will have to be repeated. Obviously, a significant energy cost can be associated with this practice. Also, this method is not effective in controlling humidity during periods with warm and humid outdoor conditions.

Condensation can be especially problematic in polyethylene houses. This is because the surface tension properties of plastics are such that condensed water rapidly forms large drops that can drip onto greenhouse crops. Covering the greenhouse with two layers of polyethylene, with an air space between the layers using inflation (see chapter 3), will reduce the temperature difference between the outside and inside. This will reduce condensation, especially when internal air circulation and heating are also used. Moreover, anticondensate products that can be sprayed on the inside of plastic, as well as improved plastics with anticondensate coatings, are commercially available. The use of a HAF system will also reduce condensation.

It should be noted that irrigation generally results in increased relative humidity due to the presence of additional water and runoff. While this occurs particularly with overhead

Photo: J. H. Lieth.

irrigation, the problem will persist even with drip irrigation, particularly if there is runoff. Clearly, under conditions of high humidity, irrigation practices should be adjusted so that there is no freestanding water. Moreover, as long as plants are transpiring, relative humidity will increase.

Fog and mist systems (described in chapters 4 and 9) can be used to raise relative humidity when plants are unable to take up the amount of water needed for transpiration or for cooling purposes. Fan-and-pad systems and shading systems used to cool greenhouses will also increase humidity. It is a common practice in propagation to raise humidity when cuttings are forming roots, thus preventing desiccation of the plant tissue. This is usually accomplished by intermittent misting or by constructing a tent over the cuttings.

Carbon Dioxide

The ambient levels of carbon dioxide are adequate for moderate rates of photosynthesis, but these rates can usually be increased with higher carbon dioxide concentrations. In greenhouses, higher levels can be maintained above ambient to as much as 1,500 parts per million to stimulate additional plant growth.

Various methods can be used to increase carbon dioxide levels in greenhouses. For example, carbon dioxide gas can be released from a compressed carbon dioxide gas tank or from a liquid carbon dioxide tank. It can also be released by burning clean natural gas with specially designed burners. Each method has advantages and disadvantages. With any carbon dioxide distribution method, it is important that no harmful contaminant gases such as ethylene are released at the same time. Distribution units should be installed properly and monitored regularly to avoid safety issues. These include excessive carbon dioxide levels of 5,000 parts per million, which can cause dizziness or lack of coordination in humans and injury in plants. Carbon dioxide levels can be monitored by sampling air and passing it through an infrared analyzer that is compared with a reference sample. Growers who monitor and control the greenhouse environment with sensors linked to a computer can integrate a carbon dioxide controller to monitor and maintain optimal carbon dioxide levels (Blom et al. 2002; Both 2003).

Periods of carbon dioxide enrichment are generally limited to times when the greenhouse vents are closed. The extra carbon dioxide should be delivered as close to the foliage as possible. Clearly carbon dioxide enrichment is only feasible when there is adequate light for photosynthesis (i.e., during the day, unless supplemental night lighting is used); otherwise, light will be the factor limiting photosynthesis. Generally during the night, decomposing organic matter in the root media and respiring plant tissues can generate sufficient levels of carbon dioxide so that there is no need to enrich levels until this buildup is drawn down after sunrise. On a sunny day, however, when plants are actively photosynthesizing, the carbon dioxide concentration could go from ambient levels to "starvation" levels within minutes.

Low carbon dioxide levels are especially a problem during the winter when a greenhouse might be completely closed for considerable periods. Opening the vents will allow outside air at ambient carbon dioxide concentrations to replenish the depleted supply. Unfortunately there is no clear, noticeable symptom when carbon dioxide levels are low; if low levels persist, there will be a significant delay or size reduction in the crop. Suboptimal levels of carbon dioxide can be avoided by monitoring with a carbon dioxide sensor.

Air Pollutants

The best method of managing air pollution is to locate the nursery away from polluted areas, or at least upwind from potential pollution sources. Mitigation of air pollution is very difficult and expensive. One possibility in greenhouse production is to pass air through a filtering system before it enters the greenhouse, but the extra energy needed to do this may make crops too expensive. It is important to have good air circulation and ventilation in greenhouses and storage areas where gas-powered equipment is used. Furthermore, fruit should not be placed in nursery storage areas, and old or decaying plants and debris should be removed, as these may be sources of ethylene gas.

Some plants are more susceptible to air pollutants than others. If some variety is found to suffer from air pollution damage (e.g., yellowing, bleaching or bronzing of leaves, "burnt" edges, leaf spotting), then it is wise to switch to varieties that are tolerant.

Wind

Greenhouses effectively eliminate the threat of wind to the plants; shade structures also reduce plant damage if the structure itself is not damaged by wind. If plants are not grown in protective structures, the use of windbreaks should be taken into account when designing and constructing the nursery. This can be accomplished by planting tall hedges or trees along the upwind boundaries of the production area to protect sensitive nursery stock from windburn and sandblasting. Vegetation used as windbreaks should be planted perpendicular, or nearly so, to the principal wind direction (Lu et al. 2008). The height of the vegetation used as windbreaks is dictated by the range over which it is to serve as a windbreak: vegetation can effectively reduce wind speed on the leeward side up to 20 times the height of the plants. Building fences or walls can also significantly reduce airflow. Artificial windbreaks are usually constructed from plastic or fiber netting or laths.

It should be noted that windbreaks can also cause problems. A permanent windbreak reduces airflow that is needed for cooling, and it reduces usable land. Also, living windbreaks constructed from plants such as trees and hedges may require pruning and pest management, thus costing money for maintenance. Living windbreaks may serve as hosts for disease organisms or insect pests that could infest nursery crops. Conversely, plant windbreaks may serve as a refuge for beneficial insects that could help keep pests in check. Windbreak plantings can also cast undesirable shadows and can create litter from falling foliage or fruit drop. Windbreaks that create shade may provide a place for crops that do better in the shade, but care should be taken so that windbreaks do not provide excessive shade onto other crops since this will reduce production, possibly increasing the production time and reducing profits.

In addition to windbreaks, other practices can be implemented to protect plants from wind damage, such as orienting planting rows and containers at right angles to the prevailing wind (Lu et al. 2008). A heavier soil substrate can be used to prevent toppling of plants. In areas of gusty winds, plants can be tethered to a support wire or trellis. Alternatively, a pot-in-pot production system can be used to prevent plants from being blown over. Reduced spacing for plants with small canopies can also help—pots that are spaced can-tight will be less likely to fall over. However, crop spacing for wind management is rarely a solution, since spacing is generally dictated by the shape and space needed for optimal plant growth.

Where overhead irrigation is used and irrigation frequency is based on timers, the standard program may have to be augmented with additional irrigations to prevent plant desiccation. This is particularly important when the winds are hot and dry (e.g., Santa Ana conditions in southern California).

Cultural Practices

Water and Irrigation

Water is a potent growth regulator. The quantity and quality of water affect plant growth and cultural options for production. Water is required to transport nutrients from the roots to the shoots. Transpiration cools plants, water maintains plant turgidity, and water is essential for biochemical processes within the cells. Lack of water immediately arrests cell expansion and division, long before leaves wilt. Longer periods of drought can cause damage to leaves, flowers, and fruit, and it can lead to abscission.

Plant growth directly depends on the rate of irrigation that plants receive. Optimal irrigation of container crops will result in healthy, fast-growing plants. Plant height, stem diameter, leaf expansion, and root growth depend on a steady water supply. During temperatures that are favorable for growth, growers should capitalize by providing sufficient irrigation. This replaces the amount of water lost to evapotranspiration since the previous irrigation. The amount of water lost from plants is a function of temperature, sunlight, humidity, wind, and other factors. The amount of water that can be stored in a container depends on the container volume and how media composition affects moisture-holding capacity. Soil moisture-release curves and tensiometers can be useful tools to optimize irrigation management in the nursery (see chapters 5 and 7 for more information on soils and irrigation).

Frequency and amount of irrigation needed depend on the type of media, the plant species and size, development stage, and environmental conditions. When temperatures exceed the optimal range for plant growth, water becomes even more important to help cool plants (especially roots) and replenish moisture lost to transpiration. During periods of low evaporative demand, irrigation should be reduced accordingly to avoid conditions that promote diseases.

Underwatering will slow plant growth rates and development, and it can be a result of insufficient frequency or quantity of irrigation. Many plants do not tolerate drought well, and photosynthesis and nutrient uptake immediately decline. Drought-stressed plants with broad leaves wilt, and leaves can turn soft, rubbery, light green, or off-color. Leaves often become stunted and may develop marginal or interveinal necrosis, and leaves and flowers may prematurely drop. Conifers under drought stress will not exhibit symptoms until the foliage is damaged, which becomes apparent as tips turn yellow or brown.

In fall, when plants enter dormancy and cold hardiness starts to develop in response to shortening days and cooler temperatures, irrigation should be reduced to discourage new vegetative growth and to facilitate hardening. Plants cannot enter dormancy unless growth stops. Additionally, new vegetative shoots and leaves are susceptible to cold temperatures and potential damage by early frost. Another time when plants intentionally receive less water is at the end of the production cycle when they are prepared for shipping. Reduced irrigation for a few days keeps plants in a state of light water stress, which enables them to withstand the rigors of shipping and handling. This is where they often experience water stress. However, plants exposed to light water stress withstand drought better than plants never exposed to drought. Intentionally reducing water to prepare plants for postproduction drought is known as hardening plants off. This term also refers to acclimating plants to full sunlight or variable outdoor temperatures when they have grown in a protective structure or under shade.

Overwatering is a problem for plants, and it is exacerbated in poorly drained substrate. Symptoms of overwatering include taller plants with yellow or wilted foliage, stunted growth, and black or dark brown roots. Overwatered plants often have saturated substrate at the bottom of the container. When trying to lift the plant with the root ball out of the container, there may be only a small part of the upper root ball with live roots attached while dead roots with saturated substrate remain in the bottom.

Water quality affects the growth of plants, and high soluble salt concentrations in irrigation water can restrict growth. Plants vary in their tolerance to salts, but symptoms of high salt levels include marginal leaf necrosis and wilting when the substrate is moist. For further information on water quality, see chapter 5.

Nutrients and Fertilizing

Fertilization with all essential elements is required for plants to develop properly. Excess height and vegetative growth can result from too much fertilizer. The lack of one or several minerals or nutritional imbalances cause stunted growth and specific deficiencies. Visual observation of leaf color and the size of new leaves for each species by an expert is one way to monitor nutrition. However, measuring electrical conductivity in the medium or leachate throughout the production cycle can be a useful tool to track a shortage or an excess of minerals.

Fertilizer requirements differ by species, growth stage, and season. At the beginning of production, when plants are growing fast, fertilizer requirements are higher than toward the end of production, when growth is much slower and plants are ready for marketing. When temperatures are favorable for growth from spring to fall, plants need appropriate amounts of fertilizer. Fertilizer application, similar to water, is commonly reduced in fall to induce dormancy and again toward the end of a production period to harden off plants. For more information on fertilizers and plant nutrition, see chapter 6.

Spacing

Plant density has a dramatic effect on all environmental parameters. Closely spaced plants will shade each other and reduce airflow, resulting in increased relative humidity and perhaps reduced carbon dioxide concentrations. Plant spacing and increased shading as canopies fill out determine final plant quality. This is because plants require sufficient light to develop a full canopy with well-balanced branches and leaves throughout the canopy. Container spacing influences plant height, width, and caliper during nursery production.

Moving containers to increase spacing as crops mature requires careful consideration, as this practice can increase production space and labor costs. Proper spacing for a crop balances two objectives. It reduces the loss of marketable plants due to misshapen or undersize canopies, and it increases production area and labor efficiency. Recent transplants in small containers are generally spaced can-tight to maximize the use of space and overhead irrigation. Container spacing must be increased before the canopy of can-tight-spaced plants closes fully. Otherwise, loss of plant quality will occur, such as defoliation in the lower canopy and insufficient development of canopy width. Spacing plants too late can result in severe sunburn and defoliation of lower leaves when they are suddenly exposed to sunlight. Spacing should accommodate unrestricted growth of side branches. Wider spacing generally results in less height increase, greater taper, and greater trunk area increase, producing a plant of better quality. In #1 and #5 containers, greatest benefits in height and trunk area growth generally are realized when plants occupy twice or three times the space that they would in a can-tight setting.

Spacing of plants in larger containers depends on the plant size and canopy shape. Therefore, trees with a narrow, upright canopy can be spaced very closely, just to allow access for maintenance. Conversely, trees with a broad canopy that may be two to three times as wide as the container diameter will require more than two to three times as much space between containers to allow unrestricted branch growth. The distance between plants and the number of times containers for a particular crop are spaced will vary by species, growth rate, and the final canopy size desired.

Container Size and Design

Plants grow larger with greater media volume in the container. Leaving plants too long in a small container reduces shoot, leaf, and root growth and can permanently reduce plant vigor. In the case of trees, root restriction causes development of a shoot system that is far too large for the root system; and although trees with restricted roots continue to grow slowly when water and fertilizer are ample, they often languish or fail once transplanted into the landscape. Transplanting in a timely manner to the next larger container size is important for maintaining optimal growth rates, a balanced root to shoot ratio, and preventing root-bound conditions. Plants are ready for transplanting when the root ball has filled the entire substrate volume and roots are starting to circle around the bottom of the container. According to guideline specifications for nursery tree quality developed in California (see the Urban Tree Foundation website in the References at the end of this chapter), the trunk diameter of trees 6 inches above the soil surface should be as follows:

- 0.5 to 0.75 inches in a #5 container
- 0.75 to 1.5 inches in a #15 container
- 1.5 to 2.5 inches in a 24-inch box

Guidelines for plant size, the proper relationship between height and caliper (trunk diameter) or height and width, and the appropriate size of root ball or container for a particular size plant can be found for 13 plant groups (including shade and flowering trees, deciduous shrubs, coniferous evergreens, roses, and perennials) in the *American Standard for Nursery Stock* (ANLA 2004). The plant sizes in the standard are guidelines for marketable plants; however, some states have specific standards for some species or groups of plants, or customers may request the final plant dimensions and container sizes they wish to purchase.

Several container shapes are available that deviate from the conventional round design with its straight wall and diameter about equal to the height. Containers with square sidewalls, stepped pyramid designs, or vertical ribs on sidewalls have been developed to reduce or eliminate root circling, matting, or kinking. Tall, narrow containers are available for container production of small trees used primarily in reforestation. Deep 3-gallon containers with a depth almost twice that of the top diameter are used in the production of citrus and avocado trees to promote large root systems before being transplanted into larger containers or the field. Custom-shaped containers with even smaller diameters and greater depth are used to grow transplants for areas that will not receive supplemental irrigation. Low-profile containers with height to width ratios of up to 1:4 and special bottom screens to promote shallow, dense root systems on trees are also marketed. Containers with small holes in container walls facilitate air pruning and prevent root matting on the outside of the root ball. Growth response of plants grown in containers with alternative designs during nursery production often varies by species, and needs testing before alternative designs are adopted on a large scale. For additional information on containers, see chapter 1.

Pruning

General objectives

The objective of pruning in the nursery is usually to enhance the natural form of a plant. Generally this entails removing the terminal shoot to break apical dominance, which stimulates growth of side branches and results in a fuller plant canopy. However, stimulating growth of side branches by providing sufficient light, water, and nutrients to plants is a more desirable alternative to shoot pruning. Pruning is a labor-intensive task and should not be done unless absolutely necessary. Pruning shoots removes photosynthetically active leaves and wastes previous growth.

Sometimes the pruning objective is to shape plants into desired forms that are not natural. Regular pruning of topiaries is necessary to create and maintain plant shape. In addition to manual pruners, various tools or machines are available to mechanize the shearing of plants (see chapter 4).

Plants in the nursery may be used as stock plants for harvesting material for vegetative propagation. Stock plants can be pruned to increase axillary bud break, thereby increasing the number of cuttings available. For more information on the management of stock plants, see chapter 9.

A common pruning practice that is not aimed specifically at directing plant growth is to remove dead, diseased, and insect-infested tissue, which is an important component of IPM programs (see chapters 12 and 13). When pruning to remove localized diseased tissue, it is generally recommended to disinfect pruning tools after each cut and before moving to another plant. However, this practice can be cost prohibitive. To make an informed decision on the frequency of disinfection, you need to know the pathogen and its life history. Diseases that invade the vascular system or form oozing cankers (e.g., fireblight, fungal and bacterial cankers, virus diseases) are the most likely to be transmitted by contaminated pruning tools, depending on the time of year. However, bacteria associated with active cankers are embedded in a sticky mass that is often difficult to remove from pruning tool surfaces. Therefore, it is important to avoid cutting active, oozing cankers (Chalker-Scott 1999).

Before disinfecting pruning tools, dirt and debris should be removed so the disinfecting solution can reach every cutting surface. Select a disinfectant that is effective, readily available and affordable, relatively safe to handle, and not harmful to tools or clothing. Many household cleaners fit this description. Chalker-Scott (2010) provides a good review of common disinfectants that have been scientifically evaluated for sterilizing pruning tools. She does not recommend using chlorine bleach because it corrodes metal, is an acute and chronic health hazard when fumes are inhaled, is toxic to plant tissues, and bleaches clothing.

Disinfectant solution can be carried into the field in a tightly sealed plastic bottle wide enough for dipping, or solution can be applied with a clean cloth. Solution can also be poured over pruning shears held over a bucket so that disinfecting solutions do not contaminate the soil. Pruning tools that are regularly disinfected need to be kept in top condition to avoid corrosion pits that may harbor microorganisms not affected by quick sterilization. After dipping pruning tools, be sure to wipe away excess disinfectant to avoid injuring the next plant (fig. 10.6). Disinfectants should not be applied to pruning wounds, although older literature often recommended this practice (Chalker-Scott 2010).

Flaming pruning tools with a propane torch can be a quick, inexpensive, and effective alternative to the use of disinfectant solutions. Downer et al. (2009) showed that flaming pruning saws with a propane torch for 10 seconds reduces total fungal colonies and palm pathogenic fungi recovered on a selective medium by 95 percent. Furthermore, flaming eliminates the waste disposal issues associated with disinfectant solutions, and it is noncaustic. However, propane torches produce an intense open flame. To avoid injury or fire damage, use extreme caution and always follow proper operating procedures and safety guidelines when using a propane torch.

Pruning shade trees and shrubs

The goal of producing shade trees is to grow straight trees with a well-balanced crown. Training of shade trees starts soon after planting, and it becomes more labor-intensive the longer plants grow in the nursery. It is important to consider tree forms when pruning to train shade trees. Trees have many forms, but the most common types are pyramidal (excurrent) or spherical (decurrent). Excurrent trees have a strong central stem and a pyramidal canopy. There is strong apical control, and lateral branches rarely compete for dominance. Most conifers and a few hardwoods such as sweetgum and tulip tree have excurrent forms. Conversely, decurrent trees (e.g., most hardwoods) have a rounded canopy without a dominant leader. There should be only one leader in excurrent trees. The leader of excurrent trees should never be cut, unless it is damaged. Cutting immediately above a healthy bud will allow that bud to assume the role of the new leader. In case of opposite buds, one bud or the weaker of the competing leaders should be removed once they start to grow.

Branches that are weak, crossing, or attached at narrow angles should be removed. Scaffold branches should be attached at 30- to 60-degree angles. Removing or heading back lower branches on the trunk affects height, caliper development, and root growth of trees. Removing lower branches completely results in a greater increase in height but less trunk diameter and taper development. Branches on the lower part of the trunk have the vital function of supplying the trunk with carbohydrates, which help increase trunk diameter and develop taper, particularly in unstaked or lightly staked trees. A promising approach for commercial shade tree production seems to be heading back lower laterals and staking lightly or not at all, especially for trees with multiple trunks.

Shrubs are often sheared once or twice during the first or second growing season to promote a dense, well-shaped canopy. Other pruning should be minimized.

Timing of pruning

Timing of pruning depends on the growth stage of the plant, the climate, and the market objective. In areas with year-round growing climates, pruning is often performed when labor is available. In colder areas, deciduous trees are generally pruned during the dormant period, while desert trees tolerate pruning in hot weather as well. Transmission of diseases by contaminated pruning tools has been shown to be affected by season (Chalker-Scott 1999).

Root pruning

Root pruning is necessary when plants are root bound and transplanted to a larger container. Otherwise this is not a necessary practice in container production. Root balls of root-bound plants are scored with a knife ¼ to ½ inch below the bottom surface to disrupt the matted mass of roots or circling roots. The amount of scoring will increase with the severity of root matting or circling.

Staking

Compared with a rigidly staked tree, an unstaked tree will have the following attributes:

- greater basal trunk
- less height growth
- greater trunk taper
- more uniform xylem to support itself upright
- larger and stronger root system
- less wind resistance
- less stress per unit of trunk area
- no rubbing or girdling injuries from stakes or tape

Binding stems of container trees tightly to a single stake is unfortunately still practiced in nurseries (Davidson et al. 2000). While this is a useful practice for vines and some trees such as shoestring acacia or standard oleander, which require support, it is also commonly employed for many other trees. As a result of tight staking, tree height increases at a fast rate, and trees reach marketable height sooner than unstaked trees. However, stem caliper often does not develop in a ratio appropriate to height growth, and once the nursery stake is removed, the tree may not be able to stand upright on its own. Tight binding of a trunk to a stake can lead to partial girdling of the trunk by the nursery tape

FIGURE 10.6. The hand pruning shears shown here were modified to allow disinfectant to slowly drip over the blades to help sterilize them after pruning diseased tissue. However, excess disinfectant that may contact plant tissues when these modified pruning tools are used could cause plant injury. Moreover, pruning tools that are regularly disinfected may become pitted; these pits can harbor microorganisms that are unaffected by quick sterilization. Tools should also be free of dirt and debris so the disinfecting solution can reach every cutting surface. *Photo:* J. K. Clark.

FIGURE 10.7. Staking hardware often causes mechanical injury to young trees. Here, a tie was installed too tightly and left on too long, causing partial girdling of the tree trunk. *Photo:* L. R. Costello.

or physical damage from the stake rubbing against the trunk (fig. 10.7). Shading of the trunk can lead to sunburn once the stake is removed. Shaded trunks will bend away from the stake once cut loose, because xylem cells next to the stake are more elongated than those on the opposite side of the trunk. Disproportionate caliper to height ratio can lead to breaking or blow-over of the tree when transplanted into a landscape.

Staking is necessary for many single-trunk tree species growing in containers. Providing support for a stem just above the point where the tree stands straight (instead of tightly binding stakes to the trunk) prevents many of the problems detailed above, yet it ensures development of a commercially desirable tree with a sturdy trunk. Instead of tall stakes, shorter stakes or bamboo stakes can be used, and staking should enable trunks sufficient sway to develop proper taper.

Chemical Regulators

Chemical growth regulation is another tool that nursery growers can use to influence vegetative and reproductive development. Plant growth regulators (PGRs) are synthetic substances that exhibit hormonelike responses in plants. PGRs can confer benefits for nursery crop production, such as accelerating or reducing shoot growth, substituting for dormancy or cold treatments, promoting rooting of cuttings, and reducing the need for pruning.

Plant hormones are defined as organic compounds produced in one part of a plant and translocated to another part where, at very low concentrations, they cause specific physiological responses. Plant growth regulators (PGRs) are synthetic compounds that exhibit responses similar to those caused by natural hormones in plants. In order for a hormone to cause such a plant response, an adequate amount of the hormone must be present in the cells, the hormone must be recognized and bound to the cells, and receptors in those target cells must be able to amplify the hormone signal. Therefore, both hormone concentration in a tissue, such as a bud, along with the sensitivity of the tissue to the hormone, must match to cause a response. As the plant grows, hormone concentrations in the different tissues change, as does the ability of receptors to amplify those signals. Therefore, to obtain the desired effect, timing is crucial regarding both the application of a hormone or growth regulator as well as the sensitivity of the plant tissue.

Five endogenous plant hormones exist: auxins, gibberellins, cytokinins, ethylene, and abscisic acid. These hormones regulate processes such as cell expansion and division, rooting, shoot elongation, flower initiation, and fruit ripening, to name a few. In addition to these hormones, synthetic growth regulators are available to manipulate vegetative and reproductive growth of plants. Chemical growth regulators can benefit nursery crop production, and they affect plants in ways such as increased compactness, more branches causing bushier plants, increased stress tolerance, darker green foliage, increased flowering, and control of root circling (table 10.2). However, because of increasing regulations of chemicals used in the nursery, the small size of the nursery crop niche market available to companies producing and selling PGRs, and an increasing consumer intolerance toward the use of chemicals, the use of PGRs in the nursery will probably diminish further.

Many woody and herbaceous crops benefit from treatments to retard stem elongation or increase the number of lateral branches. Thus chemical growth regulators can be important in preventing or reducing the need for pruning to

Table 10.2. Plant growth regulators (PGRs) for herbaceous and woody ornamental crops in nursery production

Purpose	Active ingredient (chemical trade name)	Mode of action	Precautions and remarks
Stimulates rooting of cuttings.	K-IAA (Rhizopon A water-soluble tablet) 50 mg IBA (C-mone) 1 to 2% K-IBA (Rhizopon AA water-soluble tablet) 50 mg K-IBA (C-mone K) 1 to 2% K-IBA + K-NAA (C-mone K+) 1.5% IBA + NAA (Dip 'N Grow) 1.5% IBA (Hormex) 0.1 to 4.5% IBA (Hormodin) 0.1 to 0.8% IBA + thiram (Hormo-Root) 0.1 to 2% IBA + 15% thiram IBA (Rhizopon AA) 0.1 to 0.8% K-NAA (Rhizopon B water-soluble tablet) 25 mg NAM + thiram (Rootone) 4.25% IBA + NAA (Wood's Rooting Compound) 1.54%	IAA, IBA, and NAA stimulate production of adventitious roots.	Concentrations vary according to species and time of year when cuttings are taken.
Promotes plant growth and stem elongation of most dicot and monocot plants. Can overcome overapplication of GA-inhibiting PGRs.	(Fascination)	GA_4+GA_7 and cytokinin BA promote cell elongation and division.	Extremely active PGR. Follow label instructions carefully.
Increases stem elongation and flowering in many species Partially replaces cold treatment to break bud dormancy in azaleas. Can overcome overapplication of GA-inhibiting PGRs, though often not economical.	GA_3 (Pro-Gibb T&O)	Gibberellic acid stimulates growth.	Extremely active PGR. Follow label instructions carefully.
Reduces height on a wide variety of ornamental plants growing in containers in greenhouses or shadehouses.	uniconazole-P (Sumagic) 0.055%	Reduces internode elongation by inhibiting gibberellin biosynthesis.	Extremely active growth retardant. Follow label instructions carefully. Stem or roots must have contact with chemical. Efficacy reduced when used as drench in bark media.
Controls height on a wide variety of ornamental plants growing in containers in nurseries, greenhouses, or shadehouses.	paclobutrazol (Bonzi) 0.4%	Reduces internode elongation by inhibiting gibberellin biosynthesis.	Extremely active growth retardant. Stem or roots must have contact with chemical. Efficacy reduced when used as drench in bark media.
Produces compact herbaceous perennials and woody landscape plants.	flurprimidol (Topflor) 0.38%	Inhibits gibberellin biosynthesis.	Increased plant quality in the absence of growth reduction includes darker leaf color, thicker stems, decreased water loss.
Produces compact herbaceous perennials and woody landscape plants.	ancymidol (A-Rest) 0.0264%	Inhibits gibberellin biosynthesis.	Efficacy is reduced as drench in pine bark media.
Produces more compact plants for many herbaceous and woody flowering plants.	chlormequat (Cycocel) 11.8%	Gibberellin inhibitor reduces internode elongation for a period of 1–3 weeks.	Multiple sprays or drenches may be necessary. Contains a wetting agent. Can be combined with B-Nine.
Produces more compact plants. For use on ornamentals in containers growing in a greenhouse, shadehouse, or nursery.	daminozide (B-Nine) 85%	Gibberellin inhibitor reduces internode elongation.	Multiple sprays or drenches may be necessary. Can be combined with Cycocel.
Increases lateral branching, prevents flower initiation and development, inhibits internode elongation, initiates flowering of bromeliads, and causes earlier defoliation of some woody nursery stock.	ethephon (Florel Brand Pistill) 3.9%	Enters plant readily and causes various physiological responses on different species.	Never apply to stressed plants. Ethephon increases lateral branching of azaleas, lantana, verbena, and other plants. For defoliation apply only after buds have matured.
Used on the inside of plastic or cellulose nursery containers to control root spiraling and promote a fibrous root system of many woody and other species.	copper hydroxide (SpinOut) 7.1%	Root growth regulator.	Inside of containers are sprayed with chemical formulated in latex paint. Once dried, containers can be stacked and stored up to 1 year.

Source: Adapted and expanded from North Carolina Cooperative Extension Service 2010.

promote a fuller canopy or to produce more cuttings from stock plants. However, compared with the floriculture industry, chemical growth regulators are not used as frequently in the production of outdoor container nursery crops. This is because the final crop size of woody perennials is often not as critical, since larger plants may be transplanted to larger containers. Also, one of the most important uses of chemical growth regulators in the floriculture industry is in timing flowering so that crops can be marketed for specific holidays, and this is rarely necessary in outdoor container nursery crops. However, for specific nursery crops such as azalea, flowering may be forced to increase marketability (see earlier section, "Timing of flowering and crop maturation"). In these instances, growth regulators can be used to overcome or compensate for the lack of environmental conditions that fulfill crop dormancy requirements, such as gibberellin application to compensate for chilling.

Chemicals are applied as spray or drench, as some compounds are taken up only through the foliage while others are taken up through the roots and foliage. Other methods of application include treating bulbs or cuttings or applying the chemical to the inside of a container.

PGRs should only be applied to well-watered, vigorous plants that are not under stress. Following label instructions is critical for the safety of humans and plants. Following label instructions is also important for obtaining the desired results from a particular PGR, as the amount of chemical used, the concentration, the environmental conditions during application, and the growth stage of the plant will affect plant response (fig. 10.8). The effects of PGRs differ for species and often even for different cultivars within a species. When using a PGR for the first time on a species or cultivar, it is wise to apply it to a few plants first as a trial, following label instructions.

Auxins

Auxins are most abundant in young tissues such as shoot and root meristems and growing leaves and fruit. Auxins are responsible for tropistic plant growth, which includes the growth of shoots and leaves toward light (phototropism), the downward growth of roots, and the upward growth of shoots (gravitropism) even when shoots are placed horizontally or upside down. Auxins stimulate the growth of leaves, stems, flowers, and fruit primarily through cell division. Auxin controls apical dominance, which is the suppression of lateral bud development by the apical bud. The role of auxin in root growth is more complicated. While exogenously applied auxin stimulates root initiation and early development on stem cuttings, it inhibits root elongation when applied to an intact root system.

The use of auxins for adventitious root formation on stem cuttings is important for asexual propagation of many ornamental plants. Synthetic auxins have been used for many decades to hasten rooting or to obtain more uniform rooting. See chapter 9 for more information on the use of auxins in plant propagation.

FIGURE 10.8. Plant growth regulator (uniconazole) applied improperly to the Easter lily on the right caused internode compression, resulting in a bushy, stunted plant. Failing to strictly follow all label directions can result in severe crop damage. *Photo:* J. K. Clark.

Gibberellins

Gibberellins enhance cell and stem elongation, overcome dormancy in seeds or buds, play a major role in flowering, and can enhance maturity in juvenile conifers. Gibberellins are synthesized in large quantities in immature seeds, young leaves, and, to a lesser extent, roots. Of more than 80 gibberellins identified, only a small number are strong growth promoters. Most plants show enhanced stem elongation particularly when treated with GA_3. The major gibberellin responsible for stem elongation is GA_1. In most plants, exogenously applied GA_3 is converted to GA_1 before growth resumes.

Gibberellin-treated plants often flower sooner compared with plants forced in a greenhouse after cold treatment. Many azalea cultivars flower earlier when dormancy is broken with gibberellin sprays versus the conventional cold treatment. This is probably due to simultaneously breaking dormancy and forcing flowers with gibberellins, while plants out of cold storage require a few weeks at warmer temperatures before they bloom. GA_3 can substitute for inadequate cold treatment of hydrangeas with one or several sprays at 5 parts per million, but it cannot be used to entirely replace cold treatment.

Antigibberellins, which block the biosynthesis of gibberellins in growing tissue, are common growth retardants and cause plants to grow shorter with thicker stems and darker green foliage. A number of different chemicals, such as daminozide, chlormequat, ancymidol, and flurprimidol, are growth retardants commonly used in floriculture and nursery production. Triazole compounds such as uniconazole and paclobutrazol are extremely effective at very low concentrations. Plants treated with these antigibberellins are more resistant to different stresses, such as heat and drought, that plants experience in shipping and handling. The use of most antigibberellins is restricted to commercial greenhouses and shadehouses, as well as uncovered production areas producing container-grown ornamentals. Although synthetic gibberellins can be used as an antidote to counteract overapplication of GA-inhibiting PGRs, it is often not economical.

Cytokinins

Cytokinins promote cell division, cell expansion, and organ formation in plants. More than 200 natural and synthetic cytokinins are known. Zeatin, dihydrozeatin, and isopentenyl adenine are natural cytokinins that are abundant and physiologically active in plants. Kinetin and benzyladenine are synthetic cytokinins with high biological activity. Cytokinins are primarily produced in the root tips and transported through the xylem to other plant parts. High cytokinin levels are also found in young leaves and immature fruit and seeds. Although cytokinins delay senescence, application of cytokinins in plant production has little practical value for this purpose.

Cytokinins can be used to increase lateral bud development. Application of a cytokinin treatment increases the low cytokinin to auxin ratio, which inhibits growth of lateral buds (apical dominance), to a high ratio, where growth of lateral buds is stimulated. The mixture of GA_4+GA_7 and benzyladenine increases growth and stem elongation in most dicot and monocot ornamental plants.

Ethylene

Ethylene is a hormone that occurs naturally as a gas. Fruit ripening, senescence, and abscission of fruit, flowers, and leaves are affected by ethylene. Ethylene influences differentiation as well as shoot and root growth, and it is involved in release from dormancy. Although ethylene seems to play a role in many plant responses, it is primarily as a reaction to environmental or physiological changes, while all other hormones seem to have a more regulatory role in vegetative and reproductive development of plants. Ethylene is produced by all tissues undergoing senescence and by most tissues in response to stress such as touching, wounding, pathogen or insect attacks, flooding, and drought. Ethylene is important in flower senescence as it is produced during the process and can hasten senescence when some flowering plants are exposed to low concentrations of the hormone (see the sections "Air Quality" and "Air Pollutants" earlier in this chapter).

Ethylene-releasing agents such as ethephon break apical dominance and can be used to stimulate branching for a more compact canopy or to increase the number of cuttings from stock plants. Ethylene is also used to initiate flowering in bromeliads and to hasten defoliation in nursery stock that is dug in the field. Although ethylene is produced naturally by plants in response to stress, application of ethylene-releasing compounds in concentrations that are too high, at an improper growth stage, or to stressed plants can cause mild to severe damage. Symptoms range from abscission of flower buds, flowers, and leaves to epinasty (bending) and damage of new growth.

Abscisic Acid

Abscisic acid (ABA) is produced in mature leaves and seeds. ABA regulates stomatal closure and dormancy. ABA responds to physiological stresses by signaling to the plant and protecting it from unfavorable conditions such as drought, salt stress, cold temperature, or frost. For example, water-stressed tips of roots produce ABA, which is then transported to the shoots, serving as a signal for leaves to prepare for water stress. Stomata close in response to the ABA signal, and the plant is protected from dehydration. Dormant seeds and buds are protected from active growth before environmental conditions are favorable.

Copper Compounds

Copper compounds prevent root circling and result in more fibrous root systems in many plant species. Formulated generally in latex paint, they are applied to the inside of containers, plug trays, or propagation trays.

References

AHS (American Horticultural Society). 1997. Plant heat zone map. AHS website, http://www.ahs.org/pdfs/05_heat_map.pdf.

ANLA (American Nursery and Landscape Association). 2004. American standard for nursery stock. Washington, DC: ANLA. AmericanHort website, http://americanhort.org/documents/nursery_stock_standards_AmericanHort_2004.pdf.

Bartok, J. 2005. Grower 101: Horizontal airflow. Greenhouse Product News. 15(9):26–28. Michigan State University Department of Horticulture website, http://www.hrt.msu.edu/energy/Notebook/pdf/Sec3/Horizontal_Air_Flow_by_Bartok.pdf.

Blom, T. J., W. A. Straver, F. J. Ingratta, S. Khosla, and W. Brown. 2002. Carbon dioxide in greenhouses factsheet. Ontario Ministry of Agriculture, Food and Rural Affairs (OMAFRA). Ontario, Canada. OMAFRA website, http://www.omafra.gov.on.ca/english/crops/facts/00-077.htm.

Both, A. J. 2003. Introduction into greenhouse environment control. Horticultural Engineering. 18(6):1–5. Rutgers School of Environmental and Biological Sciences website, http://aesop.rutgers.edu/~horteng/newsletter/2003/vol18-6Dec2003.pdf.

Brenzel, K. N., ed. 2007. Sunset western garden book. 8th ed. Menlo Park, CA: Sunset.

Cameron, A., B. Fausey, R. Heins, and W. Carlson. 2000. Firing up perennials – Beyond 2000. Greenhouse Grower, July: 74–78.

Chalker-Scott, L. 1999. Sterilized pruning tools: Nuisance or necessity? PlantAmnesty Newsletter, 11 (1): 4–5. Washington State University Puyallup Research and Extension Center website, http://www.puyallup.wsu.edu/~Linda%20Chalker-Scott/Fact%20sheets.html.

———. 2010. The informed gardener blooms again. Seattle: University of Washington Press.

Davidson, H., R. Mecklenburg, and C. Peterson. 2000. Nursery management: Administration and culture. 4th ed. Upper Saddle River: NJ: Prentice-Hall.

Dole, J. M., and H. F. Wilkins. 2005. Floriculture: Principles and species. 2nd ed. Upper Saddle River, NJ: Pearson Prentice Hall.

Downer, A. J., D. R. Hodel, and M. Mochizuki. 2009. Pruning landscape palms. HortTechnology 19:695–699.

Ingels, J. E. 2001. Ornamental horticulture. 3rd ed. Albany, NY: Delmar.

Lu, J., L. Wu, and B. Faber. 2008. Erosion control and runoff management. In J. Newman, ed., Greenhouse and nursery management practices to protect water quality. Oakland: University of California Division of Agriculture and Natural Resources Publication 3508. 83–96.

Nelson, P. V. 2003. Greenhouse operation and management. 6th ed. Upper Saddle River, NJ: Prentice-Hall.

North Carolina Cooperative Extension Service. 2010. North Carolina agricultural chemical manual. NC State University IPM website, http://ipm.ncsu.edu/agchem/9-toc.pdf.

Pemberton, H. B., J. W. Kelly, and J. Ferare. 1997. Production of pot roses. Portland, OR: Timber Press.

Pittenger, D., ed. 2002. California master gardener handbook. Oakland: University of California Division of Agriculture and Natural Resources Publication 3382.

Shahak, Y., E. Gal, Y. Offir, and D. Ben-Yakir. 2008. Photoselective shade netting integrated with greenhouse technologies for improved performance of vegetable and ornamental crops. Acta Horticulturae 797:75–80.

Shahak, Y., E. E. Gussakovsky, E. Gal, and R. Ganelevin. 2004. ColorNets: Crop protection and light-quality manipulation in one technology. Acta Horticulturae 659:143–151.

Urban Tree Foundation. Guideline specifications for nursery tree quality. Urban Tree Foundation website, http://www.urbantree.org/newspecs.asp.

USDA (United States Department of Agriculture). 2012. Plant hardiness zone map. USDA Agricultural Research Service website, http://planthardiness.ars.usda.gov/PHZMWeb/.

Photo: J. H. Lieth.

Diagnosing Plant Problems

Steven T. Koike and Steven A. Tjosvold

CHAPTER ELEVEN

Photo: J. K. Clark.

Diseases, disorders, and other plant problems are critical concerns for the wholesale nursery. These include biotic problems (caused by living organisms such as pathogens, nematodes, and insects and other arthropods) as well as abiotic problems (caused by factors such as temperature and moisture extremes, mechanical damage, chemicals, nutrient deficiencies or excesses, salt damage, and other environmental factors). The perishable nature of nursery commodities and the market demand for high-quality, defect-free plants place great pressure on growers, nursery managers, pest control advisers, and other personnel to minimize damage caused by these problems. Many plant problems, especially biotic problems, if not recognized and controlled early in their development can result in significant economic damage for the producer. Therefore, timely and accurate diagnoses are required so that appropriate pest and disease management options and other corrective measures can be implemented.

Diagnosis is the science and art of identifying the agent or cause of the problem under investigation. When one renders a diagnosis, one has collected all available information, clues, and observations and then arrived at an informed conclusion as to the causal factor(s). Hence, plant problem diagnosis is an investigative, problem-solving process.

It is essential to conduct the diagnostic process in a systematic, organized way. This chapter proposes and discusses such an approach for the diagnosis process, which includes the following elements:

1. Ask and answer the appropriate questions to define the problem, and obtain information that is relevant to the case under investigation.
2. Conduct a detailed, thorough examination of the plants and production areas.
3. Use appropriate field diagnostic kits and lab tests to obtain clinical information on possible causal agents and factors.
4. Compile all the collected information, and consult additional resources and references.
5. Finally, make an informed diagnosis.

In addition to discussing the process of making plant problem diagnoses, this chapter describes special considerations for diagnosing infectious diseases, insects and invertebrate pests, and abiotic diseases. See chapter 13 for more information on biotic and abiotic diseases, and see chapter 14 for more information on problems caused by insects, mites, snails, and slugs. For information on diagnosing vertebrate pest problems, see chapter 16.

Systematic Approach to the Diagnosis Process

Asking the Right Questions

As a problem-solving process, the diagnostic strategy requires that information be obtained in three major areas of investigation:

- the nursery plant in its healthy and unhealthy state
- the presence or nature of the possible agents responsible for the problem

- the surrounding environment, growing conditions, and production practices that form the context of the problem

Such information is best obtained by asking and then answering a series of questions as completely as possible. Examples of such questions follow.

Questions about the healthy plant

What are the genus, species, and cultivar of the plant in question? Is this particular plant suited to the production area? Is the cultivar resistant or especially susceptible to diseases and other problems? Is the plant sensitive to certain environmental factors (salinity, excess or deficient soil moisture, etc.)? How does the plant normally appear when grown under various conditions (full sun, shadehouse, greenhouse, coastal versus inland production locations, winter versus summer) or at different stages of growth and development? What is the normal growth rate? What are the characteristics, appearance, and growth habits of a healthy plant, and could these features be confused with an unhealthy plant?

Examples: Sweet gum trees sometimes develop unusual looking flanges, or "wings," on the bark of young branches. Such "winged" bark is a symptom caused by some viruses that infect woody plants. However, in the case of sweet gum, such bark formation is a natural feature of this tree species (fig. 11.1). Seedling eucalyptus trees can develop swellings at the base of the sapling; such galls resemble those caused by the crown gall pathogen *(Agrobacterium tumefaciens)*. These seedling eucalyptus, however, form naturally occurring lignotubers that mimic crown gall swellings (fig. 11.2). Some viruses cause abnormal color patterns such as mosaics or streaking; however, some plant cultivars are specifically selected and propagated for their genetic variegation (fig. 11.3).

Questions about the unhealthy plant

Symptoms. What are the symptoms of the affected plant? What plant parts are affected? Are symptoms restricted to external plant surfaces, or are there also internal symptoms (vascular streaking, discolored crown tissue)? Are symptoms present only on exposed plant surfaces or also on protected, covered tissues such as unexpanded inner leaves or unopened flowers? What is the distribution of the symptoms on any one particular plant (do symptoms occur on one side

FIGURE 11.1. Sweet gum naturally has branches with both smooth bark (lower branch) and flanged, or "winged," bark (upper branch). The flanged bark appears to have symptoms similar to those caused by some tree-infecting viruses. *Photo:* S. T. Koike.

FIGURE 11.2. Eucalyptus seedlings produce natural storage organs called lignotubers; such structures can resemble galls caused by the pathogen *Agrobacterium tumefaciens. Photo:* S. T. Koike.

FIGURE 11.3. Growing points with variegated leaves can sometimes arise spontaneously from some species such as this *Origanum*. Genetic variants such as this are sometimes confused with plants with virus disease or nutrient deficiency symptoms. *Photo:* S. A. Tjosvold.

FIGURE 11.4. *Raphiolepis* is host to the pathogen *Entomosporium maculatum*, which can also infect ornamental pear, photinia, loquat, toyon, and other plants in the Rosaceae family. *Photo:* S. T. Koike.

FIGURE 11.5. If symptoms occur in a regular pattern that abruptly ends, one might suspect a noninfectious problem. In this case, bare-root rose plants were laid out and left exposed for an extended period during hot weather. The affected plants died or developed very slowly. *Photo:* S. A. Tjosvold.

of the plant, only on older or newer leaves, on secondary roots but not on primary roots, etc.)? How did the symptoms first start? How do early symptoms differ from more advanced symptoms? How long have the symptoms been present? What is the extent (distribution and severity) of symptoms in the population of plants of a given species?

Growth stage. What is the growth stage of the affected plant (seedling, cutting, newly transplanted plant, mature plant, flowering or fruiting plant, senescent plant)? Is a particular growth stage associated with the problem? What was the condition of the plant when first placed in the production area? How does the growth rate of the affected plant compare to that of a healthy plant?

Examples: The stage at which a nursery plant is affected could significantly influence the direction of a diagnostic investigation. For example, if young maple seedlings or hebe seedlings or cuttings begin to develop wilting symptoms, investigation into the Verticillium wilt or Fusarium wilt diseases, respectively, are likely to be unfruitful as both problems occur when plants are much more mature. In contrast, crown and root rot of either young or mature pansy bedding plants can be caused at any stage by *Thielaviopsis basicola* or *Rhizoctonia solani*.

Symptoms on other plants. Are symptoms restricted to one species or one cultivar of nursery plant? Do the same symptoms occur on one type of plant that is located in different areas of the nursery? Alternatively, are there different plantings of the same plant type that are symptomless? What other plant species and cultivars might be affected? Do adjacent plantings, weeds, or nearby landscape plants exhibit similar symptoms? If other plants are affected, do they belong to a common group or family of plants?

Examples: When examining the problem of a particular nursery crop, the surrounding context of adjacent and nearby plantings can be informative. If the concern under investigation is a leaf spot problem on *Raphiolepis* (fig. 11.4), and if an adjacent crop of ornamental pear also suffers from a leaf spot problem, the possibility of Entomosporium leaf spot disease should come to mind because both plant species are susceptible to *E. maculatum*. Several pathogens can cause leaf spots and lesions on the leaves of agapanthus; however, if adjacent plantings of agapanthus and clivia exhibit similar light-colored leaf spots, it is possible that both crops were damaged by a granular herbicide or other chemical because clivia is not susceptible to the known agapanthus foliar pathogens.

Patterns. How are the symptoms distributed within the specific production area of concern? Are there patterns (repeating numbers of plants or plant rows) to the symptoms, or are they completely random throughout the planting? Are symptomatic plants found in clustered groups? Do the symptoms occur in lines, streaks, or circles? Are symptomatic plants found mostly along the edges of the planting? Are affected plants next to buildings, roads, ditches, weedy areas, other crops, or other production areas? Are symptoms associated with subsets of plants within the planting, indicating an association with plants from certain transplant trays, different sources of plant material, or other production factors?

Are symptoms associated with physical features at the nursery, such as low or high spots of the field, places where water does not drain well, presence or absence of underlying gravel, or particular soil types? Do such areas become flooded after rains or receive irrigation runoff? Are these areas under sprinklers or watered by drip irrigation systems?

Examples: Patterns are very useful clues to consider in a diagnostic investigation. Biotic problems caused by fungi, bacteria, viruses, and nematodes usually occur randomly in a nursery planting, at least early in disease development. However, distinct patterns often point to abiotic problems caused by chemicals, production practices, or weather extremes (fig. 11.5). If recently pruned, potted photinia shrubs all show brown to black marginal burns on newly emerging shoots in February and later exhibit shoot dieback, these plants could be suffering from cold damage and not a disease. If a block of potted rose plants shows unusual distortion and discoloration of foliage only on the plants along one edge of the block, then herbicide drift may be a possibility. If cyclamen are potted into 4-inch pots and the pots placed into trays (16 pots per tray) and all 16 plants in certain trays appear stunted or different compared with other trays, then these affected sets of 16 plants likely share some common factor. For example, a particular nursery worker may be handling and planting these cyclamen improperly, or this worker may have somehow planted a cultivar in these trays that is different from the ones planted in other trays by other workers.

Timing. What temporal factors are important? When did the symptoms first occur? Are there various stages of symptoms indicating new infections versus older ones? Have the features or severity of the symptoms changed over time? Do symptoms appear to have developed gradually over a period of time or rapidly and all at once?

Examples: Like patterns, timing evidence is valuable in the diagnostic process. Are the symptoms associated in time with a previous event or practice? If foliage begins to discolor, burn, or die following extremely cold or hot weather, then this environmental condition is likely responsible for the damage (fig. 11.6). Rain events are very important in the development of many foliar problems; if a series of rains occur and leaf spot problems begin to show up afterward, then examination and testing for bacterial or fungal foliar diseases are warranted. If symptoms develop shortly after a pesticide or fertilizer application, then these production inputs need to be considered as possible factors causing the symptoms (fig. 11.7). The reverse situation can also occur. If two blocks of nursery stock are planted at the same time but one block appears healthier, is greener, or has no disease symptoms compared with the second block, then perhaps the lower-quality set of plants did not receive a pesticide application to protect it, or it may have been skipped when fertilizer was applied.

Biotic or abiotic. Do symptoms resemble those caused by biotic agents, such as pathogens, nematodes, arthropods, or vertebrate pests? Alternatively, are symptoms more suggestive of physiological or abiotic factors, such as nutritional problems, physiological disorders, genetic mutations (chimeras), chemical damage, or environmental extremes? (See the section "Distinguishing Abiotic and Biotic Problems" later in this chapter for information that will help you answer these questions.) Do the symptoms provide evidence that more than one factor or pathogen is involved?

Questions about possible agents and factors responsible for the problem

Biotic. Which pathogens, nematodes, arthropod pests, snails and slugs, and vertebrate pests are known to occur on the host or cause feeding damage or other injury to the host? Which biotic agents occur in the geographic area of concern? Compile a list of common biotic agents that occur in the area.

Abiotic and physiological. Compile a list of abiotic and physiological factors known to cause problems for the plant in question (e.g., iron chlorosis in ornamental pear, water spotting on African violet).

Signs. A sign is the visible presence of a causal agent or factor. Are such signs present? Examples of signs are fungal growth and spores, bacterial ooze, insect bodies or excrement (frass), and chemical residue. Are there multiple signs indicating that more than one factor may be involved? What is the distribution of signs on the affected plant (present on all or only on certain plant parts)? Are signs present on all symptomatic plants, and do they appear to be associated with the symptoms? Are signs present on adjacent plantings or nonsymptomatic plants?

Questions about the surrounding environment, growing conditions, and production practices

Timing. What time of year did the problem occur? If the problem occurred before, did it take place during the same time of year?

Environmental conditions and context. What are the current and past weather conditions? Have there been any unusual weather patterns or developments recently or in the past few weeks or months? Have there been any conditions that would hinder plant growth or favor pathogen and pest development? Is there evidence of abiotic stress factors (temperature extremes, water stress or excess, salt buildup, mineral deficiencies and toxicities, pollution, wind or other mechanical damage, etc.)? What is the general location of the nursery (coastal versus inland, next to other crops or nurseries, next to roads, etc.)?

FIGURE 11.6. Low night temperatures at a critical flower bud development period caused malformation of gerbera flowers. In this case, a boiler was not heating the greenhouse properly as gerbera flower buds were developing. *Photo:* S. A. Tjosvold.

FIGURE 11.7. An accidental application of a miticide at a higher concentration than allowed on the pesticide label resulted in damage to *Hedera*. *Photo:* S. A. Tjosvold.

FIGURE 11.8. Ornamental pear leaves infected with *Entomosporium maculatum* will fall from the tree. The resulting leaf litter acts as an inoculum source that can infect other plants. *Photo:* S. T. Koike.

Examples: See the section "Timing" under "Questions about the unhealthy plant."

Potting mixes and soils. For potted plants, what type of potting mix is used? What is the condition of the mix (pH, porosity, salinity, nutrient level, etc.)? Has the potting mix been subject to any environmental extremes (excess water or salts, water deficit conditions, high temperatures in the container, etc.)? What supplements have been added to the potting mix? Has the potting mix been recycled or exposed to contaminating factors?

For field-grown stock, what is the soil type and soil condition (heavy clay, porous sand, compacted layers, change in soil type from affected to unaffected areas, pH, salinity levels, alkalinity, etc.)?

Water. What is the water source and quality? How does the water drain off the nursery? Are there low spots where water collects or drains poorly?

What type of irrigation system is used? Does the system deliver water uniformly? What are the frequency and duration of irrigations? How does the irrigation schedule correspond to other production practices such as pesticide sprays, fertilizer applications, transplanting, and other procedures? Are there indications that there was a problem with water delivery or irrigation, such as broken pipes, washed-out areas, deposition of mud and sediment, plugged drip lines, or damaged pots and containers?

Planting histories and previous crops. For both field-grown and container-produced plants, what nursery stock was produced prior to the current crop? Were the same species grown previously? Did the previous crops have the same problem that is facing the current planting? If the previous crop was different from the current crop, did the previous crop have the same problem nevertheless? If plants or plant residues (old leaves, stems, etc.) are left over from previous crops, examine these materials for signs or symptoms of problems.

Examples: The history and record of previous plantings can provide valuable background information and assistance for diagnosis. For example, if container-grown *Pyrus kawakamii* was infected with *Entomosporium maculatum*, then diseased leaves will senesce and fall off the plant, leaving inoculum at this location even when the containers are removed (fig. 11.8). This information is pertinent if a block of container-grown *Raphiolepsis* is later put at this same location. Knowing the histories of crops and their problems can be important for those pathogens that have a broad host range. For example, *Phytophthora* pathogens such as *P. cinnamomi* infect many nursery species. If a previous block of potted nursery stock was infected by a *Phytophthora* pathogen, then subsequent plantings may also be exposed if placed at the same location, due to the presence of infested crop residues or inoculum in the soil below the pots. For field-grown nursery stock, crop and disease histories are critical to maintain because pathogens such as *Fusarium, Phytophthora, Rhizoctonia, Verticillium,* and others can persist in the ground for many seasons and infect a series of plantings.

General production practices. What are the normal production practices for the crop (propagation steps, planting technique, irrigating, fertilizing, pruning, pest management)? What is the condition of the facility, including structures, fertilizer injection systems, irrigation systems, spray equipment, and so on? What production steps were completed or omitted? What was previously planted or placed in the area of concern? What equipment, such as spray rigs, was recently used in the area? Were any new production practices recently implemented? Did the onset of symptoms correspond with any cultural practice? Compile a complete record of production practices used on and around the symptomatic plants.

What fertilizers, pesticides, or other chemicals were applied to the plants, adjacent stock, ground beneath potted stock, benches, and nonproduction areas in the vicinity? Compile information on all materials, formulations, rates, additives, and gallonage used on symptomatic, nonsymptomatic, and surrounding plants. Focus especially on herbicide (selective, nonselective, preemergence, postemergence) and pesticide tank mix applications. What weather patterns occurred before, during, and after the spray applications?

Examples: See the sections "Patterns" and "Timing" under "Questions about the unhealthy plant."

Conducting the Examination

Gather appropriate equipment in preparation for examining the nursery. Equipment includes the following: notebook, hand lens, knife, pruning shears, shovel, plastic bags, bottles for water samples, labels and marking pens, soil sampling tube, flags for tagging plants for future observations, camera, ice chest or cooler, and disinfectant for tools. Examine all relevant propagation and production areas in an attempt to answer the pertinent information-gathering questions. Tour off-site areas such as adjoining roads, fields, and landscaped areas.

Thoroughly examine symptomatic plants. Carefully examine all roots, aboveground parts of the plants, and internal tissues of stems. Examine and dissect multiple examples of affected plants. Conduct a similar examination of healthy or asymptomatic plants for comparison.

Record all observations. Draw maps to indicate patterns of symptoms and locations of plantings having the problem. Take photographs to document symptoms and distribution of patterns. Photograph asymptomatic plants for comparison.

Select and use appropriate detection tools and field test kits. Deploying and assessing yellow sticky cards or other insect traps can provide information on insect presence as well as whether virus vectors are in the area (see chapters 12 and 14). A series of quick test kits for use in the field is available for rapidly detecting various bacterial, fungal, or viral pathogens. For example, ground-up pieces of leaf can be tested for *Tomato spotted wilt virus*, *Impatiens necrotic spot virus*, and other viruses (fig. 11.9), or macerated root pieces can be tested for Phytophthora root rot pathogens. Other field test kits can be used to assess nitrogen or potassium levels in leaf tissue, helping to determine whether nutritional problems are involved.

Collect and label representative plant examples for possible laboratory analysis. Samples should include various stages of affected plants (initial and more developed symptoms) taken from multiple sites. In many cases, and if feasible, the entire plant should be collected and not just a few leaves and stems; foliar symptoms may be caused by root and crown problems, so laboratory personnel need to see the entire specimen. Include a healthy plant for comparison. Keep samples cool and transport in a timely manner to the lab.

If the plant being examined is a sample brought in from the nursery, what is its condition? Samples that are in poor condition, incomplete, old, or of limited size and number may not yield useful information. Is the sample representative of the nursery problem? Have unaffected plants, placed in separate bags or containers, also been included in the sample for comparison?

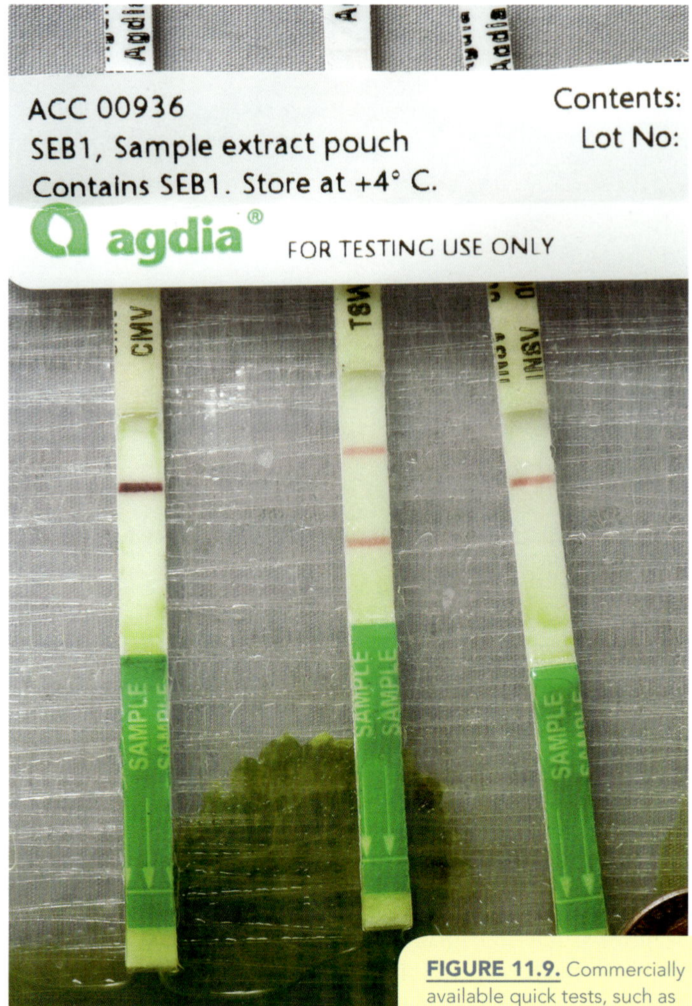

FIGURE 11.9. Commercially available quick tests, such as these test strips, can be used to rapidly assay for virus and other pathogens. *Photo:* S. T. Koike.

Incorporating Laboratory Tests

Because diagnosis based solely on symptoms is risky and may lead to inaccurate conclusions, laboratory analysis is usually highly recommended. Specific procedures for laboratory analysis of biotic and abiotic problems are presented later in the chapter, but some general guidelines follow.

If pathogens are possibly involved, send good-quality, representative samples to a plant pathology laboratory. If physiological or abiotic factors are likely involved, send samples of plant tissue, soil or potting mix, and water to appropriate laboratories. Other labs can be consulted for possible herbicide residue testing.

If warranted, collect appropriate root media, water, and leaf tissue samples for various analyses. Check with the laboratory for specific tests and their sampling requirements. In general, however, conduct analyses on soil from plants that are symptomatic and nonsymptomatic so that quantitative comparisons can be made. If possible, several small soil or potting mix samples should be taken from the root zone of several plants and combined into one sample (approximately 1 pint). Though not conclusive by itself, preliminary evidence of possible *Pythium* or *Phytophthora* problems can be gathered by conducting a pear baiting test. Soils are placed in containers, water added, and pears placed on top of the slurry; if these soilborne pathogens are present in the soil, distinctive brown lesions will develop on the pear (fig. 11.10).

Water samples should be representative of the water used to irrigate plants. Collect samples in clean plastic bottles. Rinse the bottles at least three times with the water to be sampled. About 1 quart is needed. Leaf tissue samples from symptomatic and nonsymptomatic tissue should be collected separately. Generally, a minimum of 3.5 ounces of fresh tissue is needed. Take 5 to 10 leaves from each plant and combine into one sample.

Compiling Information and Drawing a Conclusion

Before making a diagnosis, consult additional information sources such as printed references and books and online website resources, and record notes from these sources. Diagnosis may be assisted by also consulting with university and other experienced professionals. Digitally Assisted Diagnosis (DAD)—the process of acquiring digital images of plant problems and sending these images to researchers, diagnosticians, and other experts for viewing and diagnosing—can be very helpful in reducing the time required to identify the cause of the problem. Digital images can also be appropriate when distinctive symptoms or signs characterize the problem (such as powdery mildews, *Taphrina* diseases, and some herbicide injuries). However, care should be taken not to rely on such images alone. DAD can be a useful supplement to the diagnostic process, but it will not be able to completely replace it.

Compile all notes, observations, maps, laboratory results, photographs, and other information. This compilation will be the information base for the present diagnosis, and it can also be a useful resource for future diagnostic cases. Keep an open mind as the information is analyzed, and do not make unwarranted assumptions. After considering all the information, render an informed diagnosis.

Strategies for Diagnosing Abiotic and Biotic Problems

In the preceding section on the systematic process for diagnosis, we discussed some specific questions to ask and some specific procedures to follow when diagnosing abiotic and biotic problems. Different strategies that may be necessary for each type of problem are further described in this section. The first step is to determine whether the problem is caused by an infectious agent, and this can be difficult. Plant symptoms caused by biotic factors such as infectious diseases and arthropod pests are often similar to damage caused by other factors. Leaf spots, chlorosis, blights, deformities, defoliation, wilting, stunting, and plant death can be common symptoms of both biotic and abiotic problems; therefore, the presence of these symptoms does not necessarily mean the problem is a disease.

FIGURE 11.10. Pears and some other fruits can be used as baits to detect pathogens such as *Pythium* or *Phytophthora* in soil or water. *Photo:* S. A. Tjosvold.

Distinguishing Abiotic and Biotic Problems

Some general guidelines for distinguishing abiotic and biotic problems follow, and they are summarized in table 11.1.

Biotic problems

Identifying biotic problems is sometimes facilitated if signs of a pathogen, primarily the growth of a fungus, are present. The most obvious examples of such signs are the mycelium and spores produced by rusts and powdery and downy mildews. However, in other cases, nonpathogenic fungi can grow on top of damaged plant tissues and appear to be signs of a pathogen, resulting in possible misdiagnoses.

Biotic problems often affect one species or cultivar or plants of the same age, and typically they are initially observed in random or irregular locations; symptoms appear at varying times, and severity varies among affected plants. Biotic problems are infectious, spreading when environmental conditions are favorable, and they may be associated with pests that have affected the crop. This infectious aspect is important, as biotic diseases will often be progressive and continue to affect additional tissues and more plants.

Abiotic problems

In contrast to biotic factors, abiotic problems often affect several species or plants of various ages; typically, damage is relatively uniform, does not spread, and is often not progressive. Abiotic problems are not associated with pests. They are often caused by a single incident and are related to environmental or physical factors or cultural practices. Once the responsible factor has dissipated and is no longer affecting the plant, the plant may grow out of the problem and develop new, normal-appearing foliage.

Diagnosing Biotic Problems

Infectious diseases

To confirm if a problem is caused by a pathogenic fungus, bacterium, nematode, or virus, it is often necessary to have symptomatic tissues analyzed by a trained horticulturalist or plant pathologist. Such experts will attempt to microscopically observe the agent and recover it, if culturable, through isolation procedures. Lab analysis is particularly important to determine if multiple pathogens are infecting the plant.

It is worth emphasizing that diagnosing plant diseases requires careful examination of the entire plant specimen. Symptoms on leaves, stems, or other aboveground plant parts might lead one to suspect that a foliar pathogen is involved. However, these symptoms could also result if the roots are diseased. Therefore, it is important to conduct a complete examination of the symptomatic plant.

Because biotic diseases are caused by living microorganisms, the collecting and handling of samples is particularly critical. Samples that are stored for too long after collecting or allowed to dry out or become hot (if left inside a vehicle, for example) will sometimes cause the pathogen in the sample to die, making pathogen recovery and identification impossible. Plants that have been diseased for a long time and that are in the late stages of disease development will often be colonized by nonpathogenic saprophytic organisms. If these tissues are collected, the lab will have difficulty recovering the primary pathogen of concern because of the presence of these secondary decay organisms. Root samples should be collected carefully, as diseased roots are sometimes difficult to dig out of the potting mix or soil. Furthermore, they are usually colonized by the pathogen as

Table 11.1. Distinguishing abiotic and biotic problems

	Abiotic problem	Biotic problem
Hosts	often affects several species or plants of various ages	often affects one species or cultivar or plants of the same age
Pattern of plant symptoms	often related to environmental or physical factors or cultural practices; may be regular or uniform	often initially observed in random or irregular locations
Rate of symptom development	relatively uniform; extent of damage appears similar among plants	relatively uneven; time of appearance and damage severity vary among affected plants
Signs	no evidence of the kinds of pests or pathogens known to cause the current symptoms	presence of insects, mites, fungal mycelium and spore clusters, bacterial ooze, and mollusks; products produced by pests such as honeydew/sooty mold, cast skins, frass, or mollusk slime
Spread	not infectious; not progressive; commonly caused by one incident and does not spread	infectious; spreads on host over time if environmental conditions are suitable
Recurrence	possibly previously associated with current or prior environmental conditions or cultural practices	possibly caused by pests that have affected this crop during previous growing seasons or are known to commonly affect this crop species or cultivar

Source: Adapted from Costello et al. 2003.

Photo: S. A. Tjosvold.

well as secondary agents, and they are very sensitive to high temperatures and drying conditions.

Arthropod and other invertebrate pests

Insects, mites, slugs, and snails cause damage while feeding on the plant. Feeding damage is usually associated with the type of feeding characteristics and mouthparts of the insect or pest. For example, mites and insects such as whiteflies, aphids, and mealybugs have tubular sucking mouthparts that suck plant fluids, causing buds, leaves, or flowers to discolor, distort, wilt, or drop. Thrips have rasping mouthparts that result in dried-out, bleached plant tissue. Caterpillars, weevils, snails, and slugs have chewing mouthparts that make holes and cuts in foliage or flowers; they can also prune plant parts and sometimes consume entire plants.

If present, these pests are visible with the naked eye, a 10X hand lens, or a stereomicroscope, all depending on their size. An assessment must be made as to whether the identified arthropod or invertebrate matches the plant damage it is associated with. Sometimes the identified arthropod or invertebrate may not be the sole problem, or it could, in fact, be a beneficial organism or insignificant pest.

Aphids, whiteflies, thrips, leafhoppers, and some other insects that suck plant juices may vector pathogens such as viruses and phytoplasmas (and to a lesser extent, fungi and bacteria). They can first feed on infected plants, acquire the pathogen, later feed on healthy host plants, and transmit the pathogen to the new host. The insects do not necessarily have to be present in large numbers to cause a significant disease outbreak. The insect vectors are not always present at the same time the disease symptoms are being expressed.

The excrement and by-products from these pests can also provide clues that the pests have been or are actively present. Caterpillars and other chewing pests produce dark excrement or droppings. Greenhouse thrips and plant bugs produce dark, watery, or varnishlike droppings on foliage. Aphids, whiteflies, soft scales, and some other sap-sucking insects excrete excess plant fluids as honeydew, a sticky sap, which provides a medium for the growth of sooty mold.

Diagnosing Abiotic Problems

Nutrient deficiencies and toxicities

Nutrient deficiencies and toxicities reduce shoot growth and leaf size, and they cause leaf chlorosis, necrosis, and dieback of plant parts. However, nutrient deficiencies cannot be reliably diagnosed on the basis of symptoms alone because numerous other plant problems can produce similar symptoms. General symptoms exist that can be expressed by deficiencies of nutrients, but usually leaf or soil samples (or both) are needed to confirm the problem.

Herbicide, insecticide, and fungicide phytotoxicity

Herbicides used to control weeds in crops or in noncropped areas sometimes injure ornamental crops when they are not used in accordance with label instructions. Examples include when an herbicide is used in or around sensitive nontarget crops, when an herbicide rate is increased above tolerable limits, or when an applicator makes a careless application. By understanding the mode of action of the herbicide, one can determine if the symptom fits an herbicide application. Herbicide detection in affected plants is possible with the help of a specialized laboratory, but the analysis can be expensive. To minimize the cost of testing, the laboratory will need to know the suspected herbicide or its chemical group to narrow the analysis.

Insecticides and fungicides occasionally cause obvious plant damage. Symptoms can vary widely. Generally, flower petals are more susceptible to damage from pesticide applications than are leaves. The younger and more tender the leaves, the more susceptible they are to pesticide applications. Hot weather can exacerbate the damage that the chemicals cause. Pesticides that have systemic action can have a more profound effect. Some active ingredients can adversely affect the photosynthetic mechanism or other physiological processes, and they can result in a general leaf chlorosis, interveinal chlorosis, leaf curling, and stunting. Emulsifiable concentrate (EC) formulations, soaps, and oils can adversely affect the waxy surface layer that protects the leaf from desiccation. Applications with these products can result in the loss of the shiny appearance of a leaf, leaf spotting, and necrosis. Pesticides applied as soil drenches can cause poor germination, seedling death, or distorted plant growth.

Check label precautions against use on certain species, make sure the pesticide is not applied more frequently than recommended, or that the pesticide is not mixed with incompatible pesticides. When in doubt as to whether the plant species is sensitive to the pesticide, spray a few plants and observe them for several days to a week for any signs of damage before spraying any more of the plants.

Physiological and Genetic Disorders

Numerous disorders can occur because of environmental extremes—too much or too little of an environmental element such as light, temperature, water, or wind. Sunburn is damage to foliage and other herbaceous plant parts caused by a combination of insufficient moisture and too much light and heat. Beginning in areas that have been exposed to direct sunlight and heat, sunburned leaves turn yellow or brown and may eventually die. Sunscald is damage to bark caused by excessive light or heat. Damaged bark becomes cracked and sunken. Frost damage causes shoots, buds, and flowers to curl, turn brown or black, and die. Hailstones injure leaves, twigs, and, in serious cases, even the bark. Chilling damage in sensitive plants can cause wilting of foliage and flowers and development of dark, water-soaked spots on leaves that can eventually turn light brown or bleached and die. Physical and mechanical injury can occur when plants are mishandled during transport or routine cultural practices. Wounds might serve as entry sites for plant pathogens and can attract boring insects to woody stems.

Ethylene can accumulate to toxic levels in greenhouses when it is produced by the incomplete combustion of fossil fuels. This could occur when heating systems are not vented properly or when exhaust from forklifts is allowed to accumulate. Toxic levels of ethylene gas can cause premature abscission of flower buds, petals, and leaves. Other symptoms include wilted flowers, chlorosis, twisted growth or downward bending of stems and leaves, and undersize or narrow leaves. For more information on ethylene, see chapter 10.

Air pollution damage is caused by gases such as ozone, carbon monoxide, nitrous oxides, and sulfur dioxide. Typical symptoms vary widely, but they include slow growth and discolored, dying, or prematurely dropping foliage. Damage is often found where plants are located near sources of polluted air, such as freeways or industries, or where weather and topography concentrate the pollutants. For more information on air pollution damage, see chapter 2.

Sometimes plants or plant shoots exhibit an unusual and sudden change of color, producing discrete markings of variegation. For example, a plant with entirely green leaves suddenly produces a shoot that has leaves with edges lacking green pigment, stripes, or blotches (see fig. 11.3). A new shoot such as this is probably a chimera. It is produced when a genetic mutation occurs in a specific region of the growing tip, resulting in different regions of genetically different cells. The ostensible result of the genetic change depends on the arrangement of the genetically different cells in the shoot tip and their expression. This can sometimes lead to bizarre variegation forms and sometimes forms that are quite desirable. Sometimes, however, variegation can be caused by viruses. Viruses usually cause nonuniform chlorosis, such as mosaics, while chimeras usually produce patterned forms such as variegation of color on leaf margins, stripes, or complete loss of pigment. Some viroids may also cause bleaching of pigments in leaves; such symptoms, however, are generally produced throughout the plant and are not restricted to a single shoot (see chapter 13). Some nutrient disorders can cause variegation, but these disorders usually do not arise from a specific shoot as with chimeras. For more information on chimeras, see chapter 9.

References

Boger, P., and G. Sandmann. 1989. Target sites of herbicide action. Boca Raton, FL: CRC Press.

Costello, L., E. Perry, N. Matheny, M. Henry, and P. Geisel. 2003. Abiotic disorders of landscape plants: A diagnostic guide. Oakland: University of California Division of Agriculture and Natural Resources Publication 3420.

Derr, J. F., and B. L. Appleton. 1988. Herbicide injury to trees and shrubs: A pictorial guide to symptom diagnosis. Virginia Beach, VA: Blue Crab Press.

Dreistadt, S. H. 2001. Integrated pest management for floriculture and nurseries. Oakland: University of California Division of Agriculture and Natural Resources Publication 3402.

Eagle, D. J. 1981. Diagnosis of herbicide damage to crops. New York, NY: Chemical Publishing Co.

Grogan, R. G. 1981. The science and art of plant disease diagnosis. Annual Review of Phytopathology 19:333–351.

Retzinger, E. J., and C. Mallory-Smith. 1997. Classification of herbicides by the site of action for weed resistance management strategies. Weed Technology 11:384–393.

Schubert, T. S., and L. L. Breman. 1988. Basic concepts of plant disease and how to collect a sample for disease diagnosis. Plant Pathology Circular No. 307. Florida Department of Agriculture and Consumer Services, Plant Pathology Circular No. 307.

Sharma, M. P. 1986. Recognizing herbicide action and injury. Alberta Environmental Centre, Alberta Agriculture. Agdex 641–647.

Shurtleff, M. C., and C. W. Averre. 1997. The plant disease clinic and field diagnosis of abiotic diseases. St. Paul, MN: American Phytopathological Society Press.

Stewart, T. M., and V. J. Galea. 2006. Approaches to training practitioners in the art and science of plant disease diagnosis. Plant Disease 90:539–547.

Tickes, B., D. Cudney, and C. Elmore. 1996. Herbicide injury symptoms. Tucson, AZ: University of Arizona Cooperative Extension Publication No. 195021.

Integrated Pest Management

Julie P. Newman, Cheryl A. Wilen, Karen L. Robb, James A. Bethke, and John N. Kabashima

CHAPTER TWELVE

Photo: J. K. Clark.

Many pests and diseases are common in greenhouse and container nurseries due to the diversity of crops that are grown. Pests cause significant economic losses to the industry on an annual basis, and a major portion of the nursery production budget is allocated for managing them. Pesticide use can be intensive because of the low consumer tolerance for cosmetic crop damage and the regulations requiring clean plants prior to sale. Regulation requirements are stringent because nurseries ship container plants over large distances, resulting in the potential movement of pest-infested plants. Furthermore, pests attacking greenhouse and nursery crops can be resistant to one or more pesticides, leading to a cycle of increasing pesticide use. At the same time, restrictions governing pesticide use and the off-site movement of pesticides in runoff from nurseries are increasing. Moreover, the pesticides that are mandated in nursery quarantine programs and shipping protocols can be the very same pesticides that have pesticide load reduction requirements set by water quality regulators.

Many growers are able to control pests and mitigate pesticide runoff by implementing integrated pest management (IPM) programs. The use of preventive control practices and pest control actions based on monitoring and threshold levels can reduce pesticide use. Pesticide reduction results in lower production costs and prevents potential liability issues, litigation, and fines related to pesticide runoff, pesticide drift, and worker health and safety. Furthermore, the implementation of IPM practices limits the spread of pests within the nursery and reduces the potential of pest infestations on harvested plants that are sold and moved to other areas.

This chapter provides the basic principles of IPM, including pest identification, action thresholds, and monitoring. It discusses numerous preventive practices that can be implemented to reduce pest population buildup, reduce pesticide loads in runoff water, and reduce the potential of pest infestations on shipped plants. It is crucial that nursery personnel are skilled diagnosticians and use the process described in the previous chapter for diagnosing plant problems so that appropriate pest and disease management options and other corrective measures are implemented. This chapter addresses the types of control measures that may have to be applied, including cultural, mechanical, physical, biological, and chemical control methods. Furthermore, pesticide resistance management, application methods to reduce pesticide use, regulations, and other topics pertaining to the use of pesticides are presented.

The reader should refer to additional chapters in this manual for information on the management of specific types of pest problems (e.g., chapter 13 for plant pathogens; chapter 14 for insects, mites, and other invertebrate pests; chapter 15 for weeds; and chapter 16 for vertebrate pests). Selection of pesticides to minimize runoff and leaching is discussed in chapter 8. Moreover, specific pesticides may need to be applied in nurseries at intervals specified by quarantine regulations and shipping protocols. This issue is addressed in chapter 17.

Definitions and Fundamentals of IPM

Pest

The term *pest* is defined differently by diverse segments of the human population. In natural systems without humans, there are no pests; all organisms are just part of an ecosystem. In plant production systems, pests are organisms in the crop environment that cause economic or aesthetic damage to crops.

Numerous pests are found in nurseries due to the wide range of crops that are grown. Examples include insects and other arthropods (e.g., mites, snails, slugs), pathogens, nematodes, weeds, and vertebrates (e.g., rodents, birds, deer). Many pests are *exotic species* that are not native to a particular region. In container nursery production, pests also include organisms that do not necessarily cause damage to the host plant but are capable of causing damage to other plants if disseminated during shipping. Once established in new areas, these pests may be invasive. *Invasive pests* are extremely difficult to eradicate and can cause ecological disruption as well as economic problems (see chapter 17). Furthermore, sometimes outbreaks of a new pest in a nursery occur due to the use of pesticides. A *secondary pest* is a pest that is normally not a problem until pesticide applications, applied for other pests, kill its natural enemies, enabling the secondary pest to increase its populations unchecked by biological control. For example, if a pesticide applied to kill aphids also kills predatory mites, a secondary pest outbreak of spider mites could occur (Flint and Dreistadt 1998).

The level or seriousness of the pest is determined by interactions between the pest, the crop (such as susceptibility of the host plant), the environment, and time (e.g., length of exposure and time of exposure). When the host plant is young or small, pest pressure is generally more damaging than it is in larger plants, which can recover more quickly. The market also determines the level or seriousness of the pest (e.g., consumers may object to any insects on marketed plants, even beneficial ones). Furthermore, regulatory issues play a role (e.g., a pest may not be damaging to the crop but may have the potential to be a serious pest if introduced to other areas).

Two major types of plant injury are caused by pests on nursery plants. First, pests can directly affect crop quality and lower aesthetic value by damaging plants or reducing plant growth. Secondly, some pests can indirectly cause injury as vectors of plant pathogens (e.g., some insects, nematodes, and weeds).

IPM

Integrated pest management (IPM) is an ecosystem-based strategy that focuses on long-term prevention of pests or their damage through a combination of control strategies. These strategies are implemented according to established guidelines only after monitoring indicates that such treatment is appropriate. This involves knowledge of pest life cycles and an understanding of crop production. Pest control materials are selected and applied in a manner that minimizes risks to human health, beneficial and nontarget organisms, and the environment, with the goal of removing only the target organism. As shown in figure 12.1, some of the key components of IPM programs are

- monitoring crops
- accurately identifying pests
- developing action thresholds or guidelines for when to apply control measures
- implementing preventive control measures ("start clean and grow clean" methods)
- appropriate and effective use of nonchemical and chemical control tactics

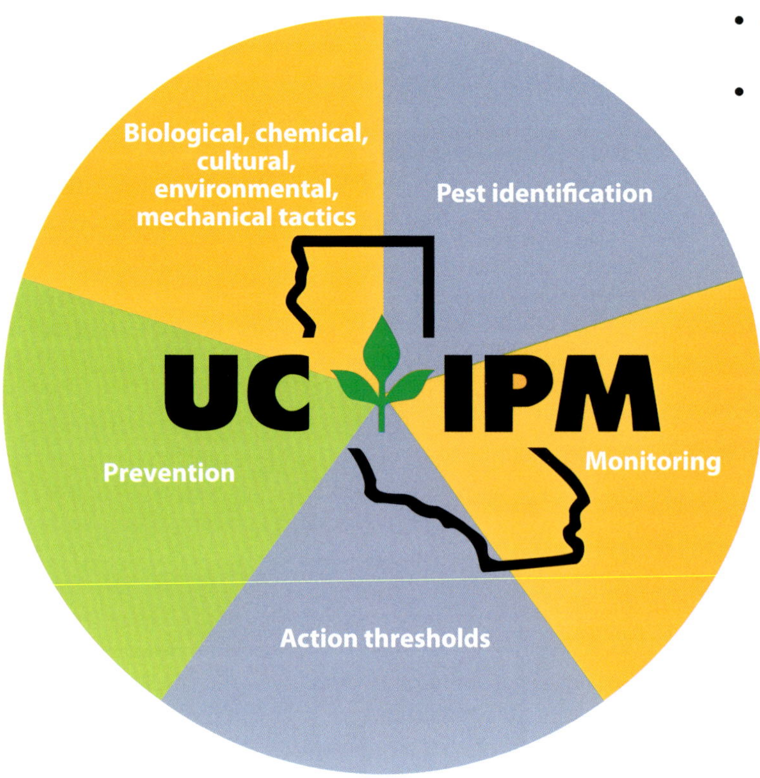

FIGURE 12.1. Components of an integrated pest management program.

Pest Identification and Diagnosis of Plant Problems

Pest identification

An important activity in IPM programs is monitoring crops. It is critical that nursery personnel involved in monitoring can identify all stages and damage symptoms of common nursery pests and relevant invasive pest species—including insects and other arthropods, pathogens, weeds, and vertebrates. This requires being aware of the pest's life cycle in relation to the growth stages of the crop. The ability to identify beneficial organisms is also important.

Diagnosing plant problems

Another important activity in IPM programs is seeking out plant problems and providing accurate diagnoses. Proper identification of the underlying cause(s) of poor plant performance is essential in determining appropriate and effective IPM decisions. Pesticides and other control measures that may be effective for one pest may not be appropriate for another. In addition, poor plant growth and damage is not always caused by pests; sometimes these problems are due to environmental and cultural conditions that cannot be remedied with pesticides. If problems are incorrectly diagnosed, inappropriate chemical treatments could be used that will be ineffective and add unnecessary pesticide loads to the environment. If the causal agent cannot easily be determined, it may be necessary to send plant samples to a diagnostic laboratory. See chapter 11 for more information on diagnosing plant problems.

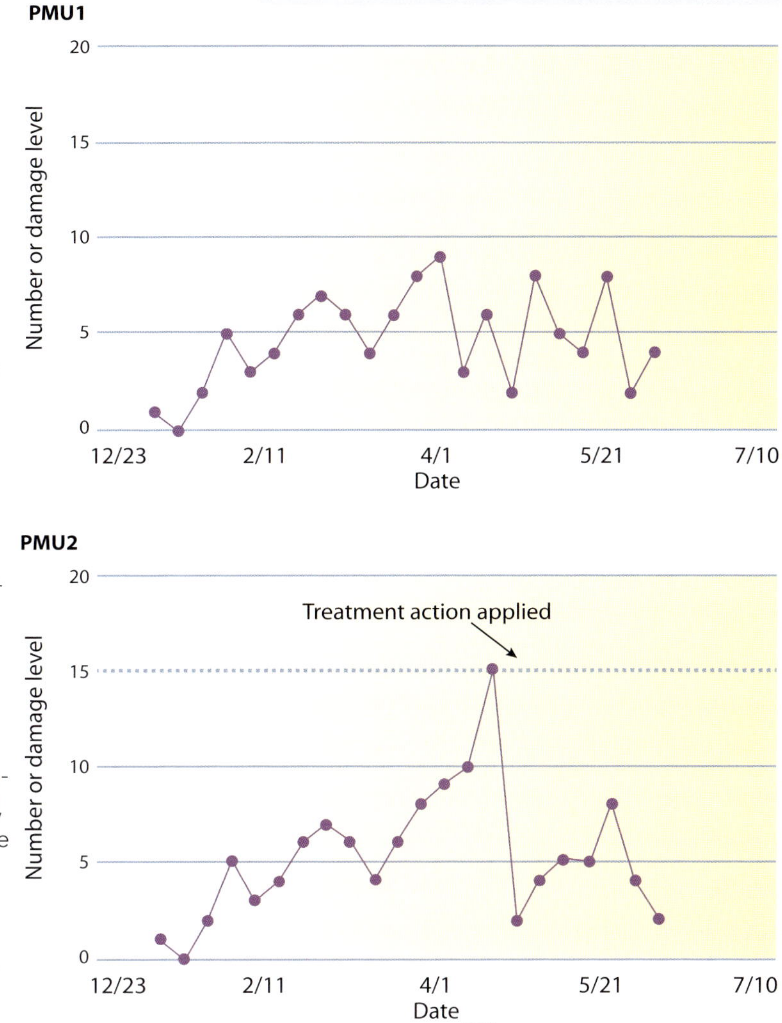

FIGURE 12.2. Example of the use of an established action threshold for pest management decisions. During the production cycle in the top graph (PMU 1), the aphid numbers varied but never reached the threshold level of 15 aphids; no control was applied. During the production cycle of the bottom graph from a different monitoring area (PMU 2), the aphid numbers reached the threshold; in this case, a control measure was taken.

Action Thresholds

Action, or treatment, thresholds are a fundamental component of IPM programs. The threshold is the level at which plant injury or pest population size is sufficient to warrant some type of control action; below this level the presence of pests and amount of damage can be tolerated. By applying remedial action only if pest populations exceed this threshold, overall pesticide use and liabilities associated with it can be reduced. (Figure 12.2 gives an example of the use of an action threshold.)

Thresholds should be quantitative (numerical) to be useful. For example, they could be based on the average number of pests per trap each week, the percentage of plants or leaves found to be damaged or infested during visual inspection, or the number or size of weeds in a container. Unfortunately, very few action thresholds have been developed for ornamental crops, in part because of the large number of crops grown, the large variety of pests, the different growing situations throughout the nursery industry, and the desire for very low levels of damage for most crops. It is therefore important that data is collected at each nursery to develop this information based on experience and economics. Even when published thresholds exist, they should be used only as a guideline or starting point for developing site-specific thresholds.

Economics and consumer tolerance of cosmetic damage are important considerations in developing treatment thresholds. In fact, action thresholds are often developed based on the cost of treating (or not treating) balanced against the price the grower will receive for the commodity. The price is affected by the buyer's willingness to purchase less-than-perfect nursery plants; because buyers typically have very low tolerance for cosmetically damaged plants, the threshold level set by nursery producers is usually low.

However, other factors should be considered besides price when setting the threshold level, such as hazards associated with making a pesticide application. Thresholds vary with crop, stage of plant development, cost of control methods, and time until harvest and market. Thresholds can often be raised if highly effective or quick-acting methods are available for controlling the problem and the plant has time to recover from the damage before being sold. Conversely, if available controls are slow acting or only partially effective, thresholds may be relatively low for that pest.

Container nurseries present unique challenges to the implementation of IPM programs because of the rapid intra- and interstate movement of potentially pest-infested plants by this industry. The ideal concept of only controlling pests when levels reach an economic threshold that could cause economic damage cannot always be practiced. This is because of the regulatory requirement that plants must be commercially clean (free of pests) and because of the increasing numbers of quarantine pests that place the nursery under quarantine protocols, mandating treatment with pesticides to meet a zero tolerance before shipping plants. Thus, regulations such as quarantines and shipping protocols may impose threshold levels of zero pests so that plants are clean when shipped for market, even when populations are low and pests do not directly damage the marketed crop. For this reason, it is especially important that plants be inspected prior to sale to ensure that they are free of pests. Treat or dispose of infected plants and remove weeds from containers.

Monitoring Programs

Monitoring (scouting) is a basic component of IPM programs. Monitoring is the regular inspection of crops and growing areas for pests, including insects, other arthropods, vertebrates, pathogens, nematodes, and weeds. Historically, nurseries applied pesticides on a calendar schedule. Today, however, growers typically use scouting data to determine appropriate pest control actions and optimal timing of pest control applications. It is likely that methods for timing these applications will further improve in the future. For example, although diseases are monitored in scouting programs, the use of preventive treatments of fungicides applied on a regular or calendar schedule is still prevalent. Research is under way to develop and refine disease models for ornamental plants. This will likely result in the ability to more accurately time preventive fungicide treatments with environmental and crop monitoring.

Benefits of Monitoring Programs

Monitoring provides many advantages. It serves as an early warning system for the presence of pests and diseases. This allows for the implementation of appropriate control strategies before pest populations and diseases escalate. Monitoring also facilitates the use of slower-acting methods that are more environment friendly and safer for workers. Furthermore, regular monitoring allows for the evaluation of pest management control actions, including the use of natural enemies.

Monitoring locates specific sites of infestation and identifies the type of pest problem, pest levels, and stage in the life cycle. Basing pest control strategies on monitoring and threshold levels reduces pesticide use, resulting in many benefits for the nursery. These include improved plant quality, lower production costs, improved worker safety, lower environmental risks, and compliance with regulations. Many of these benefits have been documented in studies conducted in various states by researchers and Extension agents. For example, in a 5-year demonstration project conducted by the University of California on a wide variety of ornamental crops that included potted plants, bedding plants, and outdoor nursery stock, scouting programs reduced pesticide use up to 40 percent and resulted in significant cost savings. Reduced insect and disease damage (from properly timed treatments) and less phytotoxicity (from reduced pesticide use) resulted in healthier crops (Newman et al. 1996, 1999).

The Scout and the IPM Team

Many nurseries involve a team of players in the overall pest management program. The IPM scout is the person who does the monitoring, records field data, and summarizes the information. The scout should have thorough knowledge of the pest complex for the specific crop varieties of the nursery and have some basic education in plant and pest biology. Ongoing training in these areas from UC Cooperative Extension personnel and industry professional groups is recommended.

However, the scout may or may not be responsible for making pest control decisions, carrying out the pest management strategies, and evaluating the effectiveness of the program. In some nurseries, the grower, pest control adviser (PCA), other employees, or consultants are responsible for these decisions. If the scout is not the decision maker, the scout should meet regularly with the person who is responsible for making pest control decisions and present summarized data for the decision-making process.

Furthermore, even the best scouts cannot see everywhere, so other nursery employees should be trained to recognize pests and contribute to the scouting effort. Trained employees can look for pests and damaged plants as they irrigate, pot, prune, and so on. A system can be developed that facilitates communication about the problems that employees see in the course of day-to-day activities. This information, passed on to the appropriate personnel, can also be used in making pest management decisions.

Requirements for a successful IPM program include good communication skills between all team members (fig. 12.3) and a shared commitment to base pest management decisions on monitoring data. Organizing monitoring information and other pest management records into a notebook kept at a place in the nursery where the data is accessible to all team members can facilitate communication. Furthermore, a plan should be developed and used to ensure that each member of the IPM team communicates with other team members on a regular basis to discuss if, when, and where control actions should be taken. The team should also evaluate the IPM program and develop strategies if pest control actions are ineffective.

Gathering Information and Eliminating Potential Problems

The scout's activities begin with gathering information concerning previous pest history and other potential pest problems. The scout should walk the entire nursery looking for sources of pests and pathogens, such as weeds that could harbor pests, and other issues that could contribute to plant problems, such as leaky irrigation lines. These issues should be addressed immediately by nursery personnel to avoid future problems.

FIGURE 12.3. Successful IPM requires good face-to-face communication on a weekly basis between the grower and the scout or pest control adviser to discuss if, when, and where control actions should be taken. *Photo:* J. K. Clark.

Developing a Monitoring Plan

Monitoring methods include the visual inspection of plant containers, crop foliage, flowers, and root systems for pests. The entire production area should be inspected for pests, such as snails, slugs, and weeds, and for potential issues that could contribute to plant problems as described in the previous section. If invasive pests are an issue, monitoring may involve inspecting plants outside the growing area that can serve as a source of pest infestations, such as landscapes, native plants, vegetative buffers, hedgerows, windbreaks, weeds in noncrop areas, and so on. In addition to the visual inspection of plants, monitoring methods include the use of tools such as insect traps, disease detection kits, and instruments for determining soil pH and conductivity. Monitoring methods may also involve collecting environmental data for forecasting disease outbreaks, sending samples to diagnostic laboratories for analysis, and reviewing records and cultural practices.

A written monitoring plan should be developed that clearly delineates what monitoring methods are to be used and how they will be employed. Monitoring methods chosen should be appropriate to the target pests and the crop (see chapters 13 to 16 and table 12.1 for monitoring methods for specific types of pests). For crops with quarantine pests, there may be protocols that must be followed.

The plot map

One of the first steps is to draw a plot map of the nursery growing area and divide it into smaller areas for sampling; nurseries are typically too large to be sampled as one unit. Each sampling area—sometimes called a pest management

Table 12.1. Monitoring methods and the types of pests they detect

Monitoring method	Pest problem				
	Pathogens	Abiotic disorders	Insects and mites	Nematodes	Weeds
visual inspection of crops	1	2	1	2	1
visual inspection of growing area and surrounding areas	1	1	1	1	1
environmental monitoring compared with optimal conditions	1	1	2	3	2
cultural care compared with recommended practices	1	1	2	2	2
on-site tools and tests (e.g., ELISA kits, traps, and water chemistry meters)	2	1	1	3	4
off-site laboratory tests	1	1	2	1	4
historic records compared with current scouting results	1	1	1	1	1

KEY: 1 = very important method for that problem; 2 = somewhat important method; 3 = not important method; 4 = not used or very rarely used.

Source: Adapted from Dreistadt 2001.

unit, or PMU (Cornell Cooperative Extension 1993; Dreistadt 2001)—is a contiguous area containing the same crop, usually at the same stage of growth. By separating a large area into smaller PMUs, scouting can be done more efficiently, and pesticide use will be limited to the PMUs that need it.

The size of the PMU depends on the number of crops and the crop age. If there is just one crop and the plants are of a similar age, PMUs may be arbitrarily divided by structure. For example, the PMU could be a shadehouse, a peak in a greenhouse range, or a nursery block in the field delineated by stakes and flagging tape. However, in most container nurseries there are numerous crops and different ages of the same crop. In this case, the PMUs should be divided by crop, variety, age, or plant size. If a crop is of very high value or is very susceptible, dividing the growing area into smaller PMUs allows for more intensive sampling and smaller treatment areas for spot treatments.

Sampling methods

Sampling is quantitative monitoring that estimates an overall pest population. Sampling often involves counting the number of each pest on a selected number of leaves of representative plants and then calculating the average number of each pest found on all plant samples to estimate the overall crop pest density for each pest. It may also involve measuring the percentage of infested plants. Sampling may include counting the number of weeds in containers or visually inspecting roots by removing the containers from a specific number of representative plants (e.g., when root pathogens or insects are a problem or the area has been overwatered).

For insects, indirect sampling (e.g., sticky traps, pheromone traps, potato disks) may be used (see chapter 14). However, these procedures must be backed by plant sampling because they are not as accurate as counting actual insects on leaves or other plant parts, and they do not ensure that the pest is actually in the crop. Sticky traps are the most common indirect sampling method in nursery monitoring programs. The number of insect pests on traps is counted, and the average number of each pest on traps is typically calculated (fig. 12.4). This data, along with the estimated pest density of the crop obtained from plant sampling, can be compared among sampling dates to determine if the pest population is increasing or decreasing. Moreover, quantitative plant sampling and counts on sticky traps can be used to estimate beneficial populations.

The monitoring plan should develop a protocol for the collection and inspection of samples for each pest of interest. The number of samples and the sampling unit should be constant over time. It is important that the site be sampled the same way each week to make results comparable among sample dates so that the status of the pest population can be assessed. Consistency is important, so ideally the same person should perform all the monitoring at a given site. If two or more people are involved, they should compare their methods in a preliminary sampling exercise to ensure that they are using the same methodology, and they should recheck periodically.

If the plants are small, the sample unit may be an entire plant; for larger plants the sample unit may be a set number of shoots and leaves or flower buds. For example, a scout may travel in a serpentine pattern through the planting blocks, stopping at regular intervals (fig. 12.5) and evaluating three leaves from the bottom, middle, and top of the randomly selected plant. The number of plants to sample depends on the value of the crop and the budget, as well as crop susceptibility to insect or disease problems.

Foliage and flowers of plants may be inspected while still attached and the number of pests immediately counted and recorded. Samples can also be pinched or pruned from the plant if the plants are large and if removal of foliage will not affect aesthetic appearance (destructive sampling). Destructive sampling will be necessary if a microscope is required to inspect samples, which is common when monitoring for parasitism in biological control programs. Microscope

FIGURE 12.4. Nursery crops are typically monitored by visually inspecting plants and plant parts for pests and damage and by setting out sticky traps in the crops. A scout (shown here) identifies and counts the number of trapped insects captured on the yellow sticky traps and records the information. Data obtained from visual plant inspections and the use of sticky traps is used to estimate pest density of the crop and to assess whether pest populations are increasing or decreasing. *Photo:* J. K. Clark.

FIGURE 12.5. Example of a path that a scout may take to systematically sample plants in a pest management unit (PMU).

work counts should preferably occur on the same day as the field sampling. Access to a microscope at the site—in a head house or office, for example—may be crucial in these instances.

The monitoring plan should also include a protocol for counting pests and beneficials on samples. Some scouts count by 10s or round off to the nearest 100 when counts are high. Another approach is to look at many plants (such as 100 plants) and record only whether pest species are present or absent on each sample. However, these shortcuts are only valid in the limited number of cases where research has obtained data using these methods and related it with the average pest density of that crop. In addition, these types of data are difficult to correlate with action treatment levels.

Implementing the Monitoring Plan

After information has been gathered and a monitoring plan has been developed, the scout is ready to begin routine inspections. It is critical that propagation areas and incoming shipments from other nurseries be intensely monitored to ensure that plants are free from pests and diseases when they are planted out (see the section "Preventive Control" later in this chapter for more details). The scout should monitor crop and noncrop areas at least once a week for pests in a consistent manner, following the procedure outlined in the monitoring plan. Foliage and other plant parts should be examined for pests by using a 10 to 15X hand lens or optivisor (a visor with a built-in magnifying lens, similar to a jeweler's magnifying visor). Be sure to examine the underside of leaves or inner plant parts, where many pests prefer to reside. Weeds in containers should also be noted and may be inspected for pests prior to their removal.

Keep in mind that pests are rarely distributed uniformly over all parts of an individual plant, and not all areas of the nursery will be infested. Although it is important to randomly select some or most plant samples, it also helps to target hot spots where pests tend to be a problem (e.g., near susceptible species or cultivars or in warmer areas in greenhouses). In addition to collecting samples using the monitoring plan protocol, be alert for any sign of pest or natural enemy presence anywhere in the nursery while scouting. Also, look for plants that are damaged by pests, adverse environmental conditions, or incorrect use of agricultural chemicals, such as pesticides and fertilizers.

Injured plants should be examined for signs of a pest. First, look at the part of the plant where the symptoms are occurring, and use a hand lens to determine if a pest is present. Look for abnormalities such as color changes, soft or necrotic areas, or spots on plant roots, leaves, and stems. Wash your hands after handling diseased plants, or wear disposable gloves to avoid spreading pathogens.

Sampling can help the grower determine where the infestation is located in the nursery to better target sprays. Record the specific location of infested plants using maps and flags so that control actions can clearly target that location. Reinspect these plants after taking action to determine if control was effective.

The scout should also inspect outgoing shipments of plants from the nursery. This is important to ensure that life stages of pests do not move to other areas along with the shipment, and it is an important tenet and regulatory requirement for managing invasive species (see chapter 17).

Data Collection and Record Keeping

The scout should keep written records of pest counts or degree of injury (fig. 12.6). Records of environmental conditions such as temperature and relative humidity are also important, as these can be used to predict the growth of pest and pathogen populations. Records should include the date, location,

FIGURE 12.6. Record keeping is essential for assessing pest and beneficial populations, conditions that promote pest problems, and for evaluating the effectiveness of pest control actions. *Photo:* J. K. Clark.

name of the person sampling, pests encountered, site plans of the nursery with monitoring areas identified, and other important information. Typically one data sheet is used for each PMU; so if you have too much data in a PMU to fit on one data sheet, consider breaking the sampling area up into smaller PMUs. An example of a data sheet that can be used as a template is shown in figure 12.7.

If the scout is not the pest management decision maker, he or she will need to summarize the data. These summarized data are presented to the grower and others involved in pest management decision making, who are often too busy to examine detailed raw data. Summarized information includes type and abundance of pests, percentage of infested plants, and location of heavy infestations. From such data collected over time, the scout may prepare a graph to illustrate pest population trends (fig. 12.8) or compare current data with data from the previous collection period.

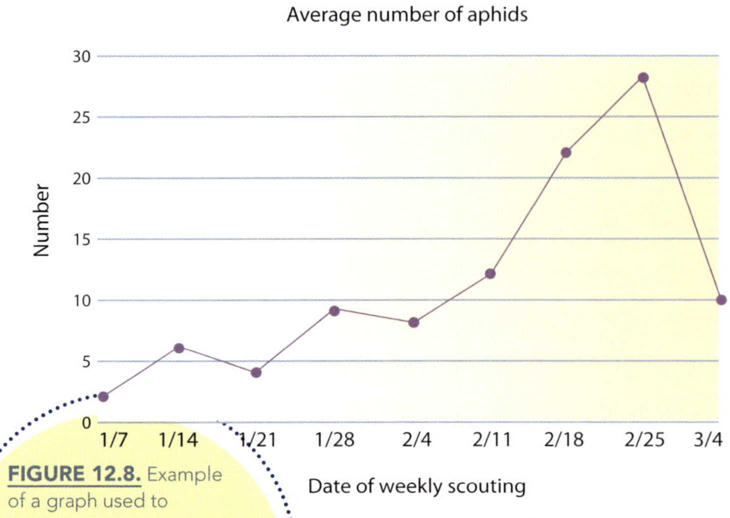

FIGURE 12.8. Example of a graph used to track pest populations over time. Graphical summaries can be used by the scout to visually present monitoring information to the grower.

Foliage inspection record **Scout's name** J. Doe			**Location** GH 20, PMU 1 **Time in** 7:15		**Date** 20 July 2000 **Time out** 7:45	
	Number of pests					
Sample no.	**Aphids**	**Mites**	**Powdery mildew**	**Thrips**	**Botrytis**	**Other**
1	55	–	–			
2	–	–	–			
3	–	–	–			
4	–	–	–			
5	–	–	–			
6	22	~50	–			
7	–	–	–			
8	–	–	–			
9	–	–	–			
10	–	–	–			
11	–	–	–			
12	–	–	–			
13	–	12	1			
14	–	–	–			
15	~100	–	–			
16	–	–	–			yellowing
17	–	–	–			yellowing
18	–	–	–			
19	–	–	–			
20	–	–	3			
Pest total	~177	~62	4			
Average	~9	~3	0.2			

Comments: – = none; ~ = approximate number, estimated
Investigate yellow plants for pathogens, nutrient pH problem
Make spot applications of soap for aphids

FIGURE 12.7. Sample recordkeeping form for a plant inspection. The percentage of plants infested or a rating scale can be substituted for actual counts of the pests. Pests and sampling units appropriate for the nursery can be substituted, such as counts of specific weed pests in sampled containers. *Source:* Dreistadt 2001.

When monitoring is also conducted to evaluate the efficacy of biological control, discussed later in this chapter, it is useful for the scout to count the number of natural enemies per sampling unit and then calculate the ratio of affected pests to healthy ones over time. If the natural enemy population is increasing faster than the pest population, the pest population will likely be controlled with the natural enemies.

Starting a Scouting Program

Important considerations when starting an IPM program are summarized in figure 12.9. Many nurseries prefer to start with existing employees as scouts rather than hire a professional scouting consultant or pest control adviser. Employees know the operation and are on site every day, which allows flexibility for changing the scouting schedule to accommodate pesticide applications and prompt inspections of incoming and outgoing plant shipments. However, if in-house employees are used, they must have extensive training in pest and natural enemy recognition and in data collection techniques. They also should have good attendance records because scouting is an activity that must be done consistently. An important limitation of in-house scouts with multiple responsibilities is that during busy periods in the nursery scouting is often neglected in lieu of more immediate duties. Therefore, hiring a professional scout is recommended if you do not have employees with a dedicated responsibility to regular monitoring. Since scouting programs reduce pesticide use and result in significant cost savings to the grower, the expense of hiring a professional scout can be offset.

Other production personnel should be trained to contribute to the scouting effort and be involved in the day-to-day IPM process as early as possible so that they understand and support the program. The success and sophistication of your scouting program will depend on the experience, skill, and enthusiasm of these employees.

When initiating a scouting program, it is best to gain experience by starting small. Keep track of the time and labor involved in scouting. This will help you determine if the program is cost effective or if you need to revise the number of samples collected. Test your methods and review your overall program regularly by evaluating records. Incorporate changes and fine-tune your program. When the program is working successfully in one area, or against one pest, expand the program to other areas or pests.

The purchase of insect identification guides and other reference materials will be critical to expanding the IPM program. Be sure you have the most recent information available. For more information on general procedures used in scouting, a good reference is *Integrated Pest Management for Floriculture and Nurseries* (Dreistadt 2001).

Preventive Control

Prevention is the best control method for all pest problems and is critical to any IPM program; it is the best way to reduce pesticide use and pesticide loads in nursery runoff. Methods to prevent the development and spread of pests include use of resistant varieties, good sanitary practices, and proper plant culture.

1. Identify all potential pests (including all life stages) for your cropping system. Train your pest management personnel to recognize all the major pests, their damage, and their natural enemies. Make sure they ask for assistance when they cannot identify the pest or problem. Provide these employees with references, resources, and tools such as hand lenses or microscopes to assist in pest identification.
2. Gather all available information on your target pests. Develop a resource library including information on pest identification, pest management practices, pesticide recommendations, and so on. Information can be collected at meetings, from your farm advisor or Extension agent, in trade journals, and many other sources.
3. Establish injury levels and action thresholds for each individual species before you get started. An injury level is the pest population size (e.g., number of whiteflies per sticky card) that is associated with unacceptable damage. Action thresholds are the set of conditions required to trigger a control action. Determine the infestation levels that will cause unacceptable damage by keeping records, taking into consideration the time of year, plant growth stage, and so on. Over time you will refine your determination of injury levels and action thresholds as you gain experience with your situation.
4. Establish monitoring procedures for each target pest. Monitoring is essential for the success of your IPM program. It provides early detection of pest problems before pests reach damaging levels, and it is important in determining the effectiveness of treatment actions. Monitoring methods vary from pest to pest, but they all require regular checking for pests and evidence of pests or pest damage. Establish a monitoring plan for detecting pests as well as natural enemies. Your monitoring methods may be rough at first, but you will refine them over time. The key is to find a simple procedure that provides the necessary information concerning your pest complex in a timely manner.
5. Establish a record-keeping system. Good records are crucial for evaluating and improving your IPM program. Your records should include information such as pest identification, size and extent of the infestation, location of the infestation, control options utilized, effectiveness of the control options, labor and material costs, and crop quality and yield.
6. Develop a list of acceptable management strategies for each pest. Methods that prevent pest problems and eliminate the need for the use of pesticides are preferred.
7. Develop specific criteria for selection of pest management methods. Criteria may include the following: least disruptive to natural enemies; least hazardous to humans; least toxic to nontarget organisms; least damaging to the general environment; logistical considerations; and short-term and long-term cost effectiveness.
8. Designate a person responsible for each step, and develop a communication plan. Monitoring should be a dedicated responsibility so that it is completed on a regular basis, but other employees can contribute to the effort if they are adequately trained. In addition to monitoring, it is important that someone be assigned the responsibility for making decisions, carrying out the pest management strategies, and evaluating the effectiveness of your program. This could be the scout, grower, PCA, or other employees or consultants. Develop a plan to ensure that all members of the IPM team communicate with one another on a regular basis and stick to it.

FIGURE 12.9. Some important considerations in starting an IPM program.

Use of Resistant Varieties

Certain plant varieties, or cultivars, have physical or biochemical characteristics that make them resistant or less affected by pests. Some varieties may be less suitable for infestation by pests, while other varieties may be able to support pests without suffering appreciable damage. Before buying plant material, especially propagation stock, check with your supplier for resistant varieties. Keep records of plants that demonstrate good resistance.

Good Sanitary Practices

Starting with clean plants (i.e., free of pests and pathogens) and implementing measures to keep them clean prevents spread and infection to other plants. It also reduces the size of areas that require pest treatment. Use the following examples of sanitation practices whenever possible and appropriate.

- Purchase certified plant material. Certified or culture-indexed stock is available for some plant species. Culture-indexed plants are vegetatively propagated and tested to confirm that they are free of specific bacterial, fungal, or viral pathogens. Certified nematode-free stock for certain crops can be obtained from participants in the California Department of Food and Agriculture Nursery Program. This is especially important when selecting propagation stock. Visit your plant supplier to become familiar with their sanitation program.
- Inspect incoming plant material. Inspect all new shipments of plants, plugs, cuttings, or transplants brought into the nursery; treat or dispose of infected plants promptly before introduction to the growing area. Avoid purchasing plants with weeds growing in the containers or liners or plants with obvious arthropod pests.
- Quarantine plants. Because disease symptoms may not be immediately expressed and arthropod pests may not be immediately evident, it is recommended that new plants be placed in a separate greenhouse or nursery section and monitored separately for pests prior to being introduced to the main production area. New plants should ideally be isolated long enough for the major potential pests of that crop to appear. For insects, plants should be isolated at least as long as it would take for the eggs to develop, which can be estimated using a degree-day model. (Information on using degree-day models for some pests found in nurseries is available at the UC IPM website listed in the References at the end of this chapter). For diseases, plants are typically quarantined for 2 to 4 weeks for symptom expression. However, if a specific disease has been coming in on plant material shipped to the nursery and there is no source of certified, pathogen-free stock from a certified propagator, the plants may need to be quarantined for a minimum of 3 months for disease expression, which would not be practical for short-cycle crops.
- Eliminate weeds and volunteer plants. Many plant disease vectors proliferate on volunteer plants and weeds (e.g., western flower thrips infected with tospoviruses). In addition, weeds can harbor other pests, increase plant competition for water and nutrients, and decrease air circulation. Install weed mat barriers if floors are bare dirt or gravel, and keep them maintained. Weeds removed from containers should be taken away from the site and not just discarded on the ground (see chapter 15 for more details).
- Eliminate pathogen reservoirs. Remove diseased plants; destroy them or treat them in an isolated area. Collect, bag, and remove crop residues, pruned plant material, and unmarketable plants. Place in a covered dumpster away from and downwind of healthy plants and production areas. Disinfect pruning tools before and after use to avoid spreading diseases.
- Treat planting areas and container media before establishing new crops. This can be done by fumigating, heat steaming (at 140°F for 30 minutes), or chemically treating media (see chapter 5). Alternatively, purchase planting media that has been pasteurized to kill plant pests. To prevent infection, all media should be stored in original bags until use or in covered containers. Reuse of untreated media increases the potential for disease and the use of fungicides.
- Disinfect hands and shoes. Hands should be washed before handling propagation material (fig. 12.10). Shoes should be cleaned of soil before going into the propagation area, and feet should never be allowed on the propagation benches.
- Use hose hangers. Do not leave hose nozzles on the ground or bench surface. This simple practice avoids transferring pathogens from the ground or bench to growing areas. It is an inexpensive practice to implement, but it is often overlooked (fig. 12.11).
- Treatment of recycled containers. Thoroughly wash containers to remove debris, soil, and plant material that cling to these items; follow with steam treatment. See chapter 13 for more details.
- Keep refuse containers tightly closed. Infested plants or potting media that are thrown into trash bins can still support insects such as whiteflies, fungus gnats, and western flower thrips. These insects can easily fly out of the container and reinfest your crop. Furthermore, pathogens on plant debris, such as *Botrytis*, can be dispersed by wind and rain.
- Avoid vectoring diseases and insect "hitch-hikers." Schedule work in infested areas of the nursery at the end of the day or after work is done in clean areas to prevent contamination. Avoid wearing clothing in colors such as yellow and blue that attract insects. Insects in infested areas that are attracted to clothing can be carried to clean areas.

Proper Cultural Practices and Environmental Management

Many pest problems can be prevented by providing good growing conditions to avoid environmental stresses. Healthy plants are more likely to avoid or withstand infection or infestation by pests than are unhealthy plants. Nurseries have numerous microenvironments—small areas that differ in temperature, humidity, and light. Select planting sites within your nursery that best match your specific crop requirements. Before planting or moving plant material into an area, check irrigation lines, heaters, and other equipment to make sure they are in proper working order. Maintain vigorous plant growth by providing plants with the proper potting mix, fertilization, irrigation, and other environmental and cultural practices that are optimal for the species.

FIGURE 12.10. Good sanitation practices include washing hands before handling propagation material to reduce disease spread. *Photo:* J. K. Clark.

FIGURE 12.11. Irrigation hose clamp to keep nozzle off the ground and reduce spread of diseases. *Photo:* J. K. Clark.

Provide adequate fertilization. Adequate fertilization is needed to avoid nutrient deficiencies and keep plants healthy enough to resist and recover from pests and diseases. Too much nitrogen, however, can encourage excessive growth of new shoots as well as overly succulent growth, which may be more susceptible to certain pests and diseases. Overfertilization can also cause an excess of soluble salts in the soil (see chapter 6).

Provide good air circulation. Provide adequate plant spacing to promote good air circulation and prevent diseases such as Botrytis. In greenhouses, use a good horizontal airflow system.

Avoid prolonged leaf wetness. Wet leaves encourage the development of foliage diseases (see chapter 13). Subirrigation or the use of a drip system is recommended. Avoid overhead watering; if this is the only method of irrigation available, then irrigate early in the day so that plants can dry as rapidly as possible.

Provide good drainage and water management. Overwatering and poor drainage promulgate root rot diseases and contribute to runoff. Never allow plants to be in or around standing water, as this is particularly damaging to plant health. Pools of water spread disease or serve as breeding areas for pests such as fungus gnats and shore flies. Also avoid water stress. For example, spider mite infestations are worse on water-stressed plants.

Nonchemical Control Practices

When monitoring indicates that pest control measures are necessary, consider implementing nonchemical control strategies as an alternative to, or in addition to, chemical control. The use of nonchemical strategies—including cultural, mechanical, and biological control—may reduce the need to use chemicals and slow the development of pesticide resistance. These practices can also help mitigate pesticide runoff.

Cultural Control

Cultural controls are modifications of normal plant care activities that reduce or prevent pests (e.g., grouping susceptible varieties together to increase scouting efficiency and better target pesticide applications; separating new plantings from older, possibly infested plantings to avoid movement of pests and pathogens to newer crops). Knowledge of the life histories of pest species is essential to the effective use of cultural control methods. Since these methods are used more to prevent pest problems than to control them, some of these methods were discussed in the section "Preventive Control."

Mechanical Control

Mechanical control tactics reduce pest abundance using methods such as hand-picking, physical barriers, or machinery. Examples include

FIGURE 12.12. Screening of greenhouse roof and fan vents to exclude insect pests. Screening used in combination with good sanitation practices in this greenhouse reduced pesticide use by 90% and resulted in better-quality plants. *Photo:* J. K. Clark, courtesy of UC IPM.

FIGURE 12.13. Double doors at greenhouse entrances used in combination with positive-pressure ventilation help to exclude pests and prevent contaminated air from flowing into greenhouses. *Photo:* J. K. Clark.

- Hand-pulling weeds or applying mulch for weed control.
- Applying reflective plastic mulches beneath container plants and using reflective mesh plant covers to repel certain insects.
- Installing insect screens over air intake vents or cooling pads in greenhouses to exclude flying and windborne pests (fig. 12.12). However, small insects such as thrips require very fine screening that can significantly reduce airflow, resulting in overheating of the greenhouse and stress on the fans. This problem can be addressed by increasing the surface area of the vent, usually by constructing a screen-covered enclosure around the vent opening. Greenhouse manufacturers and suppliers sell screening of various opening sizes, and they can provide information and guidance on proper installation of screens to achieve the necessary airflow. Also, refer to *Integrated Pest Management for Floriculture and Nurseries* (Dreistadt 2001) for more information.
- Using a double-door system in greenhouses to exclude insects (fig. 12.13). Pests can be excluded by constructing a small room onto the greenhouse with a door to the outside so that the outside door must be closed before the greenhouse door is opened. In some designs, a fan is installed that blows air back toward the outside door (positive-pressure ventilation).
- Applying sticky barriers to exclude climbing insects such as ants and weevils.
- Applying mesh covers for small containerized fruit trees and other plants to keep out large insects and birds.

Environmental Control

Environmental control methods indirectly control pests and prevent damage by manipulating the physical environment. Examples include

- Using heat treatments such as composting, solarization, and steam pasteurization to control soilborne pests.
- Heat-treating seeds by soaking them in hot water to control seedborne diseases.
- Reducing relative humidity and condensation in greenhouses by increasing ventilation and applying heat.
- Changing irrigation timing and method (such as switching from sprinklers to drip irrigation) to control foliar pathogens.
- Improving drainage and aeration of planting media to prevent abiotic and pathogenic problems.

Biological control

Biological control practices use living organisms to reduce pest populations. The use of biological control with other IPM strategies can be highly effective in keeping pests in check and reducing pesticide loads. Beneficial organisms can occur naturally in nursery environments. Often, the importance of naturally occurring biological control agents may not be appreciated until the use of a broad-spectrum pesticide aimed at primary pests results in mortality of natural enemies. The eradication of beneficial insects then results in outbreaks of secondary pests.

Types of organisms used in biological control

Organisms that provide biological control of pests are often called beneficials, natural enemies, or biological control agents. Many kinds of biological control agents exist. The most common ones used for biological control in nurseries and greenhouses are predators, parasitoids (parasites), and pathogens.

Predators are generally very mobile and larger than their prey (the pest species). Predators include a wide range of beneficial insects, such as spiders, mites, and other living organisms. A predator seizes, overpowers, or immobilizes its prey and then either consumes it entirely or sucks its body fluids. General predators such as lacewing and lady beetle larvae feed on a diverse variety of pest species that attack nursery crops. In comparison, parasitoids are more host-specific.

Parasitoids are smaller than the host pest species and spend part or all of their life cycle with their host. Insect parasitoids that attack insect pests such as aphids, scales, and whiteflies are generally tiny wasps or flies that use their ovipositor (or stinger) to lay eggs on or in their insect hosts. Insect parasitoids typically deposit one or more eggs in or on the host, and when the eggs of the parasitoid hatch, the larvae feed on or inside the host. Eventually the host dies and an adult parasitoid emerges. When a parasitoid that feeds inside the host chews its way out of the dried-out shell of the host's body, it leaves a distinct exit hole. When monitoring, these exit holes and dried-out insect bodies (e.g., aphid mummies) are characteristic signs of parasitoid activity. Ant control is important when relying on insect parasitoids for pest control because ants often attack and kill beneficial predators and parasitoid insects. Refer to chapter 14 for details on controlling ants.

Pathogens are microscopic organisms such as bacteria, viruses, and fungi that cause diseases in pest insects, mites, nematodes, or weeds. Examples of pathogens that are routinely used in nursery crops as biopesticides are *Bacillus thuringiensis* (a bacterial by-product that, once ingested, paralyzes the insect's digestive system) and *Beauveria bassiana* (a fungal organism that infects the insect through the exoskeleton).

Types of biological control practices

In nursery production systems, biological control practices generally fall into two categories, conservation and augmentation. In most nursery situations, the emphasis is on conservation of naturally occurring biological control organisms to keep insects such as aphids maintained at low populations. Conservation is accomplished by monitoring populations of natural enemies in scouting programs and considering

FIGURE 12.14. The periodic application of biological control agents in combination with good management practices and judicious use of pesticides have been successful in some commercial greenhouse crops. Here, cards containing greenhouse whitefly pupae parasitized by *Encarsia formosa* are fastened onto leaves of poinsettia plants. Photo: J. K. Clark, courtesy of UC IPM.

the effects of pesticides on beneficials when implementing pest control strategies. Whenever possible, pesticides are avoided, especially broad-spectrum pesticides that are detrimental to beneficial organisms in the environment. When the use of pesticides is necessary, the lowest effective rate is used. In addition, a pesticide that is specific to the target pest is selected when feasible, such as *Bacillus thuringiensis* for caterpillars, or a material that could be minimally disruptive to natural enemies, such as insecticidal soap or horticultural oil. (*UC IPM Pest Management Guidelines for Floriculture and Nurseries* includes ratings of relative toxicity to natural enemies for common insecticides and miticides used on these crops. See the website in the References at the end of this chapter.) Furthermore, treating only the plants or plant parts that are in immediate need of an application leaves other plant areas as reservoirs for natural enemies. For example, control strategies such as direct spraying and spot spraying, described later, can be used to conserve beneficial insects and mites. Conservation may also include the maintenance of alternative hosts (including plants) for naturally occurring beneficial organisms.

Augmentation is the second type of biological control strategy that is sometimes used in nurseries. Augmentation is the periodic release or application of biological control agents into the production system. It usually involves the purchase and release of commercially produced insect or microbial beneficial organisms (see the References for the Association of Natural Biocontrol Producers website). Augmentation in combination with good management practices and judicious use of pesticides have been highly successful with certain greenhouse crops including poinsettias (fig. 12.14), cut roses, and gerberas. More research is necessary before recommendations concerning the use of biological agents can be made for other ornamental crops. If recommendations for the crops you grow are not available and you are interested in trying augmentation, start small and evaluate effectiveness before making large economic investments. A biological control practice that can be used on many ornamental crops is the use of commercial preparations containing antagonistic bacteria or fungi. This can be incorporated into planting mixes to control root pathogens. In addition, crown gall—a disease that affects the roots and stems of a wide range of nursery crops—can be prevented by treating exposed root systems or pruned surfaces with an antagonistic strain of *Agrobacterium*. Furthermore, as previously mentioned, some biopesticides containing insect pathogens are commonly used on nursery crops.

Chemical Control

Chemical control refers to the use of pesticides, which are products for controlling, repelling, or mitigating pest problems. Chemical controls should be used only when needed and after the pest treatment threshold has been reached. Ideally, the selected pesticide is effective against the specific pest but is also less hazardous to people, the environment, and beneficial insects. Overuse of pesticides can cause a number of adverse environmental effects as well as health and safety risks for applicators and workers. Furthermore, repeated use of pesticides can adversely affect crop quality and lead to the development of chemical resistance in the target pest.

Classification of Pesticides

Pesticides are categorized in many different ways. Some examples are provided below:

- Target pests. Three common groups that are classified by the target pests that the pesticide is intended to control are insecticides (control insects); herbicides (control weeds); and fungicides (control fungal diseases). More specific pesticide categories are listed in table 12.2.
- Effect of the pesticide on the pest. For example, attractants lure pests; repellents repel pests; desiccants and defoliants remove or kill leaves and stems; and plant growth regulators alter normal growth processes.

Table 12.2. Types of pesticides and the pests they target

Pesticides	Pests controlled
acaricides	mites, ticks, spiders
avicides	birds
bactericides	bacteria
fungicides	fungi
herbicides	plants (used to control weeds)
insecticides	insects
miticides	mites
molluscicides	slugs, snails
nematicides	nematodes
piscicides	fish
rodenticides	rodents

Note: Pesticides can be classified by the target pests they are intended to control. The names of these types of pesticides end in the suffix "-cide." Examples of these pesticides are listed above.

- Source of the pesticide material. Examples include botanicals or plant extracts such as neem oil; inorganic pesticides such as sulfur and copper; synthetic pesticides such as organophosphate and chlorinated hydrocarbon insecticides, dinitroaniline herbicides, and thiophanate fungicides; and biological extracts or fermentation products.
- How the active ingredient reaches the target pest. Examples include contact pesticides (killing when the pesticide touches the target organism); stomach pesticides (must be ingested by the pest); systemic insecticides and fungicides (carried in the treated plant tissue to protect plants from insect pests and diseases); translocated herbicides (moving from the point of initial application to other parts of the plants, where they interfere with plant biochemical processes, resulting in plant death); and fumigants (killing when they are inhaled or absorbed by pests, including weeds and pathogens). How the chemical reaches the target pest can determine the type of pesticide selected. For example, chewing insects such as caterpillars are targets of stomach poisons, whereas piercing-sucking insects that feed on plant sap (e.g., aphids, mites, whiteflies, psyllids) may be killed more effectively by a systemic pesticide. A systemic fungicide may be a better choice as a preventive, while a contact fungicide would be used when trying to manage a small disease outbreak quickly.
- Chemical class or chemical family. This categorization is useful for understanding the properties of a specific compound. Many examples of chemical classes can be found. Some examples of insecticides are organophosphates, chlorinated hydrocarbons, carbamates, and synthetic pyrethroids. Common classes of herbicides include dinitroanilines, sulfonylureas, and pyridines. Two examples of fungicide classes are piperazines and strobilurins.
- Mode of action. How a pesticide works is called its mode of action (MoA) or mode of activity. It refers to the action of a pesticide at its specific target site (the physical location within an organism where the pesticide acts). In other words, the mode of action is the way the pesticide causes physiological disruption at its target site. The chemical structure of a pesticide generally defines its target site and its mode of action at that target site. This allows pesticides to be classified based on the commonality of the active ingredient's mode of action, which is useful for resistance management (see the section "Minimizing Pesticide Resistance" later in this chapter). Four specialist committees have classified MoA groups of insecticides, miticides, fungicides, herbicides, and rodenticides, based on the best available evidence of the mode of action of the active ingredients. These MoA groups can be found in the *UC IPM Pest Management Guidelines* or in the pesticide action

committee databases and publications (see the References at the end of this chapter).

- Selectivity. Pesticides can be selective or non-selective. Selective pesticides kill only certain groups or species of plants or animals and are generally preferred in IPM programs. Nonselective, or broad-spectrum, pesticides kill a wide range of plants or animals.
- Persistence. Pesticide persistence is a measure of how long the chemical resists degradation. Typically degradation by soil microbial and chemical reactions eventually detoxifies a pesticide, which is a desirable process from the perspective of environmental safety. However, fast degradation may render a pesticide ineffective for pest control, so a balance is desirable. The use of highly persistent pesticides poses the greatest risk of groundwater and surface water contamination (see chapter 8).
- Risk. Pesticides traditionally used in nursery crops are broad spectrum and persistent. They control a broad range of pests that includes natural enemies. They also typically pose hazards to the environment and can increase pesticide resistance among numerous organisms. Reduced-risk pesticides are pesticides that are IPM compatible. They usually are targeted to specific pests and pose less risk to the environment. They may provide reduced potential for groundwater contamination and offer better worker safety. In the past decade, many new reduced-risk pesticides have become available for ornamental crops, making it easier to implement IPM programs that include biological control.

Timing of Pesticide Applications

Pesticides are most effective when they are applied at the correct time in the pest's life cycle. In most cases, the pest must be present in the treated area and in a vulnerable life stage for the pesticide to have an impact. Proper timing of any pesticide requires information about the pest's life cycle and about environmental conditions that favor the pest. Timing of the control actions before pests approach damaging levels requires early detection of the pests, so monitoring on a regular basis is extremely important.

With most pests, only certain life stages are susceptible to control by pesticides. For instance, a stomach poison insecticide is only useful if applied during the life stage when the insect is feeding. Preemergence herbicides are only effective in controlling weeds as they germinate. Furthermore, the location of the life stage of the pest is also important when using pesticides. This is because the pest must, in some way, encounter the pesticide.

Many types of pesticides lose their effectiveness or cause damage to treated plants during certain weather conditions. Extremes in temperatures, for instance, may render some pesticides ineffective. Fog, rainfall, or irrigation may wash pesticides off treated surfaces so they cannot function properly. High temperatures, wind, and lack of irrigation may cause treated plants to become stressed, and the pesticides applied to them may damage these plants. This is especially true of the soaps and oils used to control insects and mites.

Proper timing may also contribute to the selectivity of certain pesticides. For example, dormant sprays with oils and insecticides help control overwintering insect and mite pests on deciduous trees without affecting natural enemies, which they would do if applied during spring or summer. Some postemergence nonselective herbicides can be applied over a crop if the crop is dormant. A preemergence herbicide may be applied to control geminating weeds without adversely affecting the already established crop.

Selecting Pesticides

Before choosing a pesticide, it is important to read the label to understand the pesticide's relative toxicity. All pesticides sold in the United States must include a product label with a signal word indicating the potential hazard of immediate or acute injury. Signal words help alert users to special hazards of a pesticide product. The signal words are CAUTION, WARNING, and DANGER. Pesticides with the word DANGER are the most hazardous, and will also have the word POISON and a skull and crossbones symbol if highly toxic due to oral or dermal toxicity. These products are typically for restricted use and are available only to certified applicators who have completed special training.

A common measure of acute toxicity is the lethal dose (LD50) or lethal concentration (LC50) that causes death (resulting from a single or limited exposure) in 50 percent of the treated animals. LD50 is generally expressed as the dose in milligrams (mg) of chemical per kilogram (kg) of body weight. LC50 is often expressed as milligram of chemical per volume (e.g., liter) of medium (i.e., air or water) that the organism is exposed to. Chemicals are considered highly toxic when the LD50 or LC50 is small and less toxic with higher numeric values. However, the LD50 or LC50 does not reflect any effects from long-term exposure (i.e., cancer, birth defects, or reproductive toxicity) that may occur at levels below those that cause death. LD50 and LC50 values are not always given on the pesticide label; rather, the relative toxicity of a pesticide product is reflected by one of the three signal words.

The pesticide label also includes information related to environmental risk such as the potential of the pesticide to contaminate surface water. It is important to read the pesticide label, including the "Environmental Hazards" section, and make sure you are familiar with any instructions to minimize the risks to the environment. Furthermore, the *UC Integrated Pest Management Guidelines* includes interactive website information that can be used to determine if a pesticide active ingredient has the potential of water-related risks. Site-specific conditions can be factored in and comparisons made with other pesticides used in nurseries. For more information, see chapter 8.

Whenever possible, select pesticides that can effectively control the pest without causing harm to the environment or human risk. As previously mentioned, many reduced-risk pesticides that are commonly used today have low toxicities to people and less detrimental impact on the environment. When selecting a pesticide, it is also important to understand the pesticide's potential for causing phytotoxicity (injury to plants). Certain active ingredients, formulations, and application methods have higher risks of phytotoxicity than others. Always check that the crop to be treated is on the label, and follow label instructions precisely. When treating a crop with a pesticide for the first time, it is wise to try a small-scale application first.

Minimizing Pesticide Resistance

Pesticides should also be selected to minimize pesticide resistance. Pesticide resistance can become a problem when chemicals with the same mode of action are used repeatedly to control a particular pest. After a time, the pest population may develop resistance to the chemical so that the chemical no longer effectively controls the pest at the same application rate. As resistance increases, higher rates and more frequent applications become necessary until eventually the chemical provides little or no control.

Minimizing pesticide use is fundamental to pesticide resistance management. Strategies to minimize pesticide use include starting with the lowest recommended rate and using the lowest effective application frequency. In addition, many management practices exist that can help reduce the number of pesticide applications, such as using sound cultural practices and nonchemical control practices (e.g., screening greenhouse openings to exclude pests, biological control). Furthermore, the pesticide application strategies described below can minimize resistance.

Rotate pesticides with the same mode of action

Resistance management strategies for insects, weeds, and fungal pathogens all include rotating pesticides with the same mode of action or group number, as described earlier. Rotation schedules should not be based only on the active ingredient or chemical class, because these may have the same mode of action. For example, organophosphates and carbamates, despite belonging to different chemical classes, have identical modes of action (interfering with the nervous system of the target insect).

Rotation strategies

Always follow label instructions for rotation schedules. When no label instructions are given, generalized rotation strategies can be used—but be aware that they differ for specific types of pests. For instance, with fungal pathogens, it is suggested that fungicides with different modes of action be rotated with every application. With weeds, herbicide rotations should be used at least once per season, along with hand or mechanical weed control to reduce the number of herbicide applications.

With arthropod pests, it is recommended that a pesticide be used for one or two generations of the target pest before it is followed by a rotation to a different mode of activity. This rotation cycle is generally every 1 to 4 weeks, depending on the length of time it takes for the pest to complete a generation. For example, at 68°F, the approximate time that it takes the following insects to complete a life cycle is 7.1 days for green peach aphid (Satar et al. 2008); 21.4 days for western flower thrips (Lewis 1997); and 34.7 days for silverleaf or sweetpotato whitefly (Butler et al. 1983). However, temperatures in outdoor nurseries fluctuate, which makes it difficult to predict pest generation time. For example, for mites, the entire life cycle can take as little as 8 days at 77° to 95°F to as many as 28 days at 50° to 68°F. Growers who monitor temperature can compute precise estimates of generation time for some nursery pests using the UC IPM Weather, Models, and Degree-Days website (see the References at the end of this chapter).

Long-term rotation through many chemical classes in successive generations will help maintain efficacy. Using a specific mode of action for longer than two generation times of the pest may enhance the chance of resistance, since the survivors of the first generation and the next will most likely be tolerant to that mode of action. On the other hand, if insecticides are switched with every application before a generation is completed, the same individual insects could be exposed to the mix of chemicals, as if the materials were tank mixed (see next section). In this case, a pest population that is resistant to pesticides with multiple modes of action could develop.

If only one chemical is effective against a pest, and other available products are only marginally effective, a good strategy is to use the marginally effective materials at times when pest pressure is less severe and reserve the effective material for periods when control must be most effective.

Tank mixes

Combining two pesticides with the same mode of action in a tank mix is not recommended. Such a "super dose" often increases the chances for the establishment of resistant pest individuals.

Combining two pesticides with different modes of action in a tank may or may not be recommended for managing resistance, depending on the target pest. For instance, with fungal diseases, fungicides with different modes of action may be tank mixed. With weeds, herbicides may be tank mixed to control both existing weeds (postemergence) and to prevent new ones from establishing (preemergence). However, with arthropod pests, tank mixes of pesticides with different modes of action should be used with caution, even when the tank mix provides superior control. This is because long-term use of these pesticide mixes may give rise to pesticide resistance to both pesticides if resistance mechanisms to both pesticides arise together in some individuals. Continued use of the mixture may select for these multiple-pesticide-resistant pests.

Tank mixing pesticides from two different modes of action is also not generally recommended for controlling multiple life stages of arthropods. Rather than tank mixing an adulticide with an insect growth regulator (a pesticide that controls the immature stages of an insect pest), it would be better to choose formulations that require different types of application. For instance, in greenhouse applications, if the insect

FIGURE 12.15. Spot spraying can significantly reduce overall pesticide use. Here, the scout identifies infested plants in the nursery with colored flags and then maps the locations. Subsequent pesticide applications are targeted to the flagged areas. *Photo:* J. K. Clark.

growth regulator is applied as a spray, it would be preferable for the adulticide to be applied as an aerosol or smoke. This provides rapid kill of the adults and little residual that might select for resistance buildup in surviving immatures.

Methods to Reduce Pesticide Use

A number of practices are typically used in IPM programs to reduce pesticide use in the nursery and pesticide loads in the environment. Some examples are described below.

Spot spraying

Monitoring identifies areas in the nursery and plant parts where pests reside. Focusing pesticide applications in these areas rather than spraying the entire crop can significantly reduce overall pesticide use. Although spot spraying can be more labor-intensive, it still results in significant economic savings for the grower due to reduced material costs (fig. 12.15).

Direct spraying

Some pests are found primarily in certain parts of the crop plant. For example, western flower thrips typically reside in the top third of the crop, in the flowers and buds. Conversely, spider mites are typically found in the lower two-thirds of the crop. Directing spray toward the part of the plant where the pest resides can result in the overall reduction of pesticides.

Use pesticide adjuvants where appropriate

Pesticide adjuvants are products used to enhance the performance of pesticide applications. In general, they cause the pesticide to spread, stick to, or penetrate the plant material, resulting in more effective pest control. The use of adjuvants can allow for reduced pesticide use by providing more uniform coverage. Wetting agents, or surfactants, reduce the surface tension of water in the pesticide spray, allowing it to spread more easily over the leaf. Stickers increase the chemical attraction of the pesticide to the plant surface. They increase the length of time the pesticide stays in contact with the pest by resisting being washed off by rain or irrigation and reducing evaporation and volatilization. There are also extenders, which work like stickers, and plant penetrants, which work like surfactants.

Before using an adjuvant, be sure to read the pesticide product label for information about the appropriate types of adjuvant, if needed, for the pesticide you are using. Some manufacturers add adjuvants to their pesticide products, so no additional adjuvants are needed in these formulations. If an adjuvant is needed, information on what and how much to use will be on the label. Some manufacturers produce adjuvants that are most compatible with their own products.

Use current IPM recommendations

To ensure that your control tactics are effective, use the most recent IPM recommendations for crops. Consult with your farm advisor, Cooperative Extension agent, or pest control adviser, as well as relevant publications such as *UC IPM Pest Management Guidelines for Floriculture and Ornamental Nurseries* (see the References at the end of this chapter).

Pesticide Application Equipment

The two main categories of pesticide sprayers used in nurseries are high-volume hydraulic and low- or ultra-low volume (LV or ULV). The simplest types of high-volume hydraulic sprayers are compression sprayers. Low- and ultra-low-volume sprayers are typically used in greenhouse production, and they deliver an active ingredient in a small volume of water or other diluents. Equipment to apply granular materials includes centrifugal spreaders and blower applicators. See chapter 4 for more information regarding pesticide sprayers.

Pesticide Regulations Impacting Nurseries in California

The Department of Pesticide Regulation (DPR) is responsible for the laws, rules, and regulations regarding pesticide use in California. Locally, the county agricultural commissioner (CAC) is the enforcing officer of all laws, rules, and regulations relative to the introduction into or the spread within the state of plant pests, and the CAC is under the supervision of the secretary of the California Department of Food and Agriculture. These organizations also serve as the primary enforcement agents for state pesticide laws and regulations. DPR and the CAC have the authority to enter and inspect or sample sites, equipment, and activities related to pesticide use (e.g., greenhouses, crops, records, equipment, storage facilities) to determine compliance with the Food and Agricultural Code, which pertains to pesticides and pest control operations.

Licensing

DPR's Licensing and Certification Program is responsible for testing and licensing qualified applicators, aircraft pilots, agents designated as pest control dealers, and agricultural pest control advisers; it is also responsible for certifying pesticide applicators who use or supervise the use of restricted pesticides. Additionally, the program licenses businesses that sell or apply pesticides or use pest control methods or devices for hire (i.e., pest control business, maintenance gardener pest control business, pest control dealer, and pesticide broker). All applicants for a license or certificate are required to pass an examination on the laws and regulations governing pesticide use and the safety precautions necessary to prevent injury. License or certificate holders must complete the required continuing education hours within each 2-year license or certificate period before they can receive a renewed license or certificate. Types of licenses required and application information is available at the DPR website (see the References at the end of this chapter).

Restricted-entry intervals

Restricted-entry intervals (REI) are the required number of hours or days after application during which entry into treated areas is prohibited. The purpose of the REI regulations is to minimize employee exposure to pesticides. REI information is listed on the pesticide label.

Treatment notification requirements

The operator of the property must notify all workers of the pesticide application. All workers likely to enter the treated area prior to the expiration of the re-entry interval must be informed as well. Notifications must include the date of the scheduled application; the location and description of the field to be treated; the pesticide product name(s), U.S. EPA registration number(s), and active ingredient(s); spray adjuvant product name(s) and California registration number(s); the restricted entry interval if applicable; information on whether the pesticide product labeling requires the posting of treated fields; and any other precautions printed on the pesticide product labeling or included in applicable laws and regulations related to the protection of employees or other persons during or after application. Notifications of completed applications must include the location and description of the treated field; the time during which entry is restricted; and instructions not to enter the treated field until the restricted entry interval has expired. For each pest control operation, the records must include the date of application; the name of the operator of the property treated; the location of property treated (by county, section, township, range, base, and meridian), crop commodity, or site treated; the total acreage or units treated at the site; the name of the pesticide, including the U.S. Environmental Protection Agency (U.S. EPA) or state registration number, which is on the pesticide label, and amount used; location of the property treated; hour the treatment was completed; the operator identification number issued to the operator of the property treated; the site identification number issued to the operator of the property treated; total acreage (planted) or units at the site; and name or identity of the person(s) who made and supervised the application if the pesticide application was made by an agricultural pest control business.

Mandatory reporting

California requires full pesticide use reporting in response to demands for more realistic and comprehensive pesticide use data. Under the program, all agricultural pesticide use must be reported monthly to county agricultural commissioners, who, in turn, report the data to DPR. For more information about pesticide use reporting, see the DPR and UC IPM websites in the References at the end of this chapter.

Restricted-use pesticides

Restricted materials are pesticides deemed to have a higher potential to cause harm to public health, farmworkers, domestic animals, honey bees, the environment, wildlife, or other crops compared with other pesticides. Restricted-use pesticides can only be bought, possessed, or used under permit (restricted-use permit) from the local CAC. Permits for use must be in the name of the operator of the property. The agricultural commissioner may allow an authorized person to sign the permit, but the authorized person must have a valid operator license or certificate.

For a list of the products considered "restricted use," see the DPR website listed in the References at the end of this chapter or contact your local county agricultural commissioner's office.

Worker safety training

Employers are required to ensure that all workers who apply, handle, or mix pesticides have been properly trained. Furthermore, the Federal Worker Protection Standard requires that all employees working in greenhouses and nurseries that have been treated with a pesticide must receive pesticide safety training regardless of whether they work directly with pesticides. Employees should be familiar with safety information found on the materials safety data sheet (MSDS) for each pesticide used. The MSDS details any potential hazards that are known, including long-term or chronic health effects that are not required to be stated on a pesticide label.

Photo: J. K. Clark.

References

Association of Natural Biocontrol Producers website, http://www.anbp.org.

Boodley, J. W., and S. E. Newman. 2009. Insects, mites, and other invertebrate greenhouse pests. In The commercial greenhouse. Clifton Park, NY: Delmar Cengate Learning.

Butler, G. D., T. J. Henneberry, and T. E. Clayton. 1983. *Bemisia tabaci* (Homoptera: Aleyrodidae): Development, oviposition, and longevity in relation to temperature. Annals of the Entomological Society of America, 76(2): 310–313.

California Department of Pesticide Regulation (DPR) website, http://www.cdpr.ca.gov/.

Cloyd, R. A., and R. S. Cowles. 2009. Resistance management: Resistance, mode of action, and pesticide rotation. Manhattan: Kansas State University Agricultural Experiment Station and Cooperative Extension Service Publication MF-2905.

Cornell Cooperative Extension. 1993. IPM for poinsettias in New York: A scouting and pest management guide. New York State IPM Program Publication No. 403.

Dreistadt, S. H. 2001. Integrated pest management for floriculture and nurseries. Oakland: University of California Division of Agriculture and Natural Resources Publication 3402.

Flint, M. L., and S. H. Dreistadt. 1998. Natural enemies handbook: The illustrated guide to biological pest control. Oakland: University of California Division of Agriculture and Natural Resources Publication 3386.

Flint, M. L., and P. Gouveia. 2001. IPM in practice: Principles and methods of integrated pest management. Oakland: University of California Division of Agriculture and Natural Resources Publication 3418.

Fungicide Resistance Action Committee website, http://www.frac.info/.

Herbicide Resistance Action Committee website, http://hracglobal.com/.

Insecticide Resistance Action Committee website, http://www.irac-online.org.

Kabashima, J. N., J. D. MacDonald, S. H. Dreistadt, and D. E. Ullman. 1997. Easy on-site tests for fungi and viruses in nurseries and greenhouses. Oakland: University of California Division of Agriculture and Natural Resources Publication 8002. ANR CS website, http://anrcatalog.ucdavis.edu/pdf/8002.pdf.

Lewis, T. 1997. Pest thrips in perspective. In T. Lewis, ed., Thrips as crop pests. Wallingford, UK: CAB International. 1–13.

Newman, J., K. Robb, and S. Tjosvold. 1999. Training scouts and developing sites to promote floriculture IPM programs. CDPR Pest Management Demonstration Project Final Report 97-0224. CDPR website, http://www.cdpr.ca.gov/docs/pestmgt/grants/97-98/finlrpts/97-0243.pdf.

Newman, J., S. Tjosvold, K. Robb, and A. King. 1996. Implementation of IPM scouting programs for ornamental crop production. IOBC/WPRS Bulletin 19(1): 111–113.

O'Connor-Marer, P. 2000. The safe and effective use of pesticides. 2nd ed. Oakland: University of California Division of Agriculture and Natural Resources Publication 3324.

Rodenticide Resistance Action Committee website, http://www.rrac.info/.

Satar, S., U. Kersting, and N. Uygun. 2008. Effect of temperature on population parameters of *Aphis gossypii* Glover and *Myzus persicae* (Sulzer) (Homoptera: Aphididae) on pepper. Journal of Plant Diseases and Protection 115 (2): 69–74.

UC IPM overview of pesticide use reporting, UC IPM website, http://www.ipm.ucdavis.edu/PUSE/overview.html.

UC IPM pest management guidelines for floriculture and ornamental nurseries, UC IPM website, http://www.ipm.ucdavis.edu/PMG/selectnewpest.floriculture.html.

UC IPM weather, models, and degree-days, UC IPM website, http://www.ipm.ucdavis.edu/WEATHER/index.html.

Photo: J. K. Clark

Plant Diseases

Deborah M. Mathews, J. Ole Becker, and Steven A. Tjosvold

CHAPTER THIRTEEN

Photo: S. A. Tjosold.

Container nurseries and greenhouses typically produce a large number of plant species and varieties that can be hosts of many disease-causing pathogens. Tight plant spacing, used to maximize available growing space, coupled with overhead irrigation create environments where disease development can be especially swift. Furthermore, nurseries are faced with the constant threat of new introductions of exotic, invasive pathogens from around the world on infested plant material or in potting media. Quarantines for pathogens causing sudden oak death and chrysanthemum white rust have contributed to significant economic losses in California nurseries. For these reasons, it is critical that nurseries implement a disease management program that emphasizes prevention and early diagnosis of pathogens to limit damage.

Diseases are caused by pathogens such as fungi, water molds, bacteria, viruses, phytoplasmas, and nematodes. They are also caused by abiotic factors such as adverse environmental conditions or poor cultural practices. Diseases caused by pathogens are infectious; they gain entry into the plant and develop and multiply on or within the plant. Diseases caused by abiotic factors are often called disorders to distinguish them from those caused by pathogens. Abiotic problems are caused by too much or too little water, poor water drainage in soil, extremes in light intensity, excess or deficient nutrients, salinity, improper temperature, pesticides, physical injury, and adverse weather. Abiotic disorders can directly injure plants, or they can make crops more susceptible to pathogens and insects.

This chapter introduces the fundamental concepts for managing plant pathogens and disorders in container nursery production. First, we will discuss the conditions that cause infectious diseases, as well as general practices that can be implemented in the nursery to prevent and control pathogens. Next, we will discuss common plant pathogens, along with basic biology, disease symptoms, and management of fungi, water molds, bacteria, nematodes, viruses, viroids, and phytoplasmas. Abiotic problems such as nutrient deficiencies and air pollution are presented in the last part of this chapter. The basic skills needed to identify plant diseases are discussed in chapter 11.

Disease Conditions

An important underlying concept of plant pathology is the disease triangle, illustrating that three components are necessary before a disease can be caused by pathogens:

- A host plant must be susceptible.
- A pathogen that is able to infect a given host must be present in sufficient quantity.
- The first two must exist in an environment conducive to disease development over time.

If all three parts of this triangle coexist, infectious disease will result (fig. 13.1). Many people also consider the effects of timing and human involvement to be integral to the development of disease, but generally these are included in the environmental aspect of the triangle. Important environmental factors include temperature, light, humidity, water

FIGURE 13.1. Diagram of the disease triangle showing the relationship between the host, pathogen, and environment required for the development of an infectious disease. If one element or more is absent, disease will not develop.

(quantity and quality), soil quality, nutrients, and wind. The environment also includes interactions with vectors, animals, and humans, including cultural practices applied by humans that alter the plant (e.g., fertilizer and chemical applications). Extremes at either end of the spectrum for each of these factors can make a plant more (or sometimes less) susceptible to a given pathogen. For example, some pathogens require fairly specific temperatures or humidity in order to infect plants and complete their life cycle; otherwise, they cannot effectively cause disease, regardless of initial inoculum concentrations or the presence of a suitable host.

Cultural Practices to Prevent Disease Development

Cultural approaches to managing diseases are extremely important, and they fit the old adage that "an ounce of prevention is worth a pound of cure." In order to prevent disease, or at least to minimize its impact, it is necessary to maintain good sanitation and sound cultural practices during all stages of plant culture, from plant selection and propagation through production and postharvest.

Plant Selection

In selecting plants for nursery production, always consider the use of resistant plant varieties to prevent diseases. Natural resistance genes exist for various diseases, including water molds, powdery mildew, Verticillium, and Fusarium, and for some bacteria, nematodes, and viruses. Check with suppliers for varieties resistant to diseases that are found in your operation. Tested and certified disease-free stock plants should be used when available.

Propagation

Plant material used in the nursery should start from good-quality and ideally pathogen-tested plant stocks. Any stock plants showing indications of disease should not be used in propagation. Stock plants should be maintained under conditions of excellent drainage, good nutrient sources, and freedom from disease-producing organisms. In general, previously used rooting media should not be recycled and used for propagation purposes.

Propagation areas should be designed so that all plant debris and media particles are cleaned out between each batch of plants or after each crop. Diseased plant material should never be taken into propagation areas. Tools should, at a minimum, be washed clean with water between crops or plant batches and allowed to dry. Tools used for cutting valuable crops (such as orchids) or crops highly susceptible to crown gall or viruses should be treated in alcohol, diluted chlorine bleach, quaternary ammonium compounds, or other disinfectant materials to inactivate such pathogens. Containers used for carrying propagative materials should not be allowed to sit on soil or growing medium before being carried into the propagation area. Plants in flats containing diseased plant materials should be dumped even though some plants look normal, and the containers should be disinfected or discarded. Shoes should be cleaned of soil before going into the propagation area, and feet should never be allowed on the propagation benches. Weeds should be controlled or eradicated so as not to act as reservoirs of insects or pathogens. Water used in propagation facilities should be chemically treated or filtered to remove pathogens.

Sanitation Practices during Production

Maintaining good sanitation is one of the most important ways of preventing and managing diseases. Sanitation should be practiced during all phases of production. Unfortunately, simple measures for the preventive control of diseases are often overlooked.

In outdoor production, no containers should sit directly on the soil. Low benches can be used or, if this is too expensive, the soil can be covered with gravel or weed cloth. Smaller containers can also be put in plastic flats that will lift them from the surfaces of the benches or the gravel. Flats with cups or wells for holding pots are also available. Before using, they should be checked to make sure they are not the type where the pots fit all the way to the bottom. These may block the drainage holes and cause problems. Some are available that have ridges across the bottoms of the cups, and some are made so the pots do not fit all the way to the bottom. Residual planting mix between crops should be removed from gravel, weed cloth, or benches.

In indoor production, the use of concrete or other impervious surfaces on at least the walkways is preferable for minimizing disease. Floor surfaces should be kept clean of debris, soil, or planting medium. After a crop cycle, wooden or metal benches should be washed clean of soil or planting mix between crops and allowed to dry before use. Wooden benches in particular should be regularly painted with copper-containing paints to eliminate any fungal or other pathogens. Care should be taken to avoid using excessive water in the washing process so that media particles are not carried in runoff, where they can potentially contribute to nutrient and pesticide loads (see chapter 8 for more information on runoff management). If disease was present in the previous crop, benches should be sprayed with diluted chlorine bleach or other suitable disinfectant. This should be done after the benches have been thoroughly cleaned of all organic materials such as potting media and plant debris that will inhibit the activity of most disinfectants. Benches should be allowed to dry before they are used again. Walking on benches, if necessary, should be done with disposable shoe covers or with boots or shoes that are used only for that purpose.

In both indoor and outdoor production, pots, flats, or other containers should never be reused without thoroughly washing them to remove all planting mix particles and plant debris. They can then be treated with a disinfectant such as chlorine. If disease problems were present, additional treatment with a disinfecting chemical should be applied, rinsed, and dried. Potting media should not be reused, as the risk of pathogen carryover is too great and not worth the savings in most cases. Even some new mixes, notably those containing peat, may benefit from steam treatment prior to use in order to eliminate or at least reduce the amount of viable disease-causing agents. For most situations, aerated steam treatment of 140° to 160°F for 30 minutes will kill the most problematic pathogens while preserving the natural microflora of the growing medium.

Carefully inspect all new shipments of plant material brought into the nursery and treat or dispose of infected plants promptly before introduction to the growing area. If space is available, it is a good idea to keep newly arrived plants isolated from main production areas and observe them for a week or so to check for developing diseases.

Any indication of poor growth should be investigated immediately to determine the cause. Old, dead, unhealthy, or sick plants should be removed from the growing areas as soon as recognized. If handling diseased plants or any soils, workers should wash their hands or use disposable gloves before handling plants. Piles of discarded plants (cull piles) should be kept covered, as far from production areas as possible and especially away from greenhouse air vents to avoid fungal spores or insects from contaminating healthy plant stocks.

Cultural and Environmental Practices during Production

Where possible, crops should be rotated in growing areas so that a disease of any one crop does not have a chance to become established. Plant spacing should allow good airflow between plants to keep humidity down and allow irrigation water to dry off of plant surfaces quickly. This will help inhibit fungal and bacterial pathogens that require high humidity or standing water for survival and multiplication. Combined with a good ventilation system, air circulation can go a long way to help prevent or minimize many diseases.

One of the most important cultural management practices is irrigation. Too much water and overhead irrigation are probably responsible for more fungal and bacterial pathogen problems than any other factor. Overwatering predisposes plants to root rot diseases, and wet leaves encourage the development of foliar diseases. If any disease of the foliage appears (other than powdery mildew), irrigation duration and timing should be adjusted so that the foliage has minimal contact with water. This should be done at a time of day when the foliage will dry as quickly as possible. This will also help to keep foliage dry. Pots with drainage holes must be used. Hose nozzles should never be allowed to touch the ground. The practice of putting gravel, shards, or rocks in the bottoms of containers should be avoided, as it does not actually enhance drainage of excess water, and it takes up room that can be used for additional soil and eventual root volume.

Disease Management

A successful disease management program should start by incorporating as many prevention methods as possible to avoid problems, such as those cultural and sanitation practices detailed in the previous section. In addition to these cultural practices, a regular scouting program to look for disease symptoms, signs of pathogens (spores, mycelium, etc.), and insect infestation will contribute to preventing and reducing problems. Symptomatic plants should be removed from the area and possibly sent for identification of the causal agent. Proper diagnosis of the problem is critical for timely control of the disease and ensuring effective chemical use. Different pathogens can have vastly different chemical sensitivities, and accurate identification of the causal agent will prevent ineffective and wasteful chemical applications. Commercial laboratories, many counties, and some universities have diagnostic clinics to aid with diagnosing plant diseases. Detailed information on how to diagnose plant problems caused by pathogens is presented in chapter 11.

Records of disease outbreaks should be kept and should note the date, species, variety, and age of the host plant, pathogen identified, control measure used, and effectiveness of application. If patterns become apparent for a given disease problem, preemptive measures can be taken to try to prevent another occurrence, or at least minimize losses, by early intervention.

The increasing use of recycled water in agriculture has amplified the need for a management program that includes monitoring water quality. Along with salt, fertilizer, and chemical buildup comes the possibility that plant pathogens will make their way through the system to plants. All major pathogen types—fungi, water molds, bacteria, viruses, and nematodes—have been shown to be transmitted via recycled water. Filtration, ultraviolet light, or chemical methods should be employed to remove pathogens from water systems. See chapter 8 for details regarding disinfecting recycled water.

Despite the best prevention measures, inevitably there will be outbreaks of disease that require treatment. As stated before, an accurate diagnosis of what pathogen(s) you need to control is critical to avoid wasting chemicals and time. Biological control is one option that is sometimes available for control of plant diseases. Biological control uses natural enemies of pathogens to eradicate or control their populations. However, commercially available biocontrol products have not always worked consistently in nursery settings, and they should be evaluated carefully before use. Ongoing research and development on biological controls will hopefully result in viable options for disease control in the future.

For ongoing treatment of the same pathogens, rotation of chemicals with different modes of action is critical to avoid pathogen resistance from developing. By applying chemicals with modes of action that affect different metabolic pathways of the pathogen, development of resistance is less likely. If a fungicide seems to be ineffective where it should be giving control, have a plant pathologist check to see if it is a resistant strain of the fungus or if a different pathogen is present.

Specific chemicals for control change frequently. To prevent the information from becoming out of date, they are not addressed directly in this chapter. But you can check the online database of pesticide recommendations at www. ipm.ucdavis.edu or consult with your Cooperative Extension advisor, a licensed pest control adviser, or company representatives for current recommendations. Always be sure to check labeling for registration and proper use against a pathogen on a given plant host in your state.

Foliar and Stem Diseases Caused by Fungi and Water Molds

Fungi and water molds cause a variety of diseases and symptoms on the leaves, stems, and roots of plants. These include leaf spots, stem and crown cankers, tip blight, root rots, and wilting. Several examples of each are addressed individually below.

Powdery Mildew

This is one of the most common problems on nursery crops in California. Many different types of fungi cause powdery mildew. Some have wide host ranges and infect many different kinds of plants, usually in the same family or family group. Others infect only one or a few kinds of plants. The powdery mildews are biotrophs—parasitic fungi that rely on living plants to obtain nutrition and complete their life cycles. Thus, they rarely kill their hosts. In fact, research has shown that powdery mildew fungi can suppress host plant defense responses and prevent cell death at infection sites. Powdery mildews only infect the epidermal surface cells of the plant and do not invade systemically throughout other parts of the plant. Individual spores called conidia are moved by wind currents. The pathogen overwinters in buds and fallen plant debris as mycelium and in special spore survival structures called cleistothecia, which are black in color when mature.

Symptoms

White to gray powdery-looking patches develop mainly on the upper surface of leaves (fig. 13.2A, B), but they can also occur on the lower surfaces, stems, and flowers. For some host-pathogen-environment combinations, the symptoms never develop beyond the white spots composed of fungal spores and mycelium. Other combinations may result in leaf distortion and loss, as well as general plant decline. The powdery mildews often develop quickly with warm days, cool nights, and reduced light intensities. Some infect mainly the older foliage of plants as it tends to go senescent, while others infect the new growth. The latter are generally much more difficult to control.

Control

The powdery mildews do not need condensed moisture for their spores to germinate and infect; in fact, spores die if put in water. This can be used as a control measure for low levels of powdery mildew by sprinkling plants with water (early in the day so they dry out before nightfall), but care must be taken if water-requiring pathogens are also a problem in that crop or growing area. Good airflow (using ventilation and plant spacing) can be an effective prevention measure. Plants resistant to powdery mildew should be selected when available. Powdery mildew control begins with scouting and removal of symptomatic plants early in the disease cycle. Most powdery mildews respond to the same chemical control methods, which are best applied as a preventive before infection occurs. Powdery mildews tend to develop resistance to chemicals, so be sure to rotate routinely. The unique, superficial nature of powdery mildews makes them vulnerable to many different chemical classes. These include many unconventional, reduced-risk fungicides such as bicarbonate salts, horticultural oils, sulfur, waxes, powdery mildew parasites, and topical chemical toxicants. Phytotoxicity is common on many hosts after treatment with eradicative chemicals, so applications should be tested on a small group of plants for a given treatment before being used broadly. Biological control of powdery mildew using fungal-feeding mites or a ladybird beetle species has sometimes been successful in experimental applications.

Downy Mildew

These pathogens may look like powdery mildews; however, they are quite distinct and are not actually true fungi. Instead, they are more closely related to the water molds like *Pythium* and *Phytophthora*. Downy mildews are low-temperature fungi, and they need moisture for the spores to germinate and infect. They also need periods of 90 percent or higher relative humidity in order to produce spores. They do not survive well below 40°F or above 80°F. Consequently, in most parts of California these diseases are usually found only during the winter months, but they can occur year-round in the cool, moist coastal areas or in humid greenhouses.

Symptoms

White, lavender, or purple "downy" areas are usually found only on the undersides of the leaves (fig. 13.2C), while yellow, reddish, purplish, or brown angular areas may develop on the corresponding upper leaf surface, but these will be smooth and not fluffy. Defoliation and stunting may also occur. Downy mildew may exist on foliage without obvious signs until the environmental conditions are right—and then, sometimes apparently overnight, they begin to sporulate and cause noticeable disease. Downy mildews are common on many ornamentals but are rarely found on woody plants. In California, hosts include marguerite daisy, impatiens, pansy, snapdragon, rose, alyssum, pincushion flower, strawflower, and cornflower.

Control

Unlike powdery mildew, most downy mildew species are host-specific, and what controls disease on one host may not control it on another. Avoid wetting foliage, and use good airflow to prevent optimal conditions for this pathogen. Remove affected plants immediately, and get a diagnosis to ensure proper selection of control measures. Downy mildews are notably prone to developing fungicide resistance, so a strict rotation schedule should be used.

Gray Mold Caused by *Botrytis cinerea*

Gray mold is another of the most common plant pathogens of field and greenhouse crops. Named for the masses of gray mycelium and spores that are produced, *Botrytis* can quickly become a problem in greenhouses because of its ability to reproduce rapidly under ideal environmental conditions of high humidity (above 90%) or water contact and moderate temperatures (60° to 77° F). It can survive, however, at tem-

FIGURE 13.2. Foliar fungal diseases. Powdery mildew on periwinkle: patches start out individually (A), then coalesce to cover leaf surfaces (B); downy mildew on *Limonium*: mycelium and spores on underside of leaves (C); *Botrytis* on *Eustoma*: gray mold on stem and leaves (D); rust on iris leaves: patches of orange-brown dry spores (E); symptoms of foliar infection of *Phytophthora ramorum* on *Rhododendron* 'Nancy Evans' (F). *Photos:* D. M. Mathews (A, B); S. A. Tjosvold (C, D, E, F).

peratures from near freezing up to 90°F and then wait until favorable conditions occur.

Symptoms

Gray mold causes a variety of diseases. These include leaf spots on stem and leaf (fig. 13.2D), flower blights, damping-off (seedling death, see fig. 13.4A), and fruit and bulb rots. When the environment is not optimal for its growth, it is weak as a pathogen. However, it is very opportunistic and readily invades stressed, wounded, or susceptible plants when the environment is favorable (cool and moist). Furthermore, *Botrytis* commonly can directly infect intact and healthy flower petals of many important hosts. Common hosts of *Botrytis* are African violet, geranium, gerbera daisy, pansy, petunia, and poinsettia, among over 200 others. *Botrytis* is especially aggressive in densely spaced plantings of bedding plants or stands of plants that have reached maturity and are crowding each other in their containers or growing areas.

Control

Gray mold control is a perfect platform for all the control measures discussed earlier in this chapter for cultural and preventive disease control: start with healthy vigorous plants that can resist infection, don't use overhead irrigation (to avoid wet leaves and standing water), keep plants well ventilated so humidity stays low, scout to detect disease early, maintain good sanitation of growing areas, and remove old, dead, or infected plant material. Keep plants adequately spaced. It is imperative to prevent the fungus from producing spores, which, once present, can easily be spread by water splashing, air currents, and human activity in the area. *Botrytis*, like many fungi, is able to develop resistance to fungicides, so regular rotation of chemicals is important for effective control.

Rusts

Rust fungi prefer mild, moist conditions, and the spores only need to be wet for as short a period as 5 or 6 hours for infection to take place. Thus dew formation is frequently long enough for rusts to become established. Rusts are very specific to a given host, but over 5,000 fungi exist that can cause rust, so many plants may become infected. These include chrysanthemum, snapdragon, rose, geranium, dichondra, fuchsia, St. John's wort, coral bells, hollyhock, iris, pine, and turfgrass. The life cycle of some rusts can be quite complex, and it varies for different species. Several different spore types may be produced, and some must overwinter on an alternate host to complete their life cycle. Others only require a single host for their survival and life cycle.

Symptoms

Leaf spots containing raised pustules generally appear on the undersides of leaves or on stems (fig. 13.2E). These are often orange or reddish but may also be yellow, white, or purple. Severe infections can cause upper leaf surfaces to have yellow, orange, or brown spotting, overall plant yellowing, and defoliation. Often the dry spores can be easily dislodged from infected leaf surfaces with a swipe of a finger or cloth and examined to discern their color.

Control

Spores mature within the blisterlike pustules and, when released, move through wind and water. Overhead watering should therefore not be used, humidity should be kept low, and air circulation should be good to reduce the leaf wetness that rusts require.

Quarantine requirements

One important rust species in the United States is chrysanthemum white rust, caused by *Puccinia horiana*. It is a quarantined pathogen that is under a strict eradication order when found. It produces white, tan, or pink spore pustules on the undersides of leaves and causes distortion of plants, defoliation, and even death of chrysanthemums. If you find symptoms like those described, contact your local agricultural commissioner immediately.

Phytophthora ramorum

A recently discovered plant pathogen, *Phytophthora ramorum* has killed thousands of native oaks in coastal areas of California and Oregon, where the disease is known as sudden oak death. Many common nursery crops are susceptible to the pathogen, which can infect leaves, stems, and roots. Soil can be infested with long-lived chlamydospores and mycelium in plant debris and fallen leaves. Although the disease is usually not serious on ornamental crops, there is concern that the pathogen could be moved on nursery stock to new areas and eventually infect new landscape and forest hosts.

Symptoms

Although hosts of *P. ramorum* show a range of symptoms, in general the disease is characterized by irregular necrotic leaf lesions rather than distinct leaf spots (fig. 13.2F). Leaf infections can develop down the petiole and into twigs. Often, as in camellia, infected leaves fall off before the lesion reaches the petiole. Infections may occur initially on stems or move into stems, causing blights in which stems and associated leaves wilt, become necrotic, and die. A distinct dark line can mark the advance of the infection on some species. Root infections are not commonly seen because roots are rarely infected or symptoms are not expressed. Additional hosts and photographs of symptoms can be viewed in Tjosvold et al. (2005).

Control

Avoid irrigation practices that wet the foliage for prolonged periods. If sprinklers are used, irrigate in the morning to allow for thorough and quick drying of foliage. In general, the same fungicides that are active on other *Phytophthora* species can also help prevent infection by *P. ramorum*. For high-risk incoming shipments, unload nursery stock in an area that can be cleaned of leafy debris because infected leaves often drop from plants. Sweep debris from the receiving area or delivery trucks and bag for disposal; retain leaves that have suspicious-looking symptoms, and submit them for laboratory testing. Loading and delivery areas should be as far from production areas as possible.

Monitoring

Symptoms are not always readily apparent on nursery stock, and a weekly systematic monitoring of introduced stock or other plant material by a trained nursery scout helps ensure that the pathogen has not been introduced. If nurseries are located where sudden oak death is found, periodically inspect nearby native hosts for disease symptoms. Infected California bay laurel trees near the perimeter of nurseries may produce inoculum that can spread and infect nearby host plants. Removal of these trees may be warranted. Rain runoff coming downslope from areas with infected hosts may contain *P. ramorum*. Consider building berms to prevent water and soil from hillsides surrounding the nursery from moving into production areas. Irrigation water pumped from streams and ponds may be contaminated with *P. ramorum*. Consider having this water periodically tested to detect *P. ramorum*. If the pathogen is found to be present, use alternative irrigation sources, such as well water, or employ disinfection treatments.

Quarantine requirements

Federal and state quarantines exist that require inspections for nursery stock shipments from quarantined counties, periodic annual inspections, and sampling of hosts; if the pathogen is detected, eradication of the pathogen is necessary, and affected plants will be destroyed. Contact your Cooperative Extension advisor or county agricultural commissioner's office for information on hosts and locations affected by current quarantines.

Fungal Leaf Spots

These diseases are not common in many outdoor-grown California crops during dry periods, because the causal organisms need many hours of moisture on the leaf surface for spore germination and infection. However, in coastal areas with high relative humidity and fog, and inside greenhouses, they can become a problem. Evergreen plants and plants grown during the winter when there is rainfall may become infected. Overhead irrigation may also promote some of the leaf-infecting fungi. General control measures include keeping foliage dry, scouting for and roguing of isolated infected plants, monitoring surrounding native vegetation for possible reservoirs of infection, and practicing good sanitation methods including regular removal of plant debris that may harbor fungal spores. Several genera that cause leaf spots are briefly addressed below.

Entomosporium

Fungi in this genus commonly cause leaf spots on plants such as apple, Indian hawthorn, photinia, pyracantha, loquat, ornamental pear, hawthorn, toyon, and flowering quince. The spores are borne in a sticky, mucilaginous material and therefore are only dispersed by water. If it does not rain or the plants are not overhead irrigated, the fungus will not spread.

Symptoms

Spots can be reddish brown with or without a yellow halo and may darken over time (fig. 13.3A). The centers of the spots may have a lighter area. Foliage may turn red as spots coalesce. Infected leaves frequently drop. Defoliation is the end result of a bad infestation.

FIGURE 13.3. Leaf spot and canker diseases. *Entomosporium* on evergreen pear leaves (A); anthracnose on Chinese elm *(Ulmus parvifolia)* leaves (B) and cankers on trunk (C) caused by *Stegophora ulmea*; dead wood under bark of wild lilac branch (D) caused by the fungus *Nectria* sp.; cross section of branch showing wedge-shaped portion of dead wood caused by *Nattrassia mangiferae* on Indian laurel leaf fig *(Ficus nitida)* (E). *Photos:* J. K. Clark, courtesy of UC IPM (A, B, C, D); D. M. Mathews (E).

Control

Avoid overhead irrigation and spray with a preventive fungicide, especially prior to rain. Decrease frequency of irrigations to allow drying out periods. Remove badly infected stock that serves as inoculum sources.

Alternaria

Fungi in this genus commonly cause leaf spots found on a large number of plants, including calendula, carnation, dichondra, gerbera, scarlet pimpernel, cineraria, candytuft, kangaroo paw, marigold, and zinnia. *Alternaria* species tend to be host-specific and are more common during rainy winter months.

Symptoms

Infected areas usually are circular, dark brown or black, starting out small and then increasing up to $\frac{3}{16}$ to $\frac{1}{2}$ inch or more in diameter. Spots may appear as concentric rings (like a target pattern).

Control

Avoid overhead watering, water early in the day to allow for drying, and space plants to enhance air circulation. Remove plant debris around infected plants to eliminate future inoculum.

Septoria

Species of *Septoria* are very host-specific and are found on veronica, ceanothus, azalea, pincushion flower, chrysanthemum, ranunculus, and California toyon. These fungi are favored by cool, moist conditions, and spores are spread by water and human activities among and between rows of plants.

Symptoms

Leaf spots are characterized by small circular or angular spots that usually start out on older leaves as yellow and then darken over time to tan or dark brown. Spots may have light centers and be surrounded by a narrow yellow area. Sometimes black dots (fruiting bodies of the causal fungus called pycnidia) are visible in the center of the spots.

Control

Avoid overhead watering, water early in the day to allow for drying, and space plants to enhance air circulation. Remove plant debris around infected plants to eliminate future inoculum.

Other Leaf-Infecting Fungi

Although the main leaf-spotting and leaf-infecting fungi have been mentioned, others that might be found include black spot (of rose and other rosaceous plants) and those resulting from fungi in the genera *Ramularia*, *Heterosporium*, *Helminthosporium*, *Drechslera*, *Cercospora*, *Bipolaris*, and many others. For hosts and pathogens, see Farr et al. (1989) and Farr and Rossman (n.d.). For current chemical control recommendations, consult the *UC Pest Management Guidelines*, Cooperative Extension advisors, professional applicators, or company representatives.

Anthracnose of Leaves and Stems

Anthracnose is a rather generic term that refers to generalized leaf spotting and twig dieback. It is sometimes also called leaf or twig blight due to the nature of dying tissue that is induced. Several different fungi are capable of causing anthracnose, including *Colletotrichum* spp., *Phyllosticta* spp., *Discula* spp., *Glomerella* spp., and *Stegophora* spp. Anthracnose fungi infect a wide range of plants, but common hosts include ash, azalea, camellia, dogwood, elm, ficus, hollyhock, lily, maple, and sycamore. Spores are spread primarily by rain splash and overhead watering. Spores tend to be sticky and are not readily moved by simple wind currents. Long periods of moisture on plants encourage its development.

Symptoms

Leaves develop irregularly shaped areas of dead tissue that may coalesce and cause distortion and premature leaf drop. Lesions range in color from tan to brown to black (often referred to as tarlike in appearance) depending on the host, but some are dark at the edges and lighter in the middle (fig. 13.3B). Leaf spots are generally randomly distributed across the leaves, but those on ash, maple, oak, and sycamore tend to form around the central midvein and expand outward. Areas in the middle of lesions may be slightly sunken. Young twigs develop lesions containing dead tissue called cankers that can expand to cause girdling and death of the shoot. Cankered areas are discolored, sunken, and may develop callus tissue around the margins as the host tries to heal itself. Most infections do not cause extensive damage beyond the above-described leaf and twig symptoms; however, severe infections can expand downward to major branches and trunks, causing extensive cankers, canopy dieback, and possible death (see the section "Canker Diseases Caused by Fungi" below). Dead leaves and shoots killed by anthracnose may be mistaken for frost damage.

Control

Pruning in the fall or winter to remove overwintering spores present in stem cankers will aid in removing future inoculum sources. Selected pruning of affected limbs and removal of all pruned material as well as plant debris under and around plants will also help prevent reinfections (do not compost this material). Increase air circulation and sunlight among plants by thinning and increasing plant spacing to lower the amount of time moisture stays on the plants. Shield from rain and avoid overhead watering. Select resistant varieties when available and, if growing from seed, use sources that are certified clean.

Canker Diseases Caused by Fungi

Cankers are areas of dead tissue on stems, limbs, and trunks of woody plant species. Many species of fungi cause cankers, and some plant hosts are more susceptible than others. Common canker-causing fungi are *Botryosphaeria* spp., affecting a wide host range including eucalyptus, hibiscus, juniper, sweet gum, maple, oak, and pine; *Seiridium cardinal*, primarily causing bark canker on cypress but also able to infect juniper and arborvitae; and *Nectria* spp., which have a wide host range including many hardwood species, boxwood, ceanothus, elm, pear, and rose. Some canker diseases start out as anthracnose spotting on the leaves and young stems (see the section "Anthracnose of Leaves and Stems" earlier in this chapter), but then the infection spreads to larger limbs and branches, resulting in dieback of the canopy. One such disease progression is common in Chinese elm and is caused by *Stegophora ulmea* (fig. 13.3B, C). Most canker fungi invade through wounds caused by pruning or other damage. Other fungi and water molds capable of causing cankers include *Ceratocystis* and *Phytophthora* spp. Note that small, similar-appearing cankers can sometimes form as a host response due to wounding caused by physical damage to the plant or by invasion of boring insects.

Symptoms

External symptoms may not be obvious at first, but plants will generally show a change in bark color or a sunken area where the cankers develop (or both). Leaves may turn yellow or brown and young shoots may wilt. Exudation of sap from a canker causing a dark stain on the bark is a common symptom. Removal of the bark or pruning a section of a branch in areas showing these symptoms will reveal dead wood, which may encompass the entire width of the branch or just a portion, shaped like a wedge (fig. 13.3 D, E). Death of limbs or entire plants can result.

Control

Make sure to plant nonsusceptible hosts where canker fungi are a problem, and use resistant varieties of susceptible plants when available, such as the 'Drake' cultivar of Chinese elm. Remove infected limbs as soon as noticed, cutting back to a point at least 2 feet below the last visible section of affected wood. When pruning diseased trees, sterilize tools between cuts and more importantly between trees to avoid transmitting new infections of the fungus. Pruned plant material should be discarded off-site immediately to avoid carryover of spores. Keep trees vigorous with appropriate water and slow-release fertilizer applications.

Soilborne Fungal Diseases

Soilborne fungi involved in nursery diseases may be more important than any other group because they can dramatically affect plant health and appearance and result in plant death. The primary fungi involved are *Fusarium*, *Rhizoctonia*, *Thielaviopsis*, *Myrothecium*, *Sclerotinia*, *Sclerotium rolfsii*, *Verticillium*, and *Cylindrocladium*, as well as the water molds *Pythium* and *Phytophthora*. There are three major disease types caused by these pathogens: damping-off (seedling death), root rots, and vascular wilts. Each is addressed separately below.

Damping-Off (Seedling Death)

Damping-off is a seed and seedling disease that usually leads to death and is the result of infection by one or several soilborne pathogens. Virtually all plants are susceptible to this disease due to their succulent, tender tissues, which allow easy invasion of the pathogens. *Pythium*, *Phytophthora*, and *Rhizoctonia* commonly are involved, although other fungi such as *Alternaria*, *Botrytis*, *Fusarium*, and *Sclerotinia* also may be causal agents.

Symptoms

The most common form of damping-off results from infection of the seedlings after emerging from the soil

Photo: E. Zhukova.

(postemergence damping-off) or on cuttings, right at the soil line. Tissues collapse, and seedlings die and fall (fig. 13.4A). A dark area on the hypocotyl at the soil line is often visible and may appear pinched, and few or no roots may be present under the soil. Since more vigorous seedlings may escape severe disease development, damping-off may appear randomly in certain spots among individual cells of seedling trays. Other plants that do not succumb to initial damping-off will still have damaged roots and be stunted. Cuttings may remain standing, then eventually dry out and die. These pathogens can also cause seed decay or can infect the seedling before it emerges from the soil after germination (preemergence damping-off). Seed decay and preemergence damping-off are frequently blamed on poor seed rather than on disease. When potting mixes are kept too wet, *Pythium* is usually the causal organism. Swimming zoospores of *Pythium* are attracted to metabolites released by seeds and young developing roots. *Rhizoctonia* and the other fungi listed are more common under drier conditions.

Control

Damping-off can be controlled in part by good cultural practices. Seed should not be planted too deeply or too close together and should not be overwatered. Fertilizers should not be added before sowing seed because they can make plants more susceptible to the damping-off fungi. Virtually all seeds have enough nutrients in them to carry them to the point when they are thinned or transplanted, and at that time they should be fertilized. Prevention by pasteurizing or fumigating the planting mix, using clean seeds and cuttings as well as clean containers, will help. Watch seed and cutting beds for any problems. It is extremely important to remove any pots, trays, or flats where damping-off is established. Do not reuse them without disinfecting completely, if at all. Propagation media should not be recycled. Fungicide-treated seeds may also help to prevent damping-off problems.

Root and Crown Rot Diseases

These are caused by two major classes of plant pathogens, fungi and water molds, which will be addressed separately here. Some species of these pathogens have wide host ranges and can infect many different plants, while others have narrow host ranges. Virtually all plants are susceptible to some form of root rot disease. Overwatering and potting media that do not dry out in cool conditions are common environmental conditions among root-rotted plants, predisposing them to this disease.

In general, infected roots are usually smaller, darkened, and soft compared with the abundant, white, and firm roots of healthy plants. Plants may be stunted and leaves will wilt or turn yellow; death often occurs. When these aboveground symptoms are present, be sure to remove a few plants from their pots and compare them with those without symptoms. If the roots look equally healthy, the cause may be something nutritional or cultural and not related to a pathogen. The symptoms could be due to a different type of pathogen (e.g., a virus that may cause stunting or chlorosis), or it may be a case in which roots are visually unaffected by the pathogen causing the aboveground symptoms.

For plants with symptomatic roots, have the cause diagnosed so proper controls will be used, since a variety of pathogens cause root rots and each can have very different susceptibilities to chemicals. The metabolic pathways of true fungi that are targeted by fungicides can be quite different than those of the water molds, so fungicide effectiveness is often very different between the two. Therefore, the need for proper diagnosis of root rot pathogens is important for effective control.

The fungi and water molds can be spread by anything that comes in contact with media or plant roots, so good hygiene is important. Use of new or pasteurized planting mixes, pots, and trays will keep existing infestations from being recycled to new crops. Water management is also important, since most root rot pathogens thrive and move in overly wet soils; use well-drained planting mixes, and if recycling water, use a good treatment program to eliminate these pathogens.

Root and Crown Diseases Caused by Water Molds

Pythium and *Phytophthora*

The water molds are so called because they develop quickly in wet conditions, can tolerate low oxygen levels found in the soil, and produce motile spores (zoospores) that can swim short distances or be

FIGURE 13.4. Root and crown fungal diseases. Damping-off of *Lisianthus* seedlings caused by the fungus *Fusarium avenaceum* (A) (note random nature of leaf chlorosis and seedling collapse at surface of soil in various pots); *Pythium* infection of *Eustoma*, causing sparse, brown, decaying roots resulting in wilting (B); brown stems and necrotic roots of Oriental lilies caused by *Rhizoctonia* (C); white fluffy clumps of *Sclerotinia* mycelium on dahlia (D). *Photos:* M. Stanghellini (A); J. K. Clark, courtesy of UC IPM (B, C); S. A. Tjosvold (D).

carried by running water. They prefer weakened plants, including those that have been exposed to too much water, poor drainage, excess salts, temperature extremes, or other plant stresses. The two genera, *Pythium* and *Phytophthora*, each contain a number of species that infect a wide variety of plants. Although these pathogens share many similarities with fungi and were classified as such in the past, in recent years it has become apparent that they are more closely related to algae. They are now in a group called the Oomycetes (organisms causing downy mildew are also in this group) and are generically called water molds. This taxonomic grouping is still in flux and may change again as more knowledge is gained about these pathogens.

Symptoms

These pathogens invade roots, rhizomes, bulbs, and corms and cause death of the tissues. Wilting may occur due to destruction of the root system, which may be darkened and sparse with little branching compared with healthy roots (fig. 13.4B). These pathogens may move into the root crown area and girdle plants, where a characteristic dark area of the stem or crown may develop at the soil line. As they move up into the lower stems, they cause a darkening of the tissues underneath the epidermis or bark, particularly in woody plants. This symptom is very helpful in diagnosing water mold infections in plants. See the section "Damping-Off (Seedling Death)" earlier in this chapter for more information on these pathogens.

Control

Prevention is a key to water mold management, starting with sterile planting mix that drains well, placing plants in clean containers with drainage holes, and keeping pots up off the ground to avoid contact with underlying soil. Control is difficult because the water molds are nearly always present where plants are grown. Providing excellent growing conditions, including good drainage and aeration and avoiding overwatering or underwatering or stress of any kind, is important. A confirmed diagnosis of the presence of *Pythium* or *Phytophthora* is critical for proper chemical control, since many fungicides used for true fungi are not active against the water molds. Research has shown that increasing calcium levels and maintaining a slightly acidic pH (5.5 to 5.7) can inhibit the formation and motility of zoospores.

Root and Crown Diseases Caused by True Fungi

Several fungi routinely cause root rots in plants. The most common ones, many of which also cause crown rots, are addressed individually below.

Rhizoctonia

Fungi in this genus are very common wherever plants are grown, and they have been reported to infect more plants than any other pathogen. These fungi rarely produce a spore stage but are one of the few soil-inhabiting pathogens that can grow through the root media. Warm temperatures (70° to 90°F) and moist but not overly wet root media are favorable for disease development primarily on herbaceous plants.

Symptoms

Rhizoctonia causes root and stem rots, most commonly attacking plants right at the soil line, where it frequently girdles them. Sparse, darkened roots and stems with dark lesions are a common symptom of *Rhizoctonia* infection (fig. 13.4C). The resulting poor root system often leads to wilting and plant death. Forms of the fungus are also found as aerial pathogens, attacking cuttings of various plants when they are being rooted under a mist system or under tarps where brown mycelium may be visible (web blight). Also see the section "Damping-Off (Seedling Death)," another disease caused by *Rhizoctonia*.

Control

Standard good sanitation practices are important for control (see the section "Root and Crown Rot Diseases" earlier in this chapter). Chemical and biological controls exist that may be useful in treating *Rhizoctonia*.

Sclerotinia sclerotiorum and *Sclerotium rolfsii*

These are soilborne fungi that can cause very rapid rotting of plant tissues. Both fungi have wide host ranges but infect primarily herbaceous plants and a few woody species.

Symptoms

Both of these fungi attack at or near the soil line and cause stem lesions, wilting, and plant collapse. Both produce white fungal masses and strands that can often be seen on stems or root crowns (fig 13.4D).

With *Sclerotinia*, scattered in the white fungal masses on crowns and stems are white clumps (leading to the name cottony rot). These eventually become black, hard resting structures called sclerotia that are round to irregular in shape. These survival structures are produced in abundance inside plants that have hollow stems. Sclerotia carry over the fungus from one growing season to the next and usually are triggered to become active after a cool period. These structures can produce fungal strands that can invade susceptible plants. They can also produce ⅛- to ¼-inch, inverted mushroomlike structures that forcibly discharge their spores into the air. The spores usually can gain entrance through flowers and damaged or dead tissues. Once established there, the fungus can move into healthy tissues. Common hosts include carnation, dahlia, eucalyptus, and petunia. *Sclerotinia* notably causes flower blight on camellias, daffodil, and narcissus. Another important related pathogen that commonly causes flower blight of camellias in California is *Ciborinia camelliae (=Sclerotinia camelliae)*. Light brown, watery spots develop that merge and eventually cover the entire flower until it drops off.

Sclerotium differs from *Sclerotinia* in that the sclerotia start out white but become tan and finally brown. These sclerotia are spherical in shape and are the size of mustard seeds; hence, *Sclerotium* frequently is called the mustard seed fungus. There is no spore stage. Common hosts include dahlia, delphinium, gladiolus, and impatiens. *Sclerotium rolfsii* can invade and rot virtually all plant tissues. The fungus is favored by high temperatures, and the disease frequently is called southern blight or southern wilt.

Control

Sanitation is key for diseases caused by both *Sclerotium* and *Sclerotinia*. Floral debris should be removed from around plants to avoid carryover of sclerotia between seasons. See the general "Root Rot" control section earlier in this chapter.

Thielaviopsis

This fungus infects the roots of over 120 plant species, notably, begonia, impatiens, pansy, petunia, poinsettia, verbena, and vinca. Cooler temperatures (55° to 60°F), wet root media above pH 5.6, and excess fertilizer all enhance disease development and should be avoided.

Symptoms

Infected roots become darkened (hence the name black rot) and sometimes ridged or thickened, which lead to stunting and chlorotic foliage. Root and stem lesions under the soil line appear black, and dark spores may be visible.

Control

Standard good sanitation practices are important for control (see the general "Root Rot" section earlier in this chapter). Mixes with peat moss should be pasteurized using steam since they may harbor inoculum of *Thielaviopsis*.

Other Fungal Pathogens That Cause Root Rots

Cylindrocladium attacks the roots, stems, or lower leaves mainly of woody plants. It also can be serious in the rooting of cuttings, where it likes very high humidity and high temperatures. The best control is sanitation, using new pots and planting mixes, since this fungus produces long-lived spores that can survive for years until a suitable host is available. *Myrothecium* can infect roots and lower portions of plants, but it mainly causes crown rot. It is active at 65° to 85°F, has a broad host range, but is a weak pathogen. Like many of the root rot fungi, it is opportunistic and invades plants weakened by other stresses. Characteristic black dots surrounded by white fringes may be seen at the soil line just above the crown.

Vascular Wilt Diseases

The wilt fungi *Verticillium* and *Fusarium* are common soilborne pathogens. Both infect the roots and move through the water-conducting tissues of infected plants, partially blocking them. They produce resting structures that can remain viable in the soil for as long as 20 years in the absence of susceptible plants. *Verticillium* is favored by lower temperatures (68° to 82°F), depending on the species. The two major forms have a nonspecific host range of over 200 plant species. Plants commonly infected with *Verticillium* include maple, heather, pelargonium, pistachio, petunia, rose, salvia, strawflower, and many more. *Fusarium oxysporum* also infects many plants, but generally each form is host-specific (formae speciales), and infection may be limited to a small group of closely related plants or even a single species. For example, the *Fusarium* that infects carnations goes only to a few species of *Dianthus*, and the one that infects *Eustoma* goes only to that plant. Plants infected by different *Fusarium* forms include China aster, chrysanthemum, cyclamen, dahlia, delphinium, freesia, gladiolus, veronica, stock, narcissus, flax, and many others. *Fusarium oxysporum* prefers warm climates (77° to 85°F or above), and its ability to infect can be affected by the plant's nutrition status, especially in regards to nitrogen and calcium.

Symptoms

Wilting occurs first in the lower leaves when water cannot move up from the roots to the aerial portions of the plant through the xylem vessels. Both fungi often produce brown to black staining of the vascular system that can be seen if stems are cut in cross section, although it is more common with *Fusarium*. In addition to wilting, which may start on one side of the plant, yellowing of leaves, stunting, and death may occur, especially under higher temperatures, which cause water stress.

Control

Some plants that are resistant to *Verticillium* exist, but no viable chemical treatments are available. Since *Fusarium oxysporum* forms are host-specific, crop rotation to another plant species can be an effective control measure for field-grown crops. Control for both of these fungi is difficult. Soils can be steam treated or solarized to destroy not only the active fungus but also the long-lived resting spores, but treatment is not always 100 percent effective. Care should be taken not to move infested soils to clean areas. Infected plants should be discarded.

FIGURE 13.5. Bacterial leaf spots on ranunculus caused by *Xanthomonas.* Spots start out small and individual (lower right), then increase in size and may coalesce to cover most of the leaf (top middle). Note yellow areas around spots, which are common with bacterial leaf spots. *Photo:* H. Azad.

Bacterial Diseases

Bacteria are primitive single-celled organisms that can multiply rapidly, approximately every 20 to 30 minutes, which enables them to produce large numbers of cells in a short period of time. There are about 100 different plant pathogenic bacterial species. Bacteria enter plants through natural openings such as stomates or hydathodes as well as wounds, and some are transmitted by insects. They are favored by high temperatures and moist environments. Outdoor growers in much of California do not generally have major problems with bacterial diseases due to low relative humidities most of the year. An exception to this is the xylem-limited bacterium *Xylella fastidiosa*, transmitted by leafhoppers, which causes bacterial leaf scorch (see the section "Bacterial Leaf Scorch" later in this chapter). In greenhouses, however, environmental conditions may be satisfactory for bacteria to cause significant disease problems. Overhead irrigation and rain splashing are common ways bacteria move from plant to plant.

Plant pathogenic bacteria cause symptoms such as leaf spots, blights, wilting, soft rots, galls, and scorching. In bacterial leaf spots, the tissues are killed and the spots tend to be angular because the bacteria move intercellularly but cannot cross large veins. Dark gray or black spots caused by bacteria tend to be less uniform than those caused by fungi, and they may be surrounded by a yellow halo or have a water-soaked margin (fig. 13.5). Spots may coalesce until the entire surface of the leaf is killed. Two bacterial genera, *Xanthomonas* and *Pseudomonas*, are the main causes of leaf spots. Leaf spots are found on begonia, chrysanthemum, delphinium, gardenia, geranium, hibiscus, ivy, lavender, ornamental pepper, primrose, ranunculus, and many others.

Wilting of plants is caused when masses of bacteria, exuded slimes (extracellular polysaccharides produced by the bacteria), or plant defense mechanisms create blockages of the xylem vessels and water transport is compromised. Common wilt-causing bacteria are *Clavibacter*, *Erwinia*, *Pseudomonas*, *Ralstonia*, and *Xanthomonas*. Plants may die from lack of water uptake. Affected ornamentals include begonia, carnation, dahlia, dieffenbachia, ranunculus, and geranium.

Management strategies for bacterial diseases include

- starting with pathogen-tested stock plants
- avoiding overhead watering or exposure to falling rain (where feasible)
- watering early in the day to allow for drying of excess water
- reducing humidity
- scouting for and discarding symptomatic plants
- sanitizing tools if used for cutting and pruning
- treating recycled irrigation water

Chemicals such as copper compounds and antibiotics can be applied prophylactically; but once symptoms are present, such controls are generally not effective.

Bacterial Galls

There are two major types of galls caused by bacteria: the tumorlike crown galls caused by *Agrobacterium tumefaciens* and leafy galls, or fasciations, caused by *Rhodococcus fascians*. Minor gall diseases are olive knot and oleander gall, caused by *Pseudomonas syringae*, and carrot gall, caused by *Rhizobacter daucus*.

Symptoms

Crown gall occurs on hundreds of dicots, including many woody plants and ornamentals such as marguerite daisy, carnation, geranium, kalanchoe, marigold, rose, and snapdragon. It is caused when the bacterium injects the plant cell with a segment of DNA that induces tumor formation, usually near the soil line, but sometimes higher on the plants or on the roots, where the bacteria survive. Plants may also be stunted and more susceptible to environmental stress. Leafy galls are a proliferation of small shoots near the soil line or in leaf axils that are flattened and distorted. These symptoms, found on over 60 dicot and monocot species, including many herbaceous ornamentals, can be confused with herbicide or eriophyid mite damage. Infected plants exhibiting such symptoms should be discarded.

Control

Preventive biological control of crown gall can be achieved for seeds, seedlings, and rootstocks by treatment with *Agrobacterium radiobacter* strains K-84 or K-1026, antagonistic species of the bacterium. Success of this biological control depends on the ability to match the correct strains with each other and may require some trial and error before finding the right combination. Tools that are used to prune plants infected with gall-inducing bacteria should be dipped in 70 percent alcohol and flamed or treated with 10 percent sodium hypochlorite. Root media in pots or soil within several inches of field crops should be discarded.

Fire Blight

This bacterial disease is found only on members of the pome tribe of the rose family, which includes apple, pear, pyracantha, cotoneaster, quince, flowering quince, hawthorn, toyon, and loquat. In California, pear and pyracantha are most severely affected. Though other members of the rose family can be experimentally infected, they are not found infected naturally. The causal agent is *Erwinia amylovora*. There also is a form of this bacterium that infects *Rubus*. The bacteria gain entrance through the flowers and rapidly spread in both directions of the stem.

Symptoms

Tissues are killed so rapidly that leaf abscission zones do not form and the dead leaves hang in the plants, making them look as though they have been scorched by fire. Only a few branches may be affected, with the rest of the tree appearing normal.

FIGURE 13.6. Leaf scorch caused by the bacterium *Xylella fastidiosa*. Necrotic tissue usually starts near the leaf tip, then moves along the margin, before moving inward. Various stages of leaf scorch on ornamental plum (A), oak (B), and mulberry (C). Note yellow margin between green and necrotic areas on oak leaf. *Photo:* H. Azad.

Control

Fire blight can be controlled by pruning and removal of infected wood 7 to 8 inches beyond the dead portions of a branch or limb and spraying during flowering with fixed coppers or the agricultural grade of streptomycin.

Bacterial Leaf Scorch Caused by *Xylella fastidiosa*

The bacterium *Xylella fastidiosa* is of increasing concern in California. It is vectored by several species of leafhoppers, the most important of which is the glassy-winged sharpshooter (GWSS, *Homalodisca coagulata*). GWSS feeds on the xylem of plants and acquires and transmits the bacterium via these vessels. *X. fastidiosa* has several strains, the most well known being the Pierce's disease strain that infects grapes, among other crops. More important to ornamental landscape plants is the almond strain, which is causing major disease in sweet gum, ornamental plum, crape myrtle, and olive, among others. Oleander has been devastated in southern California by the oleander strain of *X. fastidiosa*, which also can infect daylily, jacaranda, and magnolia. There are over 200 hosts of the various strains of *X. fastidiosa*, and more are being found each year. Outside nursery stock can be at risk if GWSS is present in the area, particularly young plants.

Symptoms

Common symptoms of bacterial leaf scorch generally start with necrosis at the tips of leaves, which progresses around the edge, and there may be a yellow margin between the necrotic and green areas of the leaf (fig. 13.6). This symptom is similar to salt damage or general water stress. As the xylem vessels continue to become blocked, wilting can occur as summer heat begins.

Control

Management includes selective pruning of affected limbs early in the disease cycle and increasing irrigation during hot months. Severely affected plants should be removed. Efforts to inject antibiotics have not generally been successful in the long term. Studies are ongoing to find ways to control this emerging disease threat through resistant varieties, biological control, disabling the bacterium, and vector control.

Phytoplasma Diseases

Phytoplasmas, previously called mycoplasmalike organisms because of their physical similarity to animal-infecting mycoplasmas, are small disease-producing organisms that are surrounded only by a cell membrane, lacking the cell wall that bacteria possess. Phytoplasmas are found in the phloem cells of the plant and are transmitted primarily by leafhoppers or by vegetative propagation of infected plants. Some planthoppers and psyllids may also act as vectors. Vectors can acquire phytoplasmas at the nymph or adult stages after feeding for hours or days, but they are not able to transmit the pathogen until a 10- to 45-day incubation period has occurred (depending on temperature) to allow for maturation and circulation of the phytoplasma within the insect. Once infected, the insects can transmit for life; however, the phytoplasma is not passed on through eggs. The most important phytoplasma that infects ornamentals is aster yellows, whose primary vector in North America is the leafhopper *Macrosteles quadrilineatus*. For additional host-phytoplasma relationships, see French (1989); and for more information on phytoplasmas, see Agrios (2005) or Schumann and D'Arcy (2010).

Symptoms

Typical symptoms are yellowing, stunting, shoot proliferation, and disruption of pigmentation in leaves and flowers, including flowers that become leaflike in appearance. Until seen with the electron microscope in the 1960s, phytoplasma-induced diseases were thought to be caused by viruses, due to similar symptoms. Multiple strains of aster yellows have a host range that includes anemone, China aster, calendula, coreopsis, delphinium, carnation, gaura, lobelia, petunias, phlox, pincushion flower, ranunculus, veronica, viola, and many others.

Control

Destruction of infected plants is the best means of controlling the spread of phytoplasmas. Injection of tetracycline-class antibiotics can be used to suppress symptoms in specimen plants (nonfood species); however, continual treatment is required or symptoms will return. Control of insect vectors is only effective if total exclusion is practiced. This is because, once infected, insect vectors can transmit the phytoplasmas before any systemic pesticide kills them. Weeds that can harbor both the phytoplasmas and insect vectors should be eliminated near greenhouses or field-grown crops.

Diseases Caused by Plant-Parasitic Nematodes

Nematodes are mostly microscopic, eel-like roundworms, typically ranging in length from 0.01 to 0.08 inch. They constitute the most abundant multicellular animals on our planet, and they occur in almost every conceivable ecosystem. Nematodes should not be confused with earthworms–they are not even closely related–but most nematode species could be considered similarly beneficial since they feed on microbes, microfauna, or decaying organic matter and thus play an essential role in nutrient cycling. Other nematode species cause devastating animal or human diseases, while some species parasitize important insect pests and are commercially available as biological control agents. Most important in the context of this manual are nematode species that cause plant diseases. Although it is difficult to determine the exact economic impact of plant-parasitic nematodes on agricultural and horticultural crops, estimates based on several surveys suggest annual production value losses of about 12 percent (Barker et al. 1994).

Plant-parasitic nematodes are also responsible for the incidence and severity of various root disease complexes (Abawi and Chen 1998). Tissues weakened by nematode attack are often more susceptible to other plant pathogens such as fungi (e.g., *Fusarium*, *Rhizoctonia*, etc.) and bacteria (e.g., *Agrobacterium tumefaciens*). Some plant-parasitic nematodes (dagger, needle, or stubby root nematodes) are able to transmit plant-pathogenic viruses as well (nepoviruses and tobraviruses).

Most plant-parasitic nematodes spend at least part of their life in soil surrounding plant roots, while other species dwell most or all of their life inside plant tissues. These nematodes feed by piercing plant cells with a mouth stylet and withdrawing cell contents. They also inject enzymes and other metabolites that help to break down cell contents or foster changes in the plant physiology and morphology. In general, a nematode's life cycle includes egg, four juvenile stages, and the adult stages. The length of a nematode's life cycle depends on the species and the environmental conditions. Most plant-parasitic nematodes have an activity temperature range from 60° to 85°F. They survive as eggs or juveniles. In soil or on plant surfaces, they require a thin water film to move. However, for significant dissemination nematodes move as contaminants on and in water, soil, plants, wind, soiled tools, and machinery as well as infested planting material or media.

Since nematodes feed on living plant tissues, they influence the host's physiology and may cause below- and aboveground symptoms such as root galling, lesions, stunting, wilting, or chlorosis. Most of these symptoms are nonspecific and can be attributed to a number of abiotic and biotic problems. The diminutive, colorless, and transparent nematode bodies are hard to detect and even more difficult to identify. However, an accurate diagnosis is required for appropriate countermeasures. Samples from symptomatic and healthy-looking plants should be analyzed in a commercial

FIGURE 13.7. Nematode damage. Galls on cucumber roots caused by the root knot nematode *(Meloidogyne incognita)* (A). Swellings on roots are induced by feeding activity and the presence of developing females; chlorosis and angular necrosis on leaves of anemone caused by foliar nematodes (B). *Photos:* J. O. Becker (A); S. A. Tjosvold (B).

nematology laboratory. See chapter 11 for more information on diagnosing nematode problems.

Common Root and Foliar Nematodes

In container nursery production, the most commonly encountered nematode pest is the root knot nematode (*Meloidogyne* spp.). Juveniles of *Meloidogyne* spp. invade roots and move into their vascular cylinders. Plant-produced nutrients are diverted to the feeding sites, while galled roots become increasingly less efficient in water and mineral uptake. During the final developmental stages, the root knot nematode female enlarges, breaks through the root surface, and lays several hundred eggs in a protective gel sack. The eggs can hatch immediately, and the next generation of juveniles is ready to attack plant roots. Under optimal conditions the life cycle is completed in 3 to 4 weeks.

In contrast to the sedentary root knot nematodes, which remain in place once they establish a feeding site, lesion nematodes (*Pratylenchus* spp.) continuously migrate and feed throughout the cortex of an infected root. Foliar nematodes (*Aphelenchoides* spp.) occur mainly within the leaves and crowns of many herbaceous perennials. *Aphelenchoides* spp. can survive desiccation in dried plant tissues but regain activity when rehydrated.

Symptoms

While feeding, root knot nematodes induce cellular changes that lead to the formation of small- to medium-size swellings, or galls, on the roots (fig. 13.7A). Stunting and wilting may occur, along with a general lack of plant vigor.

The destructive feeding behavior of lesion nematodes leads to characteristic dark spots that later may join to form larger necrotic areas and eventually cause root death. Lesion nematodes have a wide host range and distribution. Typical symptoms of foliar nematodes include localized chlorotic or purplish lesions delineated by leaf veins (fig. 13.7B). If buds or young leaves are infested, they may not develop properly and may become deformed.

Control

Once plant-parasitic nematodes are established, they are very difficult—and quite often impossible—to eradicate. Consequently, prevention of introduction and dissemination of plant-parasitic nematodes is of utmost importance. In container nursery production, there should be a zero threshold level of tolerance for plant-parasitic nematodes. Exclusion is the only option, as there are currently no chemical or physical methods that will reliably and safely eliminate plant-parasitic nematodes from infested plant tissues.

Any planting material and media should be free of plant-parasitic nematodes. Stock plants should be periodically checked for potential nematode infestations. Containers, tools, and working benches should be clean and disinfected. Treatments approved by the California Department of Food and Agriculture (CDFA) for soil and container disinfection include certain soil fumigants, aerated steam, and tent soil solarization (CDFA 2009). Soil and containers need to be protected from infestation by plant-parasitic nematodes after the treatments. This includes using nematode-free irrigation water and avoiding placement of containers on bare ground and in flood-prone or water runoff areas.

Quarantine requirements

Of special importance are certain exotic, nonindigenous plant-parasitic nematodes that may enter the state with nursery shipments. The danger is not only related to their damage potential to any particular host crop but also to the far-reaching implications for trade should the nematode become established. For example, the burrowing nematode *(Radopholus similis)* is an endoparasite of hundreds of plant species, including more than 200 perennials. These nematodes, economically important in all tropical and subtropical areas, have been encountered in the past in soil and plant material at California border inspections. Importing countries would require quarantines if this nematode becomes established in the state (Ferris et al. 2003).

Virus Diseases

Viruses are infectious pathogens that are composed of either RNA or DNA surrounded by a protective coat of protein that the virus produces from its own genetic information. They are submicroscopic in size, requiring an electron microscope for visualization. Viruses require a living host cell in order to multiply and produce new particles, but existing particles can survive for days, months, or even years, depending on the virus, in plant debris, on hard surfaces such as greenhouse benches or walls, pots, tools, or watering implements. There are over 1,000 known plant viruses, with each usually having several different strains that may vary in the severity of their symptoms or host range (Adams and Antoniw 2006). Some viruses are able to infect hundreds of different plant species, even across plant families, while others have very narrow host ranges and can only infect a few species within a single plant family. Viruses are named for the first host plant they were found in and the type of symptom that they cause. Therefore, the name does not reflect the only host that may be infected by that virus. For example, *Tobacco mosaic virus* (TMV) can infect 200 different plants, while *Cucumber mosaic virus* (CMV) can infect over 1,000 plant species. Many ornamental plants are on the host list for both of these viruses, as well as for most plant viruses. Mixed infections of more than one virus are quite common.

All viruses can be transmitted by vegetative propagation and grafting of infected source plants. Many can be transmitted by insects (primarily aphids, whiteflies, leafhoppers, and thrips), cutting tools, irrigation water, or any action that causes wounding. A small number of viruses can be seedborne via the embryo or simply by contamination of the outer seed coat, and a few can be transmitted by pollen, fungi, or nematodes. Parasitic plants such as dodder can also transmit viruses by acting as a living bridge between virus-infected and healthy plants. Some virus-infected plants are used commercially. For instance, *Camellia yellow mottle virus* causes bright yellow variegation on camellia leaves and sometimes flower breaking, usually with no other obvious deleterious symptoms; it cannot be transmitted to other plants mechanically or by vectors, only by vegetative propagation.

Commercial labs and some universities are able to test for the most common viruses affecting many plants. Testing kits called immunostrips, similar to a home pregnancy test kit, are now available for many of the more common plant viruses. These can be used with little training in greenhouse or field operations when it is necessary to monitor for viruses that affect nursery crops. New viruses are continually being discovered, particularly in ornamental plants. This is due in part to the international nature of trade for these plants, which are primarily vegetatively propagated. This complicates testing since virus-specific detection methods must usually be used, and obviously these do not exist for new viruses. Once detected by other means, often accidentally, it can take up to 3 years to have a usable detection system. Some of the most important plant viruses are addressed individually below.

Since viruses use the plant's resources to multiply, they often disrupt the metabolism of the host, affecting things such as pigment production and growth. Viruses cause a variety of symptoms on infected plants, including leaf mosaics (light and dark green mottling), stunting, irregular growth, deformation of multiple plant parts, ring spots, and color breaks in flowers (fig. 13.8). It is important to note that chemical spray damage can mimic typical virus symptoms, since herbicides, as well as phytotoxicity due to some pesticides, also cause metabolic disruptions to plants. Symptoms caused by chemicals, however, usually occur uniformly and rapidly, whereas those caused by viruses develop more gradually and randomly among plants. Subsequent new plant growth usually returns to normal after chemical exposure, while virus symptoms persist throughout the life of the plant.

Once plants become infected with a virus, there is no cure. Therefore, the main management strategies are as follows:

- Start with tested clean stock plants or seed.
- Use proper hygiene to prevent spreading viruses that may be present.
- Scout routinely for virus symptoms.
- Discard infected materials.
- Control insect vectors.

Vector control is best practiced as exclusion instead of eradication, since most infected vectors can transmit the virus immediately upon feeding on a healthy plant before systemic pesticides can take effect. Additional precautions that should all be used in an active virus suppression program are

- changing or disinfesting pruning and propagation tools between varieties or rows of plants
- using disposable gloves when handling plants
- removing plant debris and cleaning benches, pots, and water emitters between crops

Tobacco Mosaic Virus

Tobacco mosaic virus (TMV) and its closely related tobamovirus *Tomato mosaic virus* (ToMV) are easily transmissible by simply touching an infected plant and then touching a susceptible plant. The host range of TMV extends to over 200 plants within 30 families, and members of the Solanaceae family are particularly susceptible to tobamoviruses, including petunia and tomato. These viruses are examples of those that can remain viable for over a decade in the absence of a living host, as mentioned above. TMV has been recovered from dried herbarium samples that were over 100

FIGURE 13.8. Types of symptoms caused by viruses. Mosaic on petunia caused by *Tobacco mosaic virus* (A); healthy (left) and three virus-infected, deformed leaves of portulaca caused by *Alternanthera mosaic virus* (B); healthy (left) and two *Angelonia flower break virus*-infected *Angelonia angustifolia* flowers (C). *Photos:* D. M. Mathews.

years old. There are no known insect vectors, but it has been shown experimentally that aphids feeding on infected plants can pick up virus particles on their legs and can transmit them to healthy plants mechanically through normal feeding activities.

Symptoms

Symptoms of TMV are mild to severe mosaic (patches of light and dark green areas) and blistering on leaves; ToMV also causes stunting.

Control

Tobamoviruses are extremely easy to transmit mechanically. Virus particles can be moved from infected plants to healthy ones by almost any touching activity, so care must be taken when using watering or pesticide spray implements, touching plants for inspection, or even walking between benches when clothing may come into contact with plant tissues. External contamination of the seed coat is common when seed are collected from TMV-infected plants. Transmission also occurs upon germination, and seedlings become infected as they emerge or during transplanting. Seeds can be treated before planting by soaking in 10 percent trisodium phosphate (TSP) for 15 to 30 minutes or in 1 percent sodium hypochlorite for 30 minutes. (This seed treatment can also be used to eliminate many fungal or bacterial pathogens if products containing TSP are registered for use.) A small lot of seed should be treated first and checked afterward for proper germination to ensure that a given species of seed is not adversely affected by the virus elimination treatment. Resistance genes for tobamoviruses are available in many plants, and these varieties should be selected when possible. Care must be used when workers who smoke cigarettes are allowed to handle plants, since the virus from TMV-infected tobacco can be picked up on the hands and transmitted to commercial plants. Thorough hand-washing after smoking should be mandated, or do not assign smokers to jobs with direct plant contact. Using reconstituted nonfat dry milk (20% solution of dissolved powder in water; 1 tablespoon powder in ¼ cup water) has been shown to inactivate TMV on tools used for pruning and propagating after a 1-minute contact time (Lewandowski et al. 2010). The ability of milk to inactivate other viruses is being tested. This is an alternative to standard treatments with chlorine bleach, quaternary ammonium compounds, or other disinfectants. See general virus control information in the list of virus management strategies above.

Cucumber Mosaic Virus

Cucumber mosaic virus (CMV) has many different strains that vary widely in their ability to cause disease. CMV has an extremely wide host range (1,200 plant species in 100 plant families), so it can spread easily once introduced into an operation. It is vectored by 80 different species of aphids and is mechanically transmissible but is not as easily moved as TMV by casual contact in the greenhouse. Some strains can be seed transmitted in certain hosts.

Symptoms

Mild to severe mosaic on leaves, stunting, and "shoestringing" of leaves (where the blade portion of the leaf becomes narrow and can be so severely reduced that the leaf is not much more than a midrib) are common symptoms of plants infected with CMV.

Control

As with all viruses, no cure is available (see the sections "Virus Diseases" and "*Tobacco Mosaic Virus*" earlier in this chapter). Standard avoidance of mechanical transmission, as with TMV, above, should be used. In greenhouse settings, aphid exclusion can limit large scale transmission of CMV. See the general virus control section earlier in this chapter.

FIGURE 13.9. Ring spots on nemesia caused by *Impatiens necrotic spot virus* (INSV). *Photo:* D. M. Mathews.

Tomato Spotted Wilt Virus and *Impatiens Necrotic Spot Virus*

Tomato spotted wilt virus (TSWV) and *Impatiens necrotic spot virus* (INSV) are two members of the genus *Tospovirus*, and they are increasingly important viruses in ornamental plants. TSWV can be transmitted by at least eight species of thrips, while INSV has to date only been transmitted by the western flower thrips *(Frankliniella occidentalis)*. The virus is acquired by the first- or second-stage thrips larvae, and it incubates in the insect during the third and fourth stages. When the thrips emerges as an adult, it is infectious and can transmit the virus for life. Noninfectious adults who feed on infected plants are not able to transmit the virus, even though it can be found within their bodies. The virus is not passed onto future generations by the laying of eggs. Both viruses can be mechanically transmitted, but not easily. Extensive research is ongoing on tospovirus transmission and epidemiology (Pappu et al. 2009). These two viruses can infect 1,200 different plant species in 90 different plant families, including weed and vegetable crops in addition to ornamentals.

Symptoms

Typical symptoms are ring spots on leaves (fig. 13.9), mosaic, stem cankers, and necrotic areas on leaves. The cankers and necrotic spots may make the plant more susceptible to other pathogens such as the fungus *Botrytis*.

Monitoring

Indicator plants are used in monitoring programs because they are hypersensitive to plant pathogens and show symptoms before crop plants do. Some indicator plants show early symptoms that are easy to recognize but do not act as a reservoir for the pathogen. These types of indicator plants should be selected over those that might also act as virus reservoirs. Research done in California greenhouses showed that petunias, which are nonreservoir plants because the virus does not become systemic in petunias, were reliable sentinel plants for tospovirus monitoring (Robb et al. 1999).

Control

Because of their extensive host range and the prevalence of western flower thrips in California, infections of TSWV and INSV are common, and control is difficult once they are established. Control of this virus/vector complex can be improved greatly by implementing an integrated disease and pest management program that includes buying virus-tested source plants, scouting for symptoms, eradicating infected plants, removing weeds, using resistant varieties when available, and controlling thrips with screening, chemical, and biological methods.

Viroid Diseases

Viroids are the smallest plant pathogens and are composed of naked RNA, approximately 250 to 400 nucleotides in length—and unlike viruses, they code for no proteins. They rely solely on their host for multiplication. They can be transmitted mechanically with tools and by grafting, pollen, and seed, as well as root grafting in the field, although not all viroids are transmissible by all methods. Since viroids do not have coat proteins like viruses, antibody detection methods such as ELISA cannot be used. Molecular tests such as RT-PCR or nucleic acid hybridization, which usually must be done by a commercial or university laboratory, are the most sensitive available tests for viroids.

Symptoms

Viroid symptoms are similar to those caused by viruses and include stunting, leaf bleaching (yellow to white), distortion, and fruit deformation (fig. 13.10). However, many ornamental plant species show no obvious symptoms when infected with viroids. There are about 30 different viroids, and their host ranges are generally narrow (Hadidi et al. 2003). However, some of them, such as the *Potato spindle tuber viroid*

FIGURE 13.10. Symptoms caused by viroids. Leaf mottling of chrysanthemum caused by *Chrysanthemum stunt viroid* (A); bleaching and deformation of peach leaves caused by *Peach latent viroid* (B); leaf bleaching and deformation (C) and fruit cracking and discoloration (D) caused by *Avocado sunblotch viroid*. *Photos:* D. M. Mathews (A, B, D); M. Arpaia (C).

FIGURE 13.11. Nutritional deficiencies. Iron deficiency on gardenia (A) and sweet gum *(Liquidambar styracifolia)* (B), showing interveinal chlorosis. *Photos:* J. K. Clark, courtesy of UC IPM (A); E. Martin (B).

(Pospiviroidae family), are showing expanding host ranges, moving from their established potato and tomato hosts to several ornamental plants such as marguerite daisy, angel's trumpet, cestrum, chrysanthemum, potato vine, petunia, and verbena. *Citrus exocortis viroid* has also been found in verbena, potato vine, and impatiens. Since most of these viroids are asymptomatic in their ornamental hosts, they may escape visual detection. If they are grown adjacent to other crops, transmission may occur, and those new hosts (e.g., potato and tomato) may be adversely affected.

Control

There is no cure for plants infected with viroids. Control and management strategies are similar to those recommended for viruses, above.

Noninfectious Disorders

Unfavorable environmental conditions can cause problems in plants without a pathogen being present. As described earlier in this chapter, these types of diseases or disorders are termed abiotic (nonliving) and are noninfectious in contrast to those caused by biotic (living) pathogens. These conditions stress the plant such that it cannot function optimally, and various symptoms may be observed. In addition, what starts out as a noninfectious disorder may weaken the plant enough so that it becomes susceptible to a biotic pathogen that normally would not be able to infect a completely healthy, nonstressed plant. Too much or too little light, water, or fertilizer, as well as poor soil or air quality, herbicide damage, and so on, may all create damaged plants. Several of these conditions are addressed individually below.

Nutrient Deficiencies and Excesses

Iron deficiency

Iron deficiency is common and appears as a loss of green color starting at the margins of leaves and progressing inward so that the veins or the tissues surrounding the veins are the last to lose their color (fig. 13.11A, B). These symptoms appear mainly on the new growth because when iron is available in the soil it becomes fixed in the older leaves of the plant; it cannot move to the new growth when it becomes unavailable or deficient in the soil. Iron deficiency may occur because high soil pH (above pH 7.5) prevents the iron from being available to the plant. Excessive phosphorus applications can also bind the iron and make it unavailable to the plant. Iron deficiency can occur under conditions that are not favorable for good root growth, such as heavy clay soils or overirrigated situations. Over time similar symptoms may appear in the older leaves of plants, along with brown spots and leaf drop. Depending on its cause, iron deficiency can be improved by reducing soil pH to between 6 and 6.8, applying iron chelate, and improving irrigation management and aeration of the root medium.

Manganese deficiency

Manganese deficiency looks very similar to iron deficiency. Flower buds and other plant parts may fail to develop properly and appear distorted. Remedies are the same as for iron except that manganese chelate is used instead.

Phosphorous deficiency

Phosphorus deficiency is usually seen on older leaves as red, purple, or sometimes yellow coloration. Plants may be stunted. Phosphorus deficiency is somewhat uncommon in most soils. Cool temperatures may induce phosphorous deficiency due to poor root growth, which inhibits uptake.

Copper deficiency

Copper deficiency causes yellowing, stunting, and upward curling of leaves as they fail to fully unfold, especially in younger leaves. Petioles may turn downward, causing shoots to appear wilted.

Potassium deficiency

Potassium deficiency appears as yellowing between leaf veins, browning at the tips, and scorching or spots on the margins of the older leaves. Young shoots may be thin and can show dieback at the tips. Yellow speckling may be present on palm fronds.

Nitrogen deficiency

Nitrogen deficiency may appear first as light green or yellowing in the older leaves. Later it will be seen in young tissues, which may grow poorly. Stunting and premature leaf drop also result from a deficiency of nitrogen.

Excess nutrients

Excess nutrients can be just as damaging to plants as deficiencies. Common symptoms include marginal burning of the older leaves in the plants that are dicots and burning at the tips in monocots, gymnosperms, or most ferns. If you suspect a nutrient imbalance to be the cause of such symptoms, a thorough soil analysis can determine the proper course of action to alleviate the problem.

Environmental Factors

Sunburn of leaves usually starts out as a yellowing or bleaching followed by browning as tissues die, mainly between the veins. Sunscald occurs on young bark and can cause a change in color followed by cracking and splitting. Plants should be acclimated gradually to sunny environments, and heavy pruning should not be done immediately prior to placement outside or before extremely hot weather occurs. Some plants, such as camellia, star jasmine, climbing fig, and weeping fig, are especially susceptible to too much sunlight. This will result in loss of leaf color in new growth, with no particular pattern associated with it. This is not a burn—and if given more shade, the chlorophyll will reappear.

Air pollution from chemicals such as smog, ozone, chlorine, sulfur dioxide, or ethylene can cause chlorosis, bleaching, spotting, or bronzing of leaves. Most plants are affected by high levels of these or other pollutants, but petunia, dahlia, pine, and tomato seem particularly susceptible. In enclosed greenhouses, care must be taken to avoid fumes from vehicle exhaust, equipment motors, heaters, disinfectant use, or other sources from entering without sufficient fresh airflow.

Edema appears as small raised areas, usually on the undersides of the leaves, that appear as blisters on the surface. The blisters usually start out as greenish and then turn light brown, appearing similar to cork. Edema is caused by excessive watering, particularly when humidity is high, and is more common when high soil temperatures are coupled with cooler air temperatures. This causes the roots to take up more water than is being lost by the aerial portion of the plant. Plants commonly affected by edema are begonias, geranium, Hans ivy, pansy, eucalyptus, and violet.

Abiotic disorders can be difficult to distinguish from problems caused by pathogens, making them difficult to diagnose. See chapter 11 for information on how to diagnose these disorders.

References

Abawi, G. S., and J. Chen. 1998. Concomitant pathogen and pest interactions. In K. R. Barker, G. A. Pederson, G. L. Windham, eds., Plant and nematode interactions. Madison, WI: American Society of Agronomy. 135–158.

Adams, M. J., and J. F. Antoniw. 2006. DPVweb: A comprehensive database of plant and fungal virus genes and genomes. Nucleic Acids Research 34: 382–385. Descriptions of Plant Viruses website, http://www.dpvweb.net/.

Agrios, G. N. 2005. Plant pathology. 5th ed. San Diego, CA: Elsevier Academic Press.

Barker, K. R., R. S. Hussey, L. R. Krusberg, G. W. Bird, R. A. Dunn, V. R. Ferris, D. W. Freckmann, C. J. Gabriel, P. S. Grewal, A. E. Macguidwin, D. L. Riddle, P. A. Roberts, and D. P. Schmitt. 1994. Plant and soil nematodes—Societal impact and focus for the future. Journal of Nematology 26:127–137.

Campbell, C. L. 1985. Wilts. In D. L. Strider, ed., Diseases of floral crops. Vol. 1. New York: Praeger. 141–151.

CDFA (California Department of Food and Agriculture). 2009. Approved treatment and handling procedures to ensure against nematode pest infestation of nursery stock. Nursery Inspection Procedures Manual, NIPM Item 7. Sacramento: Plant Health and Pest Prevention Services, Pest Exclusion Branch. CDFA website, http://www.cdfa.ca.gov/phpps/pe/nipm.htm.

Coyier, D. 1985. Powdery mildews. In D. L. Strider, ed., Diseases of floral crops. Vol. 1. New York: Praeger. 103–140.

Dreistadt, S. H. 2001. Integrated pest management for floriculture and nurseries. Oakland: University of California Division of Agriculture and Natural Resources Publication 3402.

Farr, D. F., and A. Y. Rossman, eds. n.d. Fungal databases, systematic mycology and microbiology laboratory. USDA Agricultural Research Service (ARS). USDA ARS website, http://nt.ars-grin.gov/fungaldatabases/fungushost/fungushost.cfm.

Farr, D. F., G. F. Bills, G. P. Chamuris, and A. Y. Rossman. 1989. Fungi on plants and plant products in the U.S. St. Paul, MN: American Phytopathological Society (APS) Press.

Ferris, H., K. M. Jetter, I. A. Zasada, J. J. Chitambar, R. C. Venette, K. M. Klonsky, and J. O. Becker. 2003. Risk assessment of plant-parasitic nematodes. In D. A. Sumner, ed., Exotic pests and diseases: Biology and economics for biosecurity. Ames, IA: Iowa State Press. 99–119.

French, A. M. 1989. California plant disease host index. Sacramento: California Department of Food and Agriculture.

Gleason, M. L., M. L. Daughtrey, A. R. Chase, G. W. Moorman, and D. S. Mueller. 2009. Diseases of herbaceous perennials. St. Paul, MN: APS Press.

Hadidi, A., R. Flores, J. W. Randles, and J. S. Semancik, eds. 2003. Viroids. Collingwood, Australia: CSIRO Publishing.

Horst, R. F. 2008. Westcott's plant disease handbook. New York: Springer-Verlag.

Lewandowski, D., J., A. J. Hayes, and S. Adkins. 2010. Surprising results from a search for effective disinfectants for *Tobacco mosaic virus*–contaminated tools. Plant Disease 94:542–550.

Linderman, R. G. 1985. Root rots. In D. L. Strider, ed., Diseases of floral crops. Vol. 1. New York: Praeger. 153–172.

Pappu, H. R., R. A. C. Jones, and R. K. Jain. 2009. Global status of tospovirus epidemics in diverse cropping systems: Successes achieved and challenges ahead. Virus Research 141:219–236.

Raabe, R. D. 1985. Rust diseases. In D. L. Strider, ed., Diseases of floral crops. Vol. 1. New York: Praeger. 173–227.

Robb, K., C. Casey, A. Whitfield, L. Campbell, J. Newman, and D. Ullman. 1999. Using petunia indicator plants to monitor tospoviruses in ornamental plants. Alexandria, VA: American Floral Endowment.

Schumann, G. L., and C. J. D'Arcy. 2010. Essential plant pathology. 2nd ed. St. Paul, MN: APS Press.

Tjosvold, S. A., K. R. Buermeyer, C. Blomquist, and S. Frankel. 2005. Nursery guide for diseases of *Phytophthora ramorum* on ornamentals: Diagnosis and management. Oakland: University of California Division of Agricultural and Natural Resources Publication 8156. ANR CS website, http://anrcatalog.ucdavis.edu/Details.aspx?itemNo=8156.

Insects, Mites, and Other Invertebrate Pests

James A. Bethke, John N. Kabashima, Julie P. Newman,
Karen L. Robb, and Cheryl A. Wilen

CHAPTER FOURTEEN

Invertebrate pests cause significant economic losses in nurseries and generally require control strategies that are labor-intensive and costly. Among invertebrates, the arthropods, which include insects and mites, have the greatest number of nursery pest species. Snails and slugs, which are mollusks, are also important invertebrates. Effective management of these pests depends on the ability of nursery managers and staff to detect and accurately identify them. Further, for effective control of common nursery pests, it is very important to know their life cycles and developmental stages.

This chapter provides basic information on common invertebrate pests found in nurseries. General life cycles and development of invertebrates, types of damage symptoms, and monitoring and control strategies are presented, and they are also described specifically for each key pest species or group of species. Other subject matter includes information that is critical to identifying each nursery invertebrate pest described in this chapter. Invasive species are especially important to the nursery industry, and some of the more important species are discussed. However, invasive species and their impact on the nursery industry are more thoroughly addressed in chapter 17.

Pest management strategies will emphasize nonchemical control; because of the constant changes in availability and registration status of pesticides, only general information on chemical use is provided here. California growers can refer to the *UC Pest Management Guidelines* (see the References at the end of this chapter) for the most up-to-date information on the use of specific pesticides for controlling invertebrate pests. Growers in other states should refer to their Cooperative Extension service or to their pest control adviser for current chemical control recommendations.

Types of Invertebrate Pests

Although many invertebrates feed on plants, only a small number are considered significant pest species, and these are found within the groups of animals known as arthropods, mollusks, and nematodes (fig. 14.1). Arthropod pest species include insects (fig. 14.1A) and mites (fig. 14.1B); mollusk pest species include snails and slugs (fig. 14.1C); and nematode pests include all roundworms (fig. 14.1D). The number of pestiferous species of snails, slugs, and nematodes is dwarfed by the number of arthropod pest species. The major arthropod pests and snails and slugs

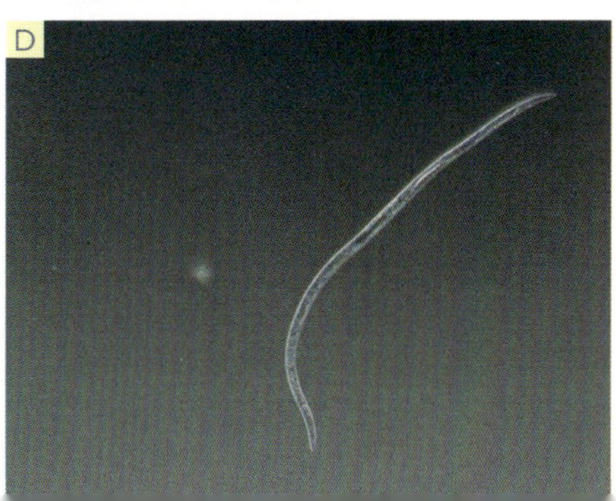

FIGURE 14.1. Invertebrate pest species are found within the groups of animals known as arthropods, mollusks, and nematodes. Arthropods include insects (A), which have three distinct body regions and three pairs of jointed legs, and mites (B), which have one body part and eight legs. Mollusks such as snails (C) are morphologically distinct from the arthropods. Nematodes (D) are small parasitic roundworms. Representative species shown in the photos are the Diaprepes root weevil (A); the two-spotted spider mite (B); the brown garden snail (C); and the foliar nematode (D). *Photos:* J. N. Kabashima (A); J. K. Clark (B, C, D).

are described below. However, nematode pests, which are parasitic organisms, are covered in chapter 13.

Arthropods

Insects

Insects are distinguished from other arthropods by having three distinct body regions and three pairs of jointed legs. Wingless and winged types of insects exist, but most adult insects have two or four wings. No other invertebrates have wings and fly—so if it has wings, it is an insect. Insects have antennae, but so do many other arthropods (centipedes or pillbugs, for example), so antennae are not a good indicator of what is an insect.

There are several types of insect mouthparts. With sucking mouthparts, for example, the mouthparts form a tube that is inserted into a food source (fig. 14.2). In chewing mouthparts (fig. 14.3), usually the most visible parts are the large mandibles on each side that move from side to side; often there are noticeable, fingerlike palps on each side of the mouth. The type of mouthpart is useful in determining what order an insect belongs to. In addition, feeding damage (see fig. 14.3) can be used as a clue to determine which insect may have caused the damage (to be discussed further in the section "Diagnosing Damage and Other Symptoms" below). Furthermore, some insecticides are more effective against insects with certain types of mouthparts, so the type of mouthparts can affect pest control management decisions.

FIGURE 14.2. An insect (southern green stink bug) feeding on a plant stem and using sucking mouthparts. *Photo:* J. K. Clark.

FIGURE 14.3. Feeding damage of an insect (the broad-winged katydid) that has chewing mouthparts. *Photo:* J. K. Clark.

Mites

Mites (arachnids) are very small and can be easily distinguished from insects because they have eight legs. Some hatchlings begin their lives with only six legs, but following their first molt, they possess eight legs. In addition to the number of legs, a mite's head, thorax, and abdomen are fused into one unsegmented body, which is easily distinguishable from a three-part body plan characteristic of insects.

Mollusks

Snails and slugs are related to oysters, clams, and other shellfish, and they are morphologically distinct from arthropods. Snails and slugs have soft, unsegmented bodies consisting of the head, neck, tail, and foot. Internal organs are found inside the shell, which slugs lack. The head possesses two pairs of tentacles; the upper pair is for vision or light sensing, and the lower pair is for taste or smell. The mouth is in the center of the head, below the lower pair of tentacles, and the mouthparts resemble a rasp. As snails and slugs feed, they rasp away at plant surfaces. Below the mouth is the opening of a large gland that produces copious amounts of slime.

Snails and slugs are problematic in nurseries and greenhouses because they can easily consume seedlings or leaves and then disappear into the potting mix or other hiding places during the day. This makes it very difficult to determine their population size and the best place to concentrate control efforts. Another problem caused by snails and slugs is that they can trigger the rejection of a shipment because they can be considered serious invasive pests. Adults and eggs may be very difficult to detect in soil and may not be found until the shipment arrives at the final destination.

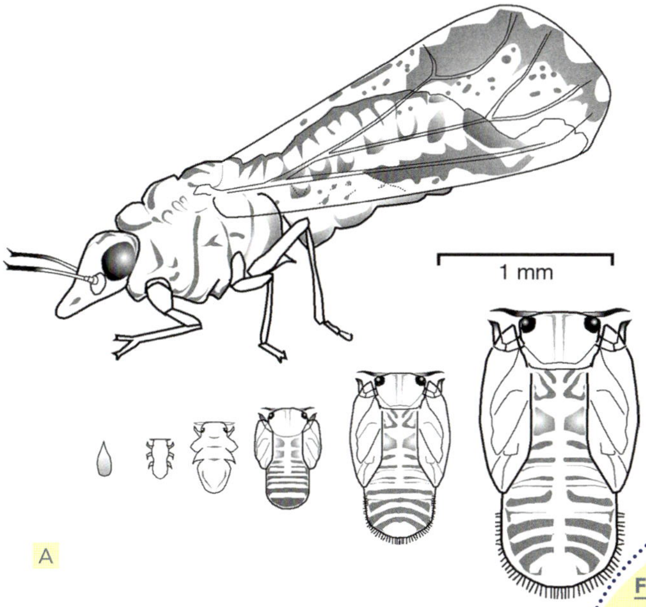

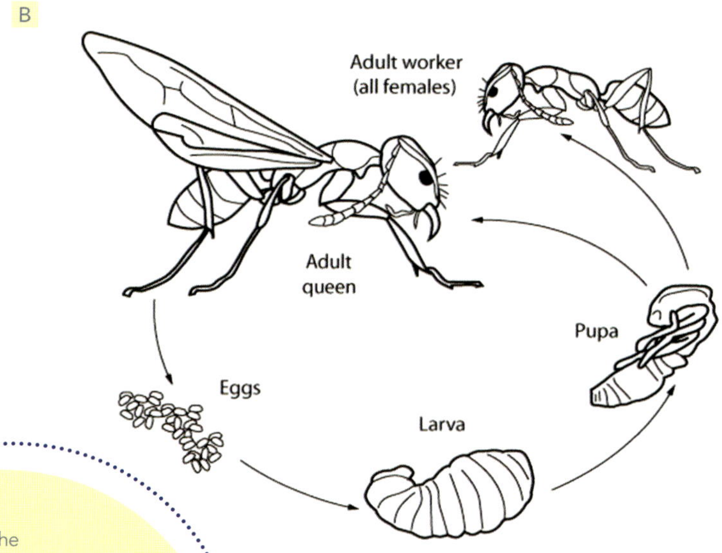

FIGURE 14.4. The number of stages in an insect life cycle depends on its type of metamorphosis. In incomplete metamorphosis, the insect lacks a pupal stage (A). In complete metamorphosis, the insect goes through a pupal stage prior to becoming an adult (B). *Sources:* after Pittenger 2002, p. 210 (A); illustration by G. O. Conville, after Catling 1970 (B).

Life Cycles and Development

Insect Life Stages

Most insects go through several life stages: an egg stage, various immature stages called instars, and an adult stage (fig. 14.4A); depending on the type of metamorphosis, some insects go through a pupal stage prior to becoming an adult (fig. 14.4B).

The vast majority of insects lay eggs. However, some insects such as aphids can give birth to live young. Some insects lay their eggs singly and others lay them in batches. The egg stage is difficult to control because of the egg's size and protective corion (shell).

In order for an insect to grow or pass through the different stages, it must molt or shed its skin (exoskeleton). This process generally occurs 3 to 5 times during the development of the immature stages. Once an insect reaches the adult stage, the life cycle is complete and it grows no further. Insects cannot reproduce until they reach the adult stage.

Spider Mite Life Stages

Spider mites go through several life stages before becoming reproductive adults (fig. 14.5). There is an egg, a larval stage, and two nymphal stages prior to the adult stage. The larva possesses only six legs, but all other life stages have eight legs. The two nymphal stages are described as the protonymph and the deutonymph, and the deutonymph will undergo a sessile stage in the transition to becoming an adult. Under warm conditions, spider mites can complete an entire life cycle from egg to egg in less than a week, which means populations can rapidly build to damaging levels during the summer months.

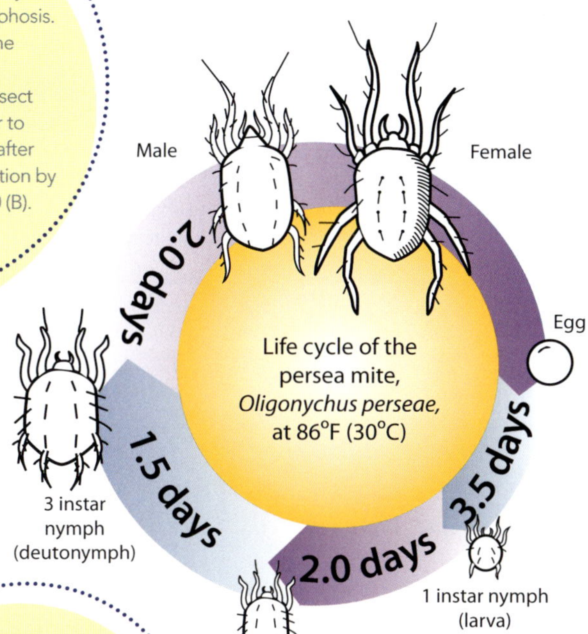

FIGURE 14.5. The life cycle of spider mite species such as the persea mite includes an egg, a six-legged larva, two nymphs, and adult. *Source:* Vince D'Amico, USDA Forest Service, and Mark Hoddle.

Snail and Slug Life Cycle

Snails and slugs do not transition through any stages or metamorphosis like the arthropods, but they do begin with an egg stage. Their growth to the reproductive stage is gradual. A snail's shell will grow slowly and progressively as it ages and increases in size. The length of time it takes for snails and slugs to reach the reproductive stage varies depending on the species. For example, it takes about 2 years for the brown garden snail to mature, whereas slugs can reach maturity in 3 to 6 months.

Snails and slugs are hermaphrodites, which means that when they reach the adult stage they each possess both male and female sexual organs. It also means that after mating, both individuals can lay viable eggs. Adult brown garden snails can dig a hole several inches deep into moist soil and lay an average of 80 spherical, pearly white eggs. Slugs typically lay clear, oval to round eggs in batches of 3 to 40 beneath leaves, in soil cracks, and in other protected areas. Slug eggs are often found in potting soil or the irrigation holes in potted nursery stock.

Snails and slugs are nocturnal but can also be seen on cloudy or foggy days. When conditions are poor, too hot or too cold, snails and slugs will hibernate in the topsoil or aestivate by sealing themselves off with a parchmentlike membrane attached to tree trunks, fences, or walls.

Diagnosing Damage and Other Symptoms

Although most people can recognize when a plant is unsightly, growing poorly, or damaged, determining the underlying cause of the damage can be challenging. Invertebrate pests cause many different damage symptoms on ornamental plants, and distinguishing these symptoms from those caused by abiotic factors or plant pathogens can be difficult. See chapters 11 and 13 for more discussion on the topic of diagnosing plant problems, including abiotic factors.

In diagnosing a problem caused by invertebrates, the damage observed can be grouped into several categories, such as stunting, dieback, stippling, and mining (table 14.1). The category of damage may be an indicator of the type of pest that caused the symptoms; a search for the cause of the damage is always warranted. Note that simply finding insects on a plant with damage does not necessarily mean that the detected insect is responsible for the damage. For example, if chewed leaves are observed and numerous aphids are present, realize that aphids do not have chewing mouthparts and cannot cause chewing damage. Some other absent insect is responsible.

Table 14.1. Common symptoms observed on damaged plants and their probable causes

Symptoms	Probable causes
chewed or tattered foliage or blossoms	larvae of moths and butterflies (e.g., caterpillars, worms, cutworms, armyworms, etc.) larvae and adult leaf-feeding beetles sawfly larvae (e.g., rose slug) grasshoppers, crickets, and katydids snails and slugs (e.g., brown garden snail)
holes in bark or twigs	adult beetles (e.g., bark beetles, boring beetles, etc.) larvae and adult leaf-feeding beetles larvae of boring beetles, not visible
stippled, yellowed, bleached, or bronzed foliage	spider mites, eriophyid (gall, blister, bud, or rust) mites leafhoppers plant bugs lace bugs thrips
distortion of plant parts	aphids true bugs thrips gall wasps larvae of certain moths psyllids eriophyid (gall, blister, bud, or rust) mites
dieback of plant parts	armored scales sharpshooters katydids larvae of boring moths or beetles cicadas gall wasps
foliage mines	larvae of some moths larvae of some flies sawfly larvae (a wasp)
arthropod by-products	honeydew and sooty mold: aphids, soft scales, mealybugs, psyllids, and whiteflies dark fecal specks: thrips and lace bugs tents, webs, silken mats, rolled leaves: tent caterpillar, webworms, and leafrollers spittle: spittlebugs cast skins: aphids, leafhoppers, and lace bugs pitch masses: larvae of certain moths, such as sequoia pitch moth pitch tubes: certain scolytid bark beetles flocculence (cottony, waxy material): adelgids, mealybugs, certain scales and whiteflies, and aphids large frass: caterpillars slime: snails and slugs

Stippling or flecking of leaves (fig. 14.6) results from the insertion of sucking mouthparts or the ovipositor of leafmining flies and the damage to or feeding from the cells that have been destroyed. As pest density increases, the areas of stippling appear to coalesce, giving rise to leaves that appear partly or mostly bleached, yellowed, bronzed, or silvered (fig. 14.7).

FIGURE 14.6. Stippling injury caused by a leafminer (A), thrips (B), spider mites (C), and lace bugs (D). Note the dark brown specks of thrips frass within the silvering area and the lack of webbing that can help distinguish thrips stippling injury from spider mite damage. *Photos:* J. K. Clark.

FIGURE 14.7. Insects with sucking mouthparts destroy leaf cells, causing discoloration. Shown is leaf yellowing caused by leafhopper feeding (A), bleached photinia foliage caused by lace bugs (B), bronzing on roses caused by two-spotted spider mite (C), and leaf silvering caused by eriophyid mites (D). (Most leaves in the photo are silvered, and some green healthy leaves are toward the center.) *Photos:* J. K. Clark.

Some pests with sucking mouthparts can also cause curled or cupped leaves, twisted new growth, or various types of galls on leaves, flowers, twigs, or stems (fig. 14.8). These are common examples of plant distortion that can be caused by arthropods that inject a hormonelike substance as they feed, which causes the plant distortion. In addition to distortion and gall formation, dieback (fig. 14.9) can occur on leaves, twigs, or branches. If the entire plant succumbs, however, examine the roots and crown for evidence of pest damage, which may be caused by insects such as black vine weevil or root mealybug.

Some pest larvae feed within leaves between layers of plant tissue. As they feed, they leave a visible trail or mine (fig. 14.10) in the leaf tissue. The diameter of the trail generally enlarges as the larva grows over time and continues to move through the leaf tissue. In certain species, various larval stages may also feed externally and skeletonize the leaf.

In addition to direct feeding damage, some insects reduce the aesthetic quality of the plants by producing by-products such as honeydew, sooty mold (fig. 14.11), dark fecal specks, webbing (including tents), spittle, cast skins (fig. 14.12), pitch masses or tubes, flocculence (soft fluffy or woolly appearance) (fig. 14.13), and slime.

It is important to note that more than one kind of pest may be present on a single plant, giving rise to multiple symptoms. Moreover, a single pest can cause several types of damage. For example, aphids can cause yellowing and plant distortion, and they can also be responsible for indirect damage such as honeydew, sooty mold, cast skins, and flocculence.

Damage Assessment and Thresholds

The presence of a few pests and some amount of damage can usually be tolerated, if noticed at all. The level at which the number of pests or damage is sufficient to warrant treatment is known as the action or treatment threshold, a fundamental concept in integrated pest management (IPM). Few thresholds have been established for flower and nursery crops, but

FIGURE 14.8. Galls are unnatural swellings of plant tissues that can be caused by secretions of certain plant-feeding insects and mites. Galling on oak (A) and eucalyptus (B) were caused by wasps (oak gall wasp and eucalyptus gall wasp, respectively). Galling on eugenia (C) was caused by the eugenia psyllid. Galling on the terminals of a honey locust tree (D) was caused by the honeylocust pod gall midge. *Photos:* J. K. Clark (A, C, D); Gevork Arakelian, Los Angeles County Department of Agricultural Commissioner/Weights and Measures (B).

FIGURE 14.9. Dieback and stunting of spruce terminals in this photo were caused by aphids. *Photo:* J. K. Clark.

FIGURE 14.10. The trail, or mine, in this chrysanthemum leaf was caused by the larva of a leafmining fly. *Photo:* J. K. Clark.

FIGURE 14.11. Sooty mold on bay laurel leaves caused by aphids feeding on the undersides of leaves from above. *Photo:* J. K. Clark.

FIGURE 14.12. Aphids and cast skins on a cineraria leaf. *Photo:* J. K. Clark.

FIGURE 14.13. Waxy flocculence caused by a psyllid (olive psyllid) on olive trees in the landscape. A number of different types of pests in the nursery are associated with waxy flocculence. *Photo:* D. Rosen.

growers who regularly monitor crops, keep good records, and frequently evaluate trends can establish and refine their own thresholds over the long term. For more details on treatment thresholds, see chapter 12.

Monitoring Methods and Sampling

Various methods are used to monitor invertebrate pests in commercial nurseries, including timed counts of pests within the crop, beat sampling, sticky traps, pheromone traps, black light traps, passive traps, and baits. Many of these techniques are discussed below. The type of method selected will largely depend on the potential pest complex of the plants to be protected. Some pests are somewhat universal and can be monitored using conventional methods such as sticky traps or visual monitoring of plant parts. Other pests require unique monitoring methods such as the use of a pheromone that is typically very selective for a particular pest species. In addition to monitoring for physical presence of the pest, visually surveying the crop for damage should also be used as part of any program to detect pest presence.

Monitoring should be conducted by a well-trained individual (a scout) familiar with the potential pests and their damage. Sampling methods chosen need to be appropriate to the target pests and local conditions. Standardized monitoring procedures should be used, applying the same methods each time a site is monitored. Pests are seldom evenly distributed over a single plant or within or among varieties and species, so it is important to monitor different individual plants and sections of the nursery every time a survey is performed. See chapter 12 and Dreistadt (2001) for detailed information on sampling strategies.

Visual Sampling

The most common way to determine if pests are present is to search for the presence of pests on leaves, stems, or other plant parts. This visual sampling method is important because pest populations can become large in a short period of time, especially if it is warm, and may require close observation with the aid of a hand lens. This strategy is best for aphids, spider mites, some psyllids, and other arthropods that do not fly readily when the plant part (or sampling unit) is counted or removed. This method can also be used for immature stages of pests, such as immature whiteflies. Visual sampling is also a good way to record the presence of natural enemies and evaluate biological control in conservation and augmentation programs. This includes examining plants for predators such as ladybird beetle adults and larvae, syrphid fly larvae, lacewing larvae, and spiders. In addition, evidence of parasitism, such as aphid mummies, darkened whitefly pupae, or psyllid nymphs and scales with exit holes of parasites, should be monitored. Furthermore, signs of insect pathogens, such as blackened, dead caterpillars or dead, discolored aphids showing evidence of fungal activity, should be noted. Remember to examine the underside of leaves or inner plant parts, where many pests prefer to reside.

Counting all the pests on the plant part (sampling unit)

In visual sampling, a representative sample is selected following a procedure outline in the sampling plan (see chapter 12), and all of the pests on the sampling unit are typically counted. Shortcut scouting methods exist that can be used (e.g., presence/absence sampling and rounding off). Detailed information on these shortcut methods and their limitations is described in chapter 12. Two other visual sample counting methods are described below.

Timed counts

This procedure is most useful for large insects such as caterpillars, egg masses on tree trunks or limbs, or insects spread over a relatively large area of plant surface. It does not involve destructive sampling (see chapter 12), so nonflying, sessile insects that are counted are available for recounting at the next sampling period. To perform timed counts, the scout counts the number of insects seen during a 1- or 2-minute visual search of the plant or plants. It is difficult to count insects and keep track of time simultaneously. Therefore, this system works best with two people or an electronic alarm watch.

Beat sampling

This method is used to sample pests such as psyllids, aphids, plant bugs, sharpshooters, thrips, and spider mites. It can also be used to sample some beneficial predators such as lady beetles and lacewing larvae. A sampling tray or beat sheet is placed just beneath plant foliage. The tray may be 1 square foot or more of surface area, or as small as a 5- or 6-inch circle. It is helpful to use a white tray to contrast with the arthropods being counted. The foliage above is struck sharply a standard number of times (usually 2 to 5) with a short stick or with the other hand (fig. 14.14). Arthropods falling to the tray or sheet are immediately counted, then removed. This process is repeated several times in the crop. If the crop is blooming, it is important to also sample flowers. It may be necessary to beat flowers over white paper to dislodge and count insects, such as western flower thrips, which may reside in the flowers.

Attractant Traps

Sticky traps

Sticky traps are efficient and important monitoring tools that can alert growers to the early presence of pests, indicate hot spots and insect migration patterns, and provide other information that can be used in control strategies. Sticky traps are used to monitor flying pests such as adult whiteflies, thrips, fungus gnats, shore flies, leafminers, psyllids, winged aphids, and leafhoppers. In general, yellow sticky traps are recommended. Blue traps are attractive to western flower thrips, but yellow traps can be equally effective for thrips monitoring and have the added advantage of trapping other species of insects. Blue traps are therefore not recommended unless thrips is the only significant pest.

The ability to correctly identify insect pests and beneficials on sticky traps is an important skill to acquire, as some insects closely resemble other types of insects on sticky traps. For example, whitefly wings tend to disappear on traps, and

IGURE 14.14. ampling for acacia syllids using a beat tick and a collection ray. *Photo:* J. K. Clark.

whiteflies may be misidentified as thrips. Springtails have also been mistaken for thrips on sticky cards. High-quality color photographs and line drawings of commonly trapped insects are available in *Sticky Trap Monitoring of Insect Pests* (Dreistadt et al. 1998), *Integrated Pest Management for Floriculture and Nurseries* (Dreistadt 2001), or at the UC IPM website.

Traps can provide a relative measure of insect density; comparisons of the number of adults caught among sample dates may indicate whether pest density is changing or remaining relatively constant over the long term. For instance, the first generation of an insect pest with multiple generations detected early in the year will possess the fewest number of individuals; later in the year, however, this pest will have a much larger population. These population trends can be monitored with the use of yellow sticky traps.

However, sticky traps are not as accurate as counting actual insects on leaves or other plant parts, and they have their limitations. For example, traps do not assure that the pest is actually in the crop. It is possible to have high numbers of the pest on traps but not in the crop, because traps catch both immigrating insects as well as adults that emerge from the crop. Foliage disturbances, such as pesticide treatments, sprinkling with water from irrigation systems, or shaking plants to promote pollination or to monitor for pests, may also increase trap catches. Moreover, the number of adults trapped may temporarily drop after a pesticide application, even if there has been relatively little change in immature populations on foliage. For these reasons, always use traps in combination with foliage inspection to confirm presence of an economically damaging population, location of the population, the dominant life stage, and the relative density within the crop so that appropriate control measures can be taken.

In greenhouses, one trap per 10,000 square feet of growing area is recommended, unless other guidelines are advised for specific pests. For example, when monitoring whiteflies, use about one trap per 1,000 square feet of growing area (Bethke et al. 2010). A lower density of traps is commonly used outdoors. Put additional traps by greenhouse vents and doors or by field edges to detect population immigration. Place additional traps in very susceptible crops and do not locate these crops near doors or field edges.

Traps should be oriented perpendicular to the ground to catch the most insects. Traps can be attached to a bamboo post or wood dowel using a clothespin (fig. 14.15), then embedded in the growing media or attached to a stand. Alternatively, traps can be hung from above using wire or string with a binder clip (fig. 14.16) tied to the end at bench level, and the height of the trap can be adjusted by tying a loop in the string or bending the wire. This method is recommended because it has the advantage of maximizing the number of insects caught while they are traveling on prevailing air currents.

FIGURE 14.15. Yellow sticky trap on a bamboo stake held by a clothespin. Predrawn line divisions make sampling from sticky cards easier when high numbers of pests are present. *Photo:* J. K. Clark.

FIGURE 14.16. Yellow sticky trap hung by a string just above poinsettia plants. *Photo:* J. K. Clark.

Number each trap and map its location in your growing area. Inspect each trap at least once or twice weekly. It is easiest to replace traps each time after you inspect them. Traps can be counted on-site (usually the fastest and the most economical method) or wrapped in clear plastic and taken to a more comfortable location for counting, which may result in more accurate data, especially if dissecting microscopes are used rather than hand lenses. Microscopes are recommended when it is important to identify and distinguish between small insect species or examine samples for parasitism in biological control programs.

When the number of pests on a sticky card is excessive, it is not necessary to count all insects on the entire trap. Counting the insects in a center vertical, 1-inch-wide column on both sides of the trap gives results that are representative of the entire trap. Note that the vertical strip is based on the orientation of the trap in the field, not necessarily on the longest side of the trap. Be sure to orient the strip to be counted in the same direction that it was when mounted. This rapid method of counting sticky traps is especially useful when insect densities on traps are high and the labor involved in counting is costly. It is important, however, not to reduce the size of the traps as a cost-saving measure, since smaller traps may be less attractive to insects.

Pheromone traps

Pheromone traps can be very useful in predicting egg-laying periods, based on peak flights of males of the targeted species. These traps typically consist of a sticky surface and a dispenser containing a pheromone (fig. 14.17). The pheromone is a sex attractant that lures adults (usually males). Pheromones are available for numerous moths, as well as some beetles. There are also pheromone traps for some mealybugs and thrips, but these are not yet widely available for commercial use.

FIGURE 14.17. A pheromone trap consists of a cover containing a pheromone lure that is attractive to specific moth species and a lower panel containing a sticky substance to capture lured moths. *Photo:* J. K. Clark.

When the number of adult moths on traps increases, more intensive inspection of plants is warranted because this is when females are most apt to be laying eggs. Timing of pesticide applications when caterpillars are young has a tremendous impact on efficacy, especially with Bt *(Bacillus thuringiensis).* Hang traps during the season when moths are expected. Separate each trap by at least several hundred feet. Reapply sticky material or replace the traps when they are no longer sticky. Pheromone dispensers may need to be replaced about once a month, especially if the weather has been hot. Check with trap suppliers for specific types of pheromones available and for recommendations.

One advantage of pheromone traps is that one trap can be used for an extended period of time, and the traps are species-specific, so no extra time is required to identify the adults caught in the trap. An additional benefit of using pheromone traps to monitor adult flight is that the traps are only attractive to the males, which obviously do not lay eggs. Use of pheromone traps is not recommended in greenhouses if you are lighting crops at night, as they will lure moths into the greenhouse.

Potato cube baits

Soil-inhabiting pests such as fungus gnat larvae and symphylans can be monitored using small potato cubes pressed into the soil surface. If these pests are present, they can be found feeding on the potato pieces 2 or 3 days after they are placed in the soil. This method is particularly useful in monitoring for the effectiveness of pest control practices.

Passive Traps

Pitfall traps

Pitfall traps collect nonflying insects, such as black vine weevil and strawberry root weevil. Adults of these insects are only active at night, so monitoring visually is difficult. Pitfall traps are placed in the soil. The traps are designed so that the insect falls down a funnel into the trap and cannot get out. These traps are commercially available and include traps that mount in a standard golf cup cutter-size hole.

Double-sided sticky tape

Double-sided sticky tape can be wrapped around branches of trees or shrubs infested with scale insects to trap hatching crawlers (immature stages of scale insects). Tapes should be replaced at regular monitoring intervals and examined with a hand lens or microscope to detect the presence of crawlers. The newly emerged crawlers are the easiest stage to control. Monitoring tapes for peak crawler emergence will allow proper timing of pesticide applications to maximize effectiveness.

Trap boards

Small boards placed on the ground underneath plants can trap pests such as weevils, snails, and slugs. Lift the board to monitor for pests during the day, when they are resting.

Control Strategies

Nonchemical Control Strategies

The use of nonchemical strategies reduces the need to use chemicals, slows the development of pesticide resistance, and helps mitigate pesticide runoff. Nonchemical control strategies include

- good cultural practices (e.g., maintaining healthy plants, good sanitation, and using resistant varieties)
- mechanical control practices (e.g., installing insect screens in greenhouses and applying sticky insect barriers)
- environmental control (e.g., using steam pasteurization to control soilborne insects)
- biological control practices

For detailed information on the use of nonchemical control strategies and their integration into integrated pest management (IPM) programs, see chapter 12. Various nonchemical control strategies appropriate against specific pests are discussed in the section "Common Invertebrate Pests and Their Management" below.

Chemical Control Strategies

Pesticides have had, and will continue to have, an important role in IPM programs for nursery production. Pesticides should only be used, however, when alternative control strategies have been ineffective and pest populations reach damaging levels. Monitoring pest populations and choosing appropriate pesticides are crucial. Consistent and thorough monitoring provides early detection of pests and may allow spot treatment applications of materials to directly infested areas rather than treating the entire crop.

Time the application of a pesticide to target the most susceptible life stage of the pest. For example, many pests are more susceptible to pesticides when they are young, and materials should be applied when first or second instars are present. If using a stomach poison, it will only be effective on larvae that are actively feeding and not on eggs or pupae. Be sure to place foliar-applied materials in the area where the pest feeds—many pests feed on the underside of leaves, and applications must be directed there to get adequate control.

Repeated use of a single product will eventually result in loss of efficacy. Therefore, pesticides from different modes of action should be rotated to avoid problems with resistance. See chapter 12 for more information on chemical control and resistance management.

Common Invertebrate Pests and Their Management

The following section gives short descriptions of some of the most common invertebrate pests found in nursery production systems. Pest descriptions of each pest type include some life histories, typical damage observed, how to monitor for the selected pest, and selected control strategies. In general, specific pesticides are not suggested in this text because pesticide availability and labels change rapidly. For chemical control recommendations for specific pests, as well as detailed information and additional photographs, consult the *UC IPM Pest Management Guidelines for Floriculture and Ornamental Nurseries* (see the References at the end of this chapter) and *Integrated Pest Management for Floriculture and Nurseries* (Dreistadt 2001). Growers in states other than California should refer to their Cooperative Extension service or their pest control adviser for current chemical control recommendations. Quarantine pests are a special case in ornamental plant production and can have a very significant impact on production and shipment of ornamental plants. For a definition of an invasive pest, some examples, and their economic impact on nurseries, see chapter 17. Pests that are known to be quarantine nursery pests at the present time are also indicated below.

INSECT PESTS

Aphids

Common Aphid Species Found in Nurseries

Melon Aphid – *Aphis gossypii*

Green Peach Aphid – *Myzus persicae*

Description of the pests. Numerous species of aphids are attacking California ornamental crops. Two commonly encountered species are the green peach aphid, *Myzus persicae,* and the cotton or melon aphid, *Aphis gossypii.* Melon aphids are typically dark green, but color variations frequently occur. Melon aphids have red eyes and relatively short antennae, which only reach to the middle of the abdomen. Green peach aphids are characterized by a depression in the front of the head between the antennae. Green peach aphids vary in color from yellowish green to rose pink. When viewed from above, winged adults have a dark blotch in the middle of the abdomen.

Adult aphids may or may not have wings, and winged aphids usually arise as a result of crowding. Green peach aphids produce winged adults at lower population densities than do melon aphids. The optimal temperature for green peach aphid development is 75°F, whereas optimal temperatures for melon aphids are above 75°F. In greenhouse situations, adult aphids give birth to living young, but outdoor populations may have a male form and an overwintering egg stage. Generally, aphids begin giving birth in 7 to 10 days, depending on temperature.

Damage. Aphids excrete copious amounts of honeydew, causing the plants to become sticky. Sooty molds then grow on the honeydew, causing plant parts and areas under the plants to become blackened. Shed skins of the aphids are white, become stuck to plant surfaces, and also detract from the plant's aesthetic quality. Sufficient feeding can cause foliage to become yellowed, and feeding on newly developing tissues causes those parts to become twisted as they grow. Melon aphids are known to transmit more than 40 plant viruses, while green peach aphids are known to transmit more than 100 plant viruses.

Monitoring. Yellow sticky cards will capture winged adults. However, aphids produce winged individuals in response to crowding, which typically occurs when the population is increasing. Therefore, a visual inspection of the crop, in proximity to the traps, is warranted. When searching for aphids, note that depending on species, they can be found on various plant parts, including leaves, stems, and even plant roots. Melon aphids tend to have a more uniform vertical distribution on plants than green peach aphids, which tend to be clustered around growing points. Thus, infestations of melon aphids under lower leaves can easily go undetected.

Control strategies. Early detection is important because aphid populations can quickly become damaging to new growth. In addition, because these pests feed on a large variety of plant species, production areas should be kept free of weeds, which can serve as reservoirs for aphid populations.

Exclusion of winged adults can be accomplished by covering openings to the greenhouse with screens that have a pore width of 355 microns or smaller. As a general rule, but especially after screening to exclude aphids, plants that are brought in to start a new crop should be carefully inspected to ensure that they are free of aphids and other pests and disinfested if needed.

Many registered insecticides exist that will control aphids, including some of the reduced-risk insecticides such as soaps, oils, and biologicals (see the *UC IPM Pest Management Guidelines*). As with all of the piercing sucking insects, however, the systemic neonicotinoid insecticides are highly effective against aphids, especially as a drench application. Drench applications are also more persistent in plant tissues, protecting the plant for a longer period of time.

Predators such as ladybird beetles (*Hippodamia* spp.), lacewings (*Chrysoperla* spp.), and midges *(Aphidoletes aphidimyza)* are commercially available. However, commercially available ladybird beetles have the tendency to disperse once they are released, even if food is abundant. Parasitoids, such as *Aphidius* spp., *Lysiphlebus testaceipes, Diaeretiella rapae,* and *Aphelinus abdominalis,* are also commercially available. Parasites and predators often occur naturally even in heavily sprayed nurseries. However, the buildup of natural enemies usually happens too slowly for economic control of aphids in nursery situations.

Ants

Common Ant Species Found in Nurseries

Argentine Ant – *Linepithema humile*

Red Imported Fire Ant (RIFA) – *Solenopsis invicta* (quarantine pest)

Southern Fire Ant – *Solenopsis xyloni*

Description of the pests. The Argentine ant is a small, shiny, brown ant with a one-segmented petiole (the narrow waist between the thorax and abdomen). Workers are all one size, about 2.2 to 2.6 millimeters in length, and they do not sting. Argentine ants are found throughout most of southern and central California. Argentine ants are difficult to control because colonies of ants over a large area are interrelated, and individual ants can easily move from one nest site to another if a colony is disturbed. In addition, colonies have many queens.

All fire ants belong to the genus *Solenopsis* and are characterized by having a 2-segmented petiole, 10-segmented antennae, and a stinger. There are two native fire ant species likely to be encountered in California and confused with the red imported fire ant (RIFA). The more common one, the southern fire ant *(Solenopsis xyloni)*, is found in coastal and inland regions. It is very similar in appearance to the RIFA but is not a major concern for nursery stock. The southern fire ant differs from the RIFA in that it is bicolored, with a reddish head and thorax and a dark brown abdomen. By contrast, the RIFA is an almost uniform dark reddish brown. Furthermore, the southern fire ant mounds are irregular craters, whereas the RIFA mounds are frequently built up into domes. Both species have workers of mixed sizes (southern fire ant workers are 3 to more than 6 mm long; RIFA workers are 1.5 to 5 mm long).

Damage. Argentine ants are the most important ant pest in agriculture and urban areas throughout California, and they pose an additional problem for the nursery industry because infested container plants can result in delay or rejection of inter-regional shipments. Argentine ants tend colonies of homopteran pests such as aphids, scales, and whiteflies. The ants collect the honeydew produced by these insects as a food source and actively interfere with natural enemy activity against homopteran pests, resulting in larger populations of these pests when ants are present.

RIFA poses an immediate threat to the area's economy because their presence requires a quarantine of nursery products. In quarantined areas, plants cannot be shipped without inspection or treatment. RIFA can chew on soft plant tissue and growing buds. Their stinging behavior can be hazardous to fieldworkers.

Monitoring. Argentine ants are effectively monitored with bait stations containing sugar water; at times these ants are attracted to protein sources. Visual estimates or the amount of sugar solution consumed can be used to provide an index of foraging activity.

Fire ants are predominantly protein feeders. Presently, for quarantine compliance, monitoring stations baited with Spam meat are being used at 50-foot intervals to monitor for the presence of RIFA in nurseries. However, for nonquarantine monitoring, sugar water baits, potato and corn chips, and protein sources such as hot dogs will also attract RIFA.

Control strategies. In terms of chemical control methods, baits are typically the preferred method whenever feasible. Unlike pesticide sprays that only kill contacted worker ants, baits are carried into the mound and fed to other ant stages. Protein or sugar-based baits are commercially available for Argentine ant control. The attractiveness of the baits depends on the time of year. For instance, during the early spring or summer, when ant colonies are building broods or auxiliary colonies, the ants tend to be more attracted to protein baits. During the rest of the year, they tend to be more attracted to the sugar baits. Refillable bait stations such as the KMAntPro allow consumption of the various commercially available boric acid baits in sugar solutions. Control can be sporadic, and it may take several weeks to achieve noticeable results because ants tend to stick with their first carbohydrate sources such as aphids or mealybugs. If these baits are persistently used, however, the ants will eventually begin to select the bait station and deplete it. It is, therefore, very important to make sure the bait station never runs out of bait. In addition, when using bait stations in production beds, care should be taken to keep baits away from irrigation lines.

Other products can act as a barrier to ant movement and recruitment. For instance, products exist that contain a silica gel or diatomaceous earth that is sometimes mixed with pyrethrins to increase efficacy. These can be broadcast or placed as a line barrier against ants. Another use for a barrier against ants is when they disrupt biological control programs. In such cases, controlling them with the selective use of pesticides applied as a band around the base of the plant or by using sticky bands around the trunks will exclude ants from tending honeydew-producing insect pests.

Foliage-Feeding Caterpillars

Common Foliage-Feeding Caterpillar Species Found in Nurseries

Beet Armyworm – *Spodoptera exigua*

Yellowstriped Armyworm – *Spodoptera ornithigalli*

Variegated Cutworm – *Peridroma saucia*

Light Brown Apple Moth – *Epiphyas postvittana* (quarantine pest)

Description of the pests. The beet armyworm is the most frequently encountered of the species listed above. Adults are heavy-bodied moths (wingspread 25 to 28 mm) with a characteristic mustard or orange-colored, liver-shaped spot narrowly ringed with white on the forewings. The female lays egg masses on the undersides of leaves, covering the eggs with a felt made from her body hairs. The first through third instar larvae often feed gregariously, skeletonizing the undersides of leaves, or they feed on the insides of buds. Later instars disperse and feed individually, chewing entirely through leaves or flowers. Early instar larvae are small and green, while late instar larval color may be green, brown, black, or gray. In all instars, there are fine lines along the length of the body, with a more conspicuous lateral stripe and a black spot just dorsal to the lateral stripe behind the head. Beet armyworm continually develops during the winter in mild areas and builds up on weeds and cultivated cotton, lettuce, and tomato fields. One generation can take as little as 31 days at 75°F and 24 days at 80°F. The preovipositional period in the adult is an additional 1 to 2 days.

Yellowstriped armyworm larvae have a pair of black triangles on the back of most segments. Some larvae appear nearly completely black when viewed from above. The lateral stripe is bright orange or yellow. The adult (wingspread 38 mm) has a complex and highly contrasting pattern of brown, yellow, and white on the front wings.

The variegated cutworm overwinters as a naked pupa in the soil. Adults have a 40 to 50 mm wingspread, with a distinct liver-shaped outline on the front wings. Larvae have yellow or orange spots or a broken longitudinal stripe at the top of the body, which is otherwise gray. Often there is a dark triangle or "W"-shaped mark on the top of the eighth body segment.

The light brown apple moth (LBAM) is 6 to 10 mm long, with a darker area at the tip of the wings and a light brown area toward the head. However, it can be variable in color and may be confused with other common or endemic leafrollers. Females begin to lay eggs 2 to 3 days after emerging, and the eggs will turn from green to orange to black prior to hatching. Larvae are leafrollers, which means they will roll a leaf and feed from inside, and they develop through six instars before becoming a pupa. The caterpillar is medium green with a darker green central stripe and two side stripes. However, it is difficult to distinguish from other species of leafrollers.

Damage. As an adult moth, the cutworm causes no damage, but the cutworm as a caterpillar is an immature and causes chewing damage. Caterpillars are primarily a concern due to direct damage to seedlings and flowers, as well as aesthetic injury. Presence of late instar larvae in seedling flats can also cause tremendous plant loss.

LBAM causes some damage to many ornamental plants; however, they are a quarantine pest and require eradication. For information on actionable pests, see chapter 17.

Monitoring. Determine adult flight activity and mating by using pheromone traps if available. When adults are caught in traps, spray as soon as newly hatched (susceptible) larvae are found. However, larvae may be difficult to detect, and preventive applications may be necessary to avoid damage for some of these pests. As a further precaution, use regular visual inspections of plants to watch for any signs of chewing damage.

Control strategies. Exclusion of winged adults can be accomplished with exclusion screens. Individual seedling flats or seedbeds may also be covered with screens or ground cloth to exclude adults and larvae. Floating row covers can be effective in field production as long as the mesh prevents entry of adults and the row cover is held above the plant surface to eliminate oviposition through the fabric.

Pheromone and ultraviolet light trapping can be used for monitoring, but this is rarely effective as a control measure. However, pheromone mating disruption has been an effective tool for many moth species and is now being employed and studied against LBAM.

A number of parasitoids, both tachinid flies and parasitic wasps, attack moth larvae or eggs and reduce their population growth rate. However, larval parasitoids do not kill their hosts immediately; thus parasitized larvae may continue to feed through to the last instar and may still damage crops. Viruses also do not usually kill the larvae until later instars. Applying broad-spectrum insecticides will likely exclude most parasite activity. Applications of *Bacillus thuringiensis* (Bt) products to control caterpillars, however, will not harm parasitoids.

Fungus Gnats and Shore Flies

Common Fungus Gnats and Shore Fly Species Found in Nurseries

Fungus Gnats – *Bradysia coprophila, Bradysia impatiens*

Shore Fly – *Scatella stagnalis*

Description of the pests. Fungus gnats are clear-winged, small (2 to 5 mm long) mosquitolike flies with delicate legs and long antennae. They lay their eggs in soil; approximately 4 days later the eggs hatch. There are four larval instars, which can grow up to 8 mm. Larvae are clear, with visible internal organs, and have shiny black head capsules. Initially larvae feed on root hairs and algae; later, larvae may feed on the insides of roots. When populations are high, larvae may bore into larger roots or stems that are in the soil. Larvae also feed on leaves touching the soil. One generation may complete development in 21 days (at 75°F) to 40 days (at 61°F).

Shore flies breed in overwatered conditions associated with algae. The adult is small (2 mm) and dark. The overall appearance is similar to a fruit fly, having short antennae and shorter legs than fungus gnats. The pair of dark wings has three white spots on each wing. When the fly is at rest, there appear to be five spots because the wings overlap.

Damage. Larvae of fungus gnats usually feed on roots and algae within 1 inch of the soil surface. Root feeding by larvae can allow entry of plant pathogens. Direct damage through root feeding can cause wilting even though the plants are being sufficiently watered. Damage is particularly severe in propagation areas, in seedling flats, and with especially sensitive crops. Unlike fungus gnats, shore flies do not feed on plants. However, both species can cause damage through the dissemination of soil-dwelling pathogens that can adhere to their bodies as they move among the soil and plants. Further, the mere presence of adults of either species can be a nuisance when present in large numbers, and they can cause customer complaints.

Monitoring. Yellow sticky traps will capture adults of both species and can be used for monitoring populations or detecting pest presence. Traps should be oriented perpendicular to the ground to catch the most insects. Although orienting traps horizontally (parallel to the soil with the flat sticky surface facing the soil) may be more effective in capturing fungus gnats due to their behavior, there are limitations with this strategy. Horizontal traps tend to stick to nearby leaves and catch more dirt, thus must be replaced more often.

Larval populations can be monitored with cubes or slices of potatoes pressed just into the soil. Larvae are attracted to the potato piece and feed between the piece and the potting media so that when the potato piece is removed, the small white maggots are visible on the slice or nearby soil.

Control strategies. Production areas should be kept free of weeds and algal scum, which can serve as breeding sites for fungus gnat and shore fly populations. Other cultural methods that should be watched carefully are overwatering (which will cause algal growth and encourage fly reproduction), using incompletely composted organic matter or manure in potting media (which creates ideal conditions for fungus gnats), and using commercial sources of peat (which may already be infested with fungus gnats and should be sterilized before use).

There are several pesticides that are effective against the larval stages of these flies, including organophosphates, carbamates, and pyrethroids. In addition, the neonicotinoid dinotefuran is highly effective when applied as a drench. *Bacillus thuringiensis* (Bt) is a biological pesticide that is also very effective against the soilborne larvae. Consult the *UC IPM Pest Management Guidelines* for a list of specific pesticides that are currently recommended. Biological control agents for fungus gnats include nematodes *(Steinernema feltiae)* and soil-inhabiting predaceous mites *(Stratiolaelaps scimitus*, formerly known as *Hypoaspis miles)*.

Gall-Forming Insects

Common Gall-Forming Species Found in Nurseries

Pemphigus Aphid – *Pemphigus betae*

Honeylocust Pod Gall Midge – *Dasineura gleditschiae*

Eucalyptus Gall Wasp – *Epichrysocharis burwelli*

Ficus Gall Wasp – *Josephiella microcarpae*

Oak Gall Wasps – family Cynipidae

Description of the pests. Galls are unnatural swellings of plant tissues that can be caused by secretions of certain plant-feeding insects and mites (see fig. 14.8). Galls may form on leaves, flowers, twigs, stems, or branches. Hundreds of arthropods can cause galling symptoms on plants. However, most galls are caused by aphids, psyllids, gall midges, and cynipid gall wasps. Adult oak gall wasps are small, stout, shiny, purple or black insects. Females deposit eggs in plant tissue, and galls form several weeks after larvae hatch from eggs and start to feed. Adult gall midges are tiny delicate flies. They lay eggs on foliage that hatch into tiny maggots. The maggots bore into plant tissue and feed inside the galls that form.

Damage. Most galls do not harm the plant, but they are primarily viewed as aesthetic damage. Oak gall wasps create a variety of gall sizes, shapes, and colors on oak leaves and stems, some of which are attractive. However, galls can also cause significant plant damage. For instance, honey locust leaflets heavily infested with galls of the honeylocust pod gall midge can turn brown and drop prematurely, leaving parts of branches leafless, and the eucalyptus gall wasp forms small, red to purple, blisterlike galls on the leaves of lemon-scented gum *(Eucalyptus citriodora)*.

Monitoring. Adult activity can be monitored by placing sticky traps near new flush tissues and maturing galls. Another method is to collect galls over time and dissect them to track insect pest development.

Control strategies. Control of gall-making insects is difficult, and if galls are a serious problem, selecting alternative species or varieties may be warranted. Galls may be pruned out and disposed of before the adults begin to emerge, but it can be a labor-intensive way to minimize gall pests. Insecticide treatments are generally not effective, since larvae are protected inside galls. Therefore, insecticides must be timed properly to repel or kill egg-laying adult gall makers.

Most gall makers have a complex of natural enemies (parasitoids, predators, pathogens) that suppress their populations. However, while natural enemies can eventually suppress pest populations, they attack after damage has already occurred, when the gall has already formed and the organism is usually feeding on the pest within the gall.

Leafhoppers and Sharpshooters

Common Leafhopper and Sharpshooter Species Found in Nurseries

Glassy-Winged Sharpshooter – *Homalodisca vitripennis* (formerly *H. coagulata*)

Aster Leafhopper – *Macrosteles quadrilineatus (=M. fascifrons)*

Blue-Green Sharpshooter – *Graphocephala atropunctata*

Two-Spotted Leafhopper – *Sophonia rufofascia*

Description of the pests. Leafhoppers, including a subgroup called sharpshooters, are in the family Cicadellidae. All have mouthparts that allow them to pierce the plant tissue and feed on plant juices. Most leafhoppers measure about 6 mm and are relatively long and slender. Species may be brightly colored or similar in color to the host plant. They often jump away or move sideways when disturbed.

Damage. Leafhopper feeding can cause leaves to appear stippled, pale, or brown, and shoots may curl and die. Excrement exuded during feeding can discolor leaves with a white chalky film, and pale cast skins may be found on leaf surfaces.

Leafhoppers are pests primarily because some are vectors of plant pathogens. The glassy-winged sharpshooter and the blue-green sharpshooter transmit a bacterial pathogen, *Xylella fastidiosa*, that grows in the xylem, or water-conducting tissue, of certain plant species. *X. fastidiosa* can cause a number of plant diseases in a variety of hosts, including oleander leaf scorch, Pierce's disease of grapevines, almond leaf scorch, and alfalfa dwarf. Another example is the aster leafhopper, which is one of several leafhoppers that can transmit phytoplasma pathogens that cause aster yellows disease in many plant hosts.

Monitoring. Monitoring methods such as yellow sticky traps, beating samples, and visual counts can be used to detect the presence of leafhoppers. Yellow sticky cards should be placed at canopy height at densities not less than one card per ½ acre. Check cards at least once a week for adult sharpshooters.

Beat or sweep sampling for nymphs and adults is most effective when temperatures are cool (< 60°F). Place a white 2-foot square sheet of fabric or other material underneath the canopy to be sampled. Strike the canopy with a stick or shake vigorously to dislodge insects from the foliage. Count the number of insects on the sheet. At warmer temperatures, the insects will fly away before they can be counted. A sweep net may also be used to sample foliage for the presence of adults and nymphs.

Visual inspection of leaves, stems, and branches is perhaps the best method for detecting all stages. Insects may try to move to the far side of the stem to avoid detection. Placing a hand close behind the stem that is being observed will make the insects move to the front, where they can be seen. Egg masses can be easily detected by inspecting the undersides of leaves against a sunny sky.

Control strategies. Leafhoppers can be excluded from growing areas using screening or row covers. Some nurseries have used greenhouse and shadehouse structures or barriers to protect certain crops from glassy-winged sharpshooters. Cultural practices such as weed removal can be effective in reducing the incidence of leafhoppers or disease. If such sources cannot be removed, avoid planting crops susceptible to disease near these alternate hosts.

When pest pressure and disease transmission potential are high, insecticide used as a preventive is warranted. As with all of the piercing-sucking insects, the systemic neonicotinoid insecticides are highly effective against leafhoppers, especially as a drench application.

There are no commercial sources of biological control agents for leafhoppers currently available; however, egg parasitism of leafhoppers and sharpshooters is pronounced late in the season.

Leafminers

Common Leafminer Species Found in Nurseries

Serpentine Leafminer – *Liriomyza* spp.

Cypress Tip Miner – *Argyresthia cupressella*

Description of the pests. Larval (immature) stages of some flies (Diptera) and moths (Lepidoptera) can feed between layers of leaf tissue, creating mines. The most common leafminers in nursery stock are *Liriomyza* flies. The small (1.8 mm) yellow and black flies are about the same size and shape as fruit flies. In contrast to shore flies, leafminer adults are brightly colored and have clear wings. They are often found on yellow flowers or objects. Females puncture the leaf surface with the ovipositor to feed on plant sap and lay eggs. Larvae have three instars, forming larger mines inside the leaves as they grow. After completing feeding, larvae drop out from the leaves to pupate in the soil. *Liriomyza trifolii* can complete one generation in 17 days at 77°F and sooner at warmer temperatures.

The cypress tip miner adult moth appears in California in the spring. Adults are small (6 mm), silvery tan, and slim when wings are folded. Females lay scalelike eggs on tips of plants that hatch into yellow or green larvae. Larvae will feed in plant tips until late winter, then spin silken cocoons between leaflets to pupate. There is one generation per year.

Damage. When *Liriomyza* populations are high, stippling, caused by females puncturing the leaves with their ovipositor to feed and lay eggs, can be serious. However, most of the damage is caused by the larval mines, which detract from the aesthetic value of the crop and can reduce photosynthesis. Foliage infested with cypress tip miner larvae turns yellow in early winter and brown by late winter or early spring.

Monitoring. Serpentine leafminers can be monitored using yellow sticky cards at a rate of one per 10,000 square feet. Additionally, it is important to monitor plants for the presence of mines by using random sampling. Flies on traps and the presence of mines will indicate hot spots of leafminers that may need control measures.

For cypress tip miner, the tips of susceptible plants should be examined for cocoons in early spring. Monitoring for adult moth emergence will determine when insecticide applications need to be applied.

Control strategies. Because serpentine leafminers feed on many plant species, production areas should be kept free of weeds, which can serve as reservoirs for leafminer populations. Leafminers can breed in weed or crop hosts, so exclusion of immigrating adults is especially valuable as a management practice. Effective screens require a pore width of 600 microns or smaller to exclude *Liriomyza*. In addition, incoming plant material should be carefully inspected to ensure that it is free of mines; discard infested plants or leaves.

Although insecticides are a good preventive measure against serpentine leafminers, rotation of insecticides is extremely important because pesticide resistance is very common in this species. Rotation to a new mode of action every 17 days or one generation cycle is advised.

The cypress tip miner is a pest of arborvitae, cypress, juniper, and redwoods; however, the susceptibility of plant species and varieties to cypress tip miner is variable. When possible, select species that are least susceptible to the cypress tip miner. A list of susceptible and least susceptible plants is available at the UC IPM website.

If pesticide applications are necessary, *Bacillus thuringiensis* (Bt) is a biological pesticide that is very effective against caterpillars. However, the early stage is the only susceptible stage, so timing is very important.

Biological control normally keeps these species in check in outdoor situations. Outbreaks of *Liriomyza* leafminers frequently occur following the disruption of their parasites with broad-spectrum insecticides. For this reason, try to use selective insecticides through the entire production cycle so that parasites can be conserved. The parasites *Diglyphus* spp. and *Dacnusa sibirica* are commercially available and may be useful in greenhouse or screenhouse situations for management of *Liriomyza.*

Mealybugs

Common Mealybug Species Found in Nurseries

Citrus Mealybug – *Planococcus citri*

Madeira Mealybug – *Phenacoccus madeirensis*

Long-Tailed Mealybug – *Pseudococcus longispinus*

Obscure Mealybug – *Pseudococcus viberni* = *P. affinis*

Root Mealybug – *Rhizoecus* spp. (quarantine pests)

Description of the pests. Mealybugs are slow-moving sucking insects that have a loose, waxy coating on the body, which gives them their mealy appearance. Mealybug infestations often occur underneath foliage and in hidden areas within dense foliage. The citrus mealybug is heavily and evenly covered with white, powdery wax, except for a faint narrow streak down the middle. It has short, wax filaments along the margins, and the filaments at the posterior end are not much longer than the others. Both the citrus mealybug and the Madeira lay eggs in ovisacs (eggs are within masses of cottony wax). The Madeira mealybug (sometimes incorrectly referred to as the Mexican mealybug) can be distinguished from the citrus mealybug by four rows of thinly waxed depressions down the back. The long-tailed mealybug has two terminal wax filaments that are longer than the body. The obscure mealybug also has noticeably longer filaments protruding from the posterior end, but in addition it has distinct filaments protruding around the perimeter of the body.

Root mealybugs are smaller mealybugs that dwell belowground and feed on the roots of plants. These mealybugs have a thin, uniform waxy coating and lack the terminal wax filaments typical of their foliar-feeding relatives. The soil around a root mealybug infestation will appear white from the waxy droppings of the insects.

Damage. Mealybugs remove sap from plants, which can cause yellowing of leaves and decline in vigor. Mealybug ovisacs and waxy deposits are unsightly, and excreted honeydew can cause black sooty mold production on leaf and stem surfaces. In addition, honeydew attracts ants, which can carry mealybugs to uninfested plants and tend them for honeydew.

The only outward sign of root mealybug feeding may be a decline in the health of infested plants. When plants are removed from the pot, the whitish mealybugs and waxy deposits are observed on the potting media. Mealybugs can be seen feeding on the roots.

Monitoring. Propagation areas should be intensely monitored and shipments of incoming plant material should be carefully inspected to ensure that they are free of mealybugs and other pests. If necessary, treat infested plants. Regularly inspect plants for signs of honeydew (i.e., glistening, sticky leaves) and ant activity. Examine plant root balls for the presence of root mealybug if plants are declining.

Control strategies. Many insecticide options exist for mealybug control (see the *UC IPM Pest Management Guidelines* for current recommendations). However, best control will be obtained using the systemic insecticides. This is because contacting the insects can be difficult, and the systemic insecticides place the toxicant in the mealybug's food source. If a contact insecticide is selected, a surfactant may be necessary to enhance contact with the insect. Well-established infestations containing females with ovisacs are much more difficult to control with either systemic or contact insecticides than are new infestations. This is because reproducing adults usually stop feeding, and the female's body or the wax secretions help protect eggs and crawlers.

Cryptolaemus montrouzieri, the mealybug-destroyer lady beetle, is an effective predator of citrus and Madeira mealybugs. Larval mealybug destroyers look like large, faster-moving mealybugs but are readily distinguished by their chewing mouthparts. *Leptomastix dactylopii,* a parasite of citrus mealybug, is also commercially available. Effective predators or parasites of long-tailed mealybugs are not yet commercially available. Biological control has not been investigated for root mealybug.

Psyllids

Common Psyllid Species Found in Nurseries

Eugenia Psyllid – *Trioza eugeniae*

Pepper Tree Psyllid – *Calophya rubra*

Red Gum Lerp Psyllid – *Glycaspis brimblecombei*

Description of the pests. Adult psyllids, in the family Psyllidae, resemble miniature cicadas. Adults are 2.5 to 5 mm long and hold their wings rooflike over their bodies. They are related to aphids, but they have strong jumping legs and jump away when disturbed. They are sometimes confused with psocids, which feed on fungi and sooty mold and do not damage plants. Unlike psyllids, psocids will run or fly away when disturbed rather than jump. Psocids also have a "neck" between the head and thorax, and they have chewing rather than sucking mouthparts.

Psyllid immatures are generally flattened, often immobile, and frequently form pits or lerps (white, waxy secretions or crystalline sugar caps). Eugenia psyllid nymphs and pepper tree psyllid nymphs resemble trilobites, and they develop in pits on the leaf surfaces. The immature forms are sometimes confused with aphids or scales.

Damage. Psyllids suck plant juices and may produce honeydew. High populations can reduce plant growth or cause plant distortion and discoloration, dieback of terminal leaves, or severe leaf drop. A few species cause galls on leaves or buds. There are many different species of psyllids, each usually feeding on only one type of plant. Most species of psyllids never reach damaging levels because the plants can tolerate relatively high numbers of psyllids without sustaining damage. However, some psyllid species can be pests. The eugenia psyllid and the pepper tree psyllid form pits on the underside of the leaves as they feed, resulting in unsightly raised bumps on leaves or distorted foliage. Red gum lerp psyllids have been responsible for the defoliation and occasionally death of *Eucalyptus camadulensis*.

Monitoring. Adult psyllids and their parasitoids can be monitored with yellow sticky traps or by beat sampling to dislodge them. Beat sampling, as discussed earlier in this chapter, allows predators to be monitored at the same time. Sample every 1 to 2 weeks beginning before the typical damaging season begins. Monitor immature psyllids by visually inspecting and counting the number of insects on four to eight leaves, new growth tips, or branch terminals of several plants. A hand lens or microscope can be used to examine specimens for the presence of immature psyllids.

Control strategies. Maintain the health of the plants by properly fertilizing and watering as needed. Avoid overfertilizing plants, as psyllid eggs are often deposited on succulent new growth. Psyllids also appear to be attracted to stressed plants, so ensure that plants are grown with adequate (not excessive) water, sunlight, and so on.

Psyllid populations are most noticeable in the spring as new flush growth appears. As a preventive, time insecticide applications when eggs and young nymphs are present, before foliage is damaged. Monitor when adults become active, and treat soon after a sharp increase in adult numbers is observed. Psyllids are susceptible to most insecticides, including the older broad-spectrum insecticides (e.g., organophosphates, carbamates, and pyrethroids). However, as with all of the piercing-sucking insects, the systemic neonicotinoid insecticides are highly effective, especially as a drench application, and will reduce impacts on natural enemies. Nonetheless, neonicotinoids must be used sparingly to delay development of resistance. Consult the *UC IPM Pest Management Guidelines* for current chemical control recommendations, and always check the product label to confirm registration status and use.

General predators, such as lady beetles, lacewing larvae, and predaceous bugs, are important in controlling psyllids. The minute pirate bug has been observed feeding on many psyllid species, including invasive species such as the Tipu psyllid.

Soft Scales

Common Soft Scale Species Found in Nurseries

Black Scale – *Saissetia oleae*
Brown Soft Scale – *Coccus hesperidum*
Green Shield Scale – *Pulvinaria psidii*
Hemispherical Scale – *Saissetia coffeae*

Description of the pests. Soft scales are typically found on woody plants and foliage plants. The first nymphal instar is called a crawler and has functional legs, while the remaining instars are attached to the leaf or twig and (with the exception of green shield scale) do not move. Soft scales typically have a more conspicuous profile from a side view compared with armored scales, and they produce copious amounts of honeydew. The protective covering over a soft scale cannot be separated from its body. Brown soft scale adults are fairly flat in profile, range in color from yellowish green to brown, and are often spotted or mottled to uniform brown. Hemispherical scale adults are strongly convex, hard, brown, smooth, and shiny. Black scale adults are globular and hardened with ridges in the back that look like the letter "H." Green shield scale, introduced into California approximately in 1992, has a light yellow-green color as an immature; it reproduces parthenogenetically (without mating, as there are no males) and produces a mass of eggs in a cottony ovisac.

Damage. Soft scales remove sap from plants, which can cause yellowing of leaves and decline. Honeydew, excreted by all the soft scales, and the resulting black sooty mold are unsightly. Additionally, honeydew attracts ants, which then carry scales to uninfested plants.

Monitoring. As with other insect pests, visual inspection of leaves, stems, and branches is perhaps the best method for detecting soft scale infestations. Heavy populations can produce copious amounts of honeydew, and the presence of sooty mold may indicate a viable population of soft scale insects. Also watch for the presence of ants tending scale insect populations.

Double-sided sticky tape wrapped around infested stems is useful for determining when crawlers are active. Optimal treatment timing is when crawlers are active; however, this can be difficult when there are overlapping, multiple generations.

Control strategies. Prune out and discard heavily infested plants and plant parts. New plants should be carefully inspected to ensure that they are free of scales and other pests and disinfested if needed. As with most of the piercing-sucking insects, the systemic neonicotinoid insecticides are highly effective against soft scales, especially as a drench application; however, neonicotinoids must be used sparingly to delay development of resistance. Ultrafine oil sprays and insect growth regulators (IGRs) are also effective against immature stages and will allow better survival of natural enemies. Consult the *UC IPM Pest Management Guidelines* for current chemical control recommendations, and always check product labels to confirm registration status and use.

A complex of *Metaphycus* spp. parasites attack brown soft scale, and some are commercially available. In addition, the lady beetles *Rhyzobius (Lindorus) lophanthae* (commercially available), *Chilocorus orbus*, and *C. cacti* prey on brown soft scales. The black scale parasite, *Metaphycus helvolus*, has also been used for control of the closely related hemispherical scale. The mealybug destroyer, *Cryptolaemus montrouzieri*, is known to be an effective predator of green shield scale. Green shield scale, however, is considered an A-rated quarantine pest by the California Department of Food and Agriculture, so plant material must be entirely free of its presence to be shipped.

Armored Scales

Common Armored Scale Species Found in Nurseries

California Red Scale – *Aonidiella aurantii*
Cycad Scale – *Furchadaspis zamiae*
Euonymus Scale – *Unaspis euonymi*
Oleander Scale – *Aspidiotus nerii*
Boisdoval Scale – *Diaspis boisduvalii*

Description of the pests. Unlike soft scales, armored scale bodies can be separated from their protective coverings, which are composed of molted skins and secretions from the scale, and armored scales do not produce honeydew. As with soft scales, however, the first nymphal instar is called a crawler and has functional legs, while the remaining instars are attached to a leaf, stem, or fruit and do not move. Populations of armored scales can go unnoticed until they reach very high levels. Female scales can produce hundreds of eggs that are protected under their scale until hatch. Following hatch, crawlers disperse away from the adult and establish on new plant growth. High populations of these sucking insects give plant stems a crusty appearance.

Damage. Along with the unaesthetic encrustations on leaves, stems, and fruit, these scales inject toxic saliva that will cause plants to decline. Watch for tree branches that begin to die, starting at the terminal bud and moving back up the branch toward the trunk.

Monitoring. As with other insect pests, visual inspection of leaves, stems, and branches is perhaps the best method for detecting armored scale infestations. Look for encrusted or rough-looking branches and terminals. Double-sided sticky tape wrapped around infested stems will indicate when crawlers are active. Optimal treatment timing is when crawlers are active.

Control strategies. Prune out and discard heavily infested plants and plant parts. New plants should be carefully inspected to ensure that they are free of scales and other pests and disinfested if needed. Various insecticides are highly effective against armored scales. Consult the *UC IPM Pest Management Guidelines* for current chemical control recommendations, and always check the product label to confirm registration status and use. Ultrafine oil sprays, which are effective against immature stages, will allow better survival of natural enemies.

Generalist predators will reduce populations, but they will not always hold them below damaging levels. Some generalist predators are available commercially, and there is at least one, *Aphytis melinus,* that is commercially effective against California red scale.

Thrips

Common Thrips Species Found in Nurseries

Western Flower Thrips – *Frankliniella occidentalis*
Greenhouse Thrips – *Heliothrips haemorrhoidalis*
Cuban Laurel Thrips – *Gynaikothrips ficorum*
Myoporum Thrips – *Klambothrips myopori*

Description of the pests. Thrips are very small insects (0.5 to 1.4 mm). Thrips adults have four featherlike wings, each consisting of a thick supporting strut with fine hairs on the front and hind edges. There are two types of pest thrips, and they need to be addressed separately because life cycles and control measures differ for each group. The main distinction between the two groups is that one group lays its eggs within plant tissues (Terebrantia thrips), and the other group lays its eggs on plant surfaces (Tubulifera thrips), usually in protected structures such as galled or distorted tissues. In general, all thrips go through six life stages: egg, first instar, second instar, prepupa, pupa, and adult. The first two instars and the adults feed by piercing and removing the contents of individual plant cells.

Development times to complete one generation of Terebrantia thrips, such as the western flower thrips, vary from 11 days (at 77° to 87°F) to 44 days (at 50° to 60°F). Eggs hatch from leaf or stem tissues in 2 to 3 days and feed in enclosed tissues such as flowers, buds, or growing tips. However, they also feed on pollen and small arthropods. Pupal stages generally take place in the soil beneath the infested plant.

The development time to complete one generation of Tubulifera thrips, such as the Cuban laurel thrips or the myoporum thrips, is about 30 days at 80°F. In the Tubulifera thrips, the pupal stages remain on the plant in protected structures such as galled or distorted tissues. Adults of Cuban laurel thrips and myoporum thrips are black, and nymphs and pupae are yellow.

Damage. Direct feeding damage includes streaking, spotting, and tissue distortion. Stippling damage is caused by thrips feeding on individual cells and is often confused with mite stippling. However, black varnishlike flecks of frass often accompany thrips feeding, which can distinguish thrips feeding damage from damage caused by mites. Cuban laurel thrips create tightly rolled, podlike leaf terminals on *Ficus,* and form galled foliage from midsummer through fall. Myoporum thrips create tightly bunched and twisted terminal growth, where large numbers of thrips can survive and reproduce. Terminal distortion can occur year-round, but the new growth in the spring is especially susceptible.

In addition to the physical damage caused by feeding thrips, some species such as the western flower thrips pose an even greater threat to ornamental plants. They can vector *Tomato spotted wilt virus* and *Impatiens necrotic spot virus,* which can devastate many species of potted flowering crops (e.g., gloxinia, chrysanthemum, gerbera) and bedding plants (e.g., impatiens, begonias).

Monitoring. Use yellow sticky traps to monitor for adult stages. Although blue sticky traps are most attractive to western flower thrips, yellow traps are sufficient predictors of thrips populations, are easier to count, and are more commonly used for general-purpose monitoring of other insect pests.

Additionally, direct sampling can be accomplished by tapping the terminals of branches or flower heads onto or over a white sheet of paper to dislodge thrips, allowing counts of adult and immature stages. With the Tubulifera thrips, samples of leaf or terminal stems are necessary to determine the extent of the infestation. Counting the number of galls or damaged leaves may also be used to make treatment decisions. For more details on how to develop treatment thresholds, see chapter 12.

Control strategies. Propagation areas and incoming plant material must be carefully inspected to ensure that they are free of thrips and other pests and disinfested if needed. Because thrips feed on a large variety of plant species, production areas should be kept free of weeds, which can serve as reservoirs for thrips and the diseases they vector. Most commercially available screens have pore sizes slightly larger than the width of the western flower thrips thorax (145 microns), which means that some winged adults can penetrate these openings. Although thrips exclusion screens are available, greenhouses may need to be modified to adapt to the changes in environmental conditions (i.e., air exchanges, temperature and humidity controls, etc.). As further protection from disease-vectoring thrips, double entryways should be employed.

Insecticides used for controlling both the Terebrantia and Tubulifera thrips should be rotated because insecticide resistance is common in thrips. Consult the *UC IPM Pest Management Guidelines* for current insecticide recommendations and modes of action. Always read the product label to confirm registration status. For western flower thrips, rotation to a different mode of action every 15 days or one generation time (which depends on temperature) is advised.

Preventive applications of pesticides are preferred for the Tubulifera thrips, such as the Cuban laurel thrips and myoporum thrips. Significant damage to new flush growth can be caused by very few thrips. Therefore, any of the common broad-spectrum insecticides that are currently registered can be used as a preventive measure, and applications should be made when new flush growth is expected for susceptible plants. Once the galls or twisted leaf tissue occurs, these thrips species are well protected and can only be controlled by systemic insecticides (e.g., neonicotinoids). Keep in mind that contact insecticides such as spinosad will be less effective against these thrips because applications are unlikely to contact the insect or its food source. Commercially available predators include the minute pirate bugs, *Orius tristicolor* and *O. insidiosus,* and two predatory mites, *Neoseiulus cucumeris* and *Stratiolaelaps scimitus* (formerly known as *Hypoaspis miles*). Minute pirate bugs are general predators and also feed on aphids, mites, and small caterpillars. They are released at a rate of 2,000 to 4,000 per acre. *Neoseiulus cucumeris* also feed on spider mite eggs, pollen, and fungi, and they are released at a rate of 10 to 50 mites per plant for 2 to 3 weeks. *Stratiolaelaps scimitus* (formerly known as *Hypoaspis miles*) are soil-inhabiting predators that feed on thrips prepupae and pupae in the soil. These mites are generally released in the soil at planting.

Weevils

Common Weevil Species Found in Nurseries

Black Vine Weevil – *Otiorhynchus sulcatus*

Fuller Rose Beetle/Weevil – *Asynonychus godmani (Pantomorus cervinus)*

Description of the pests. Several weevils, or snout-nosed beetles, feed on nursery plants. Adult weevils are 6.3 to 12.7 mm long with the head elongated into a snout and elbowed antennae. Larvae of most species are whitish grubs that feed on roots and pupate in the soil.

Damage. Adult weevils are nocturnal and generally feed on foliage or flowers. The edges of leaves appear notched or ragged or may be clipped from the twigs. Unless populations are high, this damage does not usually harm established plants. Larvae of the black vine weevil, however, live in the soil and can seriously damage roots and girdle plants near the soil surface, resulting in a general decline or death of potted plants.

Monitoring. Regularly inspect the foliage, buds, and flowers of host plants for evidence of weevil feeding. Night-feeding weevils such as the black vine weevil can be monitored 1 to 2 hours after dark by sweeping foliage with a net or by beat sampling of branches onto a light-colored sheet or tray. Black vine weevils can also be monitored using pitfall traps, trap boards, or by trapping them under burlap bands wrapped around trunks.

Control strategies. If some foliage damage can be tolerated in larger plants, pesticide application can be delayed until 3 to 4 weeks after first detecting foliage feeding damage or adults. Few, if any, weevils will have laid eggs by then because they must feed for about a month before they start laying eggs. Use any conventional insecticide labeled for control of foliage-feeding adult weevils.

An alternative to pesticide use may be to select varieties of plants that are more resistant to weevil damage (e.g., less susceptible rhododendron species for black vine weevil listed on the UC IPM website) and to maintain vigorous plants that will be able to tolerate some damage.

Soil-dwelling immature weevils can be controlled with commercially available parasitic nematodes *Heterorhabditis* spp. or *Steinernema* spp. Apply nematodes when weevil larvae or pupae are present.

Whiteflies

Common Whitefly Species Found in Nurseries

Greenhouse Whitefly – *Trialeurodes vaporariorum*

Sweetpotato Whitefly – *Bemisia tabaci* (B and Q biotypes)

Description of the pests. Whiteflies are generally small (1 to 2 mm), and there are four life stages: egg, nymph, redeye pupa, and the adult. Whitefly adults prefer to feed and lay eggs on the undersides of the leaves. There are four nymphal instars, and the first is called a crawler because it crawls away from the eggshell and establishes permanently on the plant tissue. The fourth instar eventually becomes a pupa. Eggs are partially inserted into the leaf; initially they are yellowish, but they darken over time to a purplish brown just before hatching. In general, whiteflies can complete a life cycle in about 25 days at 80°F. They can reproduce throughout the year in greenhouses and in warmer climates, and they can have several overlapping generations even in nurseries.

Damage. Whiteflies feed from the vascular system of plants much like related species (aphids, mealybugs, and scale) and excrete honeydew on plant surfaces, which then promotes the production of black sooty mold. Very large populations of whiteflies cause stunting of plant growth, and leaves may senesce and die. In addition to the physical and aesthetic damage that most species of whiteflies can cause, the sweetpotato whitefly B biotype (silverleaf whitefly) is a known vector of virus disease in tomatoes and silverleaf on squash. On some ornamental plants, physiological abnormalities can occur, such as white stem on poinsettia, when sweetpotato whiteflies are feeding.

Monitoring. The most common method of monitoring for whiteflies is using yellow sticky cards. Yellow sticky cards should be placed at a density of one per 1,000 square feet in more susceptible crops or up to one per 10,000 square feet in other areas. In addition to yellow sticky traps used for adult monitoring, a random inspection in search of immature stages on the undersides of leaves of susceptible plants may be necessary. Observations should be made with a 10X hand lens due to the small size of the immature stages.

Control strategies. Control of whiteflies requires an integrated approach—for example, a good monitoring program, appropriate cultural practices, physical exclusion where possible, and preventive chemical control if necessary. Start any control practice early, when the first whiteflies are detected, because control is difficult once populations have increased to high levels. As a good cultural practice, always keep production areas free of weeds, which can serve as hosts for whitefly populations. Whitefly exclusion can be accomplished by using screens with a hole size of 405 microns or smaller; when employing exclusion screening, it is especially important to carefully check all incoming shipments to make sure they are free of all whitefly stages.

Chemical control options vary, but it is critical that insecticides with different modes of action are rotated every 21 days during warm weather or every one or two generation cycles. Although the B biotype sweetpotato whitefly can become insecticide resistant if not properly managed, the Q biotype sweetpotato whitefly is highly resistant to common insecticides. Knowledge of the chemicals that can kill B and a change to chemicals that can control Q may assist in effective control of the sweetpotato whitefly. Alternatively, you can have the biotype determined genetically by a reputable scientific source, and then you can make sure you are using the proper chemicals. For a list of chemicals that are effective against the B or Q biotype and a reputable source for biotype determination, see the whitefly management program at the Mid-Florida Research and Education (MREC) website in the References at the end of this chapter.

Encarsia formosa, a parasitic wasp, has long been recognized as an effective biological control for greenhouse whiteflies. Wasps are released at a rate of two to five parasitoids per plant for 8 to 10 weeks of the growing season. This sort of release program can be effective if long-residual insecticides have not been applied in advance of the parasitoid release and when the initial population is quite low (only a few whiteflies per plant). Greenhouse whitefly pupae turn black when parasitized by *Encarsia*, which emerge as adults through circular exit holes. *Eretmocerus eremicus* is another commercially available whitefly parasitoid for sweetpotato whitefly; it turns sweetpotato whitefly nymphs light brown when parasitized. *Delphastus pusillus* is a whitefly predator that has also been used commercially against sweetpotato whiteflies.

MITE PESTS

Spider Mites (Tetranychid Mites)

Common Spider Mite Species Found in Nurseries

Two-Spotted Spider Mite – *Tetranychus urticae*

Persea Mite – *Oligonychus persea*

Description of the pests. Spider mites are web-forming mites that pierce plant cells and remove their contents. They tend to feed and reproduce on the underside of leaves and form rudimentary webs between leaf veins and other surfaces. Spider mites are very small and difficult to see without the aid of a hand lens.

All spider mites have one body segment and four pairs of legs as adults. Two-spotted spider mite adults, as the name suggests, are yellowish green with two large dark spots on their sides. These mites lay round or onion-shaped eggs that can be laid on the webbing or on the leaf surfaces. The eggs hatch into six-legged larvae. The subsequent stages, the protonymph and deutonymph stages, are eight-legged, as are the adults. Since the entire life cycle can take as little as 8 days at 95°F, spider mites have many generations per year.

Persea mites have a similar life cycle with generation times reported at 13 to 20 days at temperatures from 77° to 86°F. Eggs are semispherical in shape, pale yellow in color, and during the later stages of development have characteristic red eyespots. Adult females have an oval-shaped body that is slightly flattened and elongated, yellowish green in color, and with several small dark spots located on the abdominal region. The male is smaller than the female and has a pear-shaped body that is slightly flattened and yellowish in color.

Damage. Removal of cell contents by spider mites changes the color of leaves from green to olive green, then to a dull brown as feeding increases. Therefore, yellowing, stippling, and bronzing can occur. Eventually, leaf drop may also occur. Undersides of leaves may have many cast skins of mites, and the webbing on foliage is unaesthetic. Feeding damage from large populations of mites can result in severe stunting of plants.

Monitoring. Scout the crop regularly, as indirect sampling methods (such as sticky traps) are ineffective. Observe the undersides of leaves with a 10X hand lens, and watch for changes in plant foliage that are characteristic of mite feeding. Foliage plants are particularly susceptible to mite damage, and some of the most susceptible varieties can be used as indicator plants—plants that should be monitored first.

Control strategies. Incoming plant material should be isolated and carefully inspected to ensure that they are free of mites. If needed, they should be cleaned or treated to eliminate mite infestations. Production areas should be kept free of mites; weeds, which harbor mites and serve as alternate hosts, should be removed. Dusty areas can encourage spider mite outbreaks, so the application of water to pathways and other dusty areas will discourage mite population increase. Midseason washing of trees and vines with water to remove dust may help prevent serious late-season mite infestations. Additionally, water-stressed trees and plants are more susceptible to spider mite damage, so be sure to provide adequate irrigation.

Most miticides only kill active stages of mites, so more than one treatment may be necessary to break the life cycle. Under high pest pressure, a tank mix of a miticide that kills motile stages and one that kills eggs and six-legged larvae will assist in breaking the life cycle. The life cycle from egg to reproductive adult can be very short—a matter of days. Therefore, during the warmer times of the year, application of miticides and rotation from different modes of action may need to occur weekly.

Many different species of predatory mites are available for use under different conditions. *Phytoseiulus persimilis* and *Galendromus occidentalis* have been used to control mite populations in greenhouses and field situations. These predators can reproduce faster than their prey; best results have been obtained when they are released into the crop well before the spider mite populations have built up.

Thread-Footed Mites (Tarsonemid Mites)

Common Thread-Footed Mite Species Found in Nurseries

Cyclamen Mite – *Phytonemus pallidus*

Broad Mite – *Polyphagotarsonemus latus*

Bulb Scale Mite – *Stenotarsonemus laticeps*

Description of the pests. Thread-footed mites are exceptionally small and cannot be seen without the aid of a microscope. These mites can usually be detected by the damage they cause. Therefore, preventive measures need to be taken for susceptible plants.

Life stages of these mites are as follows: egg, nymph, pseudopupa, and adult (one less stage than for spider mites). Cyclamen mites are generally found feeding on growing terminals, in buds, or unfolding leaflets. Their development is optimal under conditions that are moderately warm (60° to 80°F) with high humidity (80 to 90%). Eggs of the cyclamen mite are half the length of the adult, and they are smooth and oval.

The broad mite is closely related to the cyclamen mite but is found more generally on the plant on the undersides of leaves. Their eggs are distinguishable from cyclamen mite eggs, since they are studded with rows of white pegs on the eggs' upper surface. Immature stages of broad mites are whiter than those of cyclamen mites. The hind pair of legs in the adult female is threadlike, while adult males have stout legs for clasping the female. Adult males carry female pseudopupae on their back. As soon as the adult female emerges, they mate. Mites disperse between plants on air currents and through mechanical transport. These mites can complete one generation in 7 to 21 days, depending on temperature.

Female bulb scale mite adults are usually found between the scales of the bulb and the neck region. These mites overwinter in bulbs between the scales, emerging as the leaves grow. They re-enter bulbs as they dry in the field. Females lay up to 28 eggs. One generation can be completed in approximately 7 weeks under field conditions.

Damage. Feeding by cyclamen and broad mites is easily recognized on all hosts because affected leaves become characteristically cupped, dwarfed, and thickened, and the internodes are greatly shortened. Broad mite damage occurs more generally over the plant than cyclamen mite damage. Bulb scale mites feeding in developing shoots can cause longitudinal bronze streaks of discoloration, horizontal cracks, distortion, and death of leaves and flowers.

Monitoring. Visually inspect plants for typical damage symptoms as part of a weekly scouting program. A microscope is needed to confirm the diagnosis, as these mites usually cannot be identified using only a hand lens.

Control strategies. Propagation areas should be intensely monitored and shipments of incoming plant material should be carefully inspected to ensure that they are free of pests; treat infested plant material to eliminate mites or discard. Disinfestation of propagation stock can be accomplished by immersing in 110.3°F water for 30 minutes or treating at 100 percent relative humidity and 110.3°F for 1 hour.

Because these mites feed on a large variety of plant species, production areas should be kept free of weeds, which can serve as reservoirs for mite populations. There are few miticides commercially available for control of thread-footed mites. Therefore, if the plants being grown are susceptible to these mites, a preventive treatment may be necessary; damage can be quite significant if these pests are ignored. If hot spots of these mites are found in production areas, consider roguing affected plants and treating the surrounding plants. *Neoseiulus californicus* and other species of predatory mites have been used for broad mite and cyclamen mite control.

Acarid Mites

Common Acarid Mite Species Found in Nurseries

Bulb Mites – *Rhizoglyphus* spp.

Description of the pest. Acarid mites are white with large amber spots and brown legs; they are 0.75 mm long when fully grown. These mites are slow moving and oval shaped, so they are sometimes mistaken for insect eggs. They can be found in hollowed-out and decomposing portions of bulbs.

Damage. Bulb mites enter the bulb and feed in a protected cavity, where fungi and bacteria may cause extensive decomposition of the bulb. They are associated with the spread of *Fusarium*, *Stromatinia*, and *Pseudomonas* diseases of bulbs. Species of bulbs attacked include freesia, hyacinth, lilies, narcissus, and onions, as well as underground peony rhizomes. Freesias are particularly affected because they require high temperature storage to break dormancy, which allows rapid growth of bulb mite populations. Lilies are attacked both belowground, which stunts growth, and at the ground level, which causes the plant to topple.

Monitoring. No specific monitoring methods are available.

Control strategies. Bulbs may be disinfested by holding them for 24 hours at 100 percent relative humidity at 105.5°F. Fumigation with carbon dioxide may be useful.

Bulbs should be thoroughly cleaned, dipped in a sulfur fungicide solution, and dried following harvest. The sulfur will control bulb mites as well as fungal diseases if the bulbs are being held at high temperatures. However, high rates of sulfur may also be phytotoxic.

Biological control against bulb mites has not been investigated.

Eriophyid Mites

Common Eriophyid Mite Species Found in Nurseries

Fuchsia Gall Mite – *Aculops fushsiae*

Coconut Mite – *Aceria guerreronis*

Description of the pests. Eriophyids are wormlike or wedge-shaped with four legs that appear to be coming out of their head. However, these mites are exceptionally small and cannot be seen without the aid of a microscope. Eriophyid mites can usually be detected by the damage they cause. Therefore, preventive measures need to be taken for susceptible plants.

Damage. In many cases, distorted plants are the first indication of the presence of eriophyid mite damage. The fuchsia gall mite causes leaves and shoots to become thickened and distorted, sometimes forming galls. They occur on growing tips year-round and in flowers during blooming. The coconut mite is associated with meristem necrosis and death of young queen palm *(Syagrus romanzoffiana)* seedlings. *Washingtonia robusta* and *Phoenix reclinata* can also serve as hosts.

Monitoring. It is very difficult to directly monitor for eriophyid mites on plants because they are often buried deep within the folds of the growing tips or inside galls. The best way to monitor for these pests is to watch for changes in plant foliage that are characteristic of mite feeding. If possible, take a sample and observe the damage under a microscope for the presence of mites. Samples can also be taken to your local Extension agent.

Control strategies. Incoming plant material should be carefully inspected to ensure that it is free of symptoms of mites and other pests and disinfested if needed. To reduce problems with fuchsia gall mite, grow resistant varieties of fuchsia (see the UC IPM website in the References for a list of resistant or low-susceptible fuchsia species and cultivars). Prune and destroy damaged tissue. Few miticide options exist for control of eriophyid mites because they are minute and protected in host plant tissues, which make them very difficult to contact with spray applications.

The predatory mite *Neoseiulus californicus* has been associated with control of some coconut mite–infested palm seedlings.

SNAILS AND SLUGS

Common Snail and Slug Species Found in Nurseries

Brown Garden Snail – *Cornu aspersum* (formerly *Helix aspersa*)

Amber Snail – *Succinea* sp.

Gray Garden Slug – *Deroceras reticulatum* (formerly *Agriolimax meticulatus*)

Banded Slug – *Lehmannia poirieri*

Greenhouse Slug – *Milax gagates*

Description of the pests. Both snails and slugs are members of the mollusk phylum and are similar in structure and biology, except slugs lack the snail's external spiral shell. Slugs and snails constantly secrete mucus, which later dries to form the silvery slime trail that indicates the presence of these pests. Adult brown garden snails lay about 80 spherical, pearly white eggs at a time into a hole in the topsoil. Although the eggs are white initially, as they develop they turn brown and may be mistaken for some brands of slow-release fertilizer. They may lay eggs up to six times a year. It takes about 2 years for brown garden snails to mature. Slugs reach maturity in about a year.

Snails and slugs are most active at night and on cloudy or foggy days. On sunny days they seek hiding places out of the heat and sun; often the only clues to their presence are their silvery trails and plant damage. In mild-winter areas such as southern California and coastal locations, young snails and slugs are active throughout the year. During cold weather, snails and slugs hibernate in the topsoil. During hot, dry periods, snails seal themselves off with a parchmentlike membrane and often attach themselves to tree trunks, fences, or walls.

Damage. Snails and slugs feed on a variety of living plants as well as on decaying plant matter and algae. On plants, they chew irregular holes with smooth edges in leaves and can clip succulent plant parts. They can also chew fruit, flowers, and young plant bark. Because they prefer succulent foliage, they are primarily pests of seedlings, herbaceous plants, and ripening fruits such as strawberries, artichokes, and tomatoes that are close to the ground. However, they also feed on foliage and fruit of some trees.

Monitoring. Snails and slugs can be trapped under boards positioned throughout the nursery. You can make traps from 12-by-15-inch boards (or any easy-to-handle size) raised off the ground by 1-inch runners. The runners make it easy for the pests to crawl underneath. Scrape off the accumulated snails and slugs daily and destroy them. Amber snails are more difficult to detect as they tend to live on the media surface. Look for these snails in the pots and at the interface of wet and dry areas between the media and the container.

Control strategies. A good snail and slug management program relies on a combination of methods. The first step is to eliminate, to the extent possible, all places where snails or slugs can hide during the day. Boards, stones, debris, weedy areas around tree trunks, leafy branches growing close to the ground, and dense ground covers such as ivy are ideal sheltering spots. There will be shelters that are not possible to eliminate (e.g., low ledges on fences, the undersides of wooden decks, and water meter boxes). Reducing hiding places allows fewer snails and slugs to survive. The survivors congregate in the remaining shelters, where they can be more easily located and controlled.

Switching from sprinkler irrigation to drip irrigation will reduce humidity and moist surfaces, making the habitat less favorable for these pests. If possible, do not use wooden benches, as these stay moist for longer periods of time, tend to be good sites for algae to grow, and allow snails and slugs to hide between the bench and the container.

Hand-picking can be very effective if done thoroughly on a regular basis. Snails and slugs can also be trapped under boards positioned throughout the nursery, as described above, and destroyed. Crushing is the most common method of destroying snails after hand-picking or collecting from traps. Do not use salt to destroy snails and slugs; it will increase soil salinity.

Several types of barriers will keep snails and slugs out of planting beds. The easiest to maintain are those made with copper flashing and screens. Copper barriers are thought to be effective through the reaction between the copper and the slime that the snail or slug secretes, causing a flow of electricity. Vertical copper screens can be erected around planting beds. The screen should be 6 inches tall and buried several inches below the soil to prevent slugs from crawling beneath the soil. Copper screens can also be placed between the bench support legs and the bench to prevent snails and slugs under the bench from reaching the top where the pots are set. Copper foil or netting can be wrapped around planting boxes and bench legs to repel snails for several months or longer, depending on the integrity of the copper strips. Instead of copper bands, Bordeaux mixture (a copper sulfate and hydrated lime mixture) can be brushed on trunks of specimen or landscape trees to repel snails. One treatment should last about a year. Adding a commercial spreader may increase the persistence of Bordeaux mixture through two seasons. Sticky material (such as Stickem Green, which contains copper) applied to trunks excludes snails, slugs, ants, and flightless species of weevils.

Snails and slugs have many natural enemies, including ground beetles, pathogens, snakes, toads, turtles, and birds, but they are rarely effective enough to provide satisfactory control in the nursery. A predaceous snail, the decollate snail *(Rumina decollata),* is available for use in Fresno, Imperial, Kern, Los Angeles, Madera, Orange, Riverside, Santa Barbara, San Bernardino, San Diego, Ventura, and Tulare Counties. However, its release for snail control in commercial nurseries is not recommended since all plants must be snail-free (including beneficial and pestiferous species) prior to shipping.

Commercial snail and slug baits and sprays can be effective when used properly in conjunction with a cultural program incorporating the other methods discussed above. In order to use the products effectively and reduce hazards to nontarget organisms such as dogs, read and follow the label carefully. The application timing of any baiting is critical; baiting is less effective during very hot, very dry, or cold times of the year because snails and slugs are less active during these periods. As snails and slugs are more likely to travel when soil is moist, it may be beneficial to irrigate to promote snail activity before applying the bait. Make broadcast applications in areas where snails and slugs have been detected, such as in and around containers and their hiding places. Sprayable formulations of molluscicides are useful where there are high numbers of snails or slugs or where the pests, such as those that feed in the whorls of leaves, are unlikely to encounter a broadcast bait application.

References

Alford, J. 1995. A color atlas of pests of ornamental trees, shrubs and flowers. New York: Wiley.

Bethke, J. A. 2000. Leafminer management strategies. Proceedings of the 16th Conference on Insect and Disease Management on Ornamentals: 91–93.

———. 2009. Insects and mites. In S. H. Dreistadt, ed., Integrated pest management for floriculture and nurseries. University of California Division of Agriculture and Natural Resources Publication 3402.

———. 2010. Mealybug control: It's worth the wait. Ohio State Florist Association (OFA) Bulletin No. 919. Columbus: OFA.

Bethke, J. A., and T. D. Paine. 1991. Screen hole size and barriers for exclusion of insect pests of glasshouse crops. Journal of Entomological Science 26(1): 169–177.

Bethke, J. A., H. S. Costa, and R. A. Redak. 2002. Whiteflies are not flies. Greenhouse Product News 12(8): 58–61.

Bethke, J. A., R. A. Redak, and U. Schuch. 1996. Use of host plant resistance to control insect pests on chrysanthemums: 1 aphids. Proceedings of the 37th Annual Entomology Conference. 58–59.

Bethke, J. A., R. A. Redak, R. S. Cowles, and K. Kido. 1992. Biology and control of fungus gnats and shoreflies in greenhouses. Abstracts, 33rd Annual Entomology Conference, Riverside: University of California, Riverside. 78–79.

Bethke, J. A., K. L. Robb, H. S. Costa, R. S. Cowles, and M. P. Parrella. 2010. Insects and mites. UC IPM Pest Management Guidelines: Floriculture and Ornamental Nurseries. Oakland: University of California Division of Agriculture and Natural Resources Publication 3392.

Borror, D. J., C. A. Triplehorn, and N. F. Johnson. 1995. An introduction to the study of insects. 6th ed. Fort Worth, TX: Harcourt Brace College Publishers.

Catling, H. D. 1970. Distribution of the psyllid vectors of citrus greening disease with notes on the biology and bionomics of *Diaphorina citri.* FAO Plant Protection Bulletin 18:8–15.

Dietrick, E. J. Dietrick papers: Published talks and articles by Everett J. "Deke" Dietrick. Rincon-Vitova Insectaries website, http://www.rinconvitova.com/dietrick_papers.htm.

Dreistadt, S. H. 2001. Integrated pest management for floriculture and nurseries. Oakland: University of California Division of Agriculture and Natural Resources Publication 3402.

Dreistadt, S. H., J. K. Clark, and M. L. Flint. 1994. Pests of landscape trees and shrubs: An integrated pest management guide. Oakland: University of California Division of Agriculture and Natural Resources Publication 3359.

Dreistadt, S. H., J. P. Newman, and K. L. Robb. 1998. Sticky trap monitoring of insect pests. Oakland: University of California Division of Agriculture and Natural Resources Publication 21572.

Flint, M. L. 2009. Pest notes: Snails and slugs. Oakland: University of California Division of Agriculture and Natural Resources Publication 7427.

Flint, M. L., and S. H. Dreistadt. 1998. Natural enemies handbook: The illustrated guide to biological pest control. Oakland: University of California Division of Agriculture and Natural Resources Publication 3386.

———. 2005. Interactions among convergent lady beetle *(Hippodamia convergens)* releases, aphid populations and rose cultivar. Biological Control 34:38–46.

Gill, S. A. 2012. Monitoring techniques for insects, mites and other arthropods. University of Maryland Extension Bulletin 359. College Park: University of Maryland Extension.

Gill, S., and J. Sanderson. 1998. Ball identification guide to greenhouse pests and beneficials. Batavia, IL: Ball Publishing.

Greenberg, L., J. Klotz, and J. N. Kabashima. 2001. Red imported fire ants. Pest notes. Oakland: University of California Division of Agriculture and Natural Resources Publication 7487.

Johnson, W. T., and H. H. Lyon, eds. 1991. Insects that feed on trees and shrubs. 2nd ed. Ithaca, NY: Cornell University Press.

Kabashima, J. N. 1992. Control of ash whitefly, *siphoninus phillyreae*, in wholesale ornamental production nurseries. Proceedings of the 1992 First Annual Southwest Ornamental Pest Management Workshop. 62–65.

Kabashima, J., G. Gutman, R. Redak, and J. Bethke. 2008. Glassy-winged sharpshooter nursery approved treatment best management practices. In J. Kabashima and G. Morris, eds., Glassy-winged sharpshooter nursery stock approved treatment manual. 4–23.

Kabashima, J. N., R. A. Redak, J. A. Bethke, and T. D. Paine. 1992. Black cutworm insecticide and biorational efficacy trials using a turfgrass bioassay. Proceedings of the 1992 First Annual Southwest Ornamental Pest Management Workshop. 58–61.

Parrella, M. P., and J. A. Bethke. 1988. Whiteflies: A search for answers. Grower Talks 52(7): 64–70.

Pittenger, D., ed. 2002. California master gardener handbook. Oakland: University of California Division of Agriculture and Natural Resources Publication 3382.

Powell, C. C., and R. K. Lindquist. 1997. Ball pest and disease manual: Disease, insect and mite control on flower and foliage crops. 2nd ed. Batavia, IL: Ball Publishing.

Redak, R. A., J. A. Bethke, and U. Schuch. 1996. Use of host plant resistance to control insect pests on chrysanthemums: 2 leafminers and other major pests. Abstracts, 37th Annual Entomology Conference. Riverside: University of California, Riverside. 60–61.

Robb, K. L. 1989. Analysis of *Frankliniella occidentallis* (Pergande) as a pest of floriculture crops in California greenhouses. Ph.D. dissertation, University of California, Riverside.

Robb K. L., and J. A. Bethke. 1991. Biology and control of whiteflies in floriculture greenhouses. 32nd Annual Entomology Conference. Riverside: University of California, Riverside. 36.

Robb, K. L., H. S. Costa, J. A. Bethke, R. Cowles, and M. P. Parrella. 2000. Insects, mites and other invertebrates. In S. H. Dreistadt, Integrated pest management for floriculture and nurseries. Oakland: University of California Division of Agriculture and Natural Resources Publication 3402. 422.

UC IPM pest management guidelines for floriculture and ornamental nurseries. UC IPM website, http://www.ipm.ucdavis.edu/PMG/selectnewpest.floriculture.html.

UC Statewide Integrated Pest Management Program. UC IPM website, http://www.ipm.ucdavis.edu.

University of Maryland IPM for Commercial Horticulture (IPMnet) website, http://ipmnet.umd.edu/publications.htm.

Vehrs, S. L., J. A. Bethke, J. Garcia, and M. P. Parrella. 1988. Aphids common problems on numerous ornamentals. California Florists (January):1–14.

Whitefly Management Program, Mid-Florida Research and Education Center (MREC) website, http://mrec.ifas.ufl.edu/lso/bemisia/WhiteflyManagementProgram_2011.pdf.

Integrated Weed Management

Cheryl A. Wilen and Clyde L. Elmore

CHAPTER FIFTEEN

Photo: C. A. Wilen.

Introduction

Weeds are found in most nurseries and, in general, they are persistent problems. This is due to the year-round cropping system of nursery plants and the effective dispersal and movement of weed seeds. Weeds in containers reduce plant growth by competing with the crop plants for water, nutrients, and light; in some cases weeds actually prevent water and fertilizer from reaching the crop. Weeds growing in and around containers reduce crop marketability by harboring insect pests, plant pathogens, and viruses. Weed control is costly and can be very labor-intensive. While there are direct weed control expenses such as personnel, equipment, and materials including herbicides and mulches, there is also the indirect cost of diverting labor away from other tasks in order to have time for weeding. For these reasons, it is critical that an IPM program that includes weed management be implemented in the nursery.

Several options are available for weed control in container nurseries (fig. 15.1). Cultural control methods (e.g., sanitation and other prevention practices), mechanical control methods (e.g., mulching, hand-weeding), physical control methods (e.g., steaming or heat-treating growing media), and chemical control methods are described in this chapter, along with their use in integrated weed management programs. How to select and develop a nursery site to minimize weed problems is also discussed. In addition, weeds species commonly found in container nurseries and management recommendations are addressed.

It is important that runoff management practices be implemented in the nursery to protect water resources. Factors that affect herbicide movement in the environment and runoff management strategies to mitigate herbicide runoff are discussed in this chapter.

Specific herbicides for control are not provided in this chapter, as pesticide labels change frequently. For help in selecting which herbicides to use for controlling weeds in container nursery crops, growers can refer to *UC IPM Pest Management Guidelines for Floriculture and Nurseries* (see the UC IPM website in the References at the end of this chapter). Growers located outside of California should consult with their Cooperative Extension office for specific herbicide recommendations.

Sources and Types of Weeds

Some weed seeds are spread within the nursery by wind dispersal; other seeds can be forcefully ejected from pods (e.g., bittercress, woodsorrel), or carried into the nursery by water or in soil. Weeds also can be spread on clothing or equipment and supplies used in the nursery, such as irrigation tubing and recycled containers. Sources of infestation include weeds growing in, between, or near pots in production areas and weeds in nonproduction sites, such as service roads and fence lines.

In California, certain winter annuals germinate in the fall, flower and seed by spring, and normally die in the summer when the temperature rises. Examples of winter annuals in the nursery include annual bluegrass *(Poa annua)*, common chickweed *(Stellaria media)*, common sowthistle *(Sonchus*

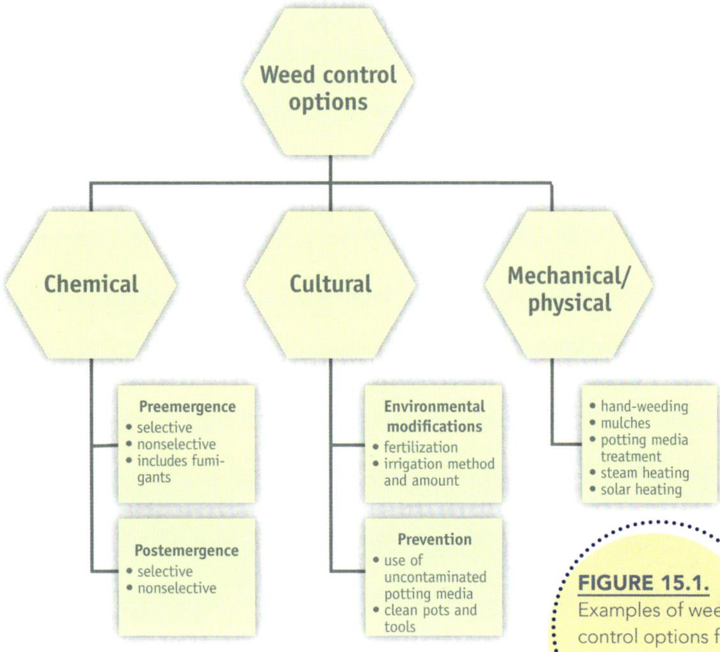

FIGURE 15.1. Examples of weed control options for container nurseries.

oleraceus), prickly lettuce *(Lactuca serriola),* common groundsel *(Senecio vulgaris),* pearlwort *(Sagina procumbens),* petty spurge *(Euphorbia [=Chamaesyce] peplus),* and bittercress (*Cardamine* spp.).

Summer annuals germinate in the spring, grow during the summer, and normally die in the fall or early winter. Examples of summer annuals include prostrate, creeping, and spotted spurge (*Euphorbia [=Chamaesyce]* spp.), crabgrass (*Digitaria* spp.), hairy fleabane and horseweed (*Conzya* spp.), and common purslane *(Portulaca oleracea).* In areas where adequate moisture and cooler summer temperatures exist, such as shaded sites, winter annuals often germinate and grow at the same time as summer annuals.

Other weeds frequently found in nurseries include bryophytes, such as mosses and liverworts, and the perennial weeds purple and yellow nutsedge (*Cyperus* spp.) as well as creeping and yellow woodsorrel *(Oxalis corniculata* and *O. stricta).* Woodsorrels are problematic because their seeds can be expelled from seedpods to other nearby pots, and the stolons or rhizomes are difficult to remove once established.

Nursery Site Selection and Preparation

Choosing a Nursery Site

Several factors should be considered for reducing weed problems when planning a nursery. If there is a choice of the nursery site, choose one that is free of perennial weeds and away from areas where windblown weed seed contamination is probable. Choose a site that is well drained; if drainage is poor, prepare the site to improve drainage or select a location that is sloped so water drains away from the containers. This is done to avoid standing water or areas that remain moist for extended periods of time; moist soil between and around containers provides excellent sites for weed seeds to germinate and become established. The area should also be designed to allow for roadways or space between the rows of plants so that equipment can be moved through the nursery to apply herbicides if necessary.

Site Preparation

It is important that the site starts out as weed-free as possible. Perennial weeds must be controlled before installing the planting bed base, or else they will grow through the gravel bed or nursery cloth. Controlling perennial weeds may take more than a year and is best done before nursery production starts. This allows more options for weed management, such as cultivating more often and in larger areas, leaving the ground fallow, and choosing from a wider range of herbicides that may be legally used.

Annual weeds should be controlled at the nursery site and surrounding areas to reduce sources of weed seeds. Nursery managers should work with neighbors by offering to clean up weeds in surrounding areas. In some cases, fine-mesh fabric fencing placed downwind from weed seed sources can be helpful, but it will not provide as much control as removing the weeds.

Postemergence nonselective herbicides can be applied in the nursery area before the container plants are placed in the beds. These should be selected to control perennial weeds, which may be more difficult to eradicate after valuable nursery stock is in place. Care must be taken if these herbicides have a long soil residual; crop roots that grow out of the pot may take up the herbicide, resulting in crop injury or death.

Weed Management in Established Container Nurseries

Weed management involves the integration of various strategies including mechanical, physical, cultural, and chemical control techniques (see fig. 15.1). Moreover, it is important that preventive practices such as sanitation be used consistently throughout the production cycle. Typical strategies used in weed programs for established outdoor container nurseries are described in the next sections.

Prevention Practices

The best weed management methods are those that prevent weeds from establishing in the first place. These include removing sources of weed infestations in production and nonproduction areas, avoiding use of contaminated soil, and starting with clean plant material.

Remove sources of infestations

Established weeds in production and nonproduction areas should be removed prior to seed head maturity, preferably before flowering, to prevent weed seed distribution by wind. In addition, a systematic program of weed control should be set up and maintained to prevent reintroduction of weeds from surrounding areas in the nursery. Often a preemergence herbicide (one that controls weed seedlings before or just as they emerge from the soil) is tank mixed with a postemergence herbicide

Photo: C. A. Wilen.

(one that controls weeds after they emerge) and used on the periphery of the nursery to reduce weeds and prevent weed seeds from reinfesting the area. Where reservoir or pond water is used for irrigation, areas around these water sources should be as weed-free as possible. If necessary, a weed screen can be placed over the water inlets to screen out seeds and debris.

Use clean potting mix

Many weed control problems can be reduced by avoiding the use of contaminated growing media. If the potting media contains soil, it should be treated to kill any weed seeds and reduce the possibility of introducing weeds from this source. The common methods to kill weeds in soil are soil solarization and steam pasteurization (see the section "Physical Control" later in this chapter) and fumigation (see the section "Chemical Control" later in this chapter). Many nurseries use soilless mixes, but even soilless media should be checked periodically for weed seeds. This can be done by placing soil mix samples in a flat or two and keeping them moist for 1 or 2 weeks. If weeds emerge, the media should be treated or not used. Organic matter such as compost should only be used if the grower is certain that it has been composted properly in order to kill all weed seeds.

Start with clean plants

An important preventive control method for managing weeds is to produce or purchase liners, transplants, and plugs that are free of weeds and weed seeds. If liners or plugs contain weeds, do not use them unless you remove all the weeds before transplanting; contaminated liners are often the primary source of weed introduction into otherwise weed-free containers. Weeds should also be removed from gallon-size plants before transplanting into larger containers. Use of preemergence herbicides in and between the containers can reduce contamination or reinfestation.

Modifying the Plant Growing Environment

In addition to preventive practices, other cultural control methods that modify the plant growing environment can be used to reduce weeds in the nursery. For example, top-dressed fertilizers should be carefully applied to keep granules in the pot and minimize weed growth in areas outside the pot where fertilizer has spilled. Irrigation type and irrigation application practices can have a significant effect on the intensity of weed pressure.

Irrigation Method and Application

Subirrigation

Subirrigation is an effective cultural control method for managing weeds. It can be used in covered structures such as greenhouses or outdoors where there is little rainfall. Pots are placed in trays, troughs, or mats, and water is drawn up through the potting mix via capillary action. There are also a few nursery pots designed as individual subirrigation systems. For effective weed control, only enough water is supplied to moisten the crop's root zone so that the top layer of the potting mix remains dry and is not a good substrate for weed seed germination.

There are limitations to this method of irrigation. First, care must be taken that plants do not sit in water for an extended period. Second, techniques must be developed that will allow growers to use this methodology on a large scale outdoors. Sand beds have been tested to try to adapt this technique to a larger scale, but problems still exist, such as algae growing on the sand. In addition, the cost of mats or specialized pots may limit use to small-scale growers. For more details on subirrigation, see chapter 7.

Irrigation application management

Even when irrigation is applied to the media surface by overhead irrigation or spot spitters, weed germination and establishment can be managed by modifying the amount and timing of the irrigation to allow the media surface to thoroughly dry between irrigations. This can be accomplished by placing soil moisture sensors in the root zone in representative pots to determine when to irrigate and turning off the irrigation when the soil moisture has reached a predetermined level. More information on irrigation management is provided in chapter 7.

Mechanical Control

Mechanical weed control options include hand-weeding as well as the use of protective barriers between plants and weeds such as mulches; additionally, fine-mesh fabric fencing and weed screens over water inlets to exclude weed seeds are occasionally used.

Hand-Weeding

Hand-pulling weeds and cultivation are widely used in container nurseries, and the labor for hand-weeding is a large part of the overall nursery production costs. The cost of labor for hand-weeding in container nurseries commonly ranges from several hundred to several thousands of dollars per acre; in field plantings, costs average about $200 to $800 per acre. In protected structures such as greenhouses, hand-weeding may be the only option due to limited registration of herbicides.

Hand-weeding is not always an effective management option. For example, weeds can quickly regrow from creeping woodsorrel stolons and yellow woodsorrel rhizomes left in the pot after hand-weeding.

Weeds that are removed by hand should be removed from the production area and disposed of rather than left to dry out near the pots. This is because some species will reroot or seed and may continue to mature even after the plant is cut down or pulled up (e.g., common purslane and bermudagrass stolons).

FIGURE 15.2. Fabric or paper collars on the surface of containers reduce weeds (left) except around the cut at the base of the plant. The cut is required to place the collar around the plant. *Photo: J. K. Clark.*

Mulches

Many annual weed species and some perennial weeds can be found growing under or between containers. An option for preventing weed establishment in these areas is to use polypropylene weed cloth or gravel mulches underneath containers in nursery beds. Another option is to use synthetic or organic mulches directly in the containers.

Weed cloth and gravel mulches

Overlapping layers of polypropylene mulches having ultraviolet light protection (landscape fabrics or geotextiles) can be used under containers in nursery beds. Potting mix that spills on the fabric should be brushed or shoveled off so weeds do not grow on the mix and into the fabrics. Weeds must be removed by hand before they establish to avoid creating tears in the fabric by pulling the entwined roots from the fabric.

Gravel mulch can also be used as a base underneath pots to prevent weed establishment. However, better weed control is achieved if fabric mulches are first installed and then covered with 2 to 5 inches of gravel. Weeds that establish on top of the gravel mulch can be pulled. As with the fabric mulch, weed seeds can grow where soil or potting media accumulate. When this occurs, the gravel will need to be scraped off and the area cleaned before replacing.

It is a good practice to schedule time to remove weeds from the mulched beds between crops. Gravel beds can be cleaned by hand-weeding or using a nonselective postemergence herbicide. They can also be cleaned by using a postemergence herbicide tank mixed with a preemergence herbicide after leveling to maintain a flat bed surface. Be aware that when postemergence herbicides are used, caution must be taken to avoid drift, which may injure nearby container plants. If the fabric mulch is worn and has many holes, it should be replaced.

Mulches used inside containers

Some mulches are used to control weeds in the container itself. Both synthetic and organic products may be used. Synthetic mulches such as fabric or paper are cut as collars that fit around the base of a plant (fig. 15.2). One brand of geotextile disk is available with copper hydroxide on one side to prevent weeds from rooting. An advantage of fabric or paper collars is that they can last up to 3 years. Disadvantages include the labor required to apply these collars to each container, as well as the growth of weeds in areas not well covered by the collar, such as along the slit or edges of the pot.

Organic products such as fir bark, pecan hulls, straw, green waste (a ground, composted, leaf-bark-branch mixture from the urban landscape), and other organic material can be used to help suppress weeds in containers. These products are applied after planting or as a topdressing after weeds are removed. It is important to use material that is small enough to completely cover the surface yet large enough so that the particles do not become a substrate for weed growth. A 1-inch layer of organic mulch is usually adequate for weed control. These mulches are especially effective in 5-gallon and larger containers.

Physical Control

Physical (environmental control) methods indirectly control pests by physical means such as the use of heat and cold. The most common methods used for weed control in established container nurseries are solarization and steam pasteurization to disinfest media.

Soil Solarization

Solarization is a nonchemical, physical control method involving trapping solar energy beneath clear plastic sheets spread over a soil surface or containers filled with potting mix. Solarization methods used in nurseries are described in chapter 5. Solarization can control many annual and perennial weeds in California, and some weed species are very sensitive to solarization. However, effective control of many weed species requires optimal conditions for control (e.g., good soil moisture, tight-fitting plastic, and high solar radiation). Detailed information on the use of solarization to control weeds is provided in the References at the end of this chapter.

Steam Pasteurization and Sterilization

Steam heating is another physical control method that kills seeds and seedlings. Details concerning this method are described in chapters 4 and 5. The steam is usually mixed with air and injected into the loose soil mix to heat the mix to at least 160°F. The mix must be kept at this temperature for 30 minutes. The steam should be carefully placed so that the soil is uniformly heated. Length of time and temperature are critical if weed seeds are to be controlled.

Chemical Control

Most growers use a regular weed management program in outdoor container nursery production that includes the use of herbicides. If used properly, herbicides are often less expensive and easier than hand-weeding, cultivation, or the use of mulches. Nevertheless, for sustainable weed management, one should not rely on herbicides alone.

Herbicide Selection

Herbicides are efficient chemicals when used to control weeds in nurseries. However, because they may injure the crop or there may be off-site movement of the herbicide, the user needs to consider the following carefully:

- the weed species that are to be controlled and their stage of growth
- whether the herbicide is registered for use to control the target weed species in the specific site location (outdoor production area, greenhouse, nonproduction area, etc.)
- whether the herbicide will adversely affect crop plants
- the herbicide residual activity and persistence of the herbicide
- the potential movement of herbicide away from the target site

The first three items listed above usually drive, or at least narrow down, the choice of which herbicide to use. Nevertheless, the last two items in the above list should be included as important factors when selecting an herbicide. Obviously, there are many other considerations when choosing a certain pesticide (e.g., toxicity class, formulation, ease of use, and cost).

Weed species and stage of growth

Herbicides are selected based on their ability to control specific weeds. Because of the broad weed spectrum found in nurseries, herbicide combinations or sequential applications are frequently used or different herbicides are used in sequential applications. For example, oxyfluorfen plus pendimethalin (a combination) could be applied in the fall for winter weeds, and oryzalin or pendimethalin could be applied in the spring for summer weeds. Rotating classes of herbicides is important because if the same herbicide or herbicide mode of action is used repeatedly year after year, herbicide-resistant weed populations may develop. Also, a weed spectrum shift toward species not controlled by the herbicide may occur.

Age and growth stage of the weeds must also be considered when selecting herbicides. Most preemergence herbicides are ineffective if the weeds have already emerged. All emerged weeds must be removed prior to the preemergence herbicide application. Most preemergence herbicides are also not effective in controlling vegetative structures, such as tubers from yellow nutsedge or rhizomes from yellow woodsorrel, regardless of whether they have sprouted or not.

Crop tolerance

Crop species vary in their tolerance for herbicides. It is critical that the user verify that the crop is listed on the pesticide label. Keep in mind that there can also be variability in susceptibility among cultivars and varieties. Before using an herbicide for the first time, test it on a small group of plants and observe them for 1 to 4 weeks to see if any injury symptoms develop. Be sure to have some untreated plants of the same type interspersed among the treated ones for comparison.

Crop age is also an important consideration when making herbicide choices because young plants are often more susceptible to root and foliar injury than are older plants. Furthermore, the growth stage and growth rate of the crop is important in predicting the tolerance of the crop to an herbicide. Newly transplanted liners or plugs are more sensitive because generally they have smaller root systems than established plants. Actively growing plants often are more sensitive to herbicide injury than dormant plants.

Additionally, the formulation of the herbicide affects crop tolerance. Granular formulations are generally safer with some products than the emulsifiable concentrate or wettable powder formulations, but injury may occur if granules are lodged in whorls or leaves.

Preplant Fumigants

Prior to planting, the media may be fumigated to control weeds and other pests. However, chemical fumigation use is quite limited, and use of alternative methods such as solarization or steaming is recommended. California growers should check with the local county agricultural commissioner prior to purchasing fumigants. Read the label thoroughly, and consult with a pest control adviser or the county agricultural commissioner as the labels and local regulations change frequently for these materials; it is important to understand where fumigants may be legally used. Growers in other states should consult with their Cooperative Extension office or state agriculture department. In all cases, buffer restrictions and location of sensitive sites such as schools, parks, hospitals, day care centers, and so on must be taken into account.

Chemical fumigation is accomplished by covering a pile of potting mix or soil with a polyethylene cover and injecting the fumigant into the media at various locations in the pile. The fumigant should be uniformly distributed throughout the pile so that all areas of the media are treated. The cover must remain on the pile for the time required as stipulated on the label. After the cover is removed, the media should be allowed to air (enabling the fumigant to dissipate) for at least 48 hours or as written on the label before it is used. Some weeds such as cheeseweed *(Malva parviflora)*, clovers, and field bindweed are not well controlled by fumigation.

Postplant Herbicides

As soon after potting as feasible, a preemergence herbicide or combination formulation should be applied, followed by hand-weeding if weeds appear before the next preemergence herbicide application. Subsequent herbicide applications are made as soon as possible after hand-weeding (unless in conflict with the label directions) to reduce additional hand-weeding costs. Preemergence herbicides will generally provide 4 to 12 weeks of control depending on the rate and environmental conditions. The same or different preemergence herbicides can be applied subsequently, depending on the weed spectrum, stage of growth, and resistance management considerations.

After plants are established, both pre- and postemergence herbicides are applied from one to five times throughout the year for managing weeds around the beds and between the pots. Hand removal of escaped weeds may also be necessary to reduce seed sources in the nursery.

Strategies to Improve Herbicide Efficacy and Avoid Crop Injury

Use of adjuvants

Tank mixing surfactants with postemergence herbicides can greatly increase herbicide activity and affect selectivity. Most of these products will decrease the surface tension of water, causing it to spread more evenly on the leaf surface or allowing the product to stick on the leaf better. Some adjuvants also help increase herbicide uptake into the plant. Herbicide labels will specify what kind of adjuvant, such as a nonionic surfactant, crop oil concentrate, or organo-silicone surfactant, to use. In some cases, the adjuvant is included in the formulation and no additional adjuvant is needed.

Application strategies

Various spray strategies can affect proper distribution of the herbicide on the target. For most herbicides, the height of the spray boom should be adjusted so that the top of the ornamental plant receives uniform spray distribution. This means that with normal spray booms equipped with fan-type nozzles, the nozzles should be at least 20 to 24 inches above the top of the plants. Spray pressure should be adjusted so that preemergence herbicides will reach the media surface, but not so high as to injure the crop. Granular materials should be applied to dry foliage. If granules remain on the leaves or other plant parts, they may be knocked off by gently shaking the plants with a stick or using a blower, taking care that the herbicide is not blown off the treated area. Overhead irrigation is often sufficient to remove herbicide residues from the crop, but herbicide may also accumulate in whorled foliage, causing injury.

Environmental conditions

To maximize herbicide performance and avoid crop injury, environmental factors such as temperature, humidity, precipitation, and wind must be considered before applying herbicides. Temperature affects rates of chemical reactions in plants; higher temperatures can greatly increase the speed of chemical reactions, which can result in greater injury to plants as well as to weeds. Higher temperatures may also increase herbicide absorption through leaves and roots. Windy conditions can move volatile herbicides off-site and potentially damage nontarget plants. Precipitation occurring before a postemergence herbicide has dried on the plant or before it is fully absorbed may wash the herbicide off the plant, resulting in reduced weed control. Very heavy rainfall may leach some preemergence herbicides into the media and cause root injury.

Herbicide Movement and Runoff Mitigation

It is important that off-site pesticide movement be limited so that the herbicide remains in place and is available for uptake by the weeds. Factors that are important in determining herbicide movement include physical and chemical properties of the herbicide, soil adsorption, water solubility, and half-life. These factors need to be considered when selecting and applying herbicides. Furthermore, management practices should be implemented to avoid herbicide movement in the environment.

Herbicide Volatility and Persistence

Herbicides that are volatile can easily drift with air currents onto nontarget plants. If a volatile pesticide is selected, it must be applied during cooler and lower wind conditions.

The herbicide's soil persistence is a measure of how long the product remains active where placed. Postemergence nonselective herbicides applied to nursery beds to control perennial weeds with a long soil residual could damage crop roots that grow out of the pot and take up the herbicide. Furthermore, pesticides with long persistence in the soil may pose environmental concerns. However, in some cases, longer soil persistence, or residual, is desirable. An herbicide that is effective over the entire cropping period will reduce or eliminate the need for additional weed control efforts. An herbicide that is active over a longer period reduces the number of times herbicide must be applied.

Herbicide persistence is influenced by the physical and chemical properties of the pesticide and the soil. Soil texture and organic matter content affect whether the herbicide can leach to the root zone. They also affect the amount of binding of herbicides on the soil particles. The more clay and organic matter content in the potting mix, the greater the binding is and, usually, the less the leaching. Microbial and

Table 15.1. Properties of various herbicides used for weed control in container nurseries

Active ingredient	Trade name	Use*	Water solubility† (ppm)	K_{oc}‡	Half-life§ (days)	Mobility#
2,4-D amine	many names	POST	796	20	10	mobile but degrades rapidly
atrazine	Aatrex	PRE	33	100	60	high, due to long half-life
dichlobenil	Casoron	PRE	21.2	400	60	low, but breakdown product is high
dimethenamid-p	Freehand Tower	PRE	1,449	105–396	31–292	high
diquat	Reward	POST	718,000	1,000,000	1,000	none; binds strongly to clay
dithiopyr	Dimension	PRE	1.4	1,638	17	low
fluazifop	Fusilade	POST	1.1	5,700	15	low
flumioxazin	Broadstar, Sureguard	PRE	1.79	336–620	20	low
glufosinate	Finale	POST	1,370,000	100	7	high, but due to short half-life and microbial breakdown, little found > 6"deep
glyphosate	Roundup	POST	900,000	24,000	47	low, due to strong adsorption to soil
halosulfuron	Sedgehammer	POST	1,630 at pH 7, 15 at pH 5	87	4–18	low to moderate
isoxaben	Gallery	PRE	1.0	190–570	50–120	very low
oryzalin	Surflan	PRE	2.6	600	20–128	low under natural rainfall
oxadiazon	Ronstar	PRE	0.7	3,200	60	low; strongly adsorbed
oxyfluorfen	Goal	PRE/ POST	0.1	100,000	35	immobile on most soils but slightly mobile in sandy soils
pelargonic acid	Scythe	POST	low	no info	not reported	not reported
pendimethalin	Pendulum	PRE	0.275	1,7200	44	very low
prodiamine	Barricade, Factor	PRE	0.01	13,000	69–120	very low
sethoxdim	Vantage	POST	257 at pH 5, 4,390 at pH 7	100	5	not reported
simazine	Princep	PRE	6.2	130	60 but persists longer in high pH soils	low
trifluralin	Treflan	PRE	0.3	7,000	45	low, due to strong soil adsorption

Notes:

*PRE = preemergence; POST = postemergence.

†Higher number indicates more soluble.

‡Measure of how strongly bound to soil; high number indicates more strongly bound.

§Time for half of active ingredient to degrade.

#Movement in soil under normal rainfall.

chemical breakdown also affect herbicide persistence. These are often affected by soil moisture, pH, and temperature. In general, herbicide breakdown occurs more rapidly when the soil is moist and warm, as microbes are more active under these conditions.

Soil Adsorption

Soil adsorption is the process by which a material associates with a surface (stickiness). It is reported as a K_{oc} value. An herbicide with a K_{oc} value of 0 is not adsorbed at all; an herbicide with a K_{oc} value greater than 1,000 is very strongly adsorbed to the soil particles. Soil texture is also important when discussing adsorption. Soils high in clay or organic matter generally will adsorb pesticides to a greater extent than will sandy soils.

Water Solubility

Water solubility is a measure of how easily the herbicide dissolves in water. Although other properties of a pesticide can affect the extent of off-target movement (see glyphosate example in the section "Factor Interaction," below), in general, more soluble herbicides are more likely to leach. Herbicides with low solubility (1 ppm or less) tend to remain at the soil surface, although they may move off-site in runoff. Pesticides with solubility greater than 30 parts per million are more likely to leach. Selecting herbicides with low water solubility can reduce leaching and protect water quality. Herbicides with high solubility must be applied with caution if site conditions are conducive to leaching.

Half-Life

Half-life is a measure of how long it takes for the herbicide to degrade by half. The longer the half-life, the more persistent the herbicide is, and therefore the more opportunity there is for the material to leach or run off when exposed to conditions that would favor off-site movement. Factors that affect half-life include microbial decomposition, degradation in light, and volatility.

Factor Interaction

It is important to remember that the interaction of soil adsorption, water solubility, and half-life, rather than the individual parameters, needs to be taken into consideration when considering how long a pesticide stays in a particular environment. For example, the water solubility of glyphosate

FIGURE 15.3. When herbicides are broadcast over container plants, some of the herbicides fall to the soil and thus control weeds between and under the container areas. (A): Preferred spacing. (Close spacing results in more herbicide landing on the media surface.) (B): Herbicide that lands on planting bed will reduce weeds between pots but may also be carried off-site in irrigation runoff. (C): Herbicide application to plants spaced farther apart is more costly because herbicide is also applied to a larger area, including where containers are not placed. *Photos:* C. Wilen.

is very high (900,000 ppm or about 1 ounce per quart; see table 15.1). Given this fact, one would expect glyphosate to move easily through soil. However, the K_{oc} value is 24,000, which indicates very strong adsorption to soil. Therefore, even though it is very water soluble, the soil holds and traps it so it is unlikely to move as an active molecule.

Table 15.1 can be used as a guide to help determine whether an herbicide will move off-site or persist in the soil. Remember that many conditions affect an herbicide's activity and that these metrics will vary according to soil type, irrigation method, weather, and other factors.

Runoff Mitigation Practices

Effects on water quality must always be considered when selecting and using herbicides. Herbicides may move with runoff water, potting mix, or sediment in the runoff. Every effort should be made to prevent soil or media from moving off-site, since the herbicide attached to these particles will be lost from the nursery and may contaminate waterways.

Management practices to reduce herbicide runoff include using the correct herbicide rate. Not only can higher herbicide dosages cause crop injury, more herbicide applied also means that more herbicide can be moved off target. It is important to know the organic matter content in the potting mix or soil to match the preemergence herbicide rate suggested on the label. It is best to use the lowest recommended herbicide for a given level of organic matter when a range of rates is given on the label.

Avoid making herbicide applications during windy conditions or when precipitation could carry herbicides that are dissolved in runoff water or adsorbed to soil or media particles off-site, where they could potentially pollute water bodies. If possible, crops should be closely spaced when herbicides are broadcast over the plants to minimize material falling between and under the containers, which could move off-site during irrigation and storm events (fig. 15.3). Little herbicide falls to the ground when gallon containers are closely spaced, but when the containers are spaced farther apart, 40 to 50 percent of the herbicide may spill onto the ground.

Spot treating and using low-volume herbicide applications can be implemented to reduce pesticide loads and minimize pesticide runoff. Furthermore, although it is important to use sufficient water to move preemergence herbicides into soil and off the crop at the first irrigation, do not overirrigate as this will result in runoff; runoff from the first irrigation following herbicide application is most likely to carry herbicide. Pulsing (applying less water per irrigation event but having more frequent events) will help reduce herbicide runoff. For more details on pulsing, see chapter 7.

Many nurseries in California are collecting and recycling irrigation water by mixing it with fresh water before reusing. Details on these runoff management practices can be found in chapter 8. If excessive applications of herbicides occur, they may be detected in the irrigation water. But, in practice, the concentration of herbicides found in the recycled water is usually not sufficient to injure most plants.

Common Weed Species and Control Methods

The overall number of weed species that are problematic in California nurseries is small but, due to the year-round nature of nursery production, includes summer and winter annuals as well as perennials (table 15.2). Annuals that propagate within the nursery or arrive as windblown seed from peripheral areas are generally the most troublesome. Bittercress, common groundsel, spurge, willowherb, and woodsorrel are the most commonly encountered weeds in California container nurseries and are described in more detail below. Other weeds found frequently in California container nurseries include common sowthistle, horseweed, pearlwort, annual bluegrass, cudweed, fireweed, eclipta, and liverwort. Pearlwort and common chickweed have become more frequent in nurseries because their tolerance to oxadiazon allows them to flourish when other weeds are controlled. As with any pest management program, correct identification of weeds and knowledge of their life cycles will help make control easier. The reference section at the end of this chapter includes publications and other sources of information for weed identification.

Bittercress

Description

Bittercress (*Cardamine* spp.) is a member of the mustard family and can complete its life cycle in as little as 6 weeks. It is a low-growing plant (2 to 6 inches) that prefers growing in shaded, moist conditions, such as those found in shadehouses and under the canopy of irrigated plants. Optimal conditions for growth are temperatures from 45° to 85°F, good moisture, and moderate to high fertility.

The tiny seed are produced in a thin capsule that shatters upon maturity, forcefully spreading the seed up to 3 feet away. Each capsule contains 18 to 28 seed. Seed do not have a dormancy period and may germinate immediately. Seed may be carried in runoff water or moved by water splashing.

Table 15.2. Common weeds found in container nurseries

Scientific name	Common name	Dicot (D) or monocot (M)
Annual winter weeds		
Cardamine spp.	bittercress	D
Cerastium glomeratum	mouse-ear chickweed	D
Euphorbia (=Chamaesyce) peplus	petty spurge	D
Lactuca serriola	prickly lettuce	D
Poa annua	annual bluegrass	M
Polygonum sp.	smartweed	D
Senecio vulgaris	common groundsel	D
Sonchus oleraceus	annual sowthistle	D
Stellaria media	common chickweed	D
Urtica urens	stinging nettle	D
Annual summer weeds		
Digitaria sanguinalis	large crabgrass	M
Echinochloa crus-galli	barnyardgrass	M
Euphorbia (=Chamaesyce) spp.	spurges	D
Gnaphalium palustre	cudweed	D
Perennial weeds		
Cynodon dactylon	bermudagrass	M
Cyperus esculentus	yellow nutsedge	M
Epilobium ciliatum	northern willowherb	D
Oxalis corniculata	woodsorrel	D

Damage

Bittercress competes with the crop for water and nutrients. The presence of bittercress in the container reduces the salability of the crop and increases the chance that the weed will spread to other containers and other areas of the nursery.

Nonchemical control

Because of the short life cycle of this weed and lack of dormancy of the seed, it is important to monitor all areas for this weed. Not only should containers be inspected but also areas surrounding containers. Water should not be allowed to flow over this weed since this will spread the seed. Also, recycled water should be treated or screened to avoid reintroduction of this weed. Irrigation tubes should be wiped before replacing in an uncontaminated container because seed could be present on the tubing. Used pots should be cleaned thoroughly to wash off soil and seed that may be sticking to them.

Chemical control

Postemergence herbicides can be used to control bittercress in areas surrounding the containers, but there is no selective postemergence treatment for use in a container. Preemergence herbicides are effective, but care must be taken that the herbicide is applied prior to seed germination. Where bittercress is known to be a problem, the herbicide must be applied as soon as possible and reapplied as needed. Over time, this will reduce the bittercress population.

Common Groundsel

Description

Common groundsel *(Senecio vulgaris)* is an early spring weed in most areas, but can grow all year in coastal areas of California. Common groundsel grows well in most shaded areas such as the environment found under the canopy of container plants. Seed can be produced very quickly and are dispersed easily by wind. Common groundsel can complete its life cycle in as little as 8 weeks. The plant habit is upright and branched. Flowers are produced in an indeterminate fashion. Because the seed is windborne, it is important to control weeds that are found around the nursery as well as those in the containers.

Damage

Common groundsel reduces the salability of plants and increases cost of production due to need for control. Infested areas must be constantly monitored and weeded to remove new plants before they produce seed.

Nonchemical control

Germination of common groundsel seed in the containers can be reduced if a layer of mulch is used to reduce the establishment of windblown seed. Mulches can effectively limit common groundsel invasion if the mulch surface dries out quickly.

Chemical control

A nonselective herbicide can be used to control weeds surrounding the nursery. However, it is important to control the plants prior to flowering since seed may still continue to mature on common groundsel even if the weed is treated or pulled after flowering. Preemergence herbicides are available that will control common groundsel in containers.

Prostrate Spurge, Spotted Spurge, Creeping Spurge, and Petty Spurge

Description

Spotted spurge *(Euphorbia [=Chamaesyce] maculata)* and prostrate spurge *(E. [=Chamaesyce] humistrata)* are similar. Creeping spurge *(Chamaesyse serpens)* has smaller, lighter-colored leaves and roots at the nodes. Petty spurge *(E.* [=Chamaesyce] *peplus)* is an upright-growing annual with light green leaves and is generally found growing under cooler temperatures. The spurges are considered some of the worst weeds in container production throughout California. These plants, except for petty spurge, are low-growing annuals that often form a dense mat on the potting media surface.

Other than petty spurge, the above spurges are summer annuals that produce numerous seeds starting within 5 weeks of germination. Seeds can remain dormant in the soil until conditions are suitable for germination. Seeds that are produced in summer germinate readily, whereas those produced in late fall are mostly dormant and will not germinate until spring. Optimal germination is from 75° to 85°F, but the range is 60° to 100°F. When moisture is available, germination can occur almost any time of the year. Light is also required for maximum germination, and seeds buried greater than ½ inch will not germinate well.

Damage

Spurge can compete with young plants and reduce crop growth. It also provides habitat for ants and snails and can be an intermediate host for fungal diseases of cultivated crops. Infested areas must be constantly monitored and new plants hand-pulled before they produce seed.

Nonchemical control

Germination can be reduced by a layer of mulch. Mulches can effectively limit spotted spurge if they prevent light from reaching the seed.

Chemical control

Preemergence herbicides will control spurge if applied before emergence of the weed.

Northern (Fringed) Willowherb

Description

Northern, or fringed, willowherb *(Epilobium ciliatum* ssp. *ciliatum)* is found in many nurseries in California. Although classified as a perennial when growing in the wild, its life history in nurseries is more like that of a summer annual as it is easily established from seed and controlled like a summer annual. It is an upright plant growing to about 2 to 3 feet when found in or around containers, with leaves longer than they are wide. The plant often has a reddish cast. Tiny pink flowers are found on the tips of a long (1 to 3 inch), tubelike structure. Seed are borne in capsules. When the seed are mature, the capsule peels back from the tip revealing four chambers, each with one row of numerous tiny seed attached to white hairs that are easily carried by wind.

Damage

Seed landing on the potting media in containers can quickly germinate and compete with young transplants. Older willowherb plants can become woody, and roots may grow out of the pot, making them difficult to remove without injuring the crop or damaging the nursery cloth. Northern willowherb can start to produce seed very quickly, so removal before capsules are developed is important to stop reinfestions. Because the plant can grow in areas surrounding the nursery, it can reinfest the nursery and piles of potting media through its windborne seed.

Photo: C. A. Wilen.

Nonchemical control

Hand-weeding young plants in and around containers and pulling plants surrounding the nursery before they go to seed is advised. Using mulch in the container will reduce seedling growth and establishment.

Chemical control

Several preemergence herbicides are available to control germinating seed in containers but some preemergence herbicides commonly used in nursery production have little to no control of this weed. Consult the label, a pest control adviser, or your local UC Cooperative Extension advisor for more information. Willowherb growing in nonproduction areas can be controlled with contact or systemic postemergence herbicides.

Woodsorrel

Description

Creeping woodsorrel, *Oxalis corniculata,* and yellow woodsorrel *(O. stricta)* are common weeds found growing in nursery containers. A purple-leafed variant also exists. Woodsorrel grows in both full sun and shaded areas that receive adequate moisture. Creeping woodsorrel is a perennial plant that grows in a prostrate manner and forms roots along its stems where nodes contact the soil. Yellow woodsorrel is also perennial but grows in a slightly more upright manner and does not root at the nodes but can reestablish from rhizomes. Flowers of woodsorrel can be found almost any time during the year, but spring is the period of heavy flowering and seed formation. Seed are borne in erect, cylindrical pods. When seedpods mature, they burst open and forcefully expel the seed, which may land 3 feet or more from the plant. Because seed are rough, they adhere to surfaces of machinery, containers that may have dirt on the outer surfaces, or clothing. Each pod contains about 10 to 50 seed, and a single plant can produce more than 5,000 seed. Light is required for germination. Optimal seed germination occurs at temperatures from 60° to 80°F, though some germination occurs at lower temperatures. The seed can germinate at any time of year, but most plant establishment takes place in fall.

Damage

Woodsorrel is a major weed in nurseries. As seedpods mature and eject seed, woodsorrel spreads from container to container. Woodsorrel reduces the salability of plants and increases cost of production due to the need for hand-weeding if it becomes established.

Nonchemical control

The primary methods of managing woodsorrel are to remove the established plant and try to control the germinating seed. Seedlings can be controlled by continual hand-weeding. Hand-weeding is of limited use, though, because the plant can quickly regrow from stolons (creeping woodsorrel) or rhizomes (yellow woodsorrel) left in the pot. Therefore, not only is it important to remove the weeds before they set seed, but they should also be removed before stolons or rhizomes become established.

Chemical control

Several preemergence herbicides are available to control germinating seed in containers.

Other methods of control

Burying seed or covering them with mulch to block their exposure to light prevents germination. Woodsorrel seedlings may grow at the base of plants, where they escape preemergence herbicide treatment or poke through mulches. It is not known how long seed remain viable in the soil; however, germination is inhibited or stopped when seed are exposed to moist, warm (97°F) conditions for 30 minutes.

References

Ayeni, A. O., B. A. Majek, J. R. Johnson, and R. G. Obal. 1999. Container nursery weed control: Bittercress, groundsel, and oxalis. Rutgers Cooperative Extension Publication FS939.

Derr, J., J. Neal, L. Kuhns, R. Smeda, L. Weston, C. Elmore, C. Wilen, J. Ahrens, A. Senesac, and T. Mervosh. 1997. Weed management in landscape and nursery plantings. In M. McGiffen, ed., Weed management in horticultural crops. Alexandria, VA: ASHS Press. 42–59.

DiTomaso, J. M., and E. A. Healy. 2007. Weeds of California and other western states. Oakland: University of California Division of Agriculture and Natural Resources Publication 3488.

Dreistadt, S. 2001. Integrated pest management for floriculture and nurseries. Oakland: University of California Division of Agriculture and Natural Resources Publication 3402.

Elmore, C. L., J. J. Stapleton, C. E. Bell, and J. E. DeVay. 1997. Soil solarization: A nonpesticidal method for controlling, diseases, nematodes, and weeds. Oakland: University of California Division of Agriculture and Natural Resources Publication 21377.

Le Strange, M., C. L. Elmore, and D. Cudney. 2010. Pest notes: Creeping woodsorrel and Bermuda buttercup. Oakland: University of California Division of Agriculture and Natural Resources Publication 7444. UC IPM website, http://ipm.ucdavis.edu/PMG/PESTNOTES/pn74145.html.

Molinar, R. H., D. W. Cudney, C. L. Elmore, and A. Sanders. 2009. Pest notes: Spotted spurge and other spurges. Oakland: University of California Division of Agriculture and Natural Resources Publication 7445. UC IPM website, http://ipm.ucdavis.edu/PMG/PESTNOTES/pn7445.html.

Neal, J. C., and J. F. Derr. 2005. Weeds of container nurseries in the United States. Virginia Tech Department of Plant Pathology, Physiology, and Weed Science IPM website, http://ipm.ppws.vt.edu/weeds_container_nurseries.php.

Rice, R. P. 1992. Nursery and landscape weed control manual. Fresno, CA: Thomson Publications.

Stapleton, J. J., C. A. Wilen, and R. H. Molinar. 2008. Pest notes: Soil solarization for gardens and landscapes. Oakland: University of California Agriculture and Natural Resources Publication 74145. ANR CS website, http://anrcatalog.ucdavis.edu/Items/74145.aspx.

UC IPM pest management guidelines for floriculture and ornamental nurseries. UC IPM website, http://www.ipm.ucdavis.edu/PMG/selectnewpest.floriculture.html.

UC IPM. Weed photo gallery. UC IPM website, http://www.ipm.ucdavis.edu/PMG/weeds_intro.html.

UC Weed Research Information Center website, http://wric.ucdavis.edu.

Wilen, C. 2009. Pests of floriculture and ornamental nurseries, Weeds section. Pest management guidelines, UC IPM website, http://www.ipm.ucdavis.edu/PMG/r280701111.html.

Wilen, C. A., U. K. Schuch, and C. L. Elmore. 1999. Mulches and subirrigation control weeds in container production. Journal of Environmental Horticulture 17:174–180.

Vertebrate Pest Management

Roger A. Baldwin

CHAPTER SIXTEEN

Photo: J. K. Clark.

Vertebrate pests are likely to be found in nearly all nurseries, although they may not always present a significant problem. Damage caused by vertebrate pests in nurseries generally results in plant injury, thereby killing the plants or causing permanent damage that lowers productivity following the initial feeding. Several rodents and rabbits eat roots, stems, or bark and can kill young plants outright. Rodent burrows and mounds (such as those of pocket gophers and ground squirrels) interfere with nursery maintenance and inflict structural damage by gnawing on drip irrigation lines. Deer strip young trees and shrubs of foliage and can stunt or even kill saplings and young shrubs through rubbing. Additionally, birds can unearth and eat seeds and seedlings, resulting in crop failure.

This chapter will first discuss some basic guidelines regarding vertebrate pest management and the legal aspects of controlling vertebrates in California. Next, it will provide details on the management of specific vertebrate pests, including ground squirrels, pocket gophers, meadow voles, rabbits, deer, and birds. The vertebrate pest species presented in this chapter are common in California. Although this chapter does not address specific vertebrate pest issues outside of California, many of the same or similar species of pests are present in other states. Therefore, some aspects of this chapter will likely be relevant also to nurseries in states other than California.

General Guidelines for Vertebrate Pest Management

The efficiency and efficacy of a vertebrate pest management program can be enhanced by adhering to the following outline:

- Correctly identify the species causing the problem.
- Alter the habitat, when feasible, to make the area less favorable to the pest species.
- Take early action and use the control methods appropriate for the nursery and time of year, with due consideration for the environment.
- Establish a monitoring system to detect reinfestation so you can determine when additional corrective measures or controls are necessary.

These points provide the framework for an effective management program and will be built upon for each species in further sections of this chapter.

Effective vertebrate pest management depends on the use of appropriate equipment and supplies (e.g., baits, fumigants, traps, deterrents). These materials can be found at local retail outlets such as farm supply and hardware stores. In addition, many county agricultural commissioners make certain rodent pesticides available to growers. For further information or sources of special control materials, consult your local Cooperative Extension advisor or county agricultural commissioner.

Legal Aspects of Vertebrate Pest Management

Under the California Fish and Game Code, if California ground squirrels, meadow voles, pocket gophers, or black-tailed jackrabbits are causing or about to cause a crop depredation, they may be lethally removed at any time by the owner or tenant of a property.

Pesticides mentioned in this chapter for vertebrate control in nurseries in California are subject to change. Only pesticides that are registered with the California Department of Pesticide Regulation (DPR), a division of the California Environmental Protection Agency (Cal/EPA), can legally be used for vertebrate pest control. Registered materials are listed in the DPR's database that is available online (see the References at the end of this chapter). You may also contact your county agricultural commissioner, Cooperative Extension advisor, or pest control adviser for information on registered materials. Pesticides for vertebrate control in gardens and landscapes are listed on the *UC IPM Pest Management Guidelines* website (see the References at the end of this chapter). These should be checked for registration status in nurseries before use. Always consult the pesticide registration label before use.

Trapping is often used to control vertebrate pests. In California, trapping most mammals, even for pest purposes, requires a trapping license issued by the California Department of Fish and Wildlife, although rats, mice, moles, voles, and gophers do not have this requirement. Additionally, you do not need a trapping license if trapping ground squirrels or rabbits on your own property for pest control purposes. However, if trapping either of these species for profit or for hire (e.g., pest control advisers or applicators), a trapping license is needed.

Because wild animals may carry infectious pathogens or parasites, do not handle them without rubber gloves. For disposal of burrowing rodents, you can bury carcasses inside their burrows. Alternatively, you can use a plastic bag slipped over your hand and arm as a glove, then grab the animal with your "bagged" hand and turn the bag inside out while slipping it off your arm and hand. Proceed to double-bag the carcass; then, dispose of it in the trash.

In some areas of California, nurseries are located within the range of federal and state protected endangered species. Species likely to be of concern include the San Joaquin kit fox, several species of rare kangaroo rats, and, where burrow fumigants are used, the blunt-nosed leopard lizard. Special guidelines apply to the use of traps, toxic baits, and fumigants for vertebrate pest control in these areas. See the composite map showing ranges of these species (fig. 16.1). Be aware that this map is not meant to be all-inclusive. Many additional species may be impacted by vertebrate pest control actions. You should contact your county agricultural commissioner if you have any doubts as to the presence or absence of threatened or endangered species associated with the area of desired application. The county agricultural commissioner's office has the latest detailed maps that show the ranges of endangered species and the latest information on restrictions that apply to pest control activities in those areas. You can also get more information on endangered species regulations from the DPR's Endangered Species Project website (see the References at the end of this chapter).

Common Vertebrate Pests

Ground Squirrels

The California ground squirrel *(Otospermophilus beecheyi)* is a medium-size rodent 14 to 20 inches long from its head to the tip of its long, slightly bushy tail (fig. 16.2). This species is responsible for major damage in nurseries throughout the state. California ground squirrels live in underground burrows and form colonies of 2 to 20 or more animals. Ground squirrels live in a variety of natural habitats. They adapt well to human activities and are found along road or ditch banks, fencerows, around buildings, and within or bordering many agricultural crops. They tend to avoid thick chaparral, dense woods, and very moist areas. Ground squirrels are active during the daytime and are easy to spot. During winter months most hibernate (fig. 16.3), but squirrels less than a year old may be active on warm, sunny winter days. Many adults go into a temporary summer sleep, called estivation, during the hottest parts of the year. Squirrels reproduce once yearly, in the early spring, and have an average litter of seven or eight young. The young are nursed in the burrow for about 6 weeks before they come aboveground to forage.

Ground squirrels are primarily herbivorous. During early spring, they consume a variety of green grasses and forbs. When these begin to dry and form seeds, the squirrels switch to seeds, grains, and nuts (see fig. 16.3).

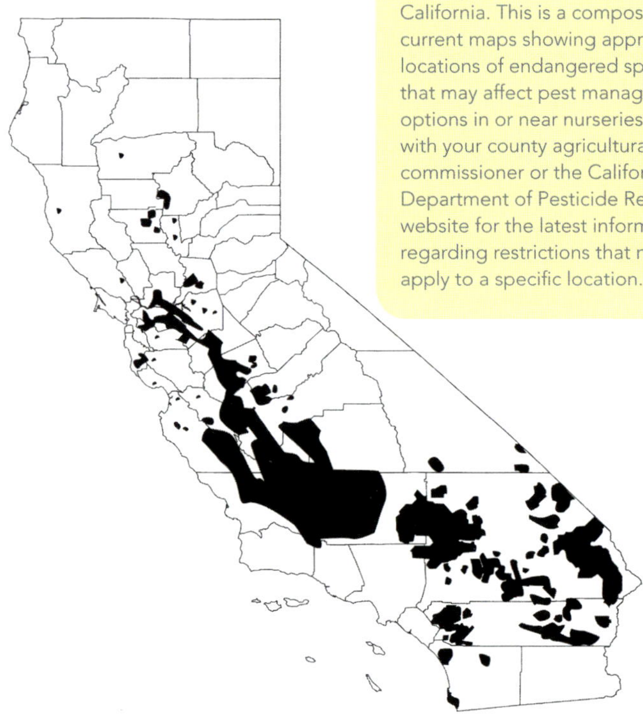

FIGURE 16.1. Distribution of endangered vertebrate species in California. This is a composite of current maps showing approximate locations of endangered species that may affect pest management options in or near nurseries. Check with your county agricultural commissioner or the California Department of Pesticide Regulation website for the latest information regarding restrictions that may apply to a specific location.

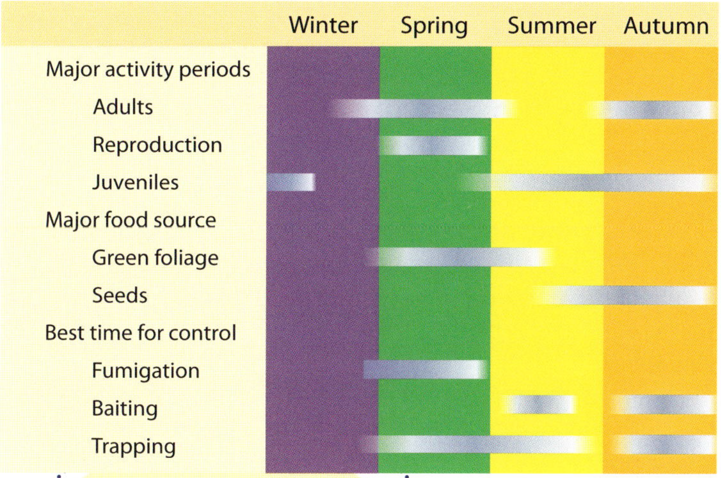

FIGURE 16.3. Activity periods and preferred food sources for the California ground squirrel. Activity periods vary somewhat from one growing area to another depending on local climate. To choose the most effective control action for ground squirrels and the proper timing, you need to know when they are active and what their preferred food sources are. *Source:* Adapted from Salmon et al. 2006.

FIGURE 16.2. California ground squirrel. *Photo:* J. K. Clark.

Damage

Ground squirrels often infest nurseries. When digging burrows, squirrels bring soil and rock to the surface and deposit it in mounds near burrow openings. They enlarge burrow systems each year by constructing new interconnecting tunnels, so the longer the squirrels occupy the burrow, the more extensive and complex it becomes. They create more entrances to serve a growing population. Large and numerous burrow openings and soil mounds are hard on nursery equipment. Ground squirrels frequently burrow around trees and damage the root systems; they can even kill trees (fig. 16.4). Bark gnawing on the trunks of young trees and on limbs of older trees is relatively rare but sometimes occurs (fig. 16.5). Squirrels also gnaw on surface-type drip irrigation pipes (fig. 16.6). They are not intimidated by people, and squirrel burrows are common beneath buildings and other structures made by humans. They are particularly fond of burrowing beneath concrete slabs.

Monitoring guidelines

Establish a plan for periodic monitoring of areas that ground squirrels are likely to invade, such as along ditch or road banks or in crops or rangelands adjacent to the nursery. The frequency of such monitoring activities depends on past problems with ground squirrels and the presence or absence of ground squirrels on adjacent properties. For example, if

FIGURE 16.4. Ground squirrel burrows at base of tree. *Photo:* R. A. Baldwin.

FIGURE 16.5. Ground squirrel gnawing on tree limbs. *Photo:* D. R. Haviland.

FIGURE 16.6. Ground squirrel gnaw marks on drip irrigation pipe. *Photo:* J. K. Clark.

you have had problems with ground squirrels in the past due to an uncontrolled population present at an adjacent property, then monitoring should be more frequent (perhaps every 1 to 2 weeks during the times of the year when ground squirrels are active). To monitor, observe squirrel haunts in midmorning when squirrels feed most actively. Where ground squirrels are a major problem, keep annual records of the dates that squirrels emerge from hibernation and when the first young are seen aboveground; changes in the general number of squirrels; and the controls used, dates of use, and their effect. Use these records as the basis for future management decisions.

Management guidelines

When even one or two ground squirrels are present in or immediately adjacent to a nursery, control them; otherwise, damage is inevitable. Fencing is practically useless against squirrels, and no feasible habitat modification within the nursery expels established animals. Unfortunately, ground squirrels are not responsive to chemical or physical repellent methods. Burrow fumigants, poison baits, and traps are the primary means of control.

Habitat modification. In natural habitat, ground squirrels generally feed in open areas where visibility is good (presumably to avoid their natural enemies), although they adapt to other situations. In nurseries, ground squirrels often burrow beneath long-standing piles of pruning refuse, wood, or rock, or use them as harborage. Removing such piles may make the area somewhat less desirable to them, but the base of trees, fence lines, and ditch banks still offer burrowing sites. As such, peripheral cleanup of cover may somewhat reduce the potential for squirrels, resulting in less expense for other more direct forms of control (i.e., trapping, baiting, and fumigation). In addition, it makes burrow detection and population monitoring easier and improves access to burrows during control operations.

Predation. Animals that prey on ground squirrels include coyotes, foxes, badgers, other mammalian carnivores, and several hawk species. Predation, however, is not a significant factor in keeping ground squirrel populations below the level that causes damage.

Trapping. Because trapping is time consuming, it is most practical for small infestations. Several types of kill traps, including modified pocket gopher box traps, tube traps, and Conibear traps, are effective. Place box-type (fig. 16.7) and tube traps on the ground near squirrel burrows or runways. Bait but do not set traps for several days so the squirrels become accustomed to them. After the squirrels are taking the bait, set the trap. Walnuts, almonds, oats, barley, and melon rinds are effective trap baits.

Set unbaited Conibear 110 traps in burrow openings so squirrels will pass through them and trip the trigger (fig. 16.8). Specially designed boxes are sometimes used with baited Conibear traps; the boxes permit the traps to be placed anywhere squirrels are active (fig. 16.9). As with all traps, take precautions to minimize capture of nontarget wildlife and pets.

Live-catch traps, such as wire-cage (fig. 16.10) and multiple-capture traps, can also be used to capture ground squirrels. As with box traps, walnuts, almonds, oats, barley, and

FIGURE 16.7. Gopher box traps set in tandem for California ground squirrels. *Photo:* J. K. Clark, courtesy of UC IPM Program.

FIGURE 16.8. Conibear 110 trap set at burrow entrance of California ground squirrel. *Photo:* R. A. Baldwin.

FIGURE 16.9. Conibear trap set in box with bait to capture California ground squirrels. *Photo:* R. E. Marsh.

FIGURE 16.10. Live-catch trap used to capture California ground squirrels. *Photo:* R. A. Baldwin.

many fruits and vegetables are all effective baits. Because these traps keep ground squirrels alive after capture, they are useful in areas where nontarget captures are a concern (e.g., areas with pets and children). However, ground squirrels must be euthanized by the trapper upon capture as translocation of ground squirrels is illegal, unloads your problem on others, and can spread disease such as bubonic plague. It is this extra step that limits the utility of live-trapping for some individuals. Methods of euthanasia considered humane by the American Veterinary Medical Association include gassing with carbon dioxide and shooting. Drowning is not an approved method of euthanasia.

Fumigation. Treating ground squirrel burrows with toxic gases can be a very effective control method when applied at the right time and in the correct manner. Late winter and early spring are the best times to fumigate for ground squirrels as moist soil is needed to hold toxic gases within the burrow system (see fig. 16.3). Conducting ground squirrel control prior to the birth of young will also dramatically increase the detrimental effect on the population. However, you must wait to fumigate until after ground squirrels have emerged from hibernation; ground squirrels wall themselves off in their burrows when hibernating, so fumigation is not effective at this time. Fumigation is also possible later in the year as long as sufficient soil moisture is present. However, it is ineffective when ground squirrels are estivating during the hottest times of the year, as ground squirrels again wall themselves off in their burrows. For safety and legal reasons, do not use fumigants in burrows that extend beneath occupied buildings; this limits the utility of fumigants in some nursery settings.

Two primary fumigants are used: gas cartridges and aluminum phosphide. Gas cartridges provide an easy and relatively safe way to fumigate ground squirrel burrows (fig. 16.11). They are available commercially and from some county agricultural commissioners' offices. Typically one or two gas cartridges are applied to each burrow that shows sign of activity; more than two may be needed for a large burrow system. For application, light the fuse on the gas cartridge and then quickly shove the ignited cartridge into the

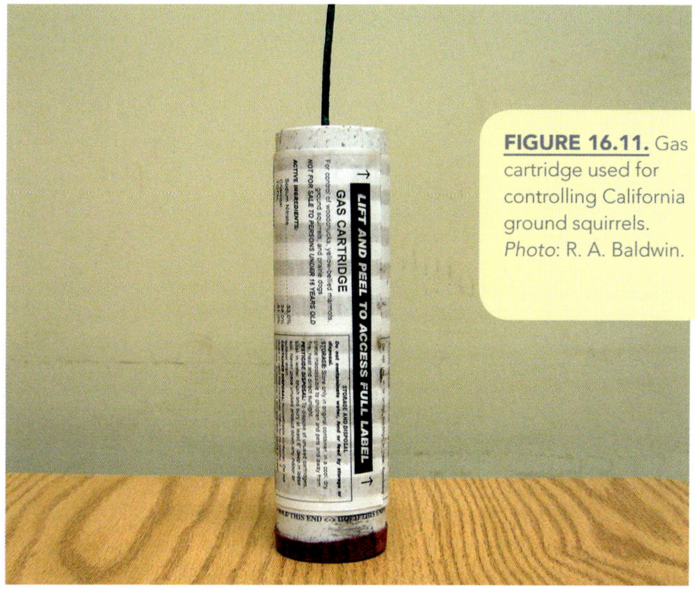

FIGURE 16.11. Gas cartridge used for controlling California ground squirrels. *Photo:* R. A. Baldwin.

burrow using a shovel handle or stick. The burrow entrance should then be sealed with soil, taking care not to bury the gas cartridge with soil. Nearby burrow entrances should be watched for escaping smoke. If other burrow openings emit smoke, they should be treated and sealed as well. The larger and more complex the burrow system, the more smoke it takes to be effective. Gas cartridges do have the tendency to flare up. Therefore, to avoid fires, they should not be used around dry vegetation.

Aluminum phosphide (fig. 16.12) is also a very effective fumigant, often outperforming gas cartridges. When aluminum phosphide tablets come into contact with moist soil in the burrow, they produce phosphine gas, which is highly toxic to any animal. When using aluminum phosphide, treat every active burrow, fill the entrance with a wad of newspaper, and cover with soil. Aluminum phosphide is a restricted-use material in California, for which a permit is required for purchase or use. Application personnel must be trained regarding the material's proper use and potential hazards. The labels for aluminum phosphide are currently being revised to limit applications in residential and landscape settings, and this will

FIGURE 16.12. Aluminum phosphide tablets used to control California ground squirrels and pocket gophers. *Photo:* T. P. Salmon.

FIGURE 16.13. Bait scattered around California ground squirrel burrow using spot treatment method. *Photo:* T. P. Salmon.

likely have some influence on applications in nurseries. Be sure to read the current label to ensure proper application; contact your local county agricultural commissioner's office for further information on label changes.

Poison baits. Poison grain baits have been developed for ground squirrel control, although they tend to be ineffective from the time ground squirrels emerge from hibernation through late spring (see fig. 16.3). During this period, ground squirrels feed extensively on green vegetation and may not accept grain baits. Hence, the time of baiting is critical. In the Sacramento region and more northerly portions of the state, ground squirrels generally switch from eating vegetation to eating seeds, grains, and nuts around the later part of May. However, farther south, this may occur 2 to 6 weeks earlier depending on the amount and timing of spring rains. The month following this dietary switch is the best time for baiting, especially with zinc phosphide baits (see following paragraph). When much of the adult population goes into estivation during the hottest part of the summer, suspend baiting until after September 15. Continue baiting through the end of October. During this time, many squirrels again feed on grain baits until they go into winter hibernation. Various grain baits with one of several poisons are available from commercial distributors or the county agricultural commissioner's office.

Zinc phosphide, a single-dose poison bait, is the most cost-effective toxin for ground squirrels and generally produces results within 48 hours. However, it is a restricted-use material in California, thereby requiring a permit for purchase and use. When applying zinc phosphide, distribute bait by spot-baiting (scattering bait by hand on bare ground to cover 2 to 3 square feet at the side or behind each active burrow) (fig. 16.13) or by broadcasting (scattering bait relatively uniformly over the entire infested area). Broadcasting can be done by hand with a belly grinder-type seeder or with a vehicle equipped with a tailgate-type seeder. Consult the product label for recommended application methods and rates.

Multiple-dose baits (anticoagulant rodenticides such as chlorophacinone and diphacinone) provide effective control when squirrels ingest them in multiple feedings for 5 days or more. Death generally occurs from 5 to 14 days following the first feeding. Eating anticoagulant bait does not immediately affect the squirrel's feeding or activity. To be effective, multiple-dose baits must remain available; effectiveness is greatly reduced if 48 hours pass between feedings. The multiple feedings usually required for a fatal dose, the slow action of the anticoagulants, and the availability of an antidote (vitamin K1) make anticoagulant rodenticides safer to livestock, pets, and children than some other rodenticides.

Use anticoagulant baits in bait stations or, if the label permits, spread it by repeated spot-baiting or broadcast baiting. Bait stations are small structures designed to hold enough bait to provide multiple feedings and to allow the squirrel to enter and feed (fig. 16.14). Bait stations safeguard larger nontarget species by excluding them from the bait.

Historically, anticoagulant baits such as chlorophacinone and diphacinone have not been restricted-use materials. However, starting in 2011, their use became federally restricted. Therefore, you will need to obtain a permit to purchase and use these anticoagulant baits. It is best to consult your local county agricultural commissioner's office before using these anticoagulant baits to ensure that you are applying them appropriately.

FIGURE 16.14. Bait station used for controlling California ground squirrels. *Photo:* J. K. Clark, courtesy of UC IPM.

FIGURE 16.15. Pocket gopher. *Photo:* J. K. Clark.

FIGURE 16.16. Pocket gopher mound with typical fan-shaped appearance. *Photo:* R. A. Baldwin.

Pocket Gophers

Pocket gophers (*Thomomys* spp.) are stout-bodied, short-legged rodents (fig. 16.15). External fur-lined cheek pouches open outside the lips, on each side of the mouth, and are used extensively for carrying food. The head and body measure 6 to 8 inches. They have a short, scantily haired tail. In nurseries and other irrigated lands, females may produce two litters in a single year, with litters averaging about five young.

Pocket gophers are most common in areas of abundant plant growth. They feed primarily on succulent underground parts of herbaceous plants, but they are capable of pulling a 2-foot-tall plant underground to consume it.

This species lives almost entirely underground. Pocket gophers are antisocial and solitary except during breeding and when the young are being raised. Burrow systems may be extensive and include deep main burrows, shallow feeding tunnels, and side tunnels to push out dirt. They create characteristic soil mounds aboveground (fig. 16.16). Main tunnels are normally 10 to 12 inches under the surface but are frequently deeper. Some lead to deeper nests or food storage chambers. The animals plug burrow openings with soil so that the tunnel system is completely enclosed. As a result, the temperature and humidity in the burrow are stable and close to optimal.

Damage

Pocket gophers often invade nurseries, feeding on many garden crops, ornamental plants, vines, shrubs, and trees. A single gopher moving down a nursery row can inflict considerable damage in a very short time. Gophers also gnaw and damage plastic water lines and lawn sprinkler systems; their tunnels can divert and carry off irrigation water, which leads to soil erosion; and mounds on lawns interfere with mowing equipment and ruin the aesthetics of well-kept turf grass.

Monitoring guidelines

Because pocket gopher damage typically occurs belowground, it often goes undetected until plants exhibit stress. By that time the plant may be beyond saving. Gopher activity, however, is readily detected; simply look for fresh mounds of soil. The animals produce these in greatest numbers in the spring and fall when the soil is amply moist.

Management guidelines

Persistent efforts can control pocket gophers and even eliminate them. The preferred control methods are baiting, trapping, and fumigation. No chemical or mechanical repellents have been effective against pocket gophers.

Habitat modification. Permanent ground covers of herbaceous (e.g., clover) and certain grassy plant (e.g., nutsedge) species favor gophers. Removal of the permanent ground cover will do much to reduce their numbers, as food is generally the factor that limits populations. Deep cultivation destroys some burrows, making the nursery less habitable; clean cultivation removes cover for gophers and allows growers to detect new mounds easily. Controlling gophers along fencerows and roadways adjacent to nurseries is important, because young gophers will disperse from there into the nursery.

When portions of nurseries are flood irrigated, very young pocket gophers may be drowned and adults may be forced to the surface, where local natural predators or the irrigator's dog can kill them. By comparison, sprinklers or drip irrigation systems generally favor pocket gophers.

Predation. A variety of predators feed on pocket gophers, although their presence does not usually keep gopher populations low enough to prevent economic damage.

Exclusion. Underground fencing might be justified for valuable ornamental shrubs or landscape trees. To protect existing plantings, bury hardware cloth or ¾-inch mesh gopher wire at least 2 feet deep with an additional 6 inches of mesh or wire bent at a 90-degree angle away from the planting. This will help keep gophers from digging around the fencing boundary. Also extend the fencing at least 1 foot aboveground to deter gophers moving overland. This method is not perfect, however, because persistent gophers can burrow below the wire; also, the wire can restrict and damage root growth of trees.

You can protect small areas such as flower or nursery beds by complete underground screening of the bed's sides and bottom (fig. 16.17). When constructing raised beds, underlay the soil with wire to exclude gophers. To protect individual plants, install wire baskets, which you can make or buy commercially, at the same time as you are putting the plants into the ground. If you use wire, use one that is light gauge and use it only for shrubs and trees that will need protection while young. Leave enough room to allow for the roots to grow. Galvanized wire provides the longest-lasting protection. Six to 8 inches of coarse gravel 1 inch or more in diameter around underground sprinkler lines or utility cables also can deter gophers.

Trapping. Trapping is safe and one of the most effective although labor-intensive methods for controlling pocket gophers. Several types and brands of gopher traps are available. The most common type is a two-pronged, pincher trap such as the Macabee, Cinch, or Gophinator, which the gopher triggers when it pushes against a flat, vertical pan (fig. 16.18). Another popular type is the choker-style box trap. (For directions on trapping gophers, see the *UC IPM Pest Notes* website in the References at the end of this chapter.)

Fumigation. The burrows of pocket gophers are extensive. Portions of these are relatively near the surface; therefore, maintaining a lethal concentration of most burrow fumigants (poison gases) is difficult. Plus, pocket gophers can escape detectable fumigants quickly by plugging their burrows. However, one registered fumigant, aluminum phosphide, has produced very acceptable results. Aluminum phosphide is a restricted-use material, for which a permit is required for purchase or use. Application personnel must be trained regarding the material's proper use and potential hazards.

Poison baits. Single-feeding poison baits (e.g., strychnine and zinc phosphide) placed in the burrow tunnels are widely used for controlling gophers, although results are sometimes erratic. Multiple-feeding anticoagulant baits (e.g., chlorophacinone and diphacinone) are also available for use on pocket gophers. Previous studies indicated less success with anticoagulant baits, given that they require multiple feedings. These baits must be reapplied after the initial bait application to maintain a constant bait supply during the baiting period; this results in greater cost for treatment than with single-dose baits. Regardless of bait used, follow label directions for application methods and amounts. The two methods of bait application are hand-baiting and mechanical baiting.

FIGURE 16.17. Raised nursery bed lined with gopher wire to exclude gophers from plants. *Photo:* R. E. Marsh.

Hand-baiting usually requires a metal probe that is used to locate one of the gopher's tunnels. With a pointed ¼-inch steel rod, probe near the fresh mounds or between two recent mounds to find the burrow. Then enlarge the probe opening with a larger rod or broomstick, and place the label-specified amount of grain-type bait in the burrow; a funnel can also be used to pour the bait into the tunnel. If gophers have infested a large area, use a hand-held bait applicator to speed treatment. Bait applicators are a combination of probe and bait reservoir (fig. 16.19). Once you have located a tunnel using the probe, a trigger releases a predetermined amount of bait into the tunnel. Generally, strychnine bait is used with such an applicator because it dispenses only a

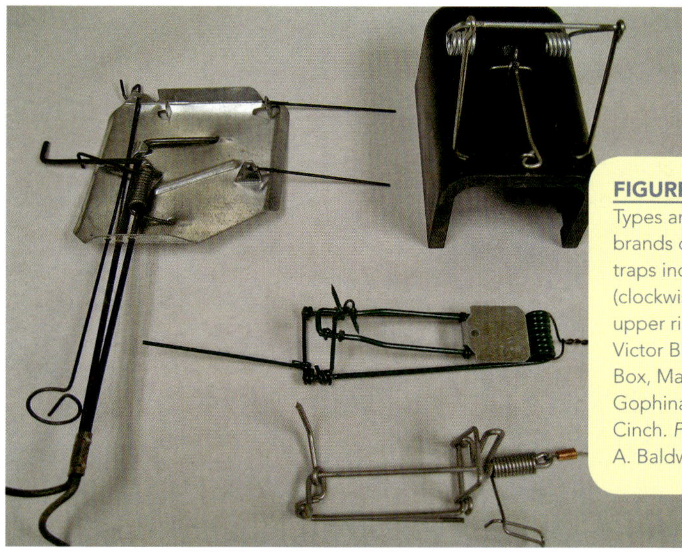

FIGURE 16.1 Types and brands of gop traps include (clockwise fron upper right) Victor Black Box, Macabee Gophinator, ar Cinch. *Photo:* A. Baldwin.

FIGURE 16.19. Combination probe and dispenser used to contro pocket gophers. *Photo:* A. Baldwin.

FIGURE 16.20. Meadow vole. *Photo:* J. K. Clark.

FIGURE 16.21. Vole runway and burrow. *Photo:* R. A. Baldwin.

small quantity of bait at a time. Whichever of these hand-baiting methods you use, be sure to cover the probe hole with a dirt clod, rock, or some other material to exclude light. However, do not cover the hole with loose soil as this will bury the bait, thereby limiting gopher access.

Mechanical bait applicators are tractor drawn and offer an excellent alternative method for controlling gophers over large areas. The device constructs an artificial burrow beneath the soil and deposits poison grain bait within it at preset intervals and quantities. The machine is driven between the rows of nursery plants where pocket gopher activity is seen. The artificial burrow intercepts some of the gopher's natural burrows, or the gopher will soon discover an artificial one and consume the bait. When using the mechanical applicator, use a shovel to occasionally open a small section of the artificial burrow and inspect its depth and condition. Soil must be moist but not overly saturated to produce a well-formed, smooth, artificial burrow.

Meadow Voles

Meadow voles (*Microtus* spp.), also called meadow mice or field mice, are mouselike rodents somewhat similar in appearance to pocket gophers (fig. 16.20). They have a compact, heavy body, short legs, short-furred tail, small eyes, and partially hidden ears. The long, coarse fur is blackish brown to grayish brown. When fully grown they can measure 5 to 8 inches long, including the tail.

Although voles spend considerable time aboveground and may occasionally be seen scurrying about, most of their time is spent belowground in their burrow system. The clearest signs of their presence are the well-traveled, aboveground runways that connect burrow openings; the runways are usually hidden beneath a protective layer of grass or other ground cover (fig. 16.21). The maze of runways leads to multiple burrow openings that are each about 1.5 to 2 inches in diameter (fig. 16.22). If abundant ground cover is present, the runways can be easily found by pulling back overhanging ground cover. Fresh clippings of green grass and greenish-colored droppings (pellets) about 3⁄16 inch long in the runways and near the burrows are further evidence of voles. With age, the droppings lose their green color and turn brown or gray (fig. 16.23).

FIGURE 16.22. Vole burrow openings. *Photo:* R. A. Baldwin.

FIGURE 16.23. Vole pellets on runway. *Photo:* R. A. Baldwin.

Females may produce from 5 to 10 litters a year. A few females breed year-round, but the principal breeding time is during late winter and spring. Because voles mature rapidly and can bear multiple litters yearly, vole populations can increase quickly. Typically the numbers peak every 6 to 8 years, with populations as high as hundreds of voles per acre.

Damage

Voles cause damage by feeding on a wide range of outdoor garden plantings and by gnawing on the bark of many tree species. Vole damage to tree trunks normally occurs from a few inches aboveground to a few inches below ground (fig. 16.24). If the damage is belowground, you will need to remove soil from the base of the tree to see it. Although voles are poor climbers, if they can climb on to low-hanging branches, they may cause damage higher up on trees as well.

Gnaw marks about ⅛ inch wide and ⅜ inch long found in irregular patches and at various angles, taken in conjunction with other signs (droppings, runways, and burrows), indicate vole damage. If voles gnaw completely around the trunk or roots, the tree's flow of nutrients and water will be disrupted; this is called girdling. Girdling damage on trunks and roots can kill trees. Signs of partial trunk or root girdling may include a prolonged time before young trees bear fruit, reduced fruit yield, abnormal yellowish leaf color, and overall poor vigor. Where snow cover is present, damage to trees may extend a foot or more up the trunk. Damage that occurs under snow cover often escapes notice until it is too late.

Monitoring guidelines

Be alert for the presence of voles. Look for fresh trails in the grass, burrows, droppings, and evidence of feeding in the nursery and surrounding area. Pay particular attention to adjacent areas that have heavy vegetation, as such areas are likely sources of invasions.

Management guidelines

To prevent vole damage, you need to manage the population in an area before it reaches high numbers. This can often be achieved by removing or reducing the vegetative cover, thus making the area unsuitable to voles. Removing cover also makes detecting voles and other rodents easier. It is important to act before vole numbers increase rapidly, as the damage these pests do to nursery plants and trees can be quite severe.

Habitat modification. One way to effectively deter vole populations is to make the habitat less suitable to them. Weeds, heavy mulch, and dense vegetative cover encourage voles by providing food as well as protection from predators and environmental stresses. If you remove this protection, their numbers will decline. You can reduce the base area from which voles can invade nurseries by regularly mowing, spraying with herbicides, grazing, or tilling grassy areas along ditch banks, rights-of-way, or field edges adjacent to planted areas. If feasible, weed-free strips can also serve as buffers around areas to be protected. The wider the cleared strip, the less apt voles will be to cross and become established in planted areas. A minimum width of 15 feet is recommended, but even that can be ineffective when vole numbers are high. A 4-foot-diameter circle around the base of young trees or vines that is free of vegetation, or a buffer strip of 4 feet or more along a row of trees, can reduce problems, because voles prefer not to feed in the open.

Predation. Predators such as coyotes, foxes, badgers, weasels, owls, and hawks feed on meadow voles; however,

FIGURE 16.24. Girdling damage caused by meadow voles. *Photo:* J. K. Clark, courtesy of UC IPM.

predation is rarely, if ever, a major factor in controlling a rapidly increasing vole population.

Exclusion. Wire fences at least 12 inches above the ground with a mesh size of ¼ inch or less will help to exclude voles from large areas. These fences can either stand alone or be attached to the bottom of an existing fence. Bury the bottom edge of the fence 6 to 10 inches to prevent voles from tunneling beneath it. A weed-free barrier on the outside of the fence will increase its effectiveness.

Cylindrical wire or plastic trunk guards can protect young trees from voles (fig. 16.25). To hinder burrowing, guards must extend at least 6 inches below the soil surface, but even then voles may dig beneath them. Meadow voles rarely climb over guards. On the other hand, trunk guards can encourage voles by giving them a sense of security. The voles often work beneath them or gnaw in seclusion behind them, where early damage goes undetected.

Trapping. When populations of voles are not overly numerous, trapping can be used to remove voles. Use a sufficient number of traps to control the population: for a

FIGURE 16.25. Tree cover used to protect trunk of tree from vole and rabbit damage. *Photo:* J. K. Clark, courtesy of UC IPM.

small area a dozen traps is probably the minimum number required, and for larger areas at least 50 or more may be needed. A simple, wooden mouse snap trap is all that is needed to capture voles, although trap placement is crucial. Voles seldom stray from their runways, so set traps along these routes. Look for burrows and runways in grass or mulch in or near the nursery. Place traps (bait is usually not necessary) at right angles to the runways with the trigger end in the runway. Examine traps daily and remove dead voles or reset sprung traps as needed. Continue to trap in one location until no further voles are caught, then move the trap to a new location 15 to 20 feet away. When possible, destroy old runways or burrows to deter immigration of new voles to the site. Bury dead voles or place them in plastic bags in the trash. Be sure to keep small children and pets out of the area where you have set traps.

Repellents. Chemical repellents are marketed for voles and have been tested, but no repellents have been effective in protecting nursery plants from voles.

Poison baits. Toxic grain baits are very effective in reducing meadow vole populations. For most effective control, apply bait in the voles' runways, where most feeding occurs. Spot-bait or broadcast bait over the entire infested area. Spot treatments involve spreading bait thinly around the burrow opening, making sure not to pile bait. For broadcasting, use a belly grinder–type seeder or a vehicle with a tailgate seeder. Broadcast application rates vary depending on estimated density of the vole population and type of toxicant. Both single-dose (e.g., zinc phosphide) and multiple-dose (e.g., chlorophacinone) poisons are used for vole control. Consult the product label for application methods and rates. For more specific instructions on vole baiting, go to the *UC IPM Pest Note* for voles on the UC IPM website (see the References at the end of this chapter).

Rabbits

Rabbits are a form of wildlife enjoyed by many people, but they can be very destructive to nurseries. Eight species of rabbits are found in California. Three of these species, the black-tailed hare, or jackrabbit, *(Lepus californicus)*, the desert cottontail *(Sylvilagus audubonii)*, and the brush rabbit *(S. bachmani)* are widespread and cause the majority of problems (fig. 16.26). Because of its greater size and abundance, the jackrabbit is the most destructive, although the cottontail may be a more frequent pest of nurseries.

The jackrabbit is about as large as a house cat, weighing 3 to 7 pounds, with a body length of 17 to 21 inches. It has a grayish brown body, long black-tipped ears, relatively long front legs, and even longer hind legs. The top of its tail is black. The desert cottontail rabbit and brush rabbit are distinguished from jackrabbits by their smaller size and shorter ears. The desert cottontail is 12 to 15 inches long, weighs 1.5 to 2.8 pounds, and has pale gray fur with yellow tints. The brush rabbit is slightly smaller at 11 to 13 inches long, weighs 1.3 to 1.8 pounds, and has brown fur.

A good sign that rabbits are present is coarse, circular fecal droppings, or pellets, found scattered over an area. Pellet

FIGURE 16.26. Common California rabbits (clockwise from top): jackrabbit, brush rabbit, and desert cottontail. *Source:* Storer and Jameson Jr. 1965.

size varies roughly with the body size of the species: jackrabbit pellets are about 0.5 inch in diameter, whereas pellets of the cottontail are closer to 0.25 inch.

Jackrabbits are usually found in open or semiopen areas of California valleys and foothills. They are seldom found in dense brush or woodlands. Within their preferred habitats, jackrabbits are quite adaptable and inhabit areas around the fringes of urban and suburban developments, green belts, golf courses, parks, airports, and agricultural lands. They make a depression in the soil, called a form, beneath a bush or other vegetation, and they use it for hiding and resting during the day.

The food habits of jackrabbits are variable, depending on location and the availability of appropriate plants. Rabbits prefer to eat succulent, green vegetation; grasses and herbaceous plants make up the bulk of their diet. In some areas rabbits eat leaves, bark, or seeds of woody shrubs. Feeding usually begins during the evening hours and continues throughout the night and into the early morning. Jackrabbits can survive without a supply of drinking water.

If food and other necessary resources are found in one place, jackrabbits exhibit no major daily movements. If food sources and areas for shelter are separated, jackrabbits will move between these areas in the morning and evening. Daily travel of 1 to 2 miles round-trip between areas is common. During dry periods, round-trips of up to 10 miles have been observed. These travels are habitually made on the same trails every day, producing noticeable paths through herbaceous vegetation.

Unlike jackrabbits, desert cottontails and brush rabbits generally inhabit places with dense cover such as brushy areas, wooded areas with some underbrush, or areas with piles of rocks or debris. Abandoned structures and sometimes cultivated fields are also used for cover. Open areas are used more at night, while dense cover is used more during the day. The brush rabbit, however, seldom feeds more than a few feet from its cover.

A good habitat, such as a nursery with a clump of low-growing junipers about 30 feet wide, may harbor 10 to 15 cottontails, but normal density is considerably less (an average of one rabbit per acre). Cottontails and brush rabbits do not exhibit the same magnitude of daily travel as seen in jackrabbits, although they do make habitual use of travel lanes within frequently used areas.

Damage

Rabbits can be very destructive in nurseries, particularly when nurseries are bordered by wild or uncultivated lands, as these wild areas provide resting and hiding cover during the day while being within easy travel distance to prime, irrigated food sources. Common foods of rabbits are wide ranging and include a variety of vegetable, tree, berry, herbaceous, and ornamental plantings. Rabbits also gnaw and cut plastic irrigation lines.

Most rabbit damage is close to the ground, except where snow allows rabbits to reach higher portions of plants. Rabbits use their incisors to make a characteristic diagonal 45-degree cut when clipping off woody twigs, buds from saplings, or flower heads. Twig clipping by rabbits is sometimes confused with deer browsing. Deer damage can be identified easily if it occurs above a height that rabbits can reach (about 2 feet) and by careful examination of the damaged twigs. Deer have no upper front teeth and must twist and pull when browsing, leaving a ragged break on the branch. Rabbits clip twigs off cleanly, as if with a knife.

Rabbits tend to gnaw the smooth, thin bark from young trees. The rough bark of older trees discourages gnawing, although old damage and gnaw marks are often present on old bark along with fresh patches of gnawing in areas of younger growth. Gnawing can completely girdle a tree, and clipping can remove the terminal shoot and lateral branches from plants. Damage by cottontails and brush rabbits is often concentrated in areas near escape cover. Jackrabbits, however, will feed far into open areas and can eat 0.5 to 1.0 pound of green vegetation each day.

Tularemia, or rabbit fever, may be carried by rabbits. This disease is relatively rare in humans but can be contracted by handling an infected rabbit with bare hands or by eating insufficiently cooked rabbit meat. Therefore, rabbits should never be handled with bare hands.

Monitoring guidelines

Rabbits are mostly nocturnal, but they are sometimes seen during the daylight hours in areas where they may simply be loafing. Rabbit signs, such as feeding damage, trails, and droppings, indicate their presence. Because few, if any, rabbits can be tolerated in a nursery setting, take appropriate action when the first sign of rabbits is observed. Rabbits that have been seen nearby frequently invade nurseries when the plantings become desirable to them. Therefore, consider exclusion methods or some other control program before serious damage occurs.

Management guidelines

Rabbit control in nurseries primarily includes exclusion, repellents, trapping, and poisoning where legal. Habitat modification can help reduce potential damage for cottontails but is less effective for jackrabbits. The choice of control method should depend on the urgency of the problem and the situation. Manage rabbit populations before severe damage occurs.

Habitat modification. Remove hiding cover to discourage cottontails and brush rabbits, especially in suburban habitats where alternate habitats may be limited. Remove brambles, piles of brush, stones, or other debris where rabbits might hide. Control vegetation along fencerows, ditch banks, or brushy areas. Keep in mind that vegetation management may affect other wildlife, notably songbirds. Removing cover will probably have little effect on jackrabbits because they can use cover that is often great distances from the feeding sites.

Exclusion. Where jackrabbits or cottontails are a constant and continuing threat to young plants, fencing the entire nursery may be the best management approach. If this is not feasible, constructing a fence around selected blocks or beds is also effective. To make an effective barrier, build a fence 36 inches high of woven wire or poultry mesh, using 1- to 1.5-inch-diameter mesh (fig. 16.27). After digging a trench 6 inches deep and 6 inches wide along the fence line, bury 48-inch-wide wire 6 inches deep, leaving a 6-inch lip turned outward at a right angle at the bottom. This will leave a 36-inch-high fence when backfilled with soil, which rabbits will normally not jump; burying the bottom prevents their digging beneath it. If you are constructing a deer-proof fence, the additional expense of a bottom roll of the smaller-mesh wire, properly buried to also exclude rabbits, may be worthwhile.

Temporary fences using silt fencing can be used to exclude rabbits in growing areas where a permanent fence is not required. As referenced above, the bottom must not have any access points where the rabbits can crawl under. Pinning with staples (e.g., those used for nursery ground cover fabric) every few inches or burying the bottom in a trench as previously described will exclude rabbits.

Electric netting (a type of electric fence) is also suitable for rabbit control. It is designed for ease of installation and frequent repositioning. Electric netting is intended for temporary use at any one site, making it ideal for small areas in nurseries.

Predation. There are a number of predators that will feed on rabbits, including coyotes, bobcats, and raptors. However, these predators are rarely able to control rabbit populations to sufficient levels to mitigate damage.

Trapping. Cottontails can be trapped fairly easily using live traps (see fig. 16.10), although capturing jackrabbits in the same manner is much more difficult given their reluctance to enter confined spaces. When using live traps, captured rabbits must be euthanized in a humane manner; relocation of captured animals is not allowable under the California Fish and Game Code. Euthanasia via a carbon dioxide chamber is the preferred method for dispatching rabbits, although

FIGURE 16.27. Woven wire fence used to exclude rabbits. *Photo:* J. K. Clark, courtesy of UC IPM.

shooting (if in an area where shooting is allowed) is another method. Often individuals are not comfortable euthanizing captured rabbits. If this is the case, consider using kill traps or hiring a wildlife pest control operator to remove the rabbits for you.

A few kinds of kill traps are effective for cottontails; Conibear 110 or 120 traps placed in a trap box are a common example. Other less common kill traps are available but may need to be specially ordered. These traps should be prebaited in target areas before setting. Common baits include fruits, vegetables, barley, or whatever foods the rabbits are consuming in the nursery. Once rabbits begin to actively consume the baits, set traps and check daily for captures. On occasion, traps may severely injure but not kill the rabbit. It is very important that the rabbit be euthanized as soon as possible. Also, coyotes or other carnivorous species may try to run off with the trap, to carry the rabbit away; checking traps daily to remove dead rabbits helps to reduce this occurrence. Multiple traps will be needed per site; one trap is rarely enough to capture rabbits. Be aware that traps may present a potential hazard to nontarget species, so use caution and good judgment when setting kill traps.

Repellents. Various chemical repellents can reduce or prevent rabbit damage. They are most useful when applied to trees, vines, or ornamentals. Repellents can be classified as area (odor), taste, or contact (sticky) repellents. Research has shown that repellents with putrescent whole-egg solids are among the best at reducing browsing by rabbits.

Apply repellents before damage occurs and reapply them frequently, especially after a rain, heavy dew, or sprinkler irrigation, or when new growth occurs. In all cases, follow the label directions for the repellent you are using. Effectiveness of repellents often depends on availability of alternative food sources. If additional food sources are abundant, repellents sprayed on target plants may be effective. If additional food sources are scarce, repellents may have little effect.

Poison baits. Poison baits offer an economical way to control large numbers of rabbits in large areas, although results are sometimes erratic. Only multiple-dose poisons (e.g., diphacinone) are registered for use against rabbits, and they come in grain or pelletized formulations. Previously, these baits were registered solely for use on jackrabbits. However, recent changes to some labels now allow for baiting of cottontails in nurseries as well. Be sure to consult the label to verify the legality of using a toxicant for rabbit control.

Place multiple-dose anticoagulant baits in a covered or enclosed self-dispensing feeder or a covered or enclosed nursery flat. Position the feeders in areas frequented by rabbits, such as trails and resting and feeding areas. If rabbits fail to feed after a few days, move feeders to where bait is readily accepted. Keep bait available until all feeding ceases, which may be from 1 to 4 weeks.

Place poisoned bait where other vertebrates—especially children—will not have access. Be aware of all wildlife in the area, such as doves or quail, and take precautions to protect them from poisoning. Carcasses of poisoned rabbits should be picked up and disposed of by deep burying, burning, or by double-bagging in a trash bag and then disposing in the garbage.

Deer

Deer (*Odocoileus hemionus*) (fig. 16.28) can cause significant damage to a nursery where the habitat adjacent to the nursery supports moderate to high populations. Foothill and coastal districts with brush or woodlands that provide cover for deer usually experience the most problems. Nurseries near stream bottoms may also suffer. State wildlife management laws limit options for control. Therefore, controlling deer damage can be difficult and expensive.

FIGURE 16.28. Mule deer. *Photo:* W. P. Gorenzel, courtesy of UC IPM.

Damage

Deer eat a variety of vegetation including woody plants, grasses, and forbs. They also consume fruit, nut, and ornamental trees, as well as shrubs, vines, and vegetables. Deer trample plants and damage young trees and shrubs by rubbing their antlers on trunks and limbs.

Monitoring guidelines

Although deer are large and easily seen, their nocturnal habits may make it necessary to check the nursery at night with a flashlight. Also look for physical signs of deer such as tracks, droppings, trails, and damage to foliage from deer browsing.

Management guidelines

The California Fish and Game Code classifies deer as game animals. If you find them damaging nursery plants, you may request a permit from your local game warden to shoot them, although this method is not generally recommended for problems around residential areas and nurseries. Traps and poisons are illegal and cannot be used. Deterrents such as fences, barriers, frightening devices, and various repellents are recommended and can all be used without a permit. Physical exclusion is by far the best and most reliable way to protect nursery plantings from deer.

Habitat modification. Eliminating suitable cover for bedding and other survival needs is rarely feasible. Deer are highly mobile and may travel ½ mile or more to reach nurseries, especially when they have become accustomed to feeding there.

Exclusion. Fencing is the most effective method for excluding deer from a nursery. A 7-foot-high wire fence usually works. A 6-foot-high mesh fence can be heightened to 7 feet by adding two or three strands of barbed or smooth wire on top (fig. 16.29). Deer may occasionally clear a 7-foot fence when being chased or if the fence is on steep, sloping ground. Electric fences have been used successfully in some areas. Check deer fences periodically to be sure they remain intact; damaged wire, broken gates, soil washout beneath fences, and so on, permit access and must be repaired immediately. Deer that manage to circumvent the fence and get inside may have to be removed by shooting if they cannot be driven out; a depredation permit would be required for this removal. The *UC IPM Pest Note* website for deer has more specific details on fencing (see the References at the end of this chapter).

In many places, protecting individual blocks of container plants or individual field-grown trees may be more practical and economical than fencing an entire area. For example, young field-grown fruit or nut trees can be individually fenced until primary branches grow above the deer's reach, usually 5 to 7 feet above the ground. Poultry wire, heavier woven wire, or strong plastic netting can be attached to two stakes to form a circle around the tree. Plastic trunk protectors are especially useful for young vines and trees. Inspect individual protectors regularly because they can restrict plant growth. In addition, care must be taken to ensure that the protector itself does not damage the vine or tree by causing an accumulation of excess heat or moisture.

FIGURE 16.29. Deer-proof fence. *Photo:* J. K. Clark, courtesy of UC IPM.

Repellents. Many chemical repellents have been tested, but deer usually adjust to them rapidly, especially when hungry. When deer populations increase, they compete fiercely for food, and repellents become totally ineffective. Taste or odor repellents can be somewhat effective, however, if applied to the foliage and treated again as new foliage develops or after rain or irrigation washes the repellent away. Noise-making devices, such as propane exploders and electronic alarms, have not been effective for more than a day or so because deer rapidly habituate to frightening devices.

Shooting. In circumstances where damage from deer is high, depredation permits may be obtained from the Department of Fish and Wildlife, but shooting is rarely a satisfactory solution to a significant deer problem. Additionally, many residential areas do not allow the discharge of firearms, thereby limiting shooting as a tool for deer management.

Birds

Many birds are frequent visitors to nurseries. Typically they cause little damage when foraging for insects and seeds. However, some species, such as white-crowned sparrows (*Zonotrichia* spp.) (fig. 16.30), horned larks *(Eremophila alpestris)*, crows (*Corvus brachyrhynchos)*, and house finches *(Carpodacus mexicanus)* (fig. 16.31), can become substantial pests when they unearth and eat seeds or feed on newly sprouted seedlings. Controlling damage in these situations is necessary to maintain nursery production.

Damage

Birds can reduce newly planted stands in field nurseries and in flats by feeding on seeds and young seedlings. They can also reduce seed germination in containers when seeds are directly sown, and they can damage newly planted plugs, liners, and transplants in containers. Young plants may be nipped off, or small holes may remain in the soil where the seedling was pulled. However, most damage occurs on young seedlings, before seedlings have two or three true leaves.

Horned larks feed on seeds of wild plants and on insects in open grasslands. They move into nurseries when natural forage is scarce or when nurseries are planted close to their habitat. They feed in flocks and can decimate newly planted areas in a few hours. They tend to feed well out into the field and do not concentrate along fencerows or wooded areas. In contrast, house finches and crowned sparrows feed along the edges of fields. Although house finches feed primarily along the field edges, they are often seen in open areas and tend to scatter to high, open perches when alarmed.

Monitoring guidelines

Closely watch newly planted or sprouted areas for bird damage (i.e., nipped seedlings and small holes where seedlings have been removed). Also, regular weekly monitoring through bird counts will help you determine when damage actually starts so you can take action early. Watch for movement of birds into or within the nursery. Keep track of species by count and location seasonally if you have had substantial damage in the past. These records will help you plan control strategies in advance and will provide information on the effectiveness of previous control actions.

Management guidelines

The U.S. Code of Federal Regulations classifies most birds that cause problems in nurseries as migratory nongame birds. As such, they are protected from indiscriminate control. Permits from either the county agricultural commissioner's office or the U.S. Fish and Wildlife Service are required for their removal from nurseries; who provides the permit depends on the depredating species. However, no permit is needed to frighten or exclude depredating or nuisance birds from the area unless they are nesting.

Of the several approaches to solving bird problems in nurseries, exclusion and frightening are the most commonly used. Anticipating the problem is critical because damage is often sudden and catastrophic. Birds may descend without warning and devour newly planted seeds or seedlings before you are aware of the birds' arrival. Preventive action is the key to success.

Habitat modification. Nurseries are attractive to birds, and not much can be done to modify the habitat to discourage them. Adequately pruning shrubs in the landscape or field to prevent overgrowth tends to reduce cover for some birds, as does eliminating brush piles, which can attract them. By observing the activity of the birds in your nursery, you may be able to identify and remove or modify the things that attract them.

Predation. A few raptor species prey regularly on crowned sparrows, horned larks, and house finches. In and around nurseries, however, the avian predators are not numerous enough to have a measurable effect on pest populations. Occasional mammalian predators capture birds when birds are feeding on the ground, but these kills mean little in terms of significant population reduction.

Exclusion. Exclusion is a consistently effective method of reducing or eliminating bird damage in the nursery. Purchase lightweight ¾-inch-mesh plastic netting for this purpose. Suspend the netting over berry vines or small trees to protect the fruit from bird damage.

FIGURE 16.30. White-crowned sparrow. *Photo:* J. K. Clark.

FIGURE 16.31. House finch. *Photo:* W. P. Gorenzel.

Build a frame out of plastic pipe or wood to support the netting above the seedbeds to keep the netting from reducing or interfering with plant growth. The netting can be loosened and folded back to permit harvest. This method can also protect crops from rabbits, squirrels, and neighborhood cats. Arches made of concrete reinforcement bar or long pliable wood also work well to support netting. These should be able to be lifted off, or the netting pulled back, to allow for attending to the plants. To keep birds from getting beneath the netting, make sure that the netting reaches the ground. Do not be careless in placing the net; keep it taut and tight fitting, as birds can become entangled in loose netting. Wire mesh or aviary wire row caps can also be effective for protecting rows of seedlings, and they are stiff enough to not require a support frame. The mesh should be secured to a wooden frame. Row caps can be lifted off as the plants mature and can be used year after year. Inverted plastic strawberry baskets can also serve the purpose. See pages 16 to 19 of Salmon et al. (2006) for additional details and diagrams of exclusionary devices for birds.

Frightening. Twisted reflective tape stretched 6 to 8 inches above the length of a row of seedlings can effectively frighten birds away when the wind causes the reflective tape to shimmer and shine. A row of tin can lids strung on a wire may also frighten birds, as can devices such as large scare-eye balloons. See pages 18 to 20 of Salmon et al. (2006) for additional details and diagrams of visual frightening devices for birds.

If a nursery is in a rural area, auditory devices such as propane cannons, electronic distress calls, shell crackers, and bird bombs can be effective at dispersing some bird species. However, these are noisy devices and not usually acceptable in residential or urban settings. When using auditory frightening devices, do not rely on any single device but rather on a combination of treatment approaches. For example, you may decide to use propane cannons for 5 to 7 days, then switch to bird bombs for the following week. Combining these auditory devices with visual repellents such as scare-eye balloons and reflective streamers may also increase their effectiveness. The most innovative and varied approach to frightening birds will usually be the most effective, as birds rapidly habituate to the same auditory and visual stimulus.

Trapping. Trapping is rarely the solution to bird problems in a nursery, as birds are highly mobile and new arrivals quickly move in if the bird population is reduced. However, if a nursery and its surrounding neighbors are all willing to cooperate on a trapping program, it can be successful at reducing house finch and crowned sparrow populations. A modified Australian crow trap works well for both species (fig. 16.32). Use a combination of canarygrass and rapeseed for bait. Once a few birds are caught, they serve as decoys to attract others. Therefore, keep a few decoy birds in the trap with ample food and water. Trapped birds should be euthanized using carbon dioxide gas from a bottle. House finch and crowned sparrow trapping must be supervised by the county agricultural commissioner.

FIGURE 16.32. Modified Australian crow trap used to capture house finches and crowned sparrows. *Photo:* J. K. Clark, courtesy of UC IPM.

References

Baldwin, R. A., T. P. Salmon, and M. Freeman. 2009. Vertebrate pest management. In L. Ferguson, ed., Pistachio production manual. 5th ed., rev. Davis: University of California, Davis, Fruit and Nut Research and Information Center. 256–269.

California Department of Pesticide Regulation (DPR). DPR website, http://www.cdpr.ca.gov/.

———. Endangered species project. DPR website, http://www.cdpr.ca.gov/docs/endspec/index.htm.

Salmon, T. P. 2009. Vertebrate pest control handbook. Sacramento: California Department of Food and Agriculture, Plant Health and Pest Prevention Services. Vertebrate Pest Control Research Advisory Committee website, http://www.vpcrac.org/about/vertebrate-pest-handbook/.

Salmon, T. P., D. A. Whisson, and R. E. Marsh. 2006. Wildlife pest control around gardens and homes. 2nd ed. Oakland: University of California Division of Agriculture and Natural Resources Publication 21385.

Storer, T. I., and E. W. Jameson Jr. 1965. Control of field rodents on California farms. Berkeley: University of California Division of Agricultural Sciences Circular 535.

UC IPM pest management guidelines. UC IPM website, http://www.ipm.ucdavis.edu/index.html.

UC IPM pest notes. Vertebrate pest notes include information on California ground squirrels, deer, pocket gophers, rabbits, and voles. UC IPM website, http://www.ipm.ucdavis.edu/PMG/menu.vertebrate.html.

University of California Statewide Integrated Pest Management Project. 2002. Integrated pest management for almonds. 2nd ed. Oakland: University of California Division of Agriculture and Natural Resources Publication 3308.

The Impact of Invasive Pests on the Nursery Industry

John N. Kabashima, James A. Bethke, and Steven A. Tjosvold

CHAPTER SEVENTEEN

Some of the most vexing issues confronting the nursery industry are related to invasive pests. With the speed and frequency of international travel and commerce today, the movement of serious agricultural pests has become a common occurrence. Prior to 1989, California gained one new potentially damaging invasive species every 60 days (Dowell and Gill1989), and one new invasive species was reported every 40 days from 1989 to 2013 (M. Hoddle, pers. comm., Dec. 9, 2013). Many introductions are unintentional as people move around with possessions such as plants and animals that can harbor undetected, potentially damaging pests. Unfortunately, there also have been documented instances of individuals smuggling fruit and plant material across borders. In addition, invasive plants and pests have been illegally transported into California by hobbyists and commercial trade.

Photo: J. N. Kabashima.

Nursery production is truly an international business, with new plants being introduced from foreign countries and plant material being imported and exported all around the world. Because pests are very difficult to detect and control and are frequently moved with plant material when plants are sold and transported, international trade can result in the introduction of invasive species that are carried with plants to new areas. Therefore, a significant share of the burden of responsibility for intra- and interstate movement of potentially infested plants lies with the nursery industry, requiring exceptional management practices and in some cases eradication and certification. Furthermore, some plants introduced into the nursery trade for use in gardens and landscapes have become pests, or noxious weeds, themselves.

Regulatory officials are challenged by new introductions of invasive pests. They are tasked with responding to the most serious ones, but many times do not have the resources to fight all of them. Priority is often given to nursery pests that threaten other agricultural commodities or the environment. Shipments of affected plant hosts from nursery operations are often restricted through regulatory actions.

In this chapter, we discuss the various types of invasive pests, the groups of people who work to minimize the introduction and spread of invasive species (e.g., regulators and university, industry, and nongovernmental organizations), and the effect of invasive plants on the production and shipping of nursery plants. We will provide examples of invasive pest species that have been problems in California, including *Phytophthora ramorum* (the pathogen that causes the disease sudden oak death), *Uromyces transveralis* (the pathogen that causes the disease gladiolus rust), invasive snail and slug species, giant reed *(Arundo donax)*, tree-of-heaven *(Ailanthus altissima)*, Diaprepes weevil, and glassy-winged sharpshooter. Nursery growers in states other than California should consult their local Cooperative Extension office, department of agriculture, or the U.S. Department of Agriculture (USDA)'s Animal and Plant Health Inspection Service/Plant Protection and Quarantine (APHIS/PPQ) for specific invasive species and state and federal regulations.

Types of Invasive Pests

The term *invasive pest* generally refers to a species that is introduced to an area, survives, reproduces, and causes economic or environmental harm or harm to human health within the new area of introduction. Many types of invasive species exist, including arthropods, plant pathogens, weeds, and mollusks. The most destructive invasive pests are plant-feeding insects and pathogens, but plants, weeds, and mollusks can also cause significant economic or environmental harm. Usually the establishment into a new area can result in uncontrolled population growth because invasive pests are often free of the natural forces such as predators and parasites that typically limit the population. Establishment of these pests may occur over a long period of time or it may be rather short, and the eventual impact varies considerably depending on the species and the area being invaded, with some species never causing significant harm. It has been estimated that in California alone, invasive pests cost at least $3 billion a year. Some estimates suggest that invasive species cost the United States $138 billion per year (Pimentel et al. 1999).

The most destructive invasive species usually have common biological characteristics, including rapid reproduction, fast growth rate, wide dispersal, tolerance of a wide range of environmental conditions, and the ability to feed on or infect many different hosts. Often the best single predictor of invasive ability is whether a species is already known to be an invasive pest somewhere else.

The unique climate and geography of California provide diverse ecosystems that are perfect for the establishment of new pests and pathogens. At the same time, ornamental nurseries offer a multitude of potential hosts and conducive environmental conditions for invasive pests and diseases that infest or infect these hosts. Moreover, some of the plants sold in the nursery trade for use in gardens and landscapes can become invasive if they escape into wildlands. As a result of the efficient marketing and distribution systems that can transport these plants long distances, the nursery industry is often implicated in the establishment of important invasive species in other ecosystems or agricultural crops.

Many invasive species infect or infest and reproduce on a great number of host plants that include ornamentals and various agricultural commodities. The implications of these invasives can therefore be far reaching. The Diaprepes root weevil, described later in this chapter, is a good example of this kind of threat. Other invasive pests that have a wide host range include the glassy-winged sharpshooter (GWSS), the red imported fire ant (RIFA), the light brown apple moth (LBAM), and *Phytopthora ramorum* (the causal agent of sudden oak death). Each of these pests has resulted in major and costly regulatory action imposed on the nursery industry in California.

Sometimes invasive species reproduce on a single plant species or type, and that may be enough to make them a serious threat to an agricultural commodity. Chrysanthemum white rust caused by the fungus *Puccinia horiana* is found on 12 chrysanthemum species including many commercial floral and garden chrysanthemums, and it is considered a serious quarantined pest in the United States. There have been outbreaks in California and other parts of the United States, and regulatory authorities have coordinated successful eradication efforts. The Asian citrus psyllid (ACP) *(Diaphorina citri)* feeds on citrus or citrus-related plants and can transmit the bacterial pathogen *Candidatus liberibacter asiaticus* that causes huanglongbing (HLB) disease, one of the world's most destructive citrus diseases. In Florida, dissemination by the nursery industry and homeowners, along with favorable weather, resulted in the rapid spread of ACP and HLB throughout the state. Regulatory agencies, the landscape industry, and nursery and citrus production industries in California have learned valuable lessons from the Florida experience with ACP and HLB.

Many agencies—including the California Department of Food and Agriculture (CDFA), USDA, Customs and Border Protection (CBP) within the Department of Homeland Security, and public universities—as well as the nursery and agriculture industries, spend a great deal of time and effort trying to predict which invasive pests will be the next threat to California. Sometimes invasive pests are downgraded because they are established and widespread, making them difficult to regulate. At this point their control becomes an IPM management strategy rather than a regulatory program. Additionally, the pest or pathogen may be downgraded (see rating system described later in this chapter) because, after research and additional time and experience, it is determined that the pest is not as dangerous and damaging as initially expected or control measures such as biological control are effective in managing it.

Many exotic pests are not serious agricultural pests so they are not regulated, even though they may be serious pests of ornamental plants. For example, various types of insects—from sucking to chewing and boring—have seriously stressed eucalyptus trees. This has created a biological threat to this important species and to the budgets of property owners and public agencies that must spend precious resources to control these pests or remove trees that have been killed by them. Moreover, tree death eliminates valuable habitat for other species such as monarch butterflies and various birds, and it poses dangerous hazards to people and structures from fallen limbs and trees. Even though these pests—which include leaf beetles, psyllids, longhorned beetles, and weevils—are not regulated because they do not pose a threat to agriculture, they are of economic importance to production nurseries, which incur the costs of controlling them in order to produce commercially clean plant stock for sale.

Regulations That Address the Invasive Pest Problem

The threats posed by the introduction of invasive species are dealt with in a number of ways. They include regulatory action or exclusion by various federal, state, and county governments, including inspection of imports that are likely to cause threats to local agricultural industries. Many times this regulatory action requires the destruction of the plants being moved. If the organism has successfully survived importation, then the action is likely to be eradication. Finally, if organisms become established, actions are required to contain or suppress the

Photo: J. N. Kabashima

population. Unfortunately, if the pest becomes established, the movement of plants may become severely limited or the regulatory actions on nurseries may become significant.

U.S. Cooperative System to Exclude Exotic Pests

Federal, state, and local governments in the United States have a cooperative system to exclude exotic pests. These agencies also collaborate with international governments and organizations to exclude and control the movement of exotic pests. The U.S. Department of Agriculture Animal and Plant Health Inspection Service (USDA-APHIS) is charged with protecting the United States and constituent states against entry of new pests. Plants and animal products are regulated by APHIS Plant Protection and Quarantine (PPQ). PPQ inspectors and Customs and Border Protection (CBP) within the Department of Homeland Security have responsibility for inspecting luggage and cargo to intercept illegal plant and animal products. The federal government also regulates interstate movement when federal domestic quarantines exist.

A person receiving or bringing nursery stock into the United States is required to have a permit from USDA in conformance with Quarantine 37 and may require clearance at a plant protection and quarantine inspection station listed on the permit. The Code of Federal Regulations (CFR) title 7, part 319, prohibits or restricts the importation into the United States of certain plants and plant products to prevent the introduction of plant pests into the country. The regulations contained in "Subpart–Nursery Stock, Plants, Roots, Bulbs, Seeds, and Other Plant Products" (§ 319.37) prohibit or restrict the importation of living plants, plant parts, and seeds intended for planting (see the Electronic Code of Federal Regulations website in the References at the end of this chapter).

All articles intended for planting or propagation must be accompanied by a phytosanitary certificate of inspection issued by the national plant protection service of the exporting country, with the exception of greenhouse-grown plants from Canada or articles accompanied by a PPQ-issued permit that specifically provides an exemption from such documentation. As an additional safeguard against the introduction and establishment of foreign plant pests and diseases, whenever any imported plant or plant part is subject to destination restrictions, the imported plant material may undergo a period of postentry quarantine, during which inspections for latent disease or pest symptoms are conducted. Federal, foreign, and domestic regulations preempt state laws and regulations; however, states are responsible for the regulation and interstate movement of domestic pest species not known to occur in the state, in addition to the regulation of intrastate movement, control, and eradication measures.

California Regulations to Exclude Pests

Prior to the California Gold Rush, California was relatively free of harmful insects and diseases of cultivated crops. The Gold Rush created a demand for agriculturally important nonnative plants from other states and countries, and with these plants came their pests and diseases. These pests and diseases spread quickly and caused significant damage to the new agriculture industry, resulting in an organized response from the agricultural community. From the first California plant pest control legislation in 1870 and the creation of the state department of agriculture in 1919, the regulatory structure has evolved to the present-day system.

State regulations

The California Department of Food and Agriculture (CDFA) Pest Exclusion Branch is responsible for keeping exotic agricultural and environmental pests out of the state of California and preventing or limiting the spread of newly discovered pests within the state. To accomplish this mission, the Branch has two roles: (1) quarantine regulatory compliance and enforcement, and (2) service to the agriculture industry and the public.

The Pest Exclusion Branch is divided into three different groups:

- the Exterior Pest Exclusion group, which oversees border station inspections of all commercial and private vehicles entering the state
- the Interior Pest Exclusion group, which oversees enforcement of quarantines, inspection of packages at parcel carrier terminals, and phytosanitary certification of exports
- the Nursery, Seed, and Quality Cotton group, which oversees licensing, pest cleanliness, registration and certification, truth in labeling, and quality of cotton

The California Food and Agricultural Code provides the authority for the rejection of plant material found to be infested or suspected of harboring pests when transiting from one state to another. Border station inspectors can reject and destroy any plant material believed to present a risk of pest or disease introduction into California.

Some nursery stock can enter California if the nursery stock has been inspected at the origin by the local state authorities and if California is satisfied that the authorities are qualified to do an inspection and ensure the stock is free from agricultural pests. Shipments coming from a state that performs "Origin Inspection" must be accompanied by the "Origin Inspection" stamp.

The secretary of agriculture of California may issue special permits (section 3154 CAC) granting permission to selected groups or states to ship nursery stock into California. QC 650 is an example of a master permit that authorizes compliance agreements to be written to allow shipment of nursery stock from Hawaii to California. Only firms that have signed a compliance agreement are allowed to ship product. The master permit describes the commodities that can be shipped and the conditions and protocols that must be followed.

County regulations

Locally, the county agricultural commissioner (CAC) is the enforcing officer of all laws, rules, and regulations relative to the introduction into or the spread within the state of plant pests, and this officer is under the supervision of the secretary of food and agriculture. The CAC inspects nurseries and incoming shipments of nursery stock for pests injurious to plants and for proper labeling and condition. The CAC also enforces plant quarantines; a hold can be placed on any nursery stock that is infested or infected with a noxious weed or other type of pest that does not meet the "commercially clean" standard, and nursery stock may be released when sufficiently cleaned. A typical standard is a "few on a few," which means only a few pests on a few plants. If that standard is exceeded, the crop is in noncompliance and may require action.

The CAC issues shipping permits, nursery stock certificates, and other required certificates to facilitate movement of nursery stock in trade. A person receiving or bringing nursery stock into a county is required to isolate and hold the stock for inspection by the CAC unless the stock is moving intrastate accompanied by a valid nursery stock certificate. The CAC must be notified of the shipment, and the plants should not be placed within the common nursery stock until inspected and released. Nursery stock certificates and shipping permits can be used for within-county and interstate shipments.

In addition, the CAC assists nursery personnel with pest control problems. The CAC has the authority to inspect or sample in order to determine compliance with the provisions that pertain to pesticides and pest control operations.

Plant industry guidelines for pest ratings

California is somewhat unique in that it has a pest policy that recognizes the potential for organisms to harm state agriculture, and it varies due to the diversity of California's agriculture and environment. The level of prevention and exclusion is determined by how serious a pest could be once it has been introduced. For instance, an aphid that is ubiquitous and a known pest may be of some harm to California agriculture and may be of moderate concern to regulators. A psyllid that vectors a disease that can kill citrus trees with certainty, however, threatens the entire state's citrus industry, and it will be taken more seriously.

This pest policy rates pests on their environmental, agricultural, and biological significance. The rating is then used by county agricultural commissioners, the CDFA, and others to determine the appropriate regulatory action to be taken against the pest. Local conditions may dictate more stringent action against selected pests at the discretion of the county agricultural commissioner because the pest may affect certain agricultural commodities in specific counties. Pest ratings represent the CDFA's policy on specific pests so that the appropriate action, such as quarantine, eradication, control, holding, or inspection, is standardized by the rating. Pests can be rated with a "Q", "A", "B", "C", or "D," and the rating system is defined as follows:

- Q. An organism or disorder requiring temporary "A" action pending determination of a permanent rating. The organism is suspected of being of economic importance, but its status is uncertain because of incomplete identification or inadequate information. In the case of an established infestation, at the discretion of the director, the Department may conduct surveys and may convene the Division Pest Study Team to determine a permanent rating.
- A. An organism of known economic importance subject to state-enforced action or commissioner-enforced action (when the commissioner is acting as a state agent) involving eradication, quarantine regulation, containment, rejection, or other holding action.
- B. An organism of known economic importance subject to eradication, containment, control, or other holding action at the discretion of the individual county agricultural commissioner. An organism of known economic importance subject to state-endorsed holding action and eradication only when found in a nursery.
- C. An organism subject to no state-enforced action outside of nurseries except to retard spread, and at the discretion of the county agricultural commissioner. An organism subject to no state-enforced action except to provide for pest cleanliness in nurseries.
- D. No action. Parasites, predators, and organisms of little or no economic importance.

The Role of Organizations and Partnerships in Providing Solutions

In addition to pest exclusion regulators, everyone has a part to play to keep exotic and invasive species from coming into the United States, including the nursery industry. Several organizations and partnerships that have formed efforts to help deal with exotic and invasive species are described below. See the References at the end of this chapter for websites providing more information.

California Association of Nurseries and Garden Centers

The California Association of Nurseries and Garden Centers (CANGC) has a standing committee, the Nursery Growers Leadership Board (NGLB), which is made up of member and nonmember grower nursery segments, as well as university and trade association representatives, that speak together as one voice for the entire industry. The Nursery Growers Leadership Board evaluates issues (e.g., plant quarantine issues), recommends positions, and takes action on matters of common concern to all or part of the nursery community. Nursery growers in states outside California are advised to work with their local grower associations on issues related to invasive species.

California Invasive Plant Council

The California Invasive Plant Council (Cal-IPC) is a nonprofit organization formed in 1992 to address one of California's top environmental threats. It works closely with agencies, industry, and other nonprofit organizations. Its active membership includes public and private land managers, ecological consultants and researchers, planners, volunteer stewards, and concerned citizens. Cal-IPC's mission is to protect California wildlands from invasive plants through restoration, research, and education.

California Horticultural Invasives Prevention partnership

In 2004, Cal-IPC helped form the California Horticultural Invasives Prevention (Cal-HIP) partnership. The partnership is facilitated by the nonprofit group Sustainable Conservation, and it includes broad representation from trade associations from the nursery and landscaping industries. The partnership develops voluntary measures to reduce the number of invasive plant species sold in California and to prevent further invasions from horticulture. Participants include industry representatives from nurseries and professional associations, as well as representatives from UC Master Gardener program.

PlantRight

PlantRight is a voluntary, proactive program for the horticultural community to prevent invasive plant introductions through horticulture. PlantRight was designed by the steering committee of California Horticultural Invasives Prevention (Cal-HIP) partnership. Its mission is to communicate the need to transition away from invasive plants in the gardening and landscaping trade. Since 2004, Cal-HIP has brought together leaders from all sectors of the horticulture industry with environmental groups, public gardens and arboreta, scientists, and government agencies to collaborate and find practical, lasting solutions. Cal-HIP brings the best information based on credible scientific research and invites the entire California horticulture industry to join in the effort. Similar programs have been developed in other areas such as Florida and Australia.

University of California Center for Invasive Species Research

The Center for Invasive Species Research (CISR), based at the University of California, Riverside, provides a forward-looking approach to managing invasions in California by exotic pests and diseases. The Center addresses inadvertent introductions of exotic insect pests, plant diseases, weeds, and other noxious organisms (e.g., exotic crabs and mussels), which are a major and continuing threat to California's agricultural, urban, and natural environments, as well as the state's precious supplies of fresh water.

Western Plant Diagnostic Network

The Western Plant Diagnostic Network (WPDN) is a regional member of the larger National Plant Diagnostic Network (NPDN). NPDN consists of five regional plant diagnostic centers located at Cornell University (Northeast); Michigan State University (North Central); Kansas State University (Great Plains); the University of Florida (Southeast); and the University of California, Davis, in partnership with the California Department of Food and Agriculture (Western). Each regional center coordinates data gathering, diagnostic collaboration, and other activities of member land grant institutions, national agencies, and state departments of agriculture. WPDN is a consortium of land grant institutions and state departments of agriculture throughout the western United States and U.S. territories in the Pacific that provide services for plant disease diagnosis, plant identification, and insect/pest identification. WPDN uses a common software interface to process diagnostic requests and share information among diagnostic laboratories.

Biological Agents and Agricultural Bioterrorism

The newest concern by the regulatory and academic community is countering agricultural bioterrorism with crop biosecurity. A "select agent list" has been developed by the USDA Animal and Plant Health Inspection Service (APHIS) in response to the Public Health Security and Bioterrorism Preparedness and Response Act signed into law in 2002. This law directs APHIS to list the plant pathogens considered as high risk for deliberate use against U.S. crops (see the References at the end of this chapter for more information).

Examples of Invasive Plants

Introduction and spread

California boasts the greatest amount of natural botanical diversity of any state in the United States, with nearly 5,000 native plant species. In addition to native species, about 1,300 non-native species have become established in the state. About 200 to 300 of these are weeds of agricultural crops, turf, or gardens. The remaining 1,000 or so are naturalized plants of wildlands or disturbed noncrop areas, some of which are important invasive plants.

Of the species listed on the California Invasive Plant Council Invasive Plant Inventory, about 37 percent were accidentally introduced to the state as contaminants of seed, clothing, equipment, vehicles, soil, ballast, animals, or packing materials. The remaining 63 percent, however, were intentionally introduced as landscape, pond, or indoor ornamentals; aquarium plants; soil stabilization species; animal forage

FIGURE 17.1. Giant reed *(Arundo donax)* is difficult to eradicate and threatens California's riparian ecosystems by outcompeting native species. *Photo:* J. M. DiTomaso.

species; or human food, fiber, or medicinal plants. The majority of these intentionally introduced plant species, such as the tree-of-heaven *(Ailanthus altissima),* came through the nursery industry as ornamental landscape species.

Impact

Invasive plants can cause significant economic and ecological damage in natural areas. From an economic standpoint, invasive species can reduce livestock forage quality and quantity, jeopardize animal and human health, increase the threat of fire or flooding, interfere with recreational activities, or lower land value. In addition, aquatic weeds can also impact the movement and navigation of private and commercial vessels, block irrigation systems, and impede livestock access to water.

Invasive plants can also cause dramatic ecological changes that impact both plant and animal communities. This is often due to landscape transformations that reduce the adaptability and competitiveness of more desired native species. Such transformation can be caused by the excessive use of resources by invasive plants. This includes an increased ability to capture light, consume water or nutrients, or deplete gases (oxygen and carbon dioxide) in aquatic systems. For example, a 10,000-acre infestation of giant reed *(Arundo donax)* on the Santa Ana River in Orange County is estimated to use 57,000 acre-feet more water per year than native vegetation (fig. 17.1).

Regulatory strategies

Some invasive plants are listed on the California Department of Food and Agriculture (CDFA) noxious weed list, allowing them to be regulated by this agency and the local county agricultural commissioner. *Noxious* is a legal term used by the CDFA and the U.S. Department of Agriculture Animal Plant Health Inspection Service (USDA-APHIS). To be considered noxious, a plant has to be listed on a noxious weed list maintained by one or both of these agencies. Listing is typically based on the threat of this weed to agriculture or noncrop areas. It allows these agencies, along with the county agricultural commissioner, to ban, quarantine, or eradicate these plants. In California, the CDFA has started to list invasive plants based on their threat to or impact on wildlands.

Nurseries that buy or receive free soil without documenting the source run the risk of finding that the soil is contaminated with unwanted organisms and noxious weeds, such as alligator weed. Noxious weeds in nursery soil will result in the soil being quarantined and fumigated or treated following strict CDFA regulatory protocol.

Most of the invasive plants that came from nursery introductions are not considered noxious and can therefore still be imported and traded in California without restriction. It is therefore important that nursery growers know which plants are invasive in their area and select appropriate alternatives for their customers. Public education efforts in retail nurseries are also important for communicating the need to transition away from invasive plants in gardens and landscapes.

Examples of Invasive Pathogens

Sudden Oak Death

Introduction and spread

Phytophthora ramorum, the pathogen that causes sudden oak death (SOD), was introduced from an as yet unknown origin. The first cases of SOD were reported in Marin County, California, in the mid-1990s, and the disease has since been detected throughout the coastal woodlands in central and northern California and southern Oregon. Thousands of coast live oak, tanoak, and related oak species have died from the trunk-girdling canker infections, while many other nonrelated hosts exhibit relatively insignificant foliar symptoms including leaf spots and blights (fig. 17.2). The first North American detection of *Phytophthora ramorum* in an ornamental nursery was reported in early 2002 in Santa Cruz County, California.

Biology

Phytophthora ramorum is a funguslike water mold (Oomycete). It has resting spores (chlamydospores) that can survive in soil and plant debris in adverse conditions, but it has other spores (sporangia and zoospores) that can be produced in rainy or similar conditions when foliar lesions are wet for several hours. They can be washed from foliar lesions, spreading and infecting nearby susceptible leaves or plants. They can also move into natural or constructed waterways, where the pathogen can become established and repeatedly detected.

FIGURE 17.2. Foliar symptoms of *Phytophthora ramorum*, the causal agent of sudden oak death, shown on rhododendron (A) and camellia (B). *Photos:* S. A. Tjosvold.

Impact

Simulations predict an expanding SOD infestation over the decade (2010 to 2020) that will likely encompass most of northwestern California and warrant treatment, removal, and replacement of more than 10,000 oak trees, with a cost of $7.5 million. In addition, the property losses to single-family homes are estimated to be $135 million (Kovacs et al. 2011). The economic impact on the nursery industry has not been analyzed, but substantial costly preventive efforts have been made by nursery operators. Moreover, if *Phytophthora ramorum* is detected, regulatory action mandates the destruction of contiguous nursery hosts from the detection locations in the nursery. There have been instances where nursery operators have decided to completely forgo growing susceptible crops.

Regulatory strategies

Federal and state officials have a regulatory program of annual monitoring of nurseries that produce listed nursery hosts or probable hosts (or products such as wreaths, wood mulch, or firewood). These commodities or products can only ship from the regulated areas if they are accompanied by a certificate affirming that they are produced and maintained in an area that has been determined to be free of the pathogen. Best management practices (BMPs) are encouraged to help nursery operators exclude the pathogen and avoid the establishment of the pathogen in nurseries. Members of industry, state departments of agriculture, universities, federal agencies, and scientists have compiled the best management practices to help the various nursery segments produce healthy nursery stock. The BMP manual, which can be accessed on the California Oak Mortality Task Force website (see the References at the end of this chapter), has been written as a living document, open to ongoing revisions as new information or research becomes available.

Gladiolus Rust

Introduction, spread, and impact

The fungus *Uromyces transversalis* causes gladiolus rust (GR) only in the members of the iris family, including *Gladiolus, Tritonia, Crocosmia,* and *Watsonia* species. Spores produced on stems and leaves disperse easily by wind or by brushing against clothing; the fungus can also be spread on surface-contaminated corms, rhizomes, and flowers.

Gladiolus rust was first detected in the United States in 2006 on a gladiolus production nursery in Manatee County, Florida. Later it was found in California in nursery production and landscapes (fig. 17.3). It has been intercepted many

FIGURE 17.3. Gladiolus rust *(Uromyces transversalis)* is occasionally found in California nurseries and landscapes. It is shown here on gladiolus in a garden near a nursery. *Photo:* H. Scheck, Santa Barbara County Agricultural Commissioner.

times on cut gladiolus flowers entering the country as commercial shipments and on passenger baggage from Mexico, Brazil, and other countries. Once detected, entire crops have been destroyed when regulatory eradication efforts are employed. This has resulted in very high, but not quantified, crop losses.

Regulatory strategies

Sanitation and fungicides for prevention are stressed, but regulatory officials have the authority to destroy infected crops. The eradication of GR is considered to be feasible because GR has a very limited distribution, and GR curative fungicides are available.

Examples of Invasive Mollusks

Introduction, spread, and impact

Invasive terrestrial mollusks can be transported on nursery stock to new locations and hosts. The European brown garden snail, *Cornu aspersum* (formerly *Helix aspersa*), is the most common snail causing problems in California gardens and economic damage in nurseries, citrus orchards, and other crops. It was introduced from France during the 1850s for use as food. Another troublesome snail is the white garden snail, *Theba pisana* (fig. 17.4). It currently is established only in San Diego County but has been found in Los Angeles and Orange Counties as well. Amber snails (Succineidae), a group of terrestrial native and non-native species, are sometimes plant feeding; but most species are nuisance pests, as they are numerous and disperse readily during and after transport. Several species of slugs also cause damage, including the gray garden slug *(Deroceras reticulatum*, formerly *Agriolimax meticulatus)*, the banded slug *(Lehmannia poirieri)*, the three-band garden slug *(L. valentiana)*, the tawny slug *(Limacus flavus)*, and the greenhouse slug *(Milax gagates)*.

FIGURE 17.4. White garden snail, with ivory white shell and brown bands, on ice plant (*Carpobrotus* sp.) in San Diego County, California. *Photo:* D. Rosen.

Regulatory strategies

A certificate of quarantine compliance (CQC) for the European brown garden snail and other plant-feeding snails and slugs must be issued by local regulatory officials. The CQC must state that compliance has been determined by official inspection immediately prior to shipment and that covered commodities were found to be free of all life stages of exotic phytophagous snails, or that commodities originate from an area determined by official inspection to be free of exotic phytophagous snails.

Examples of Invasive Arthropods

Two invasive species that have severely impacted the nursery industry in California are the Diaprepes root weevil (DRW) *(Diaprepes abbreviatus)* and the glassy-winged sharpshooter (GWSS) *(Homalodisca vitripennis* [formerly *H. coagulata*]). They are examples of invasive species with a large number of ornamental hosts. These invasive pests demonstrate the difference that government and industry funding can make in how such problems are dealt with.

Diaprepes Root Weevil

Introduction and spread

The Diaprepes root weevil (DRW) is a native of the Caribbean region. It was probably introduced into California on mature landscape stock from Florida that was taken directly to the site and planted. Two isolated populations of DRW were first found in Newport Beach (Orange County) and Long Beach (Los Angeles County) in 2005; DRW was discovered in San Diego County in 2006.

Biology

The DRW spends the majority of its life below the soil surface as a larva, pupa, or teneral (newly emerged from the pupa) adult (fig. 17.5). In citrus, the larvae initially feed on the smaller roots, then progress to large roots as they mature, eventually girdling the roots and killing the plant. For most plants, symptoms (e.g., wilting, defoliation, or yellowing) are not obvious until it is too late. The larvae complete 10 or 11 instars, with the most aggressive feeding occurring in instars 3 through 9. The mature larvae (instars 10 and 11) feed very little, as they enter a quiescent, prepupal period. Larval development can take from 6 to 18 months. The pupae remain in pupal chambers in the soil for 15 to 30 days. The adults are long-lived, and the females are very fecund. During her lifetime, an adult female can lay thousands of eggs in batches of 25 to 250, which are protected between two leaves that are glued together.

FIGURE 17.5. Late instar larva or prepupa and adult Diaprepes root weevil (*Diaprepes abbreviatus*) held in hand next to weevil-chewed root excavated from soil. *Photo:* D. Rosen.

Impact

DRW could severely impact the ornamental nursery industry because California considers it a pest that is a significant threat to the citrus industry. Unfortunately, the host range of this weevil is extensive, comprising 293 species of plants in 59 plant families; the host range is increasing as it comes in contact with plant species in California that it has never encountered before. The biggest threat to the ornamental industry has been the strict regulations on movement of nursery stock. This is due to the potential of transporting these insects as immatures in potted plants to noninfested areas.

Approximately 38 percent of nursery stock production in California occurs in the three counties infested with DRW (San Diego, Orange, and Los Angeles Counties, with most of the production in San Diego County). One of the decisions that the CDFA had to make was whether or not to impose a quarantine of whole counties affected by DRW. If a quarantine of whole counties had been imposed, this would have had a tremendous impact, as all nurseries in these counties would have then been required to make additional insecticide treatments on their entire crop whether they were infested or not, resulting in an increased insecticide load in the environment and increased costs associated with treatment applications for every nursery. Prior to the adoption of the CDFA Approved Nursery Treatment Protocol (table 17.1), untreated nursery stock was required to be held for a period that varied from 2 weeks to 6 months before movement from the facility was allowed, depending on the size of the container and length of the growth phase, which could cause significant loss of income due to the reduced marketing period.

It was estimated that the average cost to meet quarantine regulations for nurseries in a quarantine area free of DRW would be $300 per acre. The additional cost for a nursery infested with DRW was an estimated $225 per acre, and the total cost to meet quarantine regulations was $525 per acre. For nurseries with infestations, an additional foliar spray treatment with carbaryl would have been required. The per-acre cost for the foliar spray treatment was provided by the industry and was estimated to be $599 to $625 an acre. Because only a portion of the total acreage would need to be treated, the estimates of the total per-acre costs to the nursery industry were calculated assuming that 10 percent of all acreage had infestations. Applying the percentage decrease in revenues per acre to the total gross revenues for nursery crops required to meet quarantine regulations, the annual losses were estimated to be $1.05 million for the floriculture industry and $4.37 million for other nursery industries. Total annual losses are estimated to be $5.42 million.

Regulatory strategies

A Voluntary Treatment Protocol was developed by the CDFA, leading to the adoption of the Approved Nursery Treatment Protocol, which requires a mandatory soil treatment to eliminate DRW larvae in the media and a foliar treatment to eliminate the possibility of egg deposition and eclosion (the emergence of the adult weevil from its pupal case). The state interior quarantine was rescinded, largely due to lack of funding; however, detection of DRW in a nursery requires the nursery to follow the nursery treatment protocol for a B-rated pest (an organism of known economic importance subject to eradication, containment, control, or other holding action at the discretion of the individual county agricultural commissioner). The timeline illustrated in table 17.1 demonstrates the potentially serious impact that an invasion of this sort can have on ornamental plant production industries.

Table 17.1. Regulatory timeline for the Diaprepes root weevil (DRW) ***(Diaprepes abbreviatus)***

<table>
<tr><th>Date</th><th>Regulatory action</th></tr>
<tr><td>May 2, 2006</td><td>The California Department of Food and Agriculture (CDFA) issued two hold notices and eight pending hold notices in San Diego County, and it discussed the possible quarantine boundaries.</td></tr>
<tr><td>May 8, 2006</td><td>Nursery industry organizations were contacted, and discussions began among the CDFA regulatory staff about a possible treatment protocol for nurseries. The CDFA agreed to limit the quarantine boundaries in San Diego County to infested areas because the initial infestations were not near any nurseries and because of the severe economic impact it would have if all of the approximately 700 production facilities in the county were to be quarantined; other factors, such as the pest's long life cycle, limited flying ability, and tendency to build on a few plants before dispersing, were also considered when making this decision.</td></tr>
<tr><td>Sept 11, 2006</td><td>CDFA began treatments during this week in a 200-meter radius around all confirmed sites where DRW had been found in San Diego County, as well as in Los Angeles and Orange Counties. Initial treatments consisted of both foliar and soil insecticide applications. Because a treatment protocol for infested nurseries had not yet been established, a meeting with select nursery owners and managers, CDFA, UC Cooperative Extension, and CAC staff was organized by the Farm Bureau in San Diego County. Representatives from CDFA came from Sacramento to discuss possible ramifications to nurseries if they fell within a DRW quarantine area or if the weevil was found in a nursery. Since all protocols were tentative at that time, it was suggested that CDFA staff developing the protocols visit San Diego County nurseries to gain a better understanding of the diversity of the industry before developing a protocol. A 2-day tour of a dozen local ornamental producers representing a wide variety of plant and production types was arranged and conducted by the Farm Bureau.</td></tr>
<tr><td>Dec 8, 2006</td><td>The first case of an infested nursery was confirmed by the CDFA. A draft protocol was used to put the entire nursery on hold since only a draft of the nursery treatment protocol was available at the time. It took approximately 2 weeks to treat the affected area and another 2 weeks to release the hold on plants. Sales were also halted in a garden center attached to the nursery.</td></tr>
<tr><td>Feb 3, 2007</td><td>An Eradication Project Work Plan was developed by the CDFA with the help of the DRW project members, which included representatives from the University of California, University of Florida, CDFA, and the affected industries. The Work Plan included educating the public and landscapers, informing the ornamental plant industry of actions needed to comply with the quarantine order, and developing treatment protocols for backyard, commercial citrus, and landscape plants.</td></tr>
<tr><td>Mar 7, 2007</td><td>Three protocols were developed, each depending on whether DRW was present or not. Nursery stock infested with any life stage of DRW was placed on hold. The hold could be lifted only after a nursery had completed 6 months of the treatments and no life stages of DRW were detected in the nursery. Generally, all plants produced at the nursery were required to be treated in one of the following ways:<ul><li>All plants that were started from seed, cutting, or bare-root transplants could be planted in growing media with an approved pesticide incorporated at a rate to control DRW larvae and could be certified for movement immediately after treatment.</li><li>Containerized plants that were grown and shipped in 4-inch or smaller containers that had been continuously treated with an approved pesticide as part of a pest management program could be certified for movement immediately after treatment.</li><li>All containerized plants with a growing cycle of 6 months or less could be treated with approved growing media and foliar pesticides according to a table that had the pesticide treatment recommendations listed for the holding period.</li><li>All containerized plants with a growing cycle of more than 6 months needed to be treated with an approved growing media pesticide (or pesticide/nematode combination) and foliar pesticides according to a table that had the pesticide treatment recommendations listed for the holding period (which in some cases was as much as 6 months).</li></ul></td></tr>
<tr><td>Aug 2, 2007</td><td>A Voluntary Treatment Protocol was developed by the CDFA, leading to the adoption of the Approved Nursery Treatment Protocol. This treatment protocol allowed movement of long-term potted crops without an extended hold.</td></tr>
<tr><td>May 3, 2008</td><td>Infested areas that were quarantined were initially provided over $4 million for the eradication effort. However, because of the state budget crisis, the CDFA proposed a budget in May for 2008–2009 that included a $3.66 million reduction for DRW activities, which left only $590,000 for regulatory activities to control the pest.</td></tr>
<tr><td>Apr 29, 2009</td><td>The state interior quarantine was rescinded largely due to the state's financial struggles and lack of funding to continue the quarantine. However, detection of DRW in a nursery would require the nursery to follow the nursery treatment protocol for a B-rated pest.</td></tr>
</table>

Glassy-Winged Sharpshooter

Introduction and spread

The glassy-winged sharpshooter (GWSS) probably first entered California as eggs deposited into plant tissues of agricultural plants. The initial establishment of GWSS in southern California occurred sympatrically with the morphologically similar, native smoke tree sharpshooter (STSS). Neither GWSS nor STSS causes noticeable feeding damage to plants, and it was not until local GWSS populations attained extremely high nuisance levels in urban landscapes that GWSS was first collected near Irvine in 1989. However, it was not recognized as a newly introduced species until later, when the CDFA identified it as a species common in the southeastern states, from Florida through eastern Texas and as far north as Missouri. The GWSS has had a major impact on agriculture in southern states because it is the prime vector of the bacterium *Xylella fastidiosa* on peach and grape. By the time the invasive pest was correctly identified in California, it had already widely established itself and spread throughout the southern part of the state, and tremendous populations could be found on a wide variety of host material.

GWSS has been seen in high numbers in citrus along the coast of southern California since the early 1990s. Recently it has become locally abundant farther inland, in Riverside and San Diego Counties. High populations on citrus and adjacent vineyards were found in southern Kern County in 1998 and 1999. GWSS could spread north into the citrus belt of the Central Valley, becoming a permanent part of various habitats throughout northern California. Unregulated, it could become established along the coast and inland at least as far north as southern Mendocino County.

The GWSS ranges over many habitats, including agricultural crops, urban landscapes, native woodlands, and riparian vegetation. It is reported to feed on hundreds of plant species. Hosts for GWSS vary widely. They include woody plants and annual and perennial herbaceous plants, and the pest occurs in unusually high numbers on citrus. Common landscape and garden host plants include bird-of-paradise, eucalyptus, euonymus, citrus, crape myrtle, pittosporum, sunflower, hibiscus, xylosma, and cottonwood, among many others. Host preference changes according to availability and nutritional value of the host plant at any given time. Well-irrigated and well-fertilized plants may become a host when in other situations they would not.

Biology

GWSS adults are about 0.5 inch long. They are generally dark brown to black when viewed from the top or side; the abdomen is whitish or yellow; and the head is brown to black and covered with numerous ivory to yellowish spots. Immature stages (nymphs) are smaller than the adult. They are wingless, uniformly olive gray in color, and have prominent bulging eyes (fig. 17.6). As they feed on xylem tissue, they excrete a large amount of liquid substance that drops to the leaves or the ground below ("hopper rain").

FIGURE 17.6. Glassy-winged sharpshooter nymph and adult on crape myrtle *(Lagerstroemia indica)* stem. *Photo:* J. K. Clark.

Females lay their eggs in masses of up to 28 in the lower leaf surface of young leaves that have recently expanded. The egg mass initially resembles a greenish blister on the leaf, covered with a white, chalklike secretion. Shortly after egg hatch, the leaf tissue that contains the egg mass turns brown, and the dead leaf tissue remains as a permanent brown scar. Nymphs emerge in 10 to 14 days. They feed on leaf petioles, small stems, and leaves, progressing through five molts before becoming winged adults. There are two generations a year.

Impact

GWSS is an insect pest that attacks urban, nursery, agricultural, and native trees and shrubs. In the 1980s and 1990s, many nurseries moved from densely populated urban areas to citrus-producing areas of southern California, in close proximity to one of the major overwintering sites for GWSS. Citrus producers were suffering through a period of low profits and did not want to spray unnecessarily for GWSS because of the very effective biological and integrated pest management programs that would be disrupted by insecticide applications. As a result, nurseries that moved to citrus production areas were inundated with waves of GWSS. This was an extremely difficult situation because GWSS has to be controlled in order to ship to noninfested GWSS counties.

Moreover, several mutations of *Xylella fastidiosa,* the bacterial pathogen that the GWSS vectors, have developed. These could result in costly losses to a wider variety of crops and to urban landscapes. Native riparian, urban, and agricultural plantings are now potential reservoirs of GWSS and *Xylella* spp., which can potentially affect their neighbors. Thus far, strains of *X. fastidiosa* that cause oleander leaf scorch, Pierce's disease of grapevines, almond leaf scorch, and alfalfa dwarf have been identified in California.

A University of California survey was used to document the economic impact of the GWSS on nursery growers. Out of 35 nursery operators who responded to this survey, 8 operations had to destroy plants due to the presence of GWSS from 2007 to 2009. The total cost of the destroyed plants in these nurseries—including wholesale value of lost plants, labor, and materials—was $95,500. While the average loss per infested nursery was $10,611, individual loss varied widely by nursery from a low of $1,500 to a high of $21,200.

Almost 30 percent of the operators who responded to survey questions regarding pest management practices used some type of barrier method (e.g., shade cloths, insect screens, hedges) or some combination of barriers to prevent the entry of GWSS. Half of the operators conducted in-house monitoring of the traps installed by the CDFA. Monitoring varied from as frequently as once a week to as little as every other month. Over 50 percent of the nursery operators who responded to the survey applied pesticides to manage the GWSS if it was found in their nursery. Total pesticide application costs varied from $175 per acre to $2,160 per acre, for an average of $940 per acre (unpublished survey data).

The GWSS shipping protocols to the nursery industry add considerable labor, which significantly increases costs. Added labor includes

- inspecting 100 percent of the plants being shipped from southern California to northern California
- applying pesticides to control GWSS in the nursery and eradicate it from all plants being shipped
- completing paperwork required to document and comply with the regulations governing the movement of nursery stock
- spending innumerable hours on the part of dedicated nursery owners who served on committees to help shape regulations and identify research needed to keep plants moving from southern California to northern California

Regulatory strategies

When GWSS was found to be the vector of oleander scorch, a request was made by the University of California, the landscape trade, and Caltrans to have the CDFA initiate an eradication program; however, because of the wide distribution and high numbers, it was determined that eradication was not feasible. Eradication would have been a viable policy option if the CDFA, the University of California, and the grower community had an effective monitoring and interception program that could have correctly identified and recognized the threat of GWSS to the urban, nursery, and agricultural industries. It was not until the devastation of several grape vineyards in Temecula, California, and the documentation of GWSS in eight southern California counties in a CDFA preliminary field survey conducted in 1999, that GWSS was recognized as a major threat to the California wine and grape industry. The regulatory timeline presented in table 17.2 demonstrates the potential serious impact of GWSS on the ornamental plant industry.

Lessons Learned from DRW and GWSS Infestations

In the case of the Diaprepes root weevil (DRW) and the glassy-winged sharpshooter (GWSS), a considerable time delay occurred between the initial infestation and the beginning of eradication efforts (pesticide treatments). The slow response to the DRW invasion most likely allowed significant dispersal during the period it took the CDFA to develop an urban treatment protocol and a nursery treatment protocol for DRW. These examples demonstrate the need for proactive protocol development so that delays in action do not occur.

The DRW is an example of an insect pest that spreads slowly (in comparison to a pest like GWSS); it was confined to a three-county area and posed an economic threat to a small but important agricultural commodity. The GWSS is an example of an insect pest that spreads quickly; it infested an eight-county area and demonstrated an ability to devastate grape vineyards by vectoring the bacteria that causes Pierce's disease, posing a major threat to one of the most valuable agricultural commodities in California. Both pests resulted in very expensive and time-consuming regulations being imposed on nurseries within the regulated areas.

A noteworthy success shared by the DRW and GWSS projects was the cooperation among the nursery and agriculture industry representatives, university, CACs, and CDFA. In the case of DRW, the California nursery industry organized a tour for CDFA policymakers to demonstrate the great diversity of production and plant types in the ornamental production industry. The tours also demonstrated that there could not be a one-size-fits-all approach to protocol development for this industry, while the collaborative involvement of nursery, university, and regulatory personnel helped identify areas of research and funding needed to support the approved treatment protocols.

To address the daunting task of developing protocols and scientific information to allow the southern California nursery industry to ship plants to the areas of northern California not infested by GWSS, the CDFA selected nursery, wine grape, university, agricultural commissioner, and CDFA representatives to form the GWSS Nursery Task Force. The makeup of this committee rapidly facilitated the development of questions and scientific answers. Through an iterative review and validation process, a protocol was developed that has resulted in 99.9 percent of nursery stock shipped to non-infested areas of northern California arriving free of any live GWSS.

The nursery industry in California is made up of widely diverse crops and cropping systems distributed over a large geographic area, which has made it difficult for the various segments to identify common issues and threats as an industry, unlike other commodities that focus on single crops. Therefore, in spite of producing a top agricultural commodity, it has been difficult for the nursery industry to organize and address regulatory issues. However, when the nursery industry has been able to partner with other commodity groups and organizations, it has been able to raise funds needed to

Table 17.2. Regulatory timeline for the glassy-winged sharpshooter (GWSS) *(Homalodisca vitripennis)*

Year	Regulatory action
2000	In response to Pierce's disease, vectored by GWSS and killing several hundred acres of grape vineyards in Temecula, California, nurseries shipping plants from infested areas of southern California to uninfested areas of northern California were required to ship under a "Blue Tag." The restriction required that the plant material be held for inspection by the destination county's agricultural commissioner's office. Counties enforcing the Blue Tag restriction were Alameda, Contra Costa, Fresno, Lake, Mendocino, Monterey, Napa, San Joaquin, Sonoma, and Tulare. Under the Blue Tag protocol, if GWSS was detected the load was rejected and had to be reloaded on the truck and transported out of the county back to the sender or to another county not requiring Blue Tag protocol, and only with the permission of the new destination county. Shortly after the implementation of these shipping restrictions, GWSS was changed from a C-rated pest to a B-rated pest. The wine grape industry supported passage of Senate Bill 671, establishing the Pierce's Disease Control Program (PDCP). It also worked at the national level for a declaration of emergency, providing $22.3 million to counter the spread of GWSS and support research to find a cure for Pierce's disease. The Kern County pilot project to conduct an area-wide suppression and eradication program was funded by this money. CDFA then developed and implemented a statewide compliance program and the California Agricultural Commissioners and Sealers Association/Glassy-Winged Sharpshooter Advisory Group was formed.
2001	Public scoping meetings were held throughout the state to maximize opportunity for public input into the Environmental Impact Review (EIR) process that addressed the environmental impacts potentially resulting from implementation of alternatives such as pesticides for the eradication of the GWSS under the Pierce's Disease Control Program. Assembly Bill 1394 was signed, bolstering research and other program activities by providing approximately $5 million per year for 5 years from an assessment levied on wine grape producers; the Pierce's Disease and Glassy-Winged Sharpshooter (PD/GWSS) Board was established to develop recommendations on the use of collected monies. The University of California and CDFA jointly conducted research to find GWSS egg parasitoids. The CDFA organized the first annual Pierce's Disease Research Symposium, bringing together researchers from throughout the United States.
2002	The Pierce's Disease Control Program released its draft environmental impact report for the public review and comment. This was followed by news that the first localized GWSS infestation was eradicated in Brentwood (Contra Costa County). A Pierce's Disease/GWSS Science Advisory Panel was convened in Sacramento to review the CDFA program and develop recommendations. Shortly thereafter, new emergency program regulations were filed and an area-wide GWSS management program was implemented in Ventura County; the area-wide control program in Kern County was expanded.
2003	Permanent program regulations governing the movement of nursery stock, bulk grapes, and bulk citrus were adopted, and the Coachella Valley Area-Wide GWSS Program was begun. Experiments to test the efficacy of preshipment treatment of nursery stock in killing GWSS egg masses and emerging nymphs, funded by the wine grape assessment and CDFA, were conducted as part of the program to help California nurseries with their efforts at keeping nursery stock shipments free of GWSS.
2004–2005	The National Academy of Sciences (NAS) released its independent scientific review of the Pierce's Disease Control Program. GWSS egg parasitoids *Gonatocerus morrilli* and *Anagrus epos* were released, and the Nursery Treatment Pilot Program began.
2006	The assessment to fund research earmarked for PD/GWSS was renewed, and the PDCP consolidated its research proposal solicitation and review process with the University of California's Pierce's Disease Research Grants Program. Five counties participated in the Nursery Treatment Pilot Program; GWSS infestations in Butte County, as well as the cities of Fowler, Fresno, and Chico, were declared eradicated.
2007	The PD/GWSS Board set up a Research Scientific Advisory Panel (RSAP), and the first PD/GWSS Research Scientific Review was published by the RSAP. GWSS infestations in Vacaville (Solano County) and the Three Flags area (Imperial County) were declared eradicated, and biological control agents were released in over 13 sites throughout California. There were 60,600 shipments of nursery stock from infested areas to uninfested areas that were inspected; statewide, no new infestations of GWSS were found, and 35 counties were verified as being free of GWSS infestation.
2008	The Nursery Stock Approved Treatment Program (ATP) was implemented for nurseries in GWSS-infested areas, and full-time staffed border inspection stations increased from 2 to 16. SB2 passed, which extended and expanded the current Pierce's Disease Control Program if growers were to approve a referendum on the research-oriented program in 2010.
2010	The referendum was approved, extending the Pierce's Disease Control Program to do research and outreach on other invasive pests that affect wine grapes.

- influence the state and federal governments to allocate resources needed to initiate effective interception programs
- institute appropriate preventive efforts to limit introductions of exotic urban pests
- fund research to understand the biology of the pests and develop science-based prevention, eradication, and control programs
- implement quick and effective eradication efforts
- develop risk assessments for potential pests of landscape and nursery plants

The California nursery industry will likely face the same scenarios many times until it finds a way to organize and develop the resources needed to effectively deal with the inevitable new invasive pests in the future. The industry should look at the wine grape industry as an example of what it takes to initiate effective interception programs, institute appropriate preventive efforts to limit introductions of exotic urban pests, develop risk assessments for potential pests of landscape and nursery plants, and influence the state and federal governments to allocate resources needed to implement quick and effective eradication efforts.

The recession from 2008 to 2012 resulted in dramatic reductions in the budgets of regulatory and NGO agencies. It also reduced the ability of those producing agricultural commodities to deal with invasive species. Without the resources of the prerecession years, a new paradigm will need to be developed to deal with invasive pests. Development and adoption of best management practices and systems approaches, reorganization of resources to accomplish more with less, public-private partnerships, and new ways of regulating nurseries will be required to deal with the ever-present threat of invasive pests.

References

Bell, C. E., J. M. DiTomaso, C. A. Wilen, and M. L. Flint. 2007. Invasive plants. Oakland: University of California Division of Agriculture and Natural Resources Publication 74139. UC IPM website, http://www.ipm.ucdavis.edu/PDF/PESTNOTES/pninvasiveplants.pdf.

BugSpot.org website, http://www.bugspot.org/.

California Horticultural Invasives Prevention (Cal-HIP) partnership. California Invasive Plant Council website, http://www.cal-ipc.org/landscaping/calhip.php.

California Invasive Plant Council website, http://www.cal-ipc.org/about/index.php.

California Oak Mortality Task Force website, http://www.suddenoakdeath.org/.

California Oak Mortality Task Force. Sudden oak death best management practices. California Oak Mortality Task Force website, http://www.suddenoakdeath.org/diagnosis-and-management/best-management-practices/.

CANGC (California Association of Nurseries and Garden Centers). CANGC website, http://www.cangc.org/.

CDFA (California Department of Food and Agriculture) Diaprepes Root Weevil Pest Profile. CDFA website, http://www.cdfa.ca.gov/phpps/pdep/target_pest_disease_profiles/Diaprepes_PestProfile.html.

———. Division of Plant Health and Pest Prevention Services. CDFA website, http://www.cdfa.ca.gov/plant/.

———. GWSS Nursery Stock Approved-Treatment Manual (BMPs) / Glassy-Winged Sharpshooter Nursery Shipping Protocol. CDFA website, http://www.cdfa.ca.gov/pdcp/guidelines.html.

———. Noxious Weed Ratings. CDFA website, http://www.cdfa.ca.gov/plant/ipc/encycloweedia/winfo_weedratings.htm.

———. Nursery Inspection Procedures Manual. CDFA website, http://www.cdfa.ca.gov/plant/pe/Nursery/NIPM.html.

———. Nursery Services Regulatory and Inspection Activities and Certification Services. CDFA website, http://www.cdfa.ca.gov/plant/pe/Nursery.

———. Nursery Stock Approved Treatment Program. CDFA website, http://www.cdfa.ca.gov/pdcp/guidelines.html.

———. Pierce's Disease Control Program. CDFA website, http://www.cdfa.ca.gov/pdcp/.

———. Plant Health Site Map. CDFA website, http://www.cdfa.ca.gov/plant/sitemap.html.

———. Plant Quarantine Section 454: Regulations and Section 510: Counties with regulations more stringent than 3650. CDFA website, http://www.cdfa.ca.gov/pdcp/Regulations.html.

———. Statewide Sudden Oak Death Quarantine. CDFA website, http://www.cdfa.ca.gov/plant/PE/interiorexclusion/SuddenOakDeath/.

CDPR (California Department of Pesticide Regulation) website, http://www.cdpr.ca.gov/.

Center for Invasive Species Research (CISR) website, http://cisr.ucr.edu/invasive_species.html.

Citrus Research Board website, http://www.citrusresearch.org/.

Dowell, R. V., and R. Gill. 1989. Exotic invertebrates and their effects on California. Pan-Pacific Entomologist 65(2): 132–145.

Dreistadt, S. H. 2001. Integrated pest management for floriculture and nurseries. Oakland: University of California Division of Agriculture and Natural Resources Publication 3402.

Electronic Code of Federal Regulations: Government Printing Office website, http://www.ecfr.gov/cgi-bin/ECFR?page=browse.

Grafton-Cardwell, B., and T. Batkins, eds. 2003. California citrus industry guide to the glassy-winged sharpshooter and related bacterial diseases. 2nd ed. Visalia, CA: Citrus Research Board.

Grafton-Cardwell, E. E., K. E. Godfrey, J. E. Pena, C. W. McCoy, and R. F. Luck. 2004. Diaprepes root weevil. Oakland: University of California Agriculture and Natural Resources Publication 8131. ANR CS website, http://anrcatalog.ucdavis.edu/pdf/8131.pdf.

Haviland, K., M. Daane, and M. C. Battany. 2009. Integrated pest management for grapes. Oakland: University of California Division of Agriculture and Natural Resources Publication 3448.

Jetter, K. M., and K. Godfrey. 2009. Diaprepes root weevil, a new California pest, will raise costs for pest control and trigger quarantines. California Agriculture 63(3): 121–126.

Kabashima, J. N. 2006. Diaprepes root weevil. UC IPM flyer. UC IPM website, http://www.ipm.ucdavis.edu/PDF/PUBS/Diaprepes_Flyer.pdf.

Knapp, J., S. Simpson, J. Pena, and H. Nigg. 2000. Diaprepes root weevil: What Floridians need to know. ENY-640. University of Florida IFAS Extension website, http://ufdcimages.uflib.ufl.edu/IR/00/00/28/01/00001/IN11800.pdf.

Kovacs, K., T. Václavík, R. G. Haight, A. Pang, N. J. Cunniffe, C. A. Gilligan, and R. K. Meentemeyer. 2011. Predicting the economic costs and property value losses attributed to sudden oak death damage in California (2010–2020). Journal of Environmental Management 92:1292–1302.

Mannion, C. M. 2002. Interaction of an insecticide and entomopathogenic nematodes for the control of a root weevil. Proceedings of the Southern Nursery Association Research Conference 47:170–175.

———. 2003. Entomopathogenic nematodes and weevil control in ornamentals. Proceedings of the Southern Nursery Association Research Conference 48:133–135.

Mannion, C. M., A. Hunsberger, J. E. Peña, and L. Osborne. 2003. Oviposition and larval survival of *Diaprepes abbreviatus* (Coleoptera: Curculionidae) on select host plants. Florida Entomologist 86(2):165–173.

McCoy, C. W., E. D. Quintela, S. E. Simpson, and J. Fojtik. 1995. Effect of surface-applied and soil-incorporated insecticides for the control of neonate larvae of *Diaprepes abbreviatus* in container-grown citrus. Proceedings of the Florida State Horticultural Society 108:130–136.

National Agricultural Pest Information Center (NAPIC). NAPIC website, http://pest.ceris.purdue.edu/.

North American Plant Protection Organization (NAPPO) website, http://www.nappo.org/en/.

Paine, T. D., K. M. Jetter, K. M. Klonsky, L. G. Bezark, and T. S. Bellows. 2003. Ash whitefly and biological control in the urban environment. In D. Sumner, ed. Exotic pests and diseases: Biology and economics for biosecurity. Ames: Iowa State Press. 203–213.

Pimentel, D., R. Zuniga, and D. Morrison. 2005. Update on the environmental and economic costs associated with alien-invasive species in the United States. Ecological Economics 52:273–288.

Pimentel, D., L. Lach, R. Zuniga, and D. Morrison. 1999. Environmental and economic costs associated with nonindigenous species in the United States. Ithaca, NY: Cornell University, College of Agriculture and Life Sciences.

PlantRight website, http://www.plantright.org/.

Quintela, E., J. Fan, and C. McCoy. 1998. Development of *Diaprepes abbreviatus* (Coleoptera: Curculionidae) on artificial and citrus root substrates. Journal of Economic Entomology 91:1173–1179.

RSAP (Research Scientific Advisory Panel). 2007. PD/GWSS research scientific review. CDFA RSAP website, http://www.cdfa.ca.gov/pdcp/Documents/RSAP_Final_Report_08302007.pdf.

Shapiro, D. I., and C. W. McCoy. 2000. Susceptibility of *Diaprepes abbreviatus* (Coleoptera: Curculionidae) larvae to different rates of entomopathogenic nematodes in the greenhouse. Florida Entomologist 83:1–9.

Simpson, S., H. Nigg, and L. Knapp. 2000. Host plants of Diaprepes root weevil and their implications to the regulatory process. In S. Futch, ed. Diaprepes short course. Lake Alfred, FL: Citrus Research and Education Foundation. 19–37.

Simpson, S., H. Nigg, N. Coile, and R. Adair. 1996. *Diaprepes abbreviatus* (Coleoptera: Curculionidae): Host plant associations. Environmental Entomology 25:333–349.

USDA (United States Department of Agriculture) Animal and Plant Health Inspection Service (APHIS). APHIS website, http://www.aphis.usda.gov/wps/portal/aphis/home.

———. Plant Health, APHIS website, http://www.aphis.usda.gov/wps/portal/aphis/ourfocus/planthealth.

———. Agricultural Select Agent: Select Agent and Toxin List. APHIS website, http://www.aphis.usda.gov/programs/ag_selectagent/ag_bioterr_toxinlist.shtml.

USDA Natural Resources Conservation Service. Plants database: Invasive and noxious weeds, California state listed noxious weeds. USDA Natural Resources Conservation Service website, http://plants.usda.gov/java/noxious?rptType=State&statefips=06.

Van Driesche, R. G., and T. S. Bellows Jr. 1996. Biological control. New York: Chapman and Hall.

Varela, L. G., J. M. Hashim-Buckey, C. A. Wilen, and P. A. Phillips. 2007. Glassy-winged sharpshooter. Oakland: University of California Division of Agriculture and Natural Resources Publication 7492.

Woodruff, R. 1968. The present status of a West Indian weevil (*Diaprepes abbreviatus* L.) in Florida (Coleoptera: Curculionidae). Florida Department of Agriculture, Division of Plant Industry Entomology Circular No. 77.

WPDN (Western Plant Diagnostics Network) website, https://www.wpdn.org/.

Photo: Courtesy of Ever-Bloom, Inc

Business Management Basics

Bridget K. Behe, Alan W. Hodges, Julie P. Newman, Trent Teegerstrom, Phil Kenkel, and Ursula K. Schuch

CHAPTER EIGHTEEN

In today's marketplace, the nursery business requires much more than the ability to produce high-quality plants—nursery managers also must have strong business management skills. Although the nursery industry in California has experienced many years of steady growth, recently the economic climate has become increasingly competitive, with many companies experiencing depressed prices, reduced profitability, and bankruptcy. The survival of a nursery operation depends on shrewd fiscal management of all aspects of the business, including production, labor, customers, marketing, and distribution.

The goal of the nursery manager is simple in theory but quite challenging in practice: use limited space, time, and resources to produce the greatest and most sustainable economic return. To do this, effective nursery managers analyze data from financial reports and other types of information before making decisions regarding the allocation of resources. They observe, listen, ask good questions, and listen again, taking this data and their past experiences into account. Effective managers also take the time to regularly plan and strategize. This includes evaluating company performance to determine whether the business is succeeding, where it is going, or how closely it is attaining the goals of management.

Nursery managers must be customer focused to ensure that a market is available when their plants are salable. Every aspect of the nursery business should be focused on serving the needs of customers. Success depends on how well managers listen to their customers and use that information in making decisions on what to grow and what services to provide. Effective nursery managers ensure that their customers are familiar with the products and services they market, and they demonstrate how the nursery adds value to these products and services. Effective managers also have a positive relationship with nursery personnel and customers.

In addition to managing the physical and financial aspects of the business, nursery managers must be able to manage people. Understanding the individual work preferences of personnel and effectively directing them to achieve firm goals that are clear and well articulated is essential to hiring, developing, and retaining good employees. Effective managers put key people in key places. They are a bit like good movie directors who can cast the right person in the right role to get the best result from the whole group. Nursery managers also need to be good leaders as well as good stewards of their own time and money. It is nearly impossible to be a good manger of other people if you aren't a good manager of yourself.

Photo: Courtesy of Ever-Bloom Inc.

Managing risk is another aspect of management. Planning reduces risk but doesn't eliminate it. Insurance can help protect against losses associated with bad decisions or results that weren't anticipated. Risk is part of business, and profit is the reward of good decisions coupled with good luck.

This chapter provides an introduction to the myriad skills and kinds of knowledge required to manage a successful nursery business. We will review different types of legal, business, and organizational structures to help you determine the best structure for your business. We will address all aspects of the business planning process, including preparation and implementation of business plans, vision and mission statements, analysis of the business strengths and weaknesses, action plans, and financial plans. We will describe various types of records and accounting methods so that you can select record-keeping systems that work best for your business and help with decision making. We will discuss the process for evaluating your business performance using financial statements and other business records, financial performance metrics, and industry benchmark data. We will introduce many terms used in financial management that are defined in sections of this chapter and in a glossary at the end of this publication. Furthermore, we will discuss the various types of risks that nurseries are subject to, as well as insurance programs that can be used to minimize risk.

Separate chapters on marketing and labor management are provided in chapters 19 and 20, respectively, because of the complexity of these topics. Although material presented in this chapter applies to people who are starting a nursery business, this topic is presented in greater detail in chapter 1. Please note that the information presented in this chapter is not meant to substitute for sound legal and financial advice but to give the nursery professional a good basic understanding of some challenges, terminology, and approaches to managing the nursery in a profitable and sustainable manner.

Business Structures

Legal Structures

The easiest way to start a business is to just start working. In the earlier part of the last century, a person would simply hang their shingle with their credentials. Entrepreneurial activity is the backbone of the United States, but there are challenges with this form of business called *sole proprietorship*. However, sole proprietorships are still started every day, often in basements, garages, or even pickup trucks. The challenge with sole proprietorships is that they don't legally separate the person from the business, so the entrepreneur is legally and financially liable for the mistakes of the business. Sole proprietors can lose their home and possessions if the business makes a big mistake. Also, when the sole proprietor dies, so does the business.

Partnerships are essentially the same as sole proprietorships, but they have more than one person involved (two or more). The greatest challenge with partnerships is not getting into them but getting out of them. Partners are legally and financially liable for the mistakes of each other, which is not the best situation. When one partner dies, the other assumes all the responsibility.

It is usually preferable to reduce risk by legally separating the owners from businesses so that the business itself is held legally liable for the actions and debts incurred by the business. Decades ago, this was only possible if a company incorporated, but it was costly and primarily practiced by larger businesses. These "C" *corporations* often have the letters "Inc." or "Co." after the name, and as part of the legal name of the business, to show they are separate entities. They have articles of incorporation and a board of directors. The incorporated business doesn't end when the founder dies, as is the case in sole proprietorships. However, in addition to being more difficult and expensive for small businesses, corporations can pay more taxes. That is to say that the business pays tax on business income, and the people who work there (including the business owner) pay taxes on their personal income from the business. If the business were a sole proprietorship, only the owner would pay personal taxes on the sales revenue generated. Often, the sum tax paid for corporations is greater than the sum tax paid for sole proprietorships or partnerships. So, while corporations separate financial and legal responsibility from the owners and the business, they may cost more in terms of start-up costs and taxes.

More recently, options have become available for smaller businesses to separate themselves from their funders or shareholders. *S-class corporations* and (more recently) *limited liability companies* were established for smaller firms. According to the U.S. Small Business Association, a limited liability company provides the limited liability features of a corporation and the tax efficiencies and operational flexibility of a partnership. In this type of business structure, the owners are referred to as members. All profits and losses are directed to each member and are reported on their personal federal tax returns (similar to owners in a partnership). Limited liability companies may have the letters "LLC" after the company's legal name. This type of business structure is often the least expensive and easiest way to create legal separation of people from the businesses. There is no board of directors and LLCs can usually be created in most states for less than $100.

In summary, the legal structure affects the business, whether it is an existing business or one that is just starting up. The legal structure may affect the actual cost of doing business, the day-to-day operation and management, the type and amount of taxes, and the level of exposure to risks and liabilities. Legal structure affects the course of business after the business owner dies—for example, as an LLC or other corporation, business ownership can be transferred to children or spouse with fewer unwanted tax burdens (Lobo et al. 2006). Therefore, it is prudent to consult an attorney who is familiar with your type of business and the horticulture industry before deciding which legal form of business to create.

In California, the distribution of legal structures in 2007 was sole proprietors, 67 percent; partnerships, 9 percent; corporations, 22 percent; and other, 2 percent (Carman 2011). Carman's analysis of economic trends (2011) in the California nursery industry from 1982 to 2007 found that the corporate share, currently larger for nursery producers than for any other agriculture sector, is decreasingly family owned. Although many nurseries are still family-managed businesses, an increasing number of nursery owners are finding that their children are losing interest in the business. Moreover, family-operated businesses may have family-related issues that complicate the management of the business, such as sibling rivalry and difficulty in disciplining or firing family members (Avent 2003).

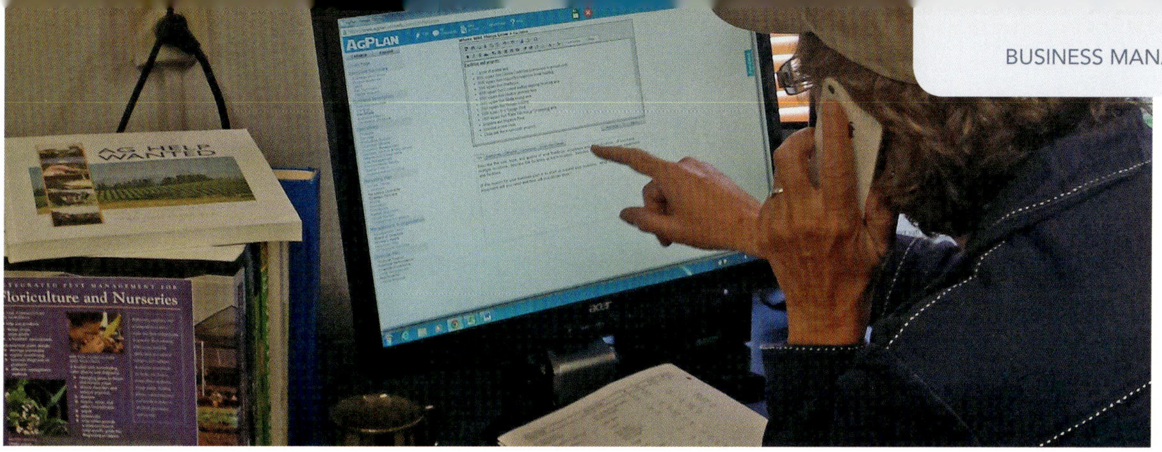

Photo: U. K. Schuch and T. Teegerstrom.

Organizational Structure

An organization chart is a useful tool to illustrate and understand the organizational structure of a business. Such a chart typically lists the title of each person's position, and, by means of connecting lines, it shows who is accountable to whom and who is in charge of what area. Actions that do not follow the structure are a common occurrence in businesses, especially in small businesses where management employees have several roles. Developing a formalized organizational structure and resisting the temptation to violate it whenever possible can facilitate smooth operation. It is essential during personnel evaluations to know who reports to whom, who makes decisions about various aspects of the firm, and which positions are responsible for which laborers.

A common organizational structure used in small businesses identifies two groups: management and workers. In this organizational structure, decision making is usually highly centralized, with the owner-manager making the important decisions, supervising employees, and being involved in every phase of the business. Workers usually have little decision-making authority. This structure provides the business with flexibility, and decisions can be made quickly. When the owner-manager decides that a change is needed, the decision can be directly and expediently communicated to the workers.

As the business grows, oversight becomes too large a task for only one person, and some responsibilities and authority need to be delegated to other individuals. A number of organizational structures are used by larger businesses for delegation of authority and responsibilities. One approach is to organize activities around common business functions (Dobbins and Ehmke 2007). With this type of structure, managers are able to specialize, which often increases efficiency and reduces duplication of effort. For example, one manager could specialize in production or operational aspects of the business, another manager could specialize in marketing, and a third could be responsible for the financial aspects of the business. Overarching these specialized managers is a general manager who concentrates on overseeing the entire business, developing and maintaining important relationships with suppliers and customers, and establishing an overall strategy for the business.

Although functional structures have advantages, there can be disadvantages. While managers in the various areas can develop expertise in their area of responsibility, they may lose sight of how the various parts fit into a whole, and they may push responsibility for overall business performance onto the general manager. Communication and efficiency are also hindered when employees have more than one supervisor.

Business Planning Process

Planning is a formal process to establish goals and develop strategies that will be used to meet them. Developing a written plan makes goals and strategies more tangible and makes the person who established them more accountable. Planning is a leveraging activity, meaning that the small investment of time put into the planning process can reap large rewards in the form of a higher return of profits (or reduction of losses). Unfortunately, because people tend to focus on immediate situations and crises, many nursery owners skip this important step in managing a business. They may spend more time planning their vacations than their business activities.

Business Plans

Just as a home is not built without blueprints, you don't start a business without a plan. If you can't make a sound and workable written business plan, how likely is it to work in reality?

Business plans are not only necessary for starting a new business—they are also vital for operating and managing a business. A business plan is necessary when an existing business applies for new loans or seeks new investments. A business plan can be an important tool used by the management to look ahead, focus on major issues, and prepare for problems and opportunities. Furthermore, a business plan is useful for making decisions regarding the allocation of resources and the pricing of products and services. To ensure that the plan is an effective tool for management, it must be interactive. This means the business plan must be reviewed, updated, adjusted and fine-tuned regularly. In a good television show, the script helps to create the next episode of the show; so, too, can a good business plan help direct the near-term future of the firm.

A business plan is a document that tells the story of a business. For example, a business plan for a nursery business describes what the nursery does, who is involved, who is included in its market, and why the nursery business should be successful against the competition. Plan formats and outlines vary, but generally a plan will include

- **The Executive Summary, or Introduction** – Provide a brief, detailed description of your business and hook readers to read the whole plan. This is the most concise version of your plan and should highlight the major points of your business, products, marketing, management, experience, and so on. If you are applying for a loan, the summary should include a statement requesting the funds and explaining the need, amount, and intended uses for the loan.
- **The Business Charter** – Define your business and the type of business you are in. Who are you? What is the purpose of your business? When are you going to begin operation, or when are you going to reach your milestones? Where are you going to sell your products? Why is your business being formed? Answers to these questions will help you develop a clear definition of your business. They will also help you identify and set short, intermediate, and long-term goals.
- **Products or Services** – Describe all crops, products, or services that you will be selling. Emphasize the unique features of your products and value-added activities that make them stand out over the competition. What makes your product different and better than the products of thousands of other nursery products? What specific need, market window, or market niche will your products fill? Describe the operating schedule, production schedule, harvest schedule, and production volume for all crops and products.
- **Markets and Marketing** – Identify and explain who the potential buyers are and who they will be in the future. Assess the need and demand for all the crops or products you will sell. Explain the nature of the market, its size, location, and important trends. Evaluate the competition, and determine how segmented the market is. What is your targeted share of the market? How will you advertise and sell your crops or products? The marketing method(s) used will impact the strategy, distribution, and type of advertising or promotion you use with your crops, products, or services.
- **Financial Information** – What resources are available? What are the sources and need for start-up funds? Financial statements needed may include an income statement, cash flow statement, and a depreciation schedule. For new businesses, personal financial statements and income tax information for the past 3 to 5 years may be required as well as projected financial statements for at least 2 years. A monthly operating budget and a statement of cash flow may be required for the first year of operation. If you are applying for a loan, a funding request should be included. Indicate the amount of funds needed, justify the need, and explain how the funds will be used.
- **Sales and Costs of Production** – Project sales and expenses for the first year of operation. Determine costs of production and break-even analysis. Estimate the cash flow of the business; this will help identify the need and timing for borrowed funds. It will also help determine the repayment capacity of the business. Regardless of the reasoning, cash flow projections should provide a road map and a true picture of the surplus or deficit of the business.
- **Legal Structure** – Choose a legal structure carefully. The choice may affect the actual cost of doing business, the operation and management of the business, taxes, and the level of exposure to risks and liabilities.
- **Management Structure, Key Personnel, and Operations** – Explain how the business will be managed on a day-to-day basis. Provide a description of the skills and experience of everyone involved with the business, including résumés for you and key employees. All duties and qualifications should be clearly stated, with emphasis placed on management or administrative responsibilities and chain of command. Hiring and personnel procedures, insurance needs, lease or rental agreements, equipment needs, production, delivery, and any other issues pertinent to the business may be included.
- **Concluding Statement** – Summarize the business goals and objectives, and reiterate your commitment to the success of the business.

FIGURE 18.1.
Sample business plan outline. *Source:* Lobo et al. 2006.

descriptions of the nursery business, products and services, market, forecasts, management team, and financial analysis. See figure 18.1 for a sample business plan outline.

A specific plan for your nursery will depend on your situation and should match its purpose. For example, description of the nursery management team is very important for investors, while financial history is most important for banks; if you're developing a plan for internal use only, you may not need to include all the background details. Many of the statements and analyses discussed later in this chapter can be part of the overall business plan (e.g., mission and vision statements, cash flow, balance sheets, income statements, SWOT analysis).

Some people hire consultants or have employees draft the business plan. While using input from experts and employees can strengthen the plan, the person or persons responsible for implementing the plan should be heavily involved in its development. If you're going to be accountable for the decisions that will be based on the plan, then you need to have a key part in its development.

Putting together a business plan requires estimates about many things that are not known with certainty. It may be difficult to estimate how many supplies such as shovels and plant labels your nursery will need, but listing all of the items needed and making educated guesses on amounts where specific information is lacking will create a better plan. Wild guesses lead to poor plans, so it is helpful to do some homework and identify reasonable costs for most items. Once established, a company can use historical records to estimate costs, but until those records are established, best cost estimates may come from colleagues in the industry, from the Internet, or from nursery catalogs. The better job a planner does with cost estimates, the better the outcome of financial success. Assumptions and guesses that are not backed by facts should be clearly identified in your plan and used cautiously in developing forecasts.

There are many resources available for preparing business plans, and a partial list is provided in the Resources at the end of this chapter. You may want to start by doing some reading. There are numerous Cooperative Extension publications with information on developing effective business plans (e.g., Ehmke and Akridge 2007; Holcomb et al. 2010; Schuch 2006). Owen et al. (1991) published a sample business plan for a greenhouse nursery business, which is a good guide for developing your own specific plan, although it is not meant to be used as an exact template.

A number of sample business plans, planning tools, and software programs (including free downloadable software) can be found on the Internet. For example, AgPlan is a free online business planning tool developed by the University of Minnesota with examples and tips on developing a strong business plan. Furthermore, business plan examples and forms that you can modify for your own use are available at the U.S. Small Business Administration's website.

Most states have small business development centers, which can be very helpful to the entrepreneur in developing a business plan and providing more information on the planning process. At some small business development centers, you can obtain legal, financial, and planning assistance.

SWOT Analysis

SWOT analysis helps the company understand how it might best make money and what challenges may prevent that from happening. SWOT stands for strength, weakness, opportunity, and threat. Strengths and weaknesses are internal to the company, while opportunities and threats are external to the company. Internal dimensions are things over which the company has, or will have, direct control—like labor, costs, prices, and what products it grows or sells. External dimensions are things beyond the control or influence of the firm, like gasoline costs, interest rates, market trends, and economic conditions.

In analyzing what strengths the company has, key partners should ask what this company can do better than others. Those are potential strengths, since the capacity to do the work comes from within, giving that company strength in that area. Weaknesses are areas that aren't strong points for the company and internal factors that may inhibit success. Most companies are started on strengths but quickly ended by weaknesses. A thoughtful, introspective approach to weaknesses is needed to keep the business from ending before it really has a chance to grow. Not all weaknesses can be overcome, but help might be purchased or out-sourced. Strength may also be acquired by partnering with other businesses. Opportunities and threats provide the context of the firm in the market. Opportunities are unmet needs or areas where the business sees an opportunity to make a profitable sale. Opportunities and strengths are the two key reasons new businesses are often started. Threats are conditions that might inhibit the success of the business, such as the inability to get start-up capital, unemployment in the local market, or many businesses of the same type in close proximity.

Vision and Mission Statements

The first step an effective planner might take is to create vision and mission statements for the business. A vision statement describes *what* you and other business leaders in the nursery want to accomplish in 3 to 5 years. Vision states what the company will look like at the end of that time frame if the plan goes as predicted. Mission statements describe *why* the workers are engaged in the business, and they often combine the values of key business leaders. Mission tells employees, customers, and suppliers why people work at your nursery over other places and why people want to do business with you. Vision and mission statements should be highly visible to employees and customers at all times. They can serve as a reminder of the direction that your nursery business is heading and why the firm is moving in that direction.

Key employees, not just managers or leaders, should take part in the development of vision and mission statements so that they have a stake in the success of the business. Otherwise, vision and mission statements may seem a bit altruistic and fuzzy to them. How much better will your workers feel about contributing their time and energy when they agree with the direction and purpose of your nursery business?

Action Plans

The action plan (fig. 18.2) can be an essential tool to help good managers translate annual sales or other large goals into workable steps that can be assigned or delegated. Action plans might take an overarching goal, like producing five hundred 1-gallon azaleas or generating $10,000 in sales, and break it down into more easily managed tasks or chronological steps. In addition to specifying steps or tasks that will need to be accomplished to achieve the goal, the action plan also allows for delegation of responsibility. Action plans are very helpful when there are activities (such as plant production, hiring, hosting an open house) that require the same steps each time (such as each year). The action plan pulls together in one place all the activities needed to achieve a particular goal.

The action plan also is well suited to transferring steps or activities to a calendar, where the steps can be integrated with other business activities. Weekly calendars are best, since not all activities take place every day. Managers take an annual or monthly calendar of activities and break it down into steps or tasks, prioritizing them in order of chronology or importance. The steps from the action plan can be recorded on the weekly calendar. Weather may cause some delays or moving up of tasks, but the action plan can provide a framework around which to make weekly crew assignments.

Financial Planning and Budgets

Financial planning may be the most important part of the planning process. Essentially, a planner should forecast income, expenses, and profits for the first 5 years. If the entrepreneur is seeking funding, then income statements and balance sheets will also be required; but it's a good idea to generate them regularly even if they're not required.

The next financial planning step is one that most people fail to take: estimating monthly cash flow. (Cash flow and other financial terms are described later in this chapter.) The most critical financial step any business can take is to estimate how much cash is needed each month to get the business through to the next month. Information for preparing

Green Side Nursery Action Plan

Goals/outcomes

1. Generate $500,000 total sales (increase 5% from last year) in hanging baskets.
2. Produce 10,000 hanging baskets, with 95% salable items.
3. Produce baskets for sales April 15.

Start Activity	Activity	Date To Be Completed
Jun 15	1. Order pots, media, tags, and fertilizer	Jun 15
Jun 30	2. Determine composition of hanging baskets.	Jun 30
Nov 1	3. Order cuttings.	Nov 1
Dec 20	4. Fill containers, prepare for cuttings.	Dec 20
Jan 1	5. Prepare greenhouse production area.	Jan 1
Jan. 15	6. Receive and stick cuttings.	Jan 16
Jan 15	7. Program lighting and irrigation requirements.	Jan 16
Jan 15	8. Begin pest management program.	Jan 16
Jan 16	9. Monitor environmental controls.	Jan 16
Apr 12	10. Price and add supplemental tag(s).	Apr 12
Apr 12	11. Sleeve and package baskets for shipment.	Apr 12
Apr 12	12. Ship or transfer to retail operation.	Apr 15
	13.	
	14.	
	15.	
	16.	
	17.	
	18.	
	19.	
	20.	

FIGURE 18.2. Example of an action plan. The action plan facilitates the manager's ability to take large or long-term company goals and divide them into steps for implementation.

cash flow projections can come from a number of sources, including records of actual cash flow, tax returns, and other nursery records from previous years.

An accurate cash flow budget is essential to the firm's success and sustainability, especially since nearly every nursery business is seasonal (meaning that sales, costs, and profits are not equally distributed across the calendar year). It is impossible to get to the end of the year financially without managing cash month by month. The bottom line of the cash flow budget shows monthly net cash flow, which is really the surplus or deficit of cash the business is facing each month. It may be tempting to spend the cash surplus as it comes in, but that cash will be needed to meet shortfalls, which inevitably occur.

The cash flow budget is helpful to show months when there will be a cash shortfall, which needs to be covered by a line of credit or some other cash source. Sustaining finances and meeting cash flow needs can be challenging. Businesses might try to get very good payment terms from suppliers to help reduce the cash needs. If a supplier will give a company 3, 6, or even 12 months to pay the bill, it will help the buyer tremendously. However, in these days of tighter credit, the likelihood of negotiating long payment terms decreases unless the business has had a stellar relationship with the supplier for many years. Another way to meet the short-term cash need might be from a line of credit. The line of credit enables a company to borrow up to the maximum amount previously agreed upon by the lender. This open line of credit helps the company to meet those short-term cash needs. The amount of the line of credit and interest rates heavily depend on the financial history of the company and the relationship it has with the lender.

Record Keeping

A successful business depends on sound record-keeping practices. Without good records it is impossible to determine the financial condition or profitability of a business. Record keeping is one of the more worthwhile time investments any company can make to improve its likelihood of success and sustainability. Moreover, some or many records are often required by law (e.g., for tax and payroll purposes, or reporting pesticide use).

The goal of any record-keeping system is to transform observations and numerical data into information that allows the manager to make better decisions. Information that nursery managers can obtain from records to improve profitability include how to grow plants most efficiently, how to set prices, where costs need to be curtailed, and whether a loan is necessary to get through periods with no cash flow.

Functions of Record Keeping

There are many reasons why complete and accurate records are crucial to the nursery business. Six common functions of record keeping are described below.

Financial planning and budgeting

Over time, your records reveal a pattern of production, sales, and expenditures. The information is useful for projecting cash flows, scheduling purchases, hiring additional help, and other activities. As described in the previous section on financial planning, all businesses should use a budget for planning purposes. A budget will help keep your business on track by forecasting your cash needs and helping you control expenditures. You must have accurate historical financial information to prepare a meaningful budget.

Financial reports

Financial records allow you to create the financial statements needed by your banker and tax preparer. Bankers and lenders require an accurate, realistic financial picture of your business. We will discuss financial reports in more detail later in the chapter, but most lenders want an income statement, a balance sheet, and a cash flow projection as part of a loan application. They may also want to see your budget as evidence that your business is well planned and stable. Your records provide the basis for these statements. Lenders are interested in the adequacy and accuracy of your record-keeping procedures. A lender may establish loan covenants that require your business to maintain target levels of working capital, accounts receivable, and other financial measures. These requirements or covenants are put in place by the bank to lower their risk in loaning money to a business. They may include things such as requiring that licenses are kept current, life insurance premiums are paid, equipment is properly insured, or that financial statements are made more often than annually. Your record-keeping system should not only supply this information but also forecast your ability to meet these requirements. Even your suppliers and other creditors may require financial information on a periodic basis.

Records are essential in the preparation of accurate tax returns. The Internal Revenue Service (IRS) and each state require that every business report income, expenses, and allowable deductions. Poor records may result in your underpaying or overpaying your taxes or filing late (and paying penalties). If your accountant prepares your income tax return, poor records will almost certainly result in your paying higher accounting fees. Most nursery managers are aware of the myriad regulations involving employees. Payroll tax deposits must be made according to strict deadlines, and payroll tax returns must be filed and reconciled with deposits every quarter. At the end of the year, you are required to give your employees and the government W-2 forms, which must agree with your quarterly payroll returns. Sound record-keeping practices are essential to comply with all these payroll rules.

Evaluating business performance

Records can provide a clear financial picture to evaluate how a nursery is performing. For example, financial records can indicate whether the business is generating a profit and show if sales are increasing or decreasing. They can also be used to determine if expenditures are above or below budget projections and identify specific areas where costs are out of control.

Projections and forecasting

Records form the basis of financial statements, which can be used as a guide for important decisions such as business expansions, financing, marketing strategies, operations planning, and product selection. They can also support the evaluation of costs and returns for individual products to determine a more profitable product mix. Business analysis can assist in identifying some common problems in wholesale plant nurseries such as low output, slow crop growth or poor pricing, excessive costs, waste or overuse, poor cash flow, overcapitalization or undercapitalization, and imbalanced debt structure.

Pricing

Records provide critical information for pricing decisions. A good set of financial records allows you to allocate direct and overhead expenses and determine if product prices cover costs and return a profit. Pricing will be further addressed in the chapter on marketing (chapter 19).

Evaluation of production systems

Managers must constantly determine if they can increase profits by growing a different set of plants, finish plants in a different period of time, or use alternative production practices. An integrated record-keeping system can provide information to help with these decisions.

Financial Record-Keeping Systems

A basic financial record-keeping system should be simple to use, timely, accurate, and detailed. It generally requires a journal to record transactions (e.g., receipts, disbursements, sales, or purchases), accounts receivable records, accounts payable records, payroll records, petty cash records, and inventory records.

Larger nursery businesses may have internal accounting personnel and sophisticated record systems to guide management. Smaller nursery businesses that rely primarily on a bookkeeper and an outside accounting firm to maintain the company's records and provide guidance should be familiar with proper record-keeping requirements and cash flow planning. Consulting with a professional accountant who has clients in horticultural businesses before establishing a financial record-keeping system can significantly improve the quality of financial record keeping by the firm.

Employees who are involved in record keeping should be trained to implement standardized methods and to regularly maintain records. A good training program will help ensure the accuracy and completeness of the data collected. Record keeping is time consuming, so it is best to delegate it to a person who is good with details and doesn't mind repetitive activity. To reduce internal financial issues, have at least one other person control the accounting ledgers for at least one week annually. Furthermore, everyone who makes purchases for the company should provide receipts and explanations for business purchases, following established company protocols.

There are numerous kinds of computer programs available for electronic record-keeping systems. Software for computer record keeping can reduce labor by facilitating repetitive tasks and streamlining record keeping. Software doesn't replace the need for professional accounting advice, but it can be adapted to meet the needs of most businesses. The real bonus is that in addition to basic bookkeeping, many of the programs can be used to generate financial statements and financial projections. The "best" software program will depend on your situation. For example, there are software programs that can provide the manager with simple to relatively complex views of the financial status of the business.

Production Record-Keeping Systems

Financial records are only one aspect of the information that managers need to operate a successful nursery. Managers also need to track physical production, cultural practices, plant environment, pesticide use, plant inventory, customer information, and other information important to the individual firm.

The level of detail depends on the complexity of the nursery operation. A more complex operation requires a more detailed system. Having the ability to drill down to detailed information improves decision making. For example, knowing that your root media costs exceed those of similar-size operations is useful. Determining root media costs at the product level allows you to identify and implement changes. In many cases it is difficult to find or customize a record-keeping system to include all of the records you find desirable. You may need to use multiple systems to form your overall record-keeping system. For example, a nursery manager might decide to keep pesticide use records in a separate system. Managers must weigh the complexity and degree of integration with the costs and their informational needs.

Creating a unique identifier for each product sold (often called a stock-keeping unit) helps you track information for decisions. A stock-keeping unit (SKU) is a unique identifier for each product sold. The SKU is the billable entity, which may be an individual item, such as a plant, or a bundle of items, such as a tray of plants. Ideally, a nursery record-keeping system should assign a unique SKU for each product and also for its variants. For example, a dracaena targeted to be sold in a 1-gallon container would have a different SKU designation than a dracaena targeted to be sold in a 2-gallon container. The units for the SKU should be whatever is logical for the plant (i.e., buds, slips, containers, etc.). The goal of the record-keeping system is to be able to track cultural, environmental, and financial information down to the SKU level. Expenses tracked to a SKU often require a coding system within the record-keeping system that allows for allocation of expenses across operations and ultimately to each crop. This is described in the section "Income and Expense Records" later in this chapter. SKUs also facilitate bar code and scanner systems.

Accounting Methods

There are two primary methods for accounting: the cash-basis method and the accrual method. The cash-basis method records income when payment is received and records expenses when the expenses have been paid; the accrual record-keeping convention records income when the sale is completed (regardless of whether the payment has been received) and records expenses at the time they are incurred (regardless of whether the expense has been paid). The IRS generally requires any firm with manufacturing, wholesale, or retail business activities to use the accrual method. However, an exemption exists for sod and nursery farms. Whether you report income (revenue) using the cash or accrual method, you must use the appropriate record-keeping system.

The cash-basis method is more commonly used for small businesses because it is a simple method for tracking cash transactions and it may avoid some tax liability for gains on unsold product. However, a disadvantage of cash-basis accounting is that records do not show all the revenue or expenses the business has incurred at the close of the fiscal year. The accrual system gives a more accurate picture of the actual financial performance of the business, since it accounts for changes in inventories, receivables, and payables. Inventories are particularly important for nursery businesses, because typically two-thirds of working capital may be tied up in plant inventories.

Types of Records

Records can be categorized into two main groups: (1) production records and (2) income and expense records. These are described below.

Production records

Production records include cultural practices, plant environment, and physical production. Cultural records should contain dates and labor used for such operations as root media preparation, planting, fertilization, scouting (monitoring), pesticide application, pinching, pruning, chemical growth regulation, disbudding, harvesting, and cleanup. The time each operation is performed, the product block, the date, and the employee name are entered on the record sheet. Pest and disease occurrence and visual observation records should also be kept. Pest management information may be integrated into the cultivar record-keeping system or maintained separately. Legally, you must keep a record of pesticide spray dates, rates, and chemicals used. Cultural records provide a plan for duplicating successful crops. They also provide a trail from which the cause of errors in the culture of the crop can be determined and then corrected in the next crop. The records can also be used to create a schedule or action plan to remind employees of the needed actions throughout the production period.

Plant environment records include solar radiation, temperature, relative humidity, and root media nutrient analysis (e.g., pH and salinity). This data can be recorded manually or automatically using data loggers or computer-controlled sensors. Plant environment information can be used to determine whether the plant environment should be altered, and it can help the manager anticipate problems such as early or late crop maturation. Good environmental records help the manager understand and correct poor quality and yield as well as reduce the occurrence of pests and disease.

Physical production records should be gathered throughout the growth period of each crop, usually on a weekly basis. The production records should include the number of plants or containers harvested, along with the date and quality. These records will be needed for cost accounting purposes. Average heights or average widths, collected regularly as appropriate for the crop, can be useful for identifying problems in early stages by monitoring smaller-than-normal growth measurements.

Income and expense records

Income or sales records should be recorded by crop or SKU. Income records should indicate date of sale, shipping date, customer, market outlet, and grade or quality.

Expenses can be either variable or fixed. *Variable (direct) expenses* are expenses that go directly to creating sales for the business. The fertilizer cost that is used to grow nursery stock is a direct cost for a production nursery. However, fertilizer applied to the demonstration beds or the office grounds is not. Typical direct expenses include seeds, cuttings, and propagules, plant media, pots, planting labor, pest control, and shipping that can be identified with a given crop or SKU. *Fixed (overhead)* costs cannot be identified with a particular segment of the business but rather are the general costs of keeping the business operating. Expenses such as the manager's time, accounting expenses, interest, depreciation, general facility maintenance, and other indirect expenses are classified as fixed costs or common overhead. Total overhead costs for a nursery operation usually range from 25 to 30 percent of total production costs or 20 to 25 percent of total net sales (total sales less returns and allowances). Managers should know each product's share of overhead costs before they set product prices.

The amounts expended for variable expense categories should be recorded as they are incurred (when using the accrual method) or on a monthly basis. These expenses should be tracked to a SKU, which should include codes allowing you to allocate the variable expenses across operations and ultimately to each crop. Some variable expenses such as containers only need a description and unit code. An $800 expense for blow-molded pots would be coded by pot size and number of units, allowing the manager to determine that a 1-gallon pot costs $0.25. The unit costs could then be easily allocated to the individual plants.

For other variable expenses, such as water, fertilizer, or pest control, the expense is coded to the operational area and then allocated to plants on a square footage basis. For example, a $2,000 fertilizer expenditure would be coded to reflect the split across the outdoor/container operation or the shade/greenhouse operation. If the entire $2,000 was allocated to outdoor/container operation with a total area of 24,000 square feet, then a fertilizer cost of $0.083 per square foot would be reflected for the outdoor/container operation. At the end of the production period, the total fertilizer cost would be allocated among plants on a square foot basis. Methods for allocating indirect costs on a square foot basis will be described in more detail in the section "Nursery Crop Cost Analysis" later in this chapter.

Direct labor expenses are variable expenses that should also be coded to allow allocation across the operational areas. The formula for allocating labor across operations is usually based on the manager's best estimate. For example, 60 percent of labor may be devoted to the outdoor/container operations while 30 percent is used by the shade/greenhouse operations and 10 percent is for miscellaneous operations. The labor allocated to the outdoor/container and shade/greenhouse operations would then be allocated among the products on the basis of square footage. The miscellaneous labor could be distributed among the operations on whatever percentage basis seems reasonable. Since labor will account for the largest single expense in any horticultural business, accurate labor records are the key to business success. Labor records are even more meaningful when the labor hours (and costs) can be associated with crops to more accurately develop direct costs. For example, a piece-rate labor system can increase the accuracy of labor cost estimates. More information on labor tracking is presented later in this chapter in the section "Nursery Crop Cost Analysis" and in chapter 20.

Marketing expenses include such items as advertising, packaging, shipping, billing, and on-site displays. Some companies even allocate salaries for salespeople as part of a marketing expense. Some marketing costs are direct costs and can be identified with an individual crop, while others must be included in overhead. Marketing costs commonly represent 5 to 15 percent of total production costs, but should be projected to the future based on the goals and objectives of the company for the next year.

Various techniques can be used to allocate overhead among departments. The allocation process is by its very nature arbitrary, but it reflects the fact that the fixed costs of the nursery must be covered by the products sold. Some managers examine the ratio of direct expenses among operations and allocate overhead in the same proportion. For example, if your outdoor/container operation represented 25 percent of the direct expenses (water, fertilizer, chemicals, labor, etc.), that operation would be allocated 25 percent of overhead expenses. Overhead costs can also be allocated on the basis of sales revenue, labor use, or square footage. If square footage is used to allocate overhead costs among crops, the fixed costs for vacant space associated with a particular crop should be included. For example, when bedding plants are produced from seedlings, vacant space may be held for the spacing of plants as they develop. The fixed costs associated with this space should be allocated to the bedding crop, even though it is not in use for the entire production period. Record keeping on actual space utilization will make this more accurate.

Evaluating Business Performance

Analysis of company financial statements and other business records is a proven approach to achieving greater efficiency, productivity, profitability, and financial security. Managers can use this information to guide financing of business expansions, developing marketing strategies, selecting an appropriate mix of products, controlling costs, and planning operations. Financial analysis can assist in identifying common problems, such as low output, poor pricing, excessive costs, waste, poor cash flow, undercapitalization, and imbalanced debt structure. When done correctly, this effort will often yield benefits in terms of increased profitability, reduced risk of business failure, better customer service, and increased job satisfaction among employees.

Financial Statements

The starting point for financial analysis of any business is to gather the most accurate and recent information available from financial statements. Using these together with other company records, one can readily calculate the standard financial ratios and other meaningful metrics.

The financial statements provided by most in-house accounting systems or professional accountants are the income statement, the balance sheet, the cash flow statement, and sometimes the statement of change in financial position, which shows how the balance sheet items change from one accounting period to the next. For a thorough analysis, it is necessary to look at all of these financial statements together because each provides a different type of information, and many of the financial measures are calculated from items on different statements.

The financial statements are supposed to conform to generally accepted accounting principles (GAAP), developed by the accounting profession and governed by the Financial Accounting Standards Board (FASB). These standards ensure that the financial information has a consistent meaning such that valid comparisons can be made over time or across different businesses. For publicly held companies, these standards ensure that there is full disclosure of information that may be important to stockholders and potential investors.

Within the standards, however, there is considerable latitude to customize the financial statements to meet your company's particular needs. For example, in a wholesale nursery it may be important to have a very detailed itemization of production expenses, whereas a retailer would have a listing of major groups of merchandise purchased for resale and the gross margin on sales. It may be necessary to work with your accountant or accounting software vendor to find the right form for your financial statements.

A description of each of the three most common financial statements and how they are used follows.

The income statement

The income statement, also known as the profit and loss, is probably the single most important financial statement. This is because it summarizes the company's revenues, expenses, and net income for the year or the accounting period of interest. An income statement has three sections: revenues, expenses, and income. Revenues are mainly the nursery sales, but recall that revenue is also composed of inventory change. This is why a manager needs to use an accrual system to really know what is going on in the business. Expenses are the cost of goods (direct expenses) and fixed costs. The difference between the revenues and expenses is called net income. A positive difference between revenues and expenses indicates a profit, and a negative value indicates a loss. The income statement is similar to a movie, illustrating changes in the costs and returns of a business over time, compared with the balance sheet, which is like a snapshot, just presenting the firm on one specific day. An example of a nursery income statement is shown in figure 18.3.

The balance sheet

The balance sheet, or statement of financial position, is a snapshot in time at the end of the accounting period that shows the company's assets, liabilities, and equity. An example of a balance sheet is shown in figure 18.4. The balance sheet has two sides and three key parts. The left side of the balance sheet shows assets, or what the business owns. The right side of the balance sheet contains the liabilities and owner's equity. Liabilities are what the company owes. The balance sheet, sometimes called the statement of financial position, shows the owner's equity, also called net worth, what the owner or entrepreneur has invested in the business. The equation used by the balance sheet is

assets = liabilities + owner's equity, or assets − liabilities = net worth

Assets and liabilities can be classified as short term and long term. Short-term assets include liquid assets; liquid means they are relatively easily converted to cash. Of course it includes cash, but also accounts receivable and inventory. Accounts receivable are the sum of the bills owed to the company. The fixed assets are durable items of value that the company owns, such as property, buildings, and equipment.

Short-term liabilities are bills owed by the company within a month or quarter, also called accounts payable. These would include an electric or telephone bill, the bill for media or containers, or unpaid wages or worker's compensation. Short-term liabilities are most often paid within 90 days of receipt. Short-term liabilities may also include lines of credit used to help with cash flow. Long-term liabilities reflect multiyear loans or mortgages taken by the company for things such as property or equipment. However, the only item recorded for these long-term loans is the remaining principal and interest. Any interest paid on a loan in the current fiscal year is recorded on the income statement as an indirect, or overhead, cost.

Owner's equity, or net worth, is the last part of the balance sheet, found in the lower right-hand corner. While the right side of a balance sheet shows what the company "owes," this is a special debt owed to the owner or investors. This item reflects the current balance of the owner's investment.

The cash flow statement

The cash flow statement, or budget, shows the inflow and outflow of cash from the business during a certain period (week, month, quarter, year). In other words, it shows how money comes into the firm and how it is paid out. It indicates the sources of cash from operations, financial lenders, and savings, and the uses of cash for operations, investments, debt service, and payments to business owners. It is one of the most useful statements in the day-to-day management of the business. In the example provided in figure 18.5, notice that the first column of the cash flow budget for Green Side Nursery is identical to the income statement when the company uses a cash accounting system. This is because the annual cash flow (or first column of the cash flow budget) should exactly match what the business has flowing

Green Side Nursery
Income Statement
Jan 1, 2013 to Dec 31, 2013

Net sales	$500,000
Cost of goods sold	$365,000
Gross profit	$135,000
Expenses	
Salary	$30,000
Wages	$10,000
Depreciation	$3000
Sales and advertising	$24,000
Utilities	$9000
Interest	$3600
Equipment rental	$6000
Insurance	$11,450
Bad debts	$1250
Supplies	$3660
Legal and accounting	$16,615
Profit before tax	$16,425

FIGURE 18.3. Example of a nursery income statement. The income statement shows income and expenses for the fiscal or calendar year.

Green Side Nursery
Balance Sheet
Dec 31, 2013

Current assets		Current liabilities	
Cash	$5,000	Accounts payable	$10,000
Accounts rec.	$500	Credit line	$0
Inventory	$12,000		
Fixed assets		**Long-term liabilities**	
Building	$75,000	Bank loan	$25,000
Vehicles	$10,000		
Equipment	$15,000	Owners equity	$82,500

FIGURE 18.4. Example of a nursery balance sheet. The balance sheet shows the financial situation of a business at one point in time, often the end of the fiscal or calendar year.

in and out on an annual basis. For Green Side Nursery, the 2013 income was $500,000, with $365,000 in cost of goods sold (or COGS) and $118,575 in expenses (or indirect costs). Sales minus COGS minus expenses yields a net income of $16,425 before taxes. Since income is not uniformly received by the company over the 12 months, sales can be distributed over the year from a historical record of the company (for forward planning) or from monthly records. The same is true for payment of COGS and expenses. The real utility of this statement is to watch the surplus or deficit accumulate as monthly cash flow (second row from bottom) and see how it compounds over the months into net cash flow (last row). December's net cash flow position should exactly match the profit before tax for a company with cash-based accounting.

As explained earlier in the chapter, the cash flow budget is essential to the planning process so that companies can strategize when and how to handle large deficits. Notice in our example for Green Side Nursery that the largest monthly cash flow deficit is in February, where a net cash flow position of −$13,541 occurs. Obtaining a line of credit, deferring payments, getting better payment terms, and shifting some payments to another month are all good strategies to help this business get through the year. Every company should construct a cash flow statement to see its history and to predict times of cash shortages.

Valuation of Company Assets

In order to correctly evaluate business performance, there must be a realistic valuation of the company's assets. Plant inventories are usually the largest class of assets in a nursery business, and this poses special challenges in accounting because growing plants are constantly changing in size, form, and value. Changes in plant inventory may be an especially significant factor in a nursery that is new or rapidly growing.

A recommended approach for dealing with changes in plant inventory is to use the degree of completion method

Green Side Nursery Cash Flow Statement, 2013

	Total	Jan	Feb	Mar	Apr	May	Jun	Jul	Aug	Sep	Oct	Nov	Dec
	($)	($)	($)	($)	($)	($)	($)	($)	($)	($)	($)	($)	($)
Income (sources of cash)	500,000	0	0	0	0	0	0	0	0	0	0	0	
		0	0	25,000	50,000	125,000	50,000	50,000	50,000	50,000	50,000	50,000	
COGS													
Wages	240,000	0	0	12,000	24,000	60,000	24,000	24,000	24,000	24,000	24,000	24,000	
FICA, unemp, 401k	68,000	0	0	3,400	6,800	17,000	6,800	6,800	6,800	6,800	6,800	6,800	
Plants	25,000	0	0	1,250	2,500	6,250	2,500	2,500	2,500	2,500	2,500	2,500	
Production supplies	32,000	0	0	1,600	3,200	8,000	3,200	3,200	3,200	3,200	3,200	3,200	
COGS	36,5000	0	0	18,250	36,500	91,250	36,500	36,500	36,500	36,500	36,500	36,500	
Expenses (uses of cash)													
Salary	30,000	2,500	2,500	2,500	2,500	2,500	2,500	2,500	2,500	2,500	2,500	2,500	2,50
Wages	10,000	833	833	833	833	833	833	833	833	833	833	833	83
Depreciation	3,000	250	250	250	250	250	250	250	250	250	250	250	25
Sales and advertising	24,000		3,600	6,000	2,400	2,400	2,400	2,400	2,400	2,400	0	0	
Utilities	9,000	750	750	750	750	750	750	750	750	750	750	750	75
Interest	3,600	0	0	900	0	0	900	0	0	900	0	0	90
Equipment rental	6,000	500	500	500	500	500	500	500	500	500	500	500	50
Insurance	11,450	954	954	954	954	954	954	954	954	954	954	954	95
Bad debts	1,250	0	0	63	125	313	125	125	125	125	125	125	
Supplies	3,660	0	0	2,745	0	0	0	0	0	915	0	0	
Legal and accounting	16,615	0	4,154	0	0	4,154	0	0	4,154	0	0	4,154	
Total expenses (total uses)	118,575	5,788	13,541	15,495	8,313	12,654	9,213	8,313	12,466	10,128	5,913	10,066	6,68
Monthly cash flow		-5,788	-13,541	-8,745	5,188	21,096	4,288	5,188	1,034	3,373	7,588	3,434	-6,68
Cumulative net cash flow		-5,788	-19,329	-28,074	-22,886	-1,790	2,498	7,685	8,719	12,091	19,679	23,113	16,42

FIGURE 18.5. Example of a nursery cash flow budget. The cash flow budget, or statement, shows the inflow and outflow of cash from the business during a certain period (week, month, quarter, year). Cash flow budgets are essential for evaluating cash flow over time as well as predicting and planning for times of cash shortages.

Photo: M. J. Dimson and J. N. Kabashima.

of valuation, based on average wholesale market prices and average production times. So, for example, if a particular crop normally requires 20 weeks to produce, and at the end of the accounting period it has been in production for 10 weeks, then this crop is 50 percent finished and would be assigned an inventory value equal to half of its ultimate wholesale value. If the firm is rapidly growing and expanding its inventory, a representative measure of financial position can be taken as an average of values at the beginning and end of the accounting period.

Owned capital in buildings, improvements, and equipment is usually assessed at original purchase cost less depreciation, known as book value. Sometimes when these assets are rapidly depreciated under IRS rules for accelerated cost recovery (ACR) over 3, 5, or 7 years, the remaining useful life and true value of the asset may be understated. For example, a nursery could have a tractor that is over 30 years old, and on the balance sheet that tractor may not have much (if any) value. Also, the value of land owned by the company is typically stated at original purchase price, which may be far below its current market value, especially during times of rapid growth in real estate prices such as in the United States during the period from 2000 to 2007. A variety of depreciation methods can be used by businesses. The choice of depreciation method can affect various financial statements such as the balance sheet, income statement, cash flow, and tax returns, as well as have implications for pricing strategies. In choosing a depreciation method, the nursery manager should consult a practicing accountant who is experienced with production nurseries in the state in which the business is operating.

One critical thought process that directly impacts the net worth, or owner's equity, has to do with equipment and depreciation. Recall that the balance sheet equation is

$$\text{assets} = \text{liabilities} + \text{owner's equity}$$

What happens to the right side of the balance sheet (liabilities) when a piece of equipment (truck or backhoe) is depreciated? While it's obvious that the left side of the balance sheet declines by the amount of value the asset lost that year, what isn't as clear is the impact on the right side of the balance sheet. The only place to make a reduction on the right side to keep the balance sheet in balance is the owner's equity or net worth. So, as assets decline in value, so does the value of the owner's equity, or net worth of the firm.

Since assets lose function over time and the value of the owner's equity declines as assets depreciate, it is necessary to rebuild or replace assets when the business generates profits. If assets are a part of the needs of the company, so is the planned replacement of them.

Performance Indicators

Another part of good financial planning is analysis of the company's financial performance. This involves more than simply constructing the income statement, balance sheet, and cash flow statement. It requires some additional calculations to create some important financial measures. The measures can be compared internally to the firm's performance last year, or they can be compared externally to the performance of other companies or published benchmark information. Most importantly, the measures should be shared within the company with key employees, because employees often overestimate how well the company for which they work is performing financially. These financial measures can be displayed in a financial dashboard, which, like the dashboard of a car, displays key information that can help the company navigate its future direction. Table 18.1 shows many of the financial metrics discussed below, as well as a typical measure that can be used as one benchmark.

Financial ratios and other metrics can be used to provide four basic analysis categories. These categories, described below, are liquidity, solvency, profitability, and efficiency.

Liquidity

Relationships among the values on a company's income statement and balance sheet express measures of financial solvency and liquidity. *Liquidity* shows a firm's ability to meet short-term debts when they become due. One indicator of liquidity is the *quick ratio*, which is calculated by dividing cash and accounts receivable by current liabilities. The quick ratio is a measure of how much the business has in liquid current assets relative to how much it owes. Cash and

Table 18.1. Common performance metrics for nursery production businesses

Metric	Calculation	Typical value*
Financial liquidity and solvency ratios		
asset to liability ratio	total assets divided by total liabilities (average of year beginning and ending values)	4.41
quick ratio	cash and accounts receivable divided by current liabilities (average of year beginning and ending values)	2.58
leverage	total assets divided by net worth (average of year beginning and ending values)	1.29
average collection period	average accounts receivable divided by the result of annual sales divided by 365 days	107 days
Profitability		
net margin	net income divided by total income	19.1%
rate of return on investment (ROI)	net income divided by total assets	15.2%
rate of return on equity (ROE)	net income divided by equity (net worth)	19.7%
earnings before interest, taxes, depreciation, and amortization (EBITDA)	net income plus interest, taxes, depreciation, and amortization added back	NA
Resource productivity and efficiency		
value produced per unit area	annual sales plus change in inventory value divided by net production area (ft^2 or acres)	$1.02/$ft^2$
value produced per employee	annual sales plus change in inventory value divided by number of full-time equivalent employees (total payroll hours divided by 2,080)	$100,720
production area per employee	net production area (ft^2) divided by number of full-time equivalent employees	98,900 ft^2
capital investment per employee	total assets divided by number of full-time equivalent employees	$206,843
capital investment per acre	total assets divided by net production area (acres)	$91,019
inventory turnover	annual sales divided by average inventory value (average of year beginning and ending values)	0.763
Cost indicators		
labor to cost ratio	expenses for production employee salaries, wages, employment taxes, fringe benefits, and bonuses divided by total annual expenses or income	33.5% of income
materials to cost ratio	expenses for plants/seeds, containers, growing media, fertilizer, agrochemicals, and packaging divided by total annual expenses or income	24.3% of income
overhead to cost ratio	expenses for management and office staff, facility repair/maintenance, equipment operation/maintenance, marketing, office, travel, entertainment, insurance, taxes, advertising, telecommunications, electric, rent, depreciation, interest, and miscellaneous other expenses divided by total annual expenses or income	28.9% of income
overhead cost per unit production area	overhead expenses divided by usable production area (ft^2 or acres)	$0.29/$ft^2$

Source: University of Florida Horticulture Business Analysis System.
Note: *Typical value is for wholesale container woody ornamental nurseries.

accounts receivable, which should be received within 90 days, are the most liquid of current assets, which are usually available on short notice; inventories are not included in this measure because they may not be immediately salable. For every dollar in bills the company has, the quick ratio shows how much cash and accounts receivable it has for paying its bills. A value for this ratio below 1 would indicate an illiquid position; that is, the firm may not have enough cash on hand to pay current bills. One of the biggest differences between the quick and current ratio is inventory. Current ratio is the ratio of current assets to current liabilities. Companies that have too much inventory on hand lower their current ratio significantly.

Solvency

Solvency is a long-range concept that shows the firm's ability to meet all debts when assets are sold. One indicator used to measure solvency is the leverage ratio. Leverage is an indication of how much the company has borrowed compared with how much it has invested.

Recall that both assets and net worth (or owner's equity) can be found on the balance sheet. Net worth is assets minus debt. Leverage is the ratio of total assets to net worth.

Many companies in the green industry have no long-term debt, so their leverage would be equal to 1. Leverage factors below 2 are generally considered to represent a relatively safe financial position. Higher values indicate

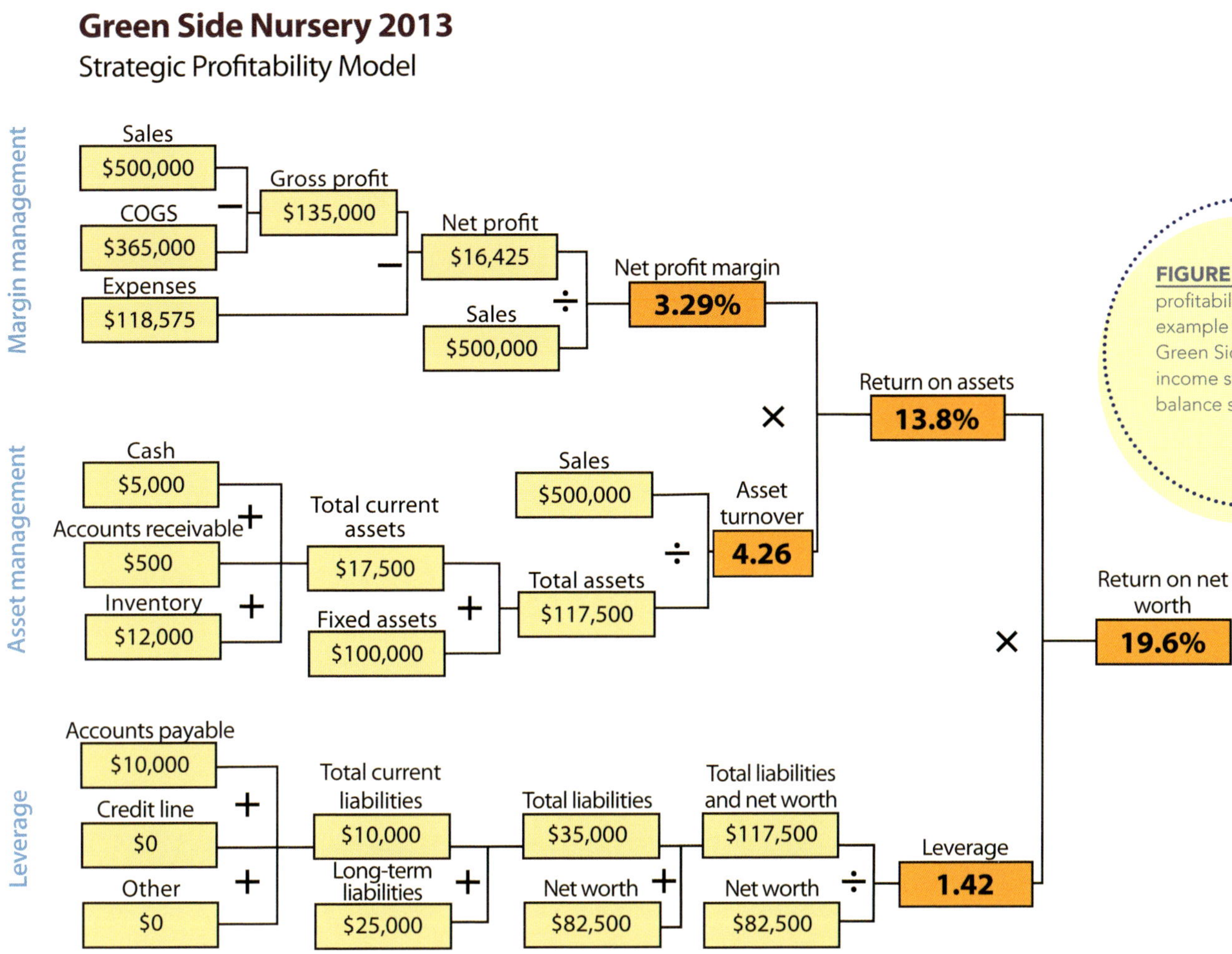

FIGURE 18.6. Strategic profitability model with example entries from Green Side Nursery's income statement and balance sheet.

greater risk, with potential for both greater returns and greater losses. A firm that has a high leverage factor may lose control of the company because others who made those loans may be more in control. Still, borrowing or leveraging may be necessary for a company to expand by purchasing more equipment or property. Purchasing assets could increase net worth if there is a good return on investment.

Profitability

The bottom line for financial performance of a company is usually *net income*, which is found on the income statement and is calculated simply as the difference between total income and total costs. A useful measure for a nursery business is profit margin, or net margin ratio. This can be calculated by dividing the net income by total income. Net margin is the ratio of net income to total income, or in other words, the share of income that is profit. Some businesses calculate *earnings before interest, taxes, depreciation,* and *amortization* (EBITDA) as a more narrow measure of profitability to exclude the effect of nonoperational changes in the business.

Some other ratios that are used to measure profitability are return on assets, rate of return on investment, and rate of return on equity. Asset turnover is the number of times a company turns assets into sales, and it is calculated as the ratio of total income to total assets. It is a measure of how effectively assets are being used to generate sales. Since assets are what a firm owns, asset turnover is a good indication of how well the business is using those things. Companies with many assets (e.g., land, equipment, buildings) often have lower asset turnover than service companies, which traditionally have substantially fewer assets. Production nurseries that own a lot of land may have an asset turnover equal to 1 or less, whereas landscape firms that rent most of their equipment may have an asset turnover greater than 15.

Rate of return on investment (ROI) is the net income divided by the total assets. *Rate of return on equity* (ROE), or *return on net worth*, is an even more refined measure of profitability, calculated by dividing net income by the net worth. Return on net worth is literally how much money or profit is returned to the owner for every dollar invested in the business (net worth, or owner's equity). This measure expresses profitability in relation to the equity of owned assets, and it is comparable to annualized yields on stocks, bonds, or savings deposits. The ROE for a business venture should be substantially higher than for a passive, risk-free investment in order to justify the considerable risk of capital investment and management effort. Typically the return on net worth ranges from 5 to 8 percent for many firms, but more profitable firms can have a return on net worth of 15 percent or more. During normal or favorable business environments, nursery managers should strive for a return on net worth of 12 percent or higher.

The *strategic profitability model*, also known as the Dupont total profitability model (see the Balancesheetwalk.com website in the Resources at the end of this chapter), is a useful tool to help understand how accounting identities among the elements of the income statement and balance sheet can be combined to calculate profitability. Figure 18.6 shows the profitability model with data from Green Side Nursery. The model contains three sections: margin management, asset management, and leverage. These three sections represent the only ways a business can make money. The business can "earn it" and make a significant net profit margin; the business can "turn it" by turning assets into sales with a high asset turnover; or the business can "leverage it" by borrowing funds and then making money with the borrowed funds. There are five key financial ratios in this model: net profit margin, asset turnover, leverage, return on assets, and return on net worth. These five key financial performance indicators can give any business a snapshot of not only how well the business is performing financially but also what the source of that performance (or lack of performance) is. The model combines two key financial statements: the income statement and the balance sheet.

Using the income statement and balance sheet from Green Side Nursery in 2013, we can complete the strategic profitability model and see how well this business performed. Using the financial information from the income statement, the top portion, or margin management section, can be completed. The net profit margin for this company was 3.29 percent. Using information from the left side of the balance sheet, the middle section, or asset management section, can be completed, showing that Green Side Nursery had an asset turnover of 4.26. Using the right side of the balance sheet to complete the bottom portion, or leverage section, we see that Green Side Nursery has a leverage of 1.42. Multiplying net profit margin by asset turnover, we see a return on assets of 13.8 percent. Multiplying that by leverage, we see that Green Side Nursery had a return on net worth of 19.6 percent. If sales decline, the metrics in the margin management and asset management sections will decline as well, since sales appears in the denominator of both.

Productivity and efficiency indicators

Beyond strictly financial information, other kinds of information on the physical and labor resources of the company may enter into an evaluation of productivity and efficiency. For nursery operations, the plant production area should be measured as the net usable growing space within growing beds and fields, excluding nonproductive space in aisles, driveways, and other service areas. This *net production area* can be used to evaluate costs and sales per square foot, or the level of *capital investment per acre* of growing area. *Value produced per unit* area is affected by nursery layout, plant selection, plant survival, vacancy rates (unused space), and plant pricing and margins. Space efficiency measures vary across types of nurseries. Efficiency measures are very useful for comparison with nurseries with similar product lines and for tracking year-to-year progress.

Plant inventory turnover is another important efficiency measure for a nursery business. Inventory turnover is calculated as the ratio of total annual sales to the average inventory of growing crops (average beginning and ending values). Inventory turnover measures the number of crops per year (or years per crop) and reflects the inherent difference in growth rates of various nursery crops. New and rapidly expanding nurseries often have lower inventory turnover because of the larger percentage of immature plants. Established nurseries should strive for an inventory turnover of 0.75 or higher.

Labor is often the most important resource in a horticultural business, and direct labor can account for 40 to 75 percent of the cost of goods sold. Labor costs include not only wages and salaries but also the benefits provided, such as worker's compensation, overtime, health and other insurances, and vacation and other leave time. It is important to track labor use in terms of physical quantity (hours) as well as monetary cost, in order to measure its productivity or efficiency. The total payroll hours of labor employed should include production, administrative costs, sales, and management personnel. Labor is often expressed in terms of *full-time equivalent* (FTE) persons, representing the number of employees working for a year at 40 hours per week, or 2,080 hours per year. Part-time employee hours should be combined to calculate the number of full-time (40-hour work week) equivalent employees in the business. It is also important to keep good labor records by crop to more accurately determine direct costs.

Common metrics to measure labor efficiency include the *value produced per employee*, (calculated as the annual sales plus change in inventory value divided by the number of FTEs), or *production area per employee* (calculated as the net production area divided by the number of FTEs). Labor efficiency can be improved by labor-saving technologies, labor management, and increasing crop turnover. For more information on improving labor efficiency, see chapter 20.

Nursery Crop Cost Analysis

Evaluation of operating costs is especially important for managers because costs are typically more subject to control than is income. For effective analysis of costs, it is important that the expenses be identified in meaningful categories. As a general rule, any cost that represents 5 percent or more of total costs should be itemized and separately tracked through the accounting system. In many nursery businesses, there may be perhaps 20 to 50 individual expense categories, which could generally be grouped into a reasonable number of major categories such as management, employee labor and benefits, materials/supplies, facility/equipment, administrative overhead, and capital (depreciation and interest). In order to evaluate meaningfully, costs should be examined as a percentage of total costs or income, as well as their absolute values.

One of the most important applications of cost analysis is to estimate the cost of producing individual products. This information is the basis for setting prices and selecting a profitable mix of products. A widely used procedure for calculating costs of individual nursery crops, known as the rental rate method, involves allocating indirect, or overhead, costs based on the time and space occupied in the nursery. This method uses the "square-foot week" concept of dividing costs by the number of square feet of production space and the number of weeks that the space is in production.

To apply the method, first determine the direct costs that can be identified, such as seed/liner, container, media, fertilizer, chemicals, and packaging, and tally these items for each crop. Some production labor expense may also be included as a direct cost if you have a piece-rate compensation system for the various nursery production tasks such as potting, pruning, fertilizing, shifting, packaging, and so on. In general, the more you can identify direct costs, the more accurate the analysis will be.

The total annual expense for the direct cost items is then deducted from the total expenses on the company's annual income statement. The remaining annual indirect expenses are divided by the net usable production area (i.e., the area occupied by plant inventory for sale) to determine the annual *overhead cost per square foot*. This figure is then expressed as an average weekly or monthly amount by dividing by 52 (weeks/year) or 12 (months/year), respectively, to arrive at the rental rate (i.e., the rate that is charged to each plant for the space and time occupied in the nursery).

For example, a crop grown in 1-gallon (6-inch-diameter) containers placed can-tight would occupy 0.25 square feet per plant. Then, if the crop is normally grown for 30 weeks, the total square-foot weeks per plant would be 7.5 (0.25 ft^2 × 30 weeks). The total cost of production is the sum of direct and indirect costs. A final adjustment is to apply a loss factor to account for the sunk costs of plants that don't survive due to pests, diseases, freeze, and so on. For example, if 5 percent of a crop is normally lost, the final cost estimate would need to be inflated by 5 percent.

A refinement of this method is to develop separate rental rates for different types of production facilities, such as greenhouses, open production fields, or propagation houses, which may have dramatically different overhead costs, and apply these rates to each crop for the time and space required in each type of facility. The rental rates should be updated annually from the annual income statement.

Example of nursery cost/profit estimator tool

The Nursery Cost and Profit Estimator is a computer spreadsheet developed cooperatively by the University of Hawaii and the University of Arizona to provide growers with a tool to track production costs and improve management of product pricing of individual items or similar product types in a nursery (see the Resources at the end of this chapter). The spreadsheet asks for input in four main categories: capital and investment, materials, machinery and equipment, and overhead. Production costs can be calculated per unit area of production space, based on how much of the available space is occupied. The spreadsheet allows comparison of different production practices, products, and time schedules, and it raises awareness of how costs are affected by maintaining or changing current practices.

The spreadsheet calculates production costs and, by using a range of high to low sales prices, can calculate whether a plant will make a profit or result in a net loss for the business. Production costs for each individual species or group of products, such as all 1-gallon pots or 5-gallon pots with similar production time and inputs, can be calculated. One feature of the computer program assesses the amount of space and time allocated for a plant to reach either the next stage of production or salable size. Using the spreadsheet tool, growers can estimate how much this production time in the nursery costs and whether this is reflected in the wholesale price. Calculations can determine which container size yields the highest profit for a species and how long the plant will remain profitable at that size.

Figure 18.7 provides an example from this computer program showing the projected profit comparison of growing plants in the nursery versus buying (brokering) stock. In this example, the grower is selling 1-year-old palms for $8.00; however, the total cost of growing this palm is $8.70, resulting in a per-plant loss of $0.70. One option available to the grower is to buy the 1-year-old palms into the operation; and in this example, the grower has a supplier who offers 1-year-old palms for $7.50 per tree. The operation can now resell the palms for the operation's original price and generate $0.50 profit. Comparing the purchase price of bought-in product at $7.50 to the own-production cost of $8.70 would result in a loss of $1.20 per plant under these condi-

	A	B	C	D	E	G	H	I	J	K	L	M	N	O
1														
2														
3					OWN PRODUCTION			PRODUCE VS BUY				SPACE USED		
4					Total Cost					Gain/Loss		12,250	12,000	Sqft used
5	#	Plant Name:	Sales		Plus Return	Expected		Purchase	Expected	from		24,000	24,000	Sqft available
6			Price		on Investment	Profit		Price	Profit	Production		Field	Shade	
7	1	Plant 1	$5.00		$4.15	$0.85		$4.50	$0.50	$0.35		-	2,000.0	
8	2	Plant 2	$9.00		$8.50	$0.50		$8.50	$0.50	$0.00		-	8,000.0	
9	3	Palm 1yr	$8.00		$8.70	($0.70)		$7.50	$0.50	($1.20)		-	1,000.0	
10	4	Palm 2yr	$15.00		$14.64	$0.36		$15.25	($0.25)	$0.61		-	1,000.0	
11	5	Plant 5			$0.00	$0.00			$0.00	$0.00		-	-	

FIGURE 18.7. Nursery Cost and Profit Estimator for comparative price analysis for palm trees.

tions. The other option would be to carry over the 1-year-old palms and sell them as 2-year-old palms. The additional cost of growing the palms amounts to $5.94 per tree; however, the grower can sell 2-year-old palms for $15.00, netting a per-plant profit of $0.36.

Without knowing the total cost of growing individual plants or groups of plants, nursery operators run the risk of underpricing their plant products. They also risk missing opportunities to buy in plant products that other firms can grow and sell cheaper than the cost of producing them in their own operation. Nursery managers can reduce risk to their operation while increasing overall profits by running a cost analysis before making these types of decisions.

Enterprise Analysis

Most nurseries have several different types of activities in their business. For example, the firm may produce both bedding plants and potted plants, and it may sell plants at different stages of production or different container sizes. In addition to opportunities to increase the efficiency of a given production mix, the nursery manager can attempt to identify a more profitable combination of crops to produce. Not all of these aspects of the firm may be equally profitable. Some segments of the operation may operate at a loss and may be supported by other operations. Enterprise analysis allows the grower to determine which items are the most profitable. The cost per unit of growing space (square feet or acres) or per unit of labor (man-hours, annual full-time equivalents) are useful measures for comparing cost efficiencies of different production systems. Designing profitable combinations and rotations of nursery crops without accurate cost information is impossible.

Trend Analysis

The changes in financial indicators from one period to the next are often as important as their absolute level. When evaluating trends in financial performance over an extended time period, say 3 or more years, it is critical that changes in values should be stated in inflation-adjusted terms, using a price index such as the consumer price index (CPI). Unless values are deflated, one can get a very misleading picture, even of the direction of trends.

For example, if a company experienced an increase in sales of 7 percent over the previous year, one might conclude that the company has healthy growth. However, if the prevailing rate of inflation is 5 percent, then the real rate of growth would be only 2 percent (7 minus 5%). As another example, consider that if you do not adjust the prices of your products over time, they are actually losing value.

The GDP implicit price deflator is a recommended macroeconomic price index that accounts for the effect of commodity price changes in the calculation of gross domestic product (GDP). The GDP deflator is published quarterly by the U.S. Commerce Department (see the Resources at the end of this chapter).

Benchmark Analysis

Financial benchmark analysis is the use of key indicators to evaluate a company's operational and financial performance in comparison with industry standards or benchmark values. Benchmark data are available for most major industries, for different types and sizes of businesses, and for regions of the country. Ideally, comparisons should be made with the leading or most profitable firms in an industry that are presumably following best practices. Some sources of information for agricultural sector industries include the Farm Credit Bank and the USDA Risk Management Agency (see the Resources at the end of this chapter), as well as land grant university research and extension programs.

The University of Florida's Horticultural Business Analysis System is a no-cost, Internet-based system for financial benchmark analysis of wholesale nursery firms (see the Resources at the end of this chapter). The system has benchmark data available for various types of nursery commodities, including woody ornamentals, flowering plants, tropical foliage, cut foliage, herbaceous perennials, and liners or plugs. In addition, information is available for large firms with $2 million or more in annual sales and for highly profitable firms with 15 percent or greater return on investment. Users may simply view the industry information available, or they can set up an account (with user name and password) and then enter their own company information for comparison with the benchmark values. Individual accounts may have data for any number of years or different types of nursery products. Industry averages are available for specific nursery commodities and locations (states or counties) having at least five valid records. Benchmark information is automatically expressed in current year dollars using the GDP implicit price deflator.

Conducting a Review

A comprehensive review of your company's financial situation should be done at least twice a year, quarterly, or, ideally, every month. It is not frequent enough to do it once a year at tax time because that is too late to spot problems and take corrective action. A good practice is to arrange for a formal meeting between the business owners, its financial advisors, and the management team to do this evaluation thoroughly and systematically. When reviewing your company's financial statements, a number of simple but important questions should be asked routinely:

- What are the total income, total expenses, net income, and profitability (net margin, rate of return on equity) of the firm in the most recent period?
- Are these values growing or declining since the previous period and over time? If the values aren't stable, is there a reason for the change? Is income growing in inflation-adjusted terms?
- What is the firm's cost structure? Are any cost items excessively high or rapidly increasing over time? What is the relationship between direct costs (cost

of goods sold) and indirect, overhead, or general operating expenses?

- What is the general financial solvency of the firm? Is the company's net worth, or equity, growing over time?
- What is the structure of capital managed in the firm? What is the condition of the company's long-term assets such as machinery, equipment, and buildings with respect to accumulated depreciation? Are any assets depleted and in need of replacement?
- What is the liquidity situation of the firm? Can current liabilities be met with expected cash flow? Is there a seasonal pattern or trend in sales or total income? What are the firm's accounts receivable in relation to sales? Is the company maintaining timely cash collections on accounts?
- What is the firm's labor situation with respect to costs for employee wages and benefits?
- Are there any changes observed in indicators for productivity or efficiency such as output per worker, output per square foot, or inventory turnover?
- How does the company's performance compare to industry standards or financial benchmarks?

Risk Management

Managing risk in nurseries is not a new concept, as the nursery industry can be volatile. Nurseries are vulnerable to both the positive and negative impacts of urban growth issues such as labor, public policy, and water and pesticide regulations, which can have tremendous impacts on the success of a nursery operation. Understanding the types of risk and the tools to manage risk is just as important as knowing how to grow the plants for a nursery business.

The Risk Management Agency, established in 1977 by the U.S. Department of Agriculture (USDA), has defined five primary categories of risk: production, finance, price or market, institutional, and human or personal. The remainder of this chapter provides an overview on each risk category and the benefits of improved risk management in these areas.

Production Risk

Production risk derives from the uncertain natural growth processes of crops and livestock. Weather, disease, pests, and other factors affect both the quantity and quality of products. Understanding production risk in the nursery industry implies an expected outcome about the yields of plant materials produced. Variability in outcomes from those that are expected poses risks to becoming a successful operation.

Enterprise diversification is an effective way of reducing income variability for an operation, and it is a common practice among nurseries. In its simplest form, it would consist of having different production processes such as container production and field production. Diversification can include different species, different types of plants (palms, shade trees, shrubs, and roses), combinations of plants and products (hanging baskets with a variety of plants, mixed plantings in large, decorative containers), or different end points in the same production process (trees sold in 15-gallon pots and larger containers such as 42-inch boxes). Diversification can also be achieved through different income sources, such as contract growing or employment outside the nursery.

Using crop insurance to cover product or price risk has the effect of transferring risk from one operation to others at a known cost for an insurance premium. Crop insurance is an example of a risk management tool that not only protects against losses but also offers the opportunity for a more predictable bottom line. When used as part of a sound management program, crop insurance can stabilize an operation after the occurrence of a catastrophic event.

Financial Risk

Financial risk results when the nursery business borrows money and creates an obligation to repay loans. Rising interest rates, the prospect of loans being called by lenders, and restricted credit availability are also aspects of financial risk. Understanding financial risk is typically the last on the list for most producers, but gaining a better understanding and control can greatly bolster the survivability of the operation. The three basic components of financial risk according to Kaan (1999) are

- the cost and availability of debt capital
- the ability to meet cash flow needs in a timely manner
- the ability to maintain and grow equity

Managing cash flow is especially important in the nursery industry, which has annual cycles of high demand for some products associated with favorable planting seasons in spring and fall. While obligations might be easy to meet during times of high cash flow, the constant turnover in inventory poses a need for purchasing inputs such as media, pots, and fertilizer, as well as making cash lease payments, tax payments, and debt repayments, which must still be met during the slower months. Controlling cash flow requires constant vigilance of current and future obligations and matching these obligations with sufficient cash inputs. Keeping and analyzing a set of well-maintained financial records is an absolute necessity for maintaining financial control of an operation. Prior sections of this chapter discuss in further detail the importance of understanding more complex financial ratios and performance indicators, as well as other metrics to evaluate liquidity, solvency, profitability, and financial efficiency.

Price or Market Risk

Marketing and pricing are discussed in chapter 19, but a brief discussion on the risks associated with these activities is presented here. Price or market risk refers to uncertainty about the prices producers will receive for commodities or the prices they must pay for inputs. The nature of price risk varies significantly from commodity to commodity. Similar to the financial and production risk areas, the unpredictability of the weather and foreign and domestic government actions affect the pricing, movement, and availability of nursery products and input supplies. The effects of regulations such as agricultural programs, embargoes, quarantines, and trade tariffs can, in extreme cases, instantly shut down the normal marketing channel for a variety of nursery products.

The key to handling market risk is understanding the market and determining one's comfort level for risk. Once it has been established how much risk an operation is willing to assume, a marketing plan can be developed to control risk exposure. An obvious way of managing market risk is to not participate in high-risk ventures or strategies. For example, a business should not jump into an unfamiliar venture because prices of a product are high. Shifting risk to others can be accomplished by executing contracts for the purchase of inputs or for the prices received for the commodity. It can also be accomplished by developing a flexible marketing plan that has the ability to take advantage of low input prices. Strategies to reduce price and market risk include the following:

- knowing current market trends and information
- spreading out sales over time
- cultivating a larger customer base to include wholesale, retail, and direct market customers
- outsourcing the production of certain products

Institutional Risk

Institutional risk results from uncertainties surrounding government rules and regulations. Tax laws, labor laws, regulations for chemical use, and rules for waste or runoff disposal are examples of government decisions that can have a major impact on nursery businesses. Most individuals refer to this area of risk as legal risk, with activities involving legal documentation such as loans. Loans and other contractual arrangements have legal implications regarding repayment specifications, product quantity, price, and delivery of plant materials. Human resource issues also have legal implications, ranging from employer/employee rules and regulations to inheritance laws. As we described earlier in the chapter, one of the critical legal issues a new nursery business faces occurs before the business begins, when the legal structure of the new business is determined.

Human Resource, or Personnel, Risk

Managing human resources is presented in detail in chapter 20, but a discussion of the risk factors is important here. Human resource risk refers to factors such as problems with employee health or personal relationships that can affect the nursery business. Accidents, illness, death, and divorce are examples of personal crises that can threaten a business. Nursery operations are acutely aware of the human resource management issues and the risks associated with family influences and employees. Human resource risk plays a role in the daily activities of a nursery operation, including coordinating work flow in plant production; recruiting and selecting employees; and supervising, training, and motivating employees. A well-documented process of correcting employees and taking disciplinary actions when necessary is an important part of reducing human risk to the operation. Another step toward reducing human risk is cultivating effective communication with employees and others. Effective communication should start when prospective employees read the help wanted ad and should continue throughout the tenure of those hired.

One often overlooked human resource risk is the loss of a key employee or family member. Having life insurance and an up-to-date estate plan for key employees is essential in helping a business through unexpected loss. Life will be difficult enough for family and business partners after a key individual's death without the extra burden of unresolved financial issues.

Rosenberg et al. (2002) published a guide for managing agricultural labor that includes ways to recognize and manage risks as well as tools and tips to help navigate through many of the common situations related to personnel management. This is a good reference to assist nursery businesses with human resource management issues.

Examples of Nursery Crop Insurance Tools

What follows are brief descriptions of two crop insurance tools administered by the USDA Risk Management Agency (RMA) that are available to most nursery operations: the Nursery Crop Insurance program and the Adjusted Gross Revenue program. The plans in both of these programs do not cover buildings and equipment, which are typically covered under the nursery owner's personal or business casualty policies. The Nursery Crop Insurance and the Adjusted Gross Revenue programs are updated on an annual basis, and more in-depth information can be found at the USDA Risk Management Agency website (see the Resources at the end of this chapter). The loss calculation examples included in the insurance program descriptions are adapted from the USDA RMA 2009 and 2010 Risk Management Agency Fact Sheets. These examples are for general illustration purposes only. Contact a private insurance agent to get additional information.

The Adjusted Gross Revenue-Lite (AGR-Lite) is a crop insurance program that we will not discuss in detail because it is not available in California. However, this program is available to nursery growers in many other states. This insurance program provides protection against low revenue due to unavoidable natural disasters and market fluctuations that affect income during the insurance year. AGR-Lite covers cut flower and field crops in addition to most nursery crops. It is similar to the Adjusted Gross Revenue program except for a few differences. More information can be found at the USDA Risk Management Agency website.

Nursery Crop Insurance Program

What's covered

The Nursery Crop Insurance program is a production insur-

Table 18.2. Distribution of premium cost and subsidies of the USDA Risk Management Agency (RMA) nursery crop insurance program*

Item	Percent					
coverage level	50	55	60	65	70	75
premium subsidy	67	64	64	59	59	55
grower's share of premium	33	36	36	41	41	45

Note: *Distribution is based on coverage levels selected for nursery crop insurance. In addition to the grower's share of the insurance premium, the program has an annual administrative fee when the policy is purchased at the catastrophic coverage level.

ance program that provides coverage of nursery products in the event of a significant damaging event. Other crop commodities that a nursery may be growing, such as cut flowers, are not covered in this program. Causes of loss covered by insurance are adverse weather conditions such as wind, hurricanes, and frost (provided adequate freeze protection measures were taken). Failure of irrigation water supply, delay in marketability of plants, fire, and wildlife damage are covered as well under the specific conditions listed in the contract.

Coverage levels

Nursery crop insurance coverage levels range from 50 to 75 percent of a nursery's plant inventory value (table 18.2). A report of the plant inventory value for all approved plants in the operation is derived from the nursery's wholesale catalog or plant listing. Eligible plants are listed in the program's Eligible Plant Listing/Plant Price Schedule. Plant prices in the Eligible Plant Listing/Plant Price Schedule are based on regional averages of plant price lists submitted by insured firms.

Cost. The cost of this crop insurance (the insurance premium) is subsidized by the USDA RMA (see table 18.2). Premium subsidies range from 67 to 55 percent of the premium.

Availability. Available to U.S. nursery growers in all states.

Loss calculation example. In a hypothetical example, a nursery operation with plant inventory originally valued at $100,000 had selected a policy with a 75 percent coverage level. The insured value in this case was $75,000. The nursery suffered a $50,000 value loss of plant material due to a disaster covered by the policy ($100,000 – $50,000). Subtracting the deductible of $25,000 leaves an indemnity (payment) of $25,000 to the policyholder. The calculations for this example are listed below.

$100,000 (plant inventory value) × 0.75 (coverage level) = $75,000 (amount of insurance)

In the event of loss:

[$100,000 (field market value before loss) – $50,000 (field market value after loss) = $50,000 (value of loss)] – $25,000 (deductible 0.25 [1-coverage level] × $100,000 [plant inventory]) = $25,000

Adjusted Gross Revenue (AGR) Program

What's covered

The Adjusted Gross Revenue (AGR) program is a whole-nursery revenue protection plan of insurance similar to the AGR-Lite plan. Revenue insurance plans—unlike traditional programs like Nursery Crop Insurance that center on yields and the production year—focus on whole-nursery income and the tax year. Crop commodities that are covered in the AGR program are far more inclusive than the Nursery Crop Insurance program. However, growers who produce only nursery crops should note that they may need to also have traditional federal crop insurance because they will be limited to the lower AGR coverage levels when more than 50 percent of the expected income is from one insurable commodity. AGR insurance is provided against loss of revenue due to unavoidable natural occurrence including, but not limited to, adverse weather, fire, insects, disease, wildlife, earthquakes, volcanic eruption, failure of irrigation water supply that causes production losses during the current or previous insurance year, or market fluctuations that cause a loss in revenue during the current insurance year. The insurance policy also lists specific conditions where no payment will be made.

Coverage levels

AGR insurance coverage level rates range from 65 to 80 percent, with two payment options on the insured revenue loss of 75 and 90 percent. The payment rate is the percentage of the insured revenue loss that will be paid to the policyholder, and the percentage of the loss that is not paid is the deductible; for example, a 90 percent payment rate means that 10 percent of the revenue loss is deductible. The payment rate in the AGR program is not comparable to the deductible amount in the Nursery Crop Insurance program, which is based on coverage level and inventory. The combination of the highest coverage and payment rate (80% coverage and 90% payment rate) requires a minimum of three separate commodities to be grown on the operation. AGR has a liability limit at $6.5 million (AGR-Lite is only $1 million).

Cost. The premium varies according to the selected coverage level. AGR can be the sole insurance for a nursery or it can be used in conjunction with other federal crop insurance plans (except for the AGR-Lite plan). When producers purchase both AGR and other federal crop insurance, the AGR premium will be reduced. Similar to other crop insurance programs, the federal government subsidizes AGR. The program currently has an annual administrative fee in addition to the producer premium ($30 in 2009). AGR must be purchased on or before January 31.

Availability. AGR is available in many states and in the following counties of California: Fresno, Kern, Tulare, Riverside,

San Diego, San Joaquin, San Luis Obispo, and Ventura.

Loss payment example. In this hypothetical example, another nursery operation with three separate crop commodities and plant inventory originally valued at $100,000 had selected an AGR policy with an 80 percent coverage level and a 90 percent payment rate. The operation in this case experienced a $50,000 shortfall in revenue ($80,000 – $30,000) due to an insured disaster. With the 90 percent payment rate option, the operation will receive an indemnity (payment) of $45,000. The calculations for this example are listed below.

liability of $100,000 (plant inventory value) × 0.80 (coverage level) × 0.90 (payment rate) = $72,000

loss inception point (when a payment is triggered) of $100,000 (plant inventory value) × 0.80 (coverage level) = $80,000

In the event of loss:

$80,000 (loss inception point) − $30,000 (revenue to count [actual income from the operation]) = $50,000 (revenue shortfall)

$50,000 (revenue shortfall) × 0.90 (payment rate [deductible]) = $45,000 (indemnity [payment to policyholder])

References

Avent, T. 2003. So you want to start a nursery. Portland, OR: Timber Press.

Carman, H. 2011. Economic aspects of the California nursery and floral industry, 2001–2009, Giannini Foundation Information Series Report 11-1. Giannini Foundation website, http://giannini.ucop.edu/InfoSeries/111-Nursery.pdf.

Dobbins, C., and C. Ehmke. 2007. Principles for structuring small businesses and farms. Purdue Extension EC-729. Purdue University Extension website, http://www.extension.purdue.edu/extmedia/EC/EC-729.pdf.

Ehmke, C., and J. Akridge. 2007. The elements of a business plan: First steps for new entrepreneurs. Purdue Extension EC-735. Purdue University Extension website, http://www.extension.purdue.edu/extmedia/EC/EC-735.pdf.

Harwood, J., R. Heifner, K. Coble, J. Perry, and A. Somwaru. 1999. Managing risk in agriculture: Concepts, research and analysis. Agricultural Economic Report No. 774. Washington, DC: USDA Market and Trade Economics Division and Resource Economics Division, Economic Research Service.

Holcomb, R., G. Muske, and P. Kenkel. Developing a business plan for value-added agricultural products. 2010. Oklahoma Cooperative Extension Service Bulletin F-909. Stillwater, OK: Oklahoma Cooperative Extension. Oklahoma Cooperative Extension Service website, http://pods.dasnr.okstate.edu/docushare/dsweb/Get/Document-981/AGEC-909web.pdf.

Kaan, D. A. 1999. An introduction to risk in agriculture: Risk and resilience in agriculture. Colorado State University Cooperative Extension Service Article 1.1.

Lobo, R. E., B. D. Wallace, K. L. Robb, and S. A. Parker. 2006. San Diego County agricultural directory and guidelines for agricultural enterprises. University of California Cooperative Extension.

Owen, L., P. Davidson, and G. Grant. 1991. British Columbia Ministry of Agriculture and Fisheries. Preparing a business plan—A guide for agricultural producers: Greenhouse floriculture. British Columbia Ministry of Agriculture and Fisheries website, http://www.agf.gov.bc.ca/busmgmt/bus_guides/flori_guide.htm.

Rosenberg, H. R., R. Carkner, J. Hewlett, L. Owen, T. Teegerstrom, J. E. Tranel, and R. R. Weigel. 2002. Ag help wanted: Guidelines for managing agricultural labor. Western Farm Management Extension Committee. AgHelpWanted.org website, www.AgHelpWanted.org.

Schuch, U. 2006. Questions to ask when planning to start a wholesale plant nursery. Arizona Cooperative Extension Publication AZ1393. University of Arizona Cooperative Extension website, http://ag.arizona.edu/pubs/garden/az1393.pdf.

USDA RMA (United States Department of Agriculture-Risk Management Agency). 1997. An introduction to risk management. Washington, DC: USDA RMA.

———. 2009. Nursery commodity insurance: A Risk Management Agency fact sheet. USDA RMA website, http://www.rma.usda.gov/pubs/rme/nursery.pdf.

———. 2010. Adjusted gross revenue-lite: A Risk Management Agency fact sheet. USDA RMA website, http://www.rma.usda.gov/pubs/rme/agr-lite.pdf.

Resources

AgPlan. Center for Farm Financial Management at the University of Minnesota. AgPlan is a free online business planning tool with resources and tips on developing business plans that are customized to different types of businesses. AgPlan allows you to share your business plan and interact with business advisors, educators, or consultants. AgPlan website, https://www.agplan.umn.edu/.

Balancesheetwalk.com website, http://www.balancesheetwalk.com/.

Bplans.com. This site offers a large online collection of free industry-specific business plans (including agricultural businesses, though none specifically for nurseries). In addition, it offers resources and information on managing businesses, including free online advice and interactive tools and calculators. Bplans website, www.bplans.com.

Business Planning Resources. Lawson, D. S., and G. Stephenson. Oregon State University Cooperative Extension. eOrganic article. List of business planning resources available, including publications, software, training courses, and technical assistance. Cooperative Extension System website, http://www.extension.org/article/18375.

Center for Business Planning. Offers samples of winning business plans, software, resources, links, and helpful articles. Center for Business Planning website, www.businessplans.org.

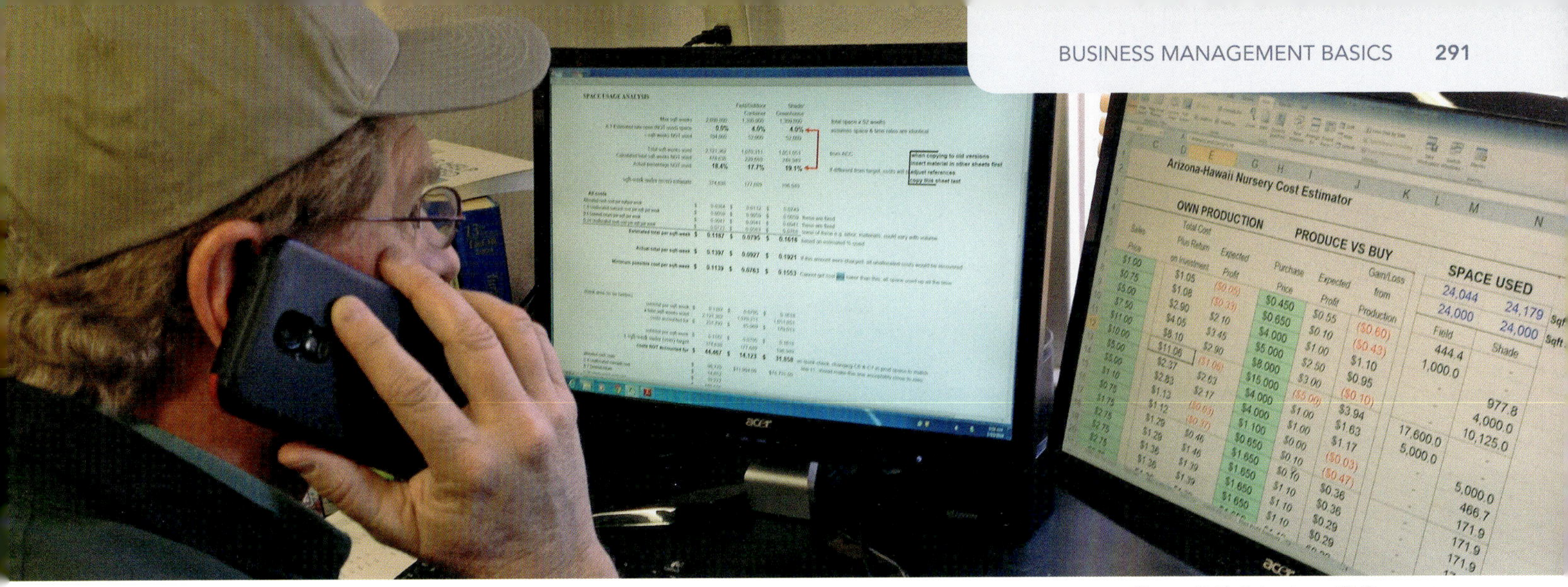

Photo: U. K. Schuch and T. Teegerstrom.

E-Myth. Gerber, M. 1988. New York: HarperBusiness. The title refers to the myth that great entrepreneurs are born from great technicians. The first third of the book is written in an entertaining manner, describing the three roles in every business: technician, manager, and entrepreneur. The last two-thirds of the book describe the business planning process and provide a framework or template for creating a business plan.

Entrepreneur Magazine. Contains topics related to business management, including starting a business, legal forms, business plans, and sample plans, and gives links to nationwide resources. Entrepreneur website, www.entrepreneur.com.

eStatement Studies. Risk Management Association (RMA). eStatement Studies is an online source of comparative data that comes directly from the financial statements of small- and medium-size business customers (including the agricultural businesses) of RMA's member institutions. Data is from two RMA Annual Statement Studies publications: *Financial Ratio Benchmarks* and *Industry Default Probabilities and Cash Flow Measures*. Risk Management Association website, http://www.rmahq.org/tools-publications/tools/estatement-studies.

Farm Credit Bank. The Farm Credit System is a nationwide network of borrower-owned lending institutions and affiliated service entities that provide credit at competitive interest rates for agriculture in every state. Funds for loans are obtained through the issuance of Farm Credit debt securities in the domestic and global capital markets. The proceeds are provided to or used for the benefit of domestic agricultural producers and cooperatives through the various Farm Credit System entities. The Farm Credit Bank can also be a source of benchmark financial data. Federal Farm Credit Banks Funding Corporation website, http://www.farmcredit-ffcb.com/.

GDP Implicit Price Deflator. U.S. Commerce Department. This is a recommended macroeconomic price index that accounts for the effect of commodity price changes in the calculation of gross domestic product. The GDP deflator is published quarterly. Federal Reserve Bank of St. Louis Economic Research website, http://research.stlouisfed.org/fred2/data/GDPDEF.txt.

Horticultural Business Analysis System. University of Florida. This is a no-cost, Internet-based system for financial benchmark analysis of wholesale nursery firms. University of Florida Horticulture Business Analysis System website, https://hortbusiness.ifas.ufl.edu/analysis/.

Nursery Cost and Profit Estimator. University of Hawaii and University of Arizona. This a computer spreadsheet that provides growers with a tool to track production costs and improve management of product pricing of individual items or similar product types in a nursery. The cost of this software is $35 to $50, and the application for purchase is available at the University of Arizona Cooperative Extension website, http://cals.arizona.edu/extension/riskmanagement/managing_finance_risk/nurserysoftwarepurchaseform.pdf.

PlanWare. Offers free downloadable business planning software, a business plan guide, financial and strategic planners, and free online advice. PlanWare website, http://www.planware.org/.

Setting Up Business in California: A Guide for Investors. California Labor and Workforce Development Agency. 2008. California has numerous government agencies, boards, and offices with which a business must interact. The number of government entities, along with their own registration and filing requirements, may seem overwhelming at first. The goal of this booklet, therefore, is to present an orderly, functional approach to the services offered or filings required by the government entities and to explain their relationship to the business investor. California Governor's Office of Business and Economic Development website,http://www.business.ca.gov/LinkClick.aspx?fileticket=sKWsQuY8Kag%3D&tabid=247.

USDA Risk Management Agency. Crop insurance programs and resources are offered by this government agency. Publications related to nursery crop insurance programs are available at the USDA Risk Management Agency website, http://www.rma.usda.gov/pubs/rme/fctsht.html.

U.S. Small Business Administration (and Small Business Development Centers). Offers many resources and much information on business plans, starting a business, and business management. Before you start your business plan, make an appointment with your local Small Business Administration center to discuss what type of information your business plan should contain. This will save you a great deal of time and energy and start you in the right direction. U.S. Small Business Administration website, www.sba.gov.

Photo: J. H. Lieth.

Marketing Considerations

Bridget K. Behe

CHAPTER NINETEEN

Marketing is an integral component of the nursery business, and knowing how to market is every bit as important as knowing how to produce plants. Yet many nursery managers neglect to develop a marketing plan, and marketing is often poorly understood by those who do try to develop marketing strategies. To better understand what marketing is, a good approach is to study the American Marketing Association (AMA) definition, which is "the activity, set of institutions, and processes for creating, communicating, and delivering value and exchanging offerings that have value for customers, clients, partners, and society at large" (AMA 2013).

Nursery owners and managers striving to develop effective marketing plans should focus on several key words in the AMA marketing definition. Perhaps the most important word to focus on is value. A business makes products, provides services, or develops experiences that its customers value and are willing to part with hard-earned cash to have. Every business must determine how or what value will be created, and this should be reflected in the business's mission statement (a statement of why they are in business). Without value there are no sales, no profits, no reason for being in business. For the most part, commercial nurseries are concerned with products, but it is important to realize that customers who sell to end users (households) may place more value on services and the experiences those products deliver or create than on the products themselves.

Another key part of the AMA marketing definition relates to who the recipients are of the created value. Customers, clients, partners, and society at large may be the primary beneficiaries. But employees, suppliers, investors, and the local community, as well as the environment, should also be considered. The wholesale nursery business provides value in many ways, not just by providing plants to its customers. For example, a wholesale nursery is a partner with manufacturers and distributors, interacting with a wide range of stores and professionals. As such, it has the purview to provide individual retail and landscape businesses with market knowledge, advertising assistance, or even product maintenance at the retail setting (as sometimes is the case in pay-by-scan retailers). Thinking more broadly about value is more likely to lead to having a more successful marketing plan.

Photo: J. K. Clark.

Finally, nursery managers should focus on three key activities described in the AMA marketing definition, which are the heart of marketing: creating, communicating, and delivering value. Marketing is a set of these processes, which makes it more than one activity. All too often, marketing is considered synonymous with advertising. Advertising is paying to communicate business value in a very public way. There is much more to marketing than just advertising. Communicating implies listening and watching what customers buy (and don't), what problems and questions they have, and where and why they are buying other products.

Another way to better understand how to market is to look at marketing models or frameworks. One common marketing model is the product approach, which is to develop and produce a product as good as possible to market and generate sales. The "Four P's" marketing model provides managers with a tool to create and implement appropriate combinations of marketing strategies and tactics. It identifies four basic variables that managers can use to fulfill their marketing objectives. The Four P's are product, price, promotion, and place (or distribution). Another more recent marketing model is the consumer approach, which studies consumer wants and needs and manufactures products in response to those needs. In the "Four C's" marketing model, the "C's" stand for customer/consumer, cost/value relationship, communication, and convenience. Many marketers consider the Four C's framework more applicable today than the original Four P's model. Times have changed, and you can no longer sell whatever you can make. The product characteristics must now match what someone specifically wants to buy. Remember the "C's" and "P's," as the material presented in this chapter is organized according to these models. Also, these frameworks can help the nursery develop a four-part marketing plan, where each dimension is addressed in terms of goals or achievements that the firm wants in each area.

Customers

The customer/consumer is the first "C" of the consumer-approach marketing model, and it is therefore a good lead topic for a discussion on marketing. Customers of producers and wholesalers can be quite diverse in their needs, preferences, and attitudes. Customers include retail garden centers and landscape professionals, whose focus is primarily plant centered, as well as supermarkets, box stores, and home centers. Customers may also be rewholesalers or other growers who seek to diversify their product mix.

In a national industry survey of eight geographic regions that closely correspond to the farm production regions delineated by the U.S. Department of Food and Agriculture (USDA) (fig. 19.1), most U.S. nurseries reported selling more than 50 percent of their product at wholesale, with the exception of nurseries in the Great Plains region, where the figure was just under 31 percent (Behe, Dennis et al. 2008). Who are the customers that buy these wholesale plants? The researchers found that the largest group of customers in all U.S. regions was landscape service professionals, who purchased 25 to 30 percent of total product sales. Rewholesalers were the second largest customer group, purchasing on average 21 percent of total sales, and single-location garden centers were typically the third largest customer group, buying an average of 14.2 percent of sales.

Keep in mind that these are the percentages across all eight regions of the United States. Customers vary in their importance (from the percentage of product they purchase) by region of the country. For example, among nurseries in the Pacific region (which includes California, Washington, Oregon, and Hawaii), sales at wholesale are greater than the national average, averaging 64.6 percent. The largest group of wholesale customers in the Pacific region in 2003 was rewholesalers, who comprised 30.1 percent of sales at wholesale (higher than any other region). Landscape firms (15.4%) and single-location garden centers (15.3%) were the second and third largest customer groups for Pacific region nurseries. Mass merchants (8.5%), multiple-location garden centers (4.4%), and home centers (3.5%) comprised the remainder of the key customer groups in the Pacific region.

For U.S. flowering plants and other floral products, box stores and other mass merchants are important customers. Yue and Behe (2008) reported that the percentage of floral transactions accounted for by box stores climbed to approximately 35 percent in 2005, while the percentage of transactions accounted for by traditional floral retailers declined to less than 25 percent. This means that box stores continue to have a strong presence in the market that is not likely to decline in the future, and they will shape the product mix that many customers see.

The studies mentioned in this section provide important information on who the customers are that are buying wholesale plants. However, nursery managers cannot rely on research studies alone, because the importance of various customer groups also varies among individual nurseries even within a region. Nursery managers who take the time to determine exactly who their key customers are often become more successful in anticipating and meeting the needs of those core customers.

Consumer Demand

At the heart of marketing is a focus on the consumer. While the direct customer may be a retail garden center, box store, or landscape professional, all eyes should be focused on the end user or consumer. The farther away from customers a business is, the more challenging it may be to connect with or identify with those customers. Growers or wholesalers may be further removed from the end user (consumer) than retailers or landscape contractors, but they still need to understand consumer demand. Savvy business managers realize that a happy retail or landscape customer is also an indirect but happy customer for the wholesaler and grower. Growers and wholesalers can help their customers tremendously by aiding with promotional efforts and collecting consumer information from a wide variety of customers and distributing that back to all customers.

Consumers aren't all alike, and that makes marketing interesting and challenging. If one plant suited everyone in every environment and situation, things would be pretty boring. The flip side of a heterogeneous demand is that business managers need to pay careful attention to markets and changes in consumers to anticipate demand for a sustainable business. Nursery crops cannot be created in a matter of hours or even days, but rather months and years are needed for lead time.

Just as consumers aren't all alike, neither are all businesses or commercial consumers. Builders, developers, and property managers all have different needs, opinions, and budgets regarding landscaping. And they have different perceptions about the benefits of good landscaping, such as improving occupancy rates, helping hospital patients and visitors relax, and improving the aesthetics of an office building. So it is also imperative for wholesalers to work with landscape contractors to best meet the needs of their commercial clients.

Furthermore, understanding the different needs of your

Photo: E. Kilmartin.

customer base will go a long way toward adding product value in addition to offering high-quality plants. For example, rewholesalers likely have similar needs, but they are somewhat different from those of landscape companies. Rewholesalers may never take physical possession of the plants, may need everything priced with scanner tags, and may be keenly aware of transportation costs (affecting the size and packaging of plants). In comparison, landscape companies won't need items priced individually but may need them delivered to the installation location.

Consumer research

One strategy that nursery managers can use to better understand what products consumers purchase is to examine consumer research reports. Unfortunately, consumer research with regard to horticultural purchases isn't abundant. However, the National Gardening Association reports annually on the percentage of consumers who make purchases of a variety of plants, including nursery trees and shrubs.

Consumer research also reveals differing purchasing habits of different consumer groups that could be used in developing a marketing strategy. For example, Dennis and Behe (2007) showed that as U.S. consumers get older, the percentage who purchase different kinds of plants (e.g., annuals, perennials, trees, and shrubs) increases, but income eliminates many of the differences observed between different age groups. The same study showed that Caucasians comprise the majority of the market for ornamental horticulture products. Differences were found among consumers of different ethnic backgrounds, but these were substantially reduced as income level rose. This means that there is greater diversity in the market for horticultural products at lower income levels than at upper income levels.

In addition to providing information about what products consumers are purchasing and differing purchasing habits among consumer groups, consumer research can provide

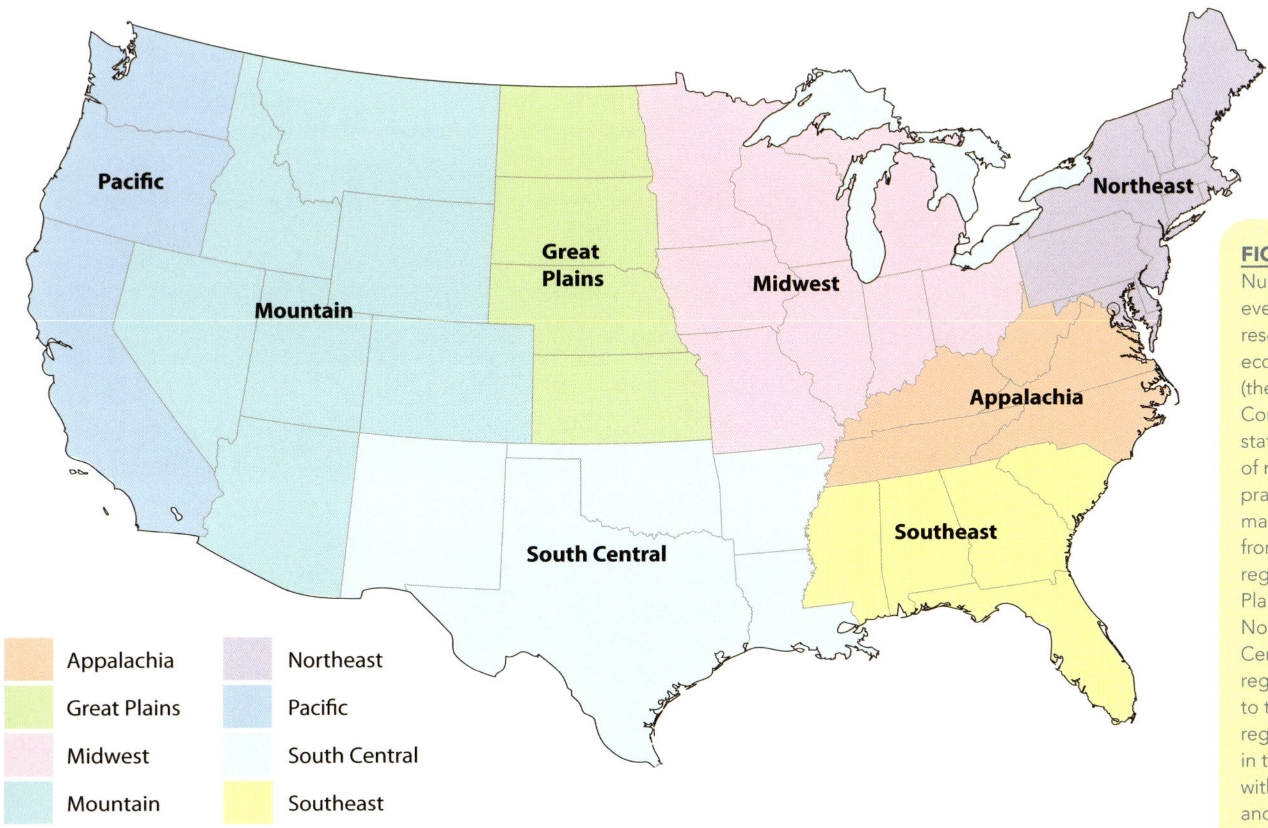

FIGURE 19.1. The National Nursery Survey is conducted every 5 years by a multistate research committee on economics and marketing (the Green Industry Research Consortium) to provide state and regional analysis of nursery production practices. This includes marketing information from eight geographic regions: Appalachia, Great Plains, Midwest, Mountain, Northeast, Pacific, South Central, and Southeast. The regions closely correspond to the USDA farm production regions. California is located in the Pacific region, along with Washington, Oregon, and Hawaii. *Source:* Behe, Dennis et al. 2008.

other helpful insights. For example, Behe et al. (2005) determined that a good landscape can add approximately 10 percent to the perceived value of a home. This study can be provided by nursery growers to customers when interacting with them to help their customers make more sales to consumers. The National Gardening Association is another example of a source of information of interest to nursery production managers regarding consumers. For example, they report that the percentage of American householders who have cared for trees has remained relatively constant over the past decade, hovering between 20 and 25 percent, but a slightly higher percentage of homeowners (30 to 35%) have participated in general landscaping (Butterfield 2006).

Trends

Many consumers today are exhibiting a greater degree of environmental awareness and concern for environmental sustainability. This has led to a desire for products that not only solve the needs of consumers but also are produced and marketed using sustainable production and business practices. Some nursery growers have used the environmentally responsible practices they have implemented as a marketing tool through various "green" eco-label certification programs. Consumers increasingly place a greater emphasis on product packaging, and this has carried over to the greenhouse/floral sector in the form of biodegradable pots. While various forms of these eco-friendly pots have been available for several years, their marketing appeal was limited due to their less-than-satisfying appearance.

However, with the recent availability of more attractive biodegradable plant containers, their consumer acceptance has grown and a renewed interest in their suitability in the green industry has emerged. Hall et al. (2010) investigated the characteristics of three biodegradable pots that consumers found most desirable, and they identified seven consumer segments. They conducted an Internet survey with 535 people from Texas, Michigan, Minnesota, and Indiana. Results showed that, on average, consumers liked rice hull pots most, followed by straw pots, and then plastic and wheat starch containers. The six market segments identified had consumer profiles that described their behaviors, attitudes, and container preferences: "rice hull likers," "straw likers," "price conscious," "environmentally conscious," "carbon sensitive," and "nondiscriminating." Diverse marketing strategies can be implemented by businesses to market biodegradable containers to specific consumer segments. Think about the pricing and packaging needs of different customers as you plan production.

Conducting Your Own Market Research

Businesses should engage in their own market research, but it doesn't need to be terribly expensive. Every business should collect customer zip codes. Because zip codes correspond to small areas geographically, the people who live within the same zip code area tend to be more similar to each other than not. Zip code collection can be part of the cash register checkout process, or it can be done by simply keeping a tally sheet at the checkout counter. If wholesale orders are taken by telephone, zip codes will probably be part of the information collected. Zip codes not only help a company understand where their customers are located geographically, but they also can provide some insight into who those customers really are. Many companies offer customer information by zip code, some for free and many for purchase. This information can be useful in planning communication activities and exploring new areas for market expansion. One of the better sites for free demographic information is the American FactFinder website (see the References at the end of this chapter).

Measures of customer satisfaction are very often used by many businesses as an indication of success with their customers. Reichheld (2003) published an easy-to-use, single-number measure for customer satisfaction. It is called the *net promoter percentage*, according to which a business can simply ask one question: "How likely are you to recommend us to a friend?" The scale used is 0 (highly unlikely to recommend) to 9 (most definitely likely to recommend). Reichheld reported that customers who respond with 8 or 9 are promoters of the business. Customers who respond with 6 or 7 are satisfied, but customers who respond with 0 to 5 are detractors. The net promoter index is calculated by comparing the ratio of promoters to detractors. In the businesses that Reichheld evaluated, firms with a net promoter percentage of 0.75 (or 3 promoters to 4 detractors) were doing a very good job and were growing in sales and profits. Many nonhorticultural companies are using this net promoter percentage as one measure of success in customer satisfaction. It is simple and reliable, so it can be easily used with confidence. Then, as changes are made, new evaluations of the net promoter percentage can help determine the impact of those changes.

One other way to keep in touch with customers is to periodically have focus groups or customer debriefing sessions. Consider hosting one session every 6 months, giving sufficient time for customers to have used, experienced, and fully interacted with the product. Plan to invite 15 to 20 customers who are similar in behavior. Perhaps they didn't make any purchases from the company this year, or perhaps they all purchased annuals or some other product. If the event is held on a weekday, estimate that 50 percent of those invited will probably attend. Begin on the half hour (5:30 or 6:30 p.m.) and serve some nonmessy, easy-to-eat finger food. Full stomachs help mouths talk more freely than hungry stomachs. Begin the questions on the hour and promise (and keep the promise) to be done by the beginning of the next hour. Offer them an incentive (free products work better than certificates for a percentage off) to stay and participate through the full session. Ask permission to record the session, and have two or three employees present to also listen to the responses firsthand. Designate one person to ask the questions.

The questions are the challenge because there are so many to be asked. Narrow the focus to five questions at most. Ask about challenges or problems; while it is tempting to ask questions about how well things went, it is better to ask about things that didn't go well. For example, ask what the single biggest challenge was facing the gardener this season (and maybe also the second and third biggest if weather was the first). Ask what they are buying elsewhere and why. Ask why they left without making a purchase. These types of questions may show some areas where the nursery business can make changes, improving customer response, increasing future sales, and perhaps boosting profitability. Discovering the answers to these questions might produce some actionable items. The cost of these sessions would include invitations, food, and employee preparation and assistance, but it will probably be well worth the investment.

Products

The product is the first "P" of the product-approach marketing model, and it is also an important topic to lead any discussion on marketing. Product mix in nurseries varies widely. Some nurseries are specialized, producing a long list of cultivars of only a few genera, or perhaps only one genus. These producers often have a climatological advantage for growing the genus (e.g., *Abies, Hydrangea, Rosa*), and they are suppliers more often for specialty retail garden centers and wholesalers than for large chain stores, box stores, or home improvement centers. Product mix is usually wider for the wholesaler who supplies the chain store, box store, or home improvement center. This producer may have several or many different types of genera with fewer cultivars being grown compared with the specialized grower. The real priority is volume, or numbers of plants. Still, most wholesale growers produce a significant number of genera and varieties, although the genera may be limited to one product line such as bedding plants or nursery stock.

Hodges et al. (2008) showed that Pacific region nurseries marketed products in three key ways: containers (60.8% of all products), bare root (12.6%), and ball and burlapped (6.5%), with "other" accounting for 15.3 percent. They also reported that whereas most U.S. regions had a significant share of total production from native American plants (approaching or exceeding 20% of total sales), this trend was not reflected in the Pacific region. On average, sales of native plants in the Pacific region was 13.2 percent of total production, a lower figure than in all regions except the Northeast. However, California is facing a dire water shortage, and there are signs that the production of native plants is increasing in this state. Government-mandated water budgets for commercial landscape plans are creating a higher demand for drought-tolerant plants, including native plants, and some nursery growers are finding a niche in this growing market.

Brands are increasingly more prevalent in the horticulture industry. Branding a product helps to differentiate it from other unbranded, generic, products. The horticulture industry has not often used the cultivar name as a brand, but it is one mechanism to distinguish one plant from another. A few plants, like holly 'Blue Prince' and 'Princess,' might be recognizable by their cultivar name. Some brands, like Proven Winners and Color Choice, have hundreds of genera under their umbrella. Others, like Flower Carpet and Knock Out roses, have a handful of cultivars and colors. More traditionally, the grower name or business name has been used as a brand. Consumers come to expect a certain level of performance across a variety of traits from their brand, and brands help customers differentiate among things that may appear very similar. Branded plants are typically sold at a premium price to help pay for future plant development, plant royalty, and supporting point-of-purchase and other advertising. Also, many branded plants can command a premium price in the marketplace due to customer demand.

One of the most common ways to help distinguish the branded plant is by use of a colored pot or container. This development has enabled plants to more readily stand out among competitors that are still typically marketed in brown, black, or dark green pots. Tags are a requirement of many branded plants, helping to enforce the royalty payment and keeping track of production numbers. Beyond containers, few brands have developed much packaging.

Cost/Value and Price

The *cost/value relationship* is the second "C" of the consumer-approach marketing model, and *price* is the second "P" of the product-approach model. The cost/value relationship can be considered synonymous with price, but there is some merit to considering the relationship between cost and value from the consumer perspective. Many horticultural professionals mistakenly base price solely on cost without considering the effects of customer perceptions. It may be termed "market value" by economists, but what is reflected in consumer perceptions is the potential to make more profits than by using simple markup percentages. Many factors contribute to setting prices, and various pricing strategies can be used. These are discussed below.

Markup on Cost

Many production nurseries set their wholesale price as a markup on the cost of producing the product. For example, if it costs $0.85 to produce a 1-gallon shrub and a nursery producer marks up 20 percent, the wholesale price for each shrub would be $1.02. To use the markup method, it is important that nursery managers know what their total cost of production is before they set product prices. This includes both the fixed and variable costs of production for each product. For a discussion on fixed and variable costs as well as cost accounting, see chapter 18. However, markup on cost is only one method of pricing and should not be considered the best.

Price Points

Price points are another pricing consideration. Typically for retail products, there are price points at which consumer behavior is likely to change. These points also usually end in 5 or 9 to show perceived value. Many consumers perceive $1.95 or $1.99 to be a better value than $2.00, even though

Photo: R. Y. Evans.

mathematically they are not more than $0.05 apart. So if a retailer's markup calculation indicates that a product should be priced at $1.67, it is relatively easy to price the product at $1.99 and have a high probability of a similar amount of sales to what would be seen with a price of $1.67. However, profit would be substantially greater. Wholesale businesses should be aware of these retail price points and consider (when prepriced product is delivered) how these points contribute to or subtract from potential profits they should be earning.

A part of price point considerations for the wholesaler should also be the question of who sets the price. In some cases, retailers may have more power (especially buying power), and they may preset or predetermine the retail and wholesale price, dictating those figures to the wholesale grower. Astute growers then need to decide if they can meet the set price, if they want to meet that price, and how much profit will be generated at that particular price point. Especially in the case of mass merchandisers, home improvement stores, and (increasingly) larger, independent retail businesses, the grower or wholesaler isn't the only one to set a price. Dictating or negotiating will likely play an increasingly larger role in price establishment as competition increases and the market for nursery plants continues to mature.

Markdowns, or Sales

A third pricing consideration is markdowns, or sales. Price reductions can help move product that is perishable in a relatively expedient manner. Most consumers will not increase their purchases until a sale offers more than 15 to 20 percent off the original price. Significant product movement can be stimulated with sales greater than 25 percent off the original price. However, the same result can be achieved with multiple-unit pricing (e.g., buy two and get one free, which is the same as 33% off). Multiple-unit pricing places the emphasis on the purchase of more than one unit so more inventory can be sold in a shorter period of time. The downside of sales is the erosion of profits, the conditioning of customers to wait for sales, and the training of customers to shop by price alone.

Discounts

Two forms of discounts should be given strong consideration for price setting: *volume discounts* and *prenegotiated orders*. The volume discount is quite common in many industries and is relatively common in the horticulture industry. But just how much of a volume is needed to qualify for a discount? Consider how much of a role transportation and freight play in a volume discount. A truck three-quarters full may be just about as expensive to ship as a full truck. What type of discount should be offered for filling the truck? Also, what percentage of the crop is purchased by one business? If a business buys a significant portion of the crop or takes the entire lot, there may be some value in that purchase that merits a discount.

Prenegotiated orders are also a consideration for price setting. This is especially the case in a maturing and highly competitive market. Prenegotiated orders allow the grower to produce crops less on speculation, which certainly makes it worth giving a discount to the buyer. Hodges et al. (2008) showed that in some U.S. regions, forward contracting accounted for a significantly higher share of total sales, perhaps indicating greater aversion to market risk. The Pacific region was one of four regions to lead the U.S. in the highest average percentage of sales negotiated for price and terms. The average percentage of the leading regions ranged from 21 to 27 percent. The other leading regions were the Appalachia, Southeast, and South Central regions. Nurseries in the Northeast, Midwest, and Mountain regions had the lowest average percentage of negotiated sales (13 to 19%).

Break-Even Point

Another key factor in price setting is the break-even point. The break-even point can, and should, be calculated for all key products or important items. It is literally the point at which all costs, including both fixed and variable costs (see chapter 18), are paid for and profits begin to accrue. While we often speak of a profit percentage on every item sold, in reality profits do not begin to accumulate until after all costs are covered. The break-even point for five hundred 1-gallon shrubs at a production cost of $0.85 each can be calculated from several wholesale prices to estimate the point at which profits would begin to accrue. In the following example, a wholesale price of $1.99 is used to determine the break-even point (BEP) because it is the most likely price to begin using as a suggested wholesale price.

Break-even point (units of shrubs) = (total units produced × cost per unit) ÷ wholesale price

BEP = (500 × $0.85) ÷ $1.99 = $425 ÷ $1.99 = 213.5678

So, the break-even point for the wholesaler would be at 214 units (always round up to cover the cost of the last unit, no matter what small fraction of a decimal it is); profits would begin to accrue after the 214th plant was sold and not before.

Price Determination Factors

Behe, Dennis et al. (2008) reported the relative importance of seven factors contributing to the determination of price. Production cost was most important, followed by plant grade, market demand, inventory levels, product uniqueness, prices of other nurseries' products, and last year's prices.

Another important contribution to price may be the amount of product lost, either on the production field or on the retail shelf. Losses on the retail shelf are becoming more of an issue with the pay-by-scan system. The pay-by-scan system refers to the payment to the wholesaler or grower when product has passed over the scanner and is purchased by the end consumer. In the pay-by-scan system, the wholesaler or grower is not paid for plants that are not sold. Moreover, pay-by-scan customers (large retailers, mass merchandisers, and home improvement stores) increasingly require producers and sometimes wholesalers to display and maintain plants in the retail setting. This additional activity may add substantially to labor costs to maintain products beyond the normal production period. If these additional maintenance costs and revenue losses from plants not sold on the retail shelf are not accounted for in the wholesaler's or grower's price negotiations with the customer, they may quickly erode the already small profit margin for the wholesaler or grower.

Another contributing factor to price may be tags. If a royalty fee is involved, it will likely be passed on to the customer. Packaging materials may add a small cost to the product, but should also be included if plants are shipped in containers. Transportation costs are usually added separately, given the volatility of petroleum pricing. Still, powerful merchandising chains may demand that growers absorb transportation costs, further reducing grower profit.

Communication and Promotion

The third "C" of the consumer-approach marketing model is communication, and the third "P" of the product approach is promotion. Communication implies a two-way conversation of speaking and listening, while promotion implies a one-way distribution of information. When a message comes from a marketer or business and is communicated in a public way to many people through mass-media channels, we consider that type of communication to be advertisement. When the message is delivered in that same public manner but to fewer individuals at once (perhaps only one), the communication form is considered to be sales or personal selling. A hybrid of personal selling and advertisement is a third type of communication called *sales promotions* or *promotional events*. This often includes activities like open houses, customer appreciation events, or special and seasonal events. When the message is delivered from an entity or business other than the marketer in a public manner or to many people, the communication form is considered publicity or public relations. When the message is delivered by someone other than the marketer in a more private manner, perhaps just to one individual, the communication form is considered word of mouth.

Estimates vary with regard to the number of advertisements consumers are exposed to on a daily basis, but most estimates are well over 10,000. Think about the number of advertisements you were exposed to today and how much attention you paid to them. It is easy to see why paid advertisements are no longer mainly how most businesses choose to communicate with customers. Paid advertisements often lack credibility and are ignored. Sales promotions may also lack credibility because they place too much emphasis on price and the need to make a purchase imminently. Messages from a marketer are quite often perceived as less credible than messages from other parties. Publicity and word of mouth are considered less biased forms of communication, even if they have identical content. One way to take advantage of this perception is to devote some resources to paid advertisements, promotions, and sales (which are needed by most businesses) but also devote some communication resources to publicity and word of mouth.

Like many planning tools in business, a communication calendar can be constructed to help allocate the resources for communication. Businesses that devote a set percentage to communication forms (e.g., 3 to 5% of sales) may not achieve their goals. New businesses, expanding businesses, or businesses facing significant market or internal changes may need to devote more resources to communication than other firms. Well-established companies with enough market demand to keep them busy may need to spend very few resources on communication. Still, the planning calendar can help management understand when and how resources will be spent. Anticipate when your customers are likely to first need information and strive to have that available just as the first demand for that information appears.

Advertising Venues

What forms of advertisement are well suited to the small grower or wholesale horticultural business? Trade magazines are one of the more common advertising venues; and depending on the market size, some national publications might also be suitable. However, regional or area advertisements are better when all customers or clientele are located in a smaller geographic region. Trade shows have traditionally been an important marketing tool in the nursery industry. Trade shows, especially those with a more regional draw, are a great place to showcase plant material and meet prospective retail and landscape customers.

Advertising on the Internet

The Internet should also be a key site for advertisements. Recent research shows that one in four persons (not necessarily gardeners) looks online for information related to horticultural products (Behe, Harte et al. 2008). Every business should have an Internet site that is updated regularly. This means weekly during peak sales periods and every other week during slower periods. Without making a commitment to regularly updating the website, your business may appear old and stagnant to potential customers. Whether you work with a professional and outsource your web page or have a committed in-house person, develop a schedule to post those updates. Pages can be crafted and reviewed during slower months and then unveiled at the scheduled time. The real challenge is that most horticultural professionals will not commit to providing the information needed for regular updates.

What should be visible on the home page, or landing page? Three key pieces of information are needed: business name, telephone number, and a physical street address (if the public is to shop there). One key consideration of the wholesaler or grower may be to keep product inventory and availability visible on the site. The retail customer may want to visit the grower's website to see how much product is available (if online inventory is available), so the demand will continue to build for current inventory levels and product availability from the wholesaler or grower. At a minimum, product mix should be current (if not quantities and prices) and as much information about the product as is feasible. Remember, too, that a retail customer may want to use images or text in advertisements or signs to help sell a product. The more information the grower or wholesaler can provide, the more value that business may provide to its customers.

The Internet will likely continue to grow in its relative importance as a sales and marketing tool. Some progressive nurseries have photographs or webcams of actual product growing in the nursery. Imagine how a prospective customer might feel if they could actually see the plants they intend to buy.

Social media has changed the marketing landscape. Social media may include Facebook, Twitter, blogs, e-mail, and other online communication means. Effective use of social media often begins with engaging a community of followers (many may be customers). Next is facilitating the conversation of those followers around meaningful points of conversation (what's your favorite flowering shrub and why?). But effective use of social media always includes a call to action. Social media should encourage discussion about the product line that the nursery sells, but imbedded in the conversation should be ways to stimulate purchases.

Advertising budget allocations

Behe, Dennis et al. (2008) showed that tremendous variability exists in the percentage of a firm's advertising budget spent on various forms of advertising. Respondents in the Pacific region (which includes California), as well as the Appalachia, Midwest, Mountain, and Northeast regions, spent more on catalog advertising than any other form of advertising (approximately 15% of total sales). In contrast, respondents in the Southeast and Great Plains regions only spent 7 to 8 percent of total sales on catalog advertising. The yellow pages were another major form of advertising for respondents in most regions, followed by newsletter advertising. Website advertising was substantially less (ranging from 2 to 5%). Radio and television advertising accounted for more than 5 percent of advertising in all regions except the Great Plains and Mountain regions. One form of advertising that did not vary statistically by geographic region was the percentage of sales allocated to gardening publications, averaging 2.8 percent for all regions. Expenditures on trade show advertising were also consistent across all eight U.S. regions, averaging 2.6 percent of sales. Billboard advertising was the category with the lowest average percentage of sales, varying from 0.1 percent in the Southeast region to 2.4 percent in the Great Plains region.

Shared advertising

Wholesalers and growers should understand the advertising venues used by retailers so that they can become partners in cooperative print advertising for the retail business. If a grower's dollars can be leveraged with the retailer's, more of an impact can be gained. The retailer's goal is to sell more product, and when aligned with the grower's or wholesaler's goal of selling more of that same type of product, it may make sense to merge advertising dollars to achieve the same goal.

A portion of the retailer's communication mix might be spent in local or area lifestyle magazines and some newspapers, especially if an advertisement is placed immediately prior to the peak sales period. Newspaper readership is declining, and those who still read a hard-copy printed newspaper are typically older. This demographic matches many businesses, but newspaper advertising still may not be the best-spent money. One idea used by many retailers is to combine the newspaper or magazine ad with a coupon to encourage a purchase; then the retailer tracks the coupons collected so that a calculation can be made of the percentage returned and the cost to pay for the ad and products.

Convenience and Place (Distribution)

The fourth "C" is convenience, and the fourth "P" is place, or distribution. These "C's" and "P's" are among the more lucrative ways adept firms can make money. Getting product from one place to another while adding value is essential in today's marketplace. For example, protective packaging usually isn't an issue when nursery plants are purchased, but ease of transportation might stimulate some manufacturers to consider other advances in packaging.

There are two extremes in the category of place: global and local. When the nursery industry began in the late 1800s, production farms were established very near population centers. The distribution channel was more local, but it changed to take advantage of climatological advantages. More production sites were located in the southeastern and western United States, and growers there shipped to other parts of the country. The tendency in the last century has been to increase the geography for the channel of distribution, and marketing of ornamental products became international and global. While the channel of distribution for food took a similar path, it is currently becoming more local.

This trend toward more local production and consumption has not returned for most of the cut flower crops, but some indications suggest that this local trend may be returning for ornamental plants.

Especially as freight costs fluctuate more upward than downward (in relation most directly to fuel costs), a trend toward more local production may ensue. For now, some climatological advantages exist for retailers buying locally grown plants because they are already acclimatized to planting in local destinations. A lower carbon footprint probably exists for locally grown plants as well. Still, few retailers and fewer customers seem to demand of their garden and yard enhancements the same eco-friendly characteristics that they are increasingly demanding of their food. Time will tell how much this trend will penetrate our part of the industry.

There are many ways that products can be distributed. Sales venues include in person, telephone, online, and mail order. Behe, Dennis et al. (2008) reported how nursery products were sold in firms across eight U.S. regions. In-person selling remains the predominant approach in the nursery industry. Nurseries in the Midwest and Great Plains regions had the highest percentage of sales made in person (approximately 60%), while nurseries in the Southeast had the lowest (45%). The Southeast region had the highest mean percentage of sales by telephone (44%), and the Midwest region had the lowest percentage of sales by telephone (30%). Nurseries in the Great Plains had the highest average percentage of sales made by mail order of any region (6%), three times more than nurseries in the Southeast. The Pacific region was one of four regions to lead the U.S. in Internet sales, although Internet sales accounted for the smallest average percentage of sales across all regions (2.3%). Other regions that were leaders in Internet sales were the Mountain, South Central, and Southeast regions (which averaged 3 to 5%).

Marketing requires an integrated effort on the part of all employees. When a business creates, communicates, and delivers value to its customers, it is effectively engaging in marketing. Marketing practices are often described within the Four P's or Four C's framework: product, price, promotion, and place, on one hand, or customer, cost/value, communication, and convenience on the other. Marketing practices vary regionally, largely because of the customer base and the distance from production that firms market their products. To be most effective in meeting customer needs, firms should develop a unique strategy based on what works for other firms but adapting that to their own unique situation.

References

AMA (American Marketing Association). http://www.marketingpower.com/aboutama/pages/definitionofmarketing.aspx.

American FactFinder website, http://factfinder2.census.gov/faces/nav/jsf/pages/index.xhtml.

Behe, B. K., B. Harte, and C. Yue. 2008. Online gardening search activities and purchases. Journal of Environmental Horticulture 26(4): 210–216.

Behe, B. K., J. H. Dennis, C. R. Hall, A. W. Hodges, and R. G. Brumfield. 2008. Regional marketing practices in U.S. nursery production. HortScience 43(7): 2070–2075.

Behe, B., J. Hardy, S. Barton, J. Brooker, T. Fernandez, C. Hall, J. Hicks, et al. 2005. Landscape plant material, size, and design sophistication increase perceived home value. Journal of Environmental Horticulture 23(3): 127–133.

Butterfield, B. 2006. Gardening in the U.S. Burlington, VT: National Gardening Association.

Dennis, J. H., and B. K. Behe. 2007. Evaluating the role of ethnicity on gardening purchases and satisfaction. HortScience 42(2): 262–266.

Hall, C., A. Hodges, and J. Haydu. 2005. Economic impact of the green industry. University of Florida IFAS Extension website, http://edis.ifas.ufl.edu/pdffiles/FE/FE63200.pdf.

Hall, C. R., B. L. Campbell, B. K. Behe, C. Yue, R. G. Lopez, and J. H. Dennis. 2010. The appeal of biodegradable packaging to floral consumers. HortScience 45(4): 1–9.

Hodges, A. W., C. R. Hall, B. K. Behe, and J. H. Dennis. 2008. Regional analysis of the U.S. nursery industry. HortScience 43(5): 1807–1812.

National Gardening Association website, http://www.garden.org/.

Reichheld, F. 2003. The one number you need to grow. Harvard Business Review 81(12): 46–54.

Yue, C., and B. K. Behe. 2008. Estimating U.S. consumers' choice of floral retail outlets. HortScience 43(3): 764–769.

Photo: J. K. Clark.

Increasing Labor Productivity

Gregorio Billikopf

CHAPTER TWENTY

Good and bad news is in store for greenhouse and nursery producers. The bad news is that one cannot transform all present employees into productive individuals through effective management principles. Research shows that, indeed, the best container nursery employee can consistently be four to eight times better than the worst. Container nursery personnel are certainly not a constant. Whatever control over production you have at your nursery or greenhouse, it is achieved through people, whether they are in production, supervisory, administrative, or management positions. How these individuals are selected and managed makes all the difference. While everyone may improve, no matter where the starting point, those with low ability potential will rarely reach half the performance of your most capable employees.

Photo: G. Billikopf.

The good news is very good, however. First, employee ability may be accurately tested before individuals are hired. Because testing has fallen out of favor and is hardly mentioned by many agricultural labor management specialists, nursery producers are in a position to hire these potentially high performers. Second, once potentially productive employees are hired, numerous management tools may be used to increase worker performance and overall employee productivity.

In this chapter we will examine those management tools that are available to help greenhouse and nursery producers of ornamentals as well as fruit and nut trees, berries, vegetables, and herbs to effectively manage their labor. Please note that our focus will be on worker productivity issues rather than legal matters. In this chapter we focus on five of the most critical aspects of labor management for your container nursery operation:

- employee selection
- performance appraisal
- compensation
- negotiating through conflict
- employee discipline

A key point in discussing productivity increases calls for measurement comparisons or benchmarking. For instance, it is ideal to measure the time taken to accomplish specific jobs. We may be able to measure the time it takes to do a particular job on a time-per-person-per-foot basis (time/person/feet) or on a square-foot basis, depending on the type of work being carried out.

One of the pioneers of performance metrics was Frederick Taylor in his *Scientific Management* approach. Indeed, with stopwatch in hand, Taylor made time and motion studies into a science. For instance, for any given type of shovel work there is the perfect size load for any given individual. Such approaches will certainly be helpful, but I suspect it does not take a time and motion study for a person to get a grip on what size shovel is needed and how full it should be for maximum efficiencies.

Since Taylor there have been numerous schemes attempting to increase worker productivity. Some of them quickly became fads. One such principle was focused on goal setting. I am all in favor of setting goals, but we ought to remember that many goals require increasing effort without necessarily showing increasing benefit beyond a certain point. For instance, Y amount of effort may be required to improve a benchmark from 60 percent success to 80 percent success, yet the amount of effort to go from 60 to 80 percent, in our hypothetical, may be well over 2 times Y.

Sometimes people who write about benchmarking talk about percentage of goals met. While managers may well be able to meet the early goals, more difficult goals will be much harder to obtain, and they are not equivalent in required effort. Even so, measuring and attempting to improve employee performance and reduce turnover, absenteeism, and injury are useful benchmarks for comparison. Some employers are constantly looking at their work flow order and trying to reduce or eliminate bottlenecks or downtime for employees who cannot do certain jobs until others are accomplished. These are also important steps to be taken. Sometimes there are mechanical tools that can be developed to increase efficiencies. In a greenhouse in Canada, for instance, I observed a mini railway that had been developed by the nursery owner. It permitted employees to sit on a cart and push themselves back while they worked in a low position that would have otherwise been quite uncomfortable. Employees, as well as their daily supervisors who are most acquainted with any particular chore, may discover great ideas. Some employees may be hesitant to offer suggestions if these translate into reduced hours and pay. Nevertheless, greenhouse and nursery growers may put in place ideas that protect jobs while increasing efficiencies. And when employees leave through natural turnover, there is no need to replace them.

Others measure labor productivity and compare it on a monthly or yearly basis. These are good ideas, but we need to remember that each employee's performance will potentially vary substantially from that of others. Just as important as keeping track of the number of full-time equivalents (FTEs) required to do specific jobs on a square-foot basis, for instance, is also keeping track of which employees are doing better. To the degree that individual performance can be measured over small group or large group performance, workers may be motivated through a number of interesting incentive pay programs. The idea is to reward the most productive employees and make sure they are earning more and are more likely to stay.

Numerous electronic devices exist that may also be used to gauge performance and quality in order to stay on top of these issues and reduce the time that it takes to record such matters in the field. For example, in some nurseries plants are specially tagged electronically so that only the very best are shipped. If in addition these plants may be associated with an individual or small group of employees, then there is also an increased possibility to use these results for feedback, training, employee discipline, and pay for performance. When carrying out quality control, it is vital for the evaluator not to know who he or she is evaluating, and the use of these electronic devices, once again, is very useful.

For a principle of productivity to be sustainable, it must be beneficial to both manager and worker in the long run. As a labor productivity specialist, I will write about tried and true principles I have applied in different types of agriculture, which, interestingly, have attracted much interest outside of agriculture.

Employee Selection

Hiring the right person for the job may be the most vital management decision you will make. One might even say that if you hire the right people, they pretty much manage themselves. Yet most employers—including greenhouse and nursery container growers—fail to spend the necessary effort up front, which requires giving applicants a practical test (also called a job sample). Here is a systematic selection process that will greatly increase the odds of hiring the right person for the job.

Decide What You Need

Before you can start hiring management, supervisors, laborers, equipment operators, office staff, or sales individuals, you first need to determine whether a temporary employee is a viable option, what the job requirements are, which specifications are most important, and how you are going to recruit.

Step 1: Determine whether a temporary employee is needed

Necessity can often be the catalyst for haphazard selection. Hiring a temporary employee is a good alternative to hiring a less suitable replacement under pressure. Exceptional temporary personnel can be encouraged to apply for permanent positions. You can share with such applicants the criteria that will be used to make the final selection decision and offer additional help and training.

A temporary employee has the advantage of having one foot in the door and the opportunity to learn what is important to you. Management also benefits by having the opportunity to better evaluate the individual's performance and personality. The downside is having to disappoint the temporary employee who does not get the job—or the coworkers who were rooting for him.

The statistical chances are not high that a temporary employee turns out to be the best candidate once the position is opened. Clear communication will help alleviate possible disappointment but is unlikely to eliminate it totally. In the end, the responsibility for qualifying for the job needs to be the employee's.

Seasonal employees, hired without the benefit of a careful selection process, can also be evaluated for future employment. The best workers can be invited to return for the next season.

Step 2: Understand the job requirements

A frequent sentiment among container nursery employers is that a good attitude and a lack of bad habits are the most important ingredients in the personal makeup of container nursery personnel. "Give me someone with a good attitude," they argue, "and someone who will learn my bad habits—rather than those of a previous container nursery!" It is indisputable that a good attitude is essential, but attitude

Photo: J. A. Bethke.

alone does not make up for poor skills any more than good skills make up for a poor attitude. You can hire people who have both the needed skills and attitude.

Successful employee selection depends on a clear understanding of a job's components, both physical and intellectual. Sit down and make a list of required skills, not taking any of them for granted. The employee selection requirements may emphasize skills and knowledge not easily learned on the job. Horticulturalists often say "The plants speak to me." Do you need to hire an employee who is observant and can tell when plants need watering, fertilizing, or some other care and thus can "listen" to those plants? Or do you need someone who is just going to follow instructions and is not required to think?

Step 3: Weight the job specification items

Weighting job duties can help the nursery manager assess the qualifications of competing candidates. Each skill, knowledge area, and ability is rated according to its importance to the job. A skill may be given less importance, for instance, if it can be easily acquired or is seldom used. For instance, when hiring office personnel, how do you rate the importance of phone skills and ability to deal with clients compared with computer skills or writing skills?

Step 4: Determine the recruitment strategy

Among the most frequent objections to the systematic selection approach discussed in this chapter are concerns about getting enough applicants. How many people apply partly depends on your recruitment efforts, the type of job, labor market, pay, and the reputation of your container nursery. The larger the applicant pool, the greater the chance of finding qualified applicants. The most thorough selection approach cannot make up for a poor candidate pool.

Sources to help you advertise the position include present employees, other container nursery employers, previous applicants, trade journals, newspapers, vocational schools, universities, and employment agencies. Nursery and greenhouse employers who are trying to attract Hispanic applicants often find great success through Spanish radio stations.

An excellent source of potential candidates is people who come looking for work when you may not have any job openings. Rather than having these individuals fill out an application form, request just the bare essentials from them: applicant name, desired job, and phone number or other information on how to contact them in the event a position becomes open.

Sometimes it may be useful to dedicate one phone line to the process. Rather than answering the questions of each applicant, you may wish to record a detailed description of the job, pay and benefits offered, working conditions, and answers to other questions potential applicants may have. For some jobs, it is helpful to create a web page with all of this information.

Design the Selection Process

A well-designed selection process will yield information about a candidate's skills and weaknesses, enabling the container nursery manager to make an informed choice.

Step 1: Determine which selection tools to use

Applicant skills can be evaluated through applications, interviews, tests, reference checks, letters of recommendation, and physicals. Some selection tools are more effective than others, but a combination of tools is usually best as long as job sample tests (i.e., practical tests) are incorporated. Some growers feel strongly about using a 1- or 2-week trial period. A trial phase in conjunction with the rest of the tools described in this chapter can be very effective. A trial period, however, makes a poor substitute for a systematic selection approach. All too often, if a person is barely good enough for the job, he or she is allowed to stay on. The chances of selecting the right individual for the job based solely on a trial period are greatly diminished.

Factors reflecting worker motivation, such as punctuality and attendance, may be elicited within the interview, but contacting previous employers may give more reliable information. If possible, try to verify evidence of specific skills, knowledge, and abilities at more than one point in the selection process. Job sample tests are the most effective tools to test skills and abilities.

Step 2: Prepare questions and situations for written and practical tests, the interview, and reference checks

At this point, the manager converts important skill areas into specific questions or activities for the application, interview, and practical tests. Also, questions for the reference check may be drawn up.

Areas of inquiry can help determine an applicant's aptitude for interpreting plant health distress signs, capability with measuring instruments, command of another language, understanding of supervisory principles, or lifting strength. Results are used to assess a candidate's technical knowledge, general problem-solving ability, interest in the operation, and other job-related attributes. Many queries or activities will elicit responses that can be judged objectively, such as how much pesticide should be mixed into a given number of gallons of water. Practical tests are even effective to see how applicants for a supervisory position deal with an employee who is not wearing appropriate safety gear or needs correction. These practical tests are the most valid approaches to employee selection—that is, the most predictive of worker performance. So, whether hiring individuals who will be irrigating, handling pest control, pruning, potting, preparing media, fertilizing, propagating, spacing plants, harvesting, maintaining equipment, or working in sales, you can easily design a job sample or practical test.

Practical tests can be classified as power versus speed tests. Speed tests require applicants to perform repetitive tasks in a limited amount of time. They are typically used for skills such as picking, packing, pruning, and sorting. Power tests require applicants to demonstrate depth of ability rather than speed, such as in the diagnosis of a mechanical malfunction or plant disease or ability to handle a difficult disciplinary situation in a role play. You will want to place reasonable time limits, however, to make the practical tests resemble the reality of time pressures on the job.

Generally, the goal of the speed tests is for applicants to show if they can follow instructions and perform well rather than see how much they remember or how much they know. At higher levels in the organization, how much individuals know is given more weight, of course, and these tests tend to be more of the power test type. But even then, how well applicants can think and reason through challenges is important.

So, let us go over some of the steps for designing a grafting speed test, as a sample of what may be done with any type of other work. After you have attracted a number of applicants, it is helpful to have them all show up for a test on the same day and at the same time. Some growers may be somewhat apprehensive about potential candidates meeting each other. Let me reassure you that the managers who conduct the job sample will be more anxious than the applicants. After doing this once, it will seem more natural.

The applicants are shown the why's and how's of what they are to do and encouraged to ask lots of questions. The manager who is giving the test may say something like "I am sure there are lots of questions" and then permit a long pause until someone asks the first question. When possible, allow applicants to do a job for a few minutes and provide individualized feedback before starting the official test period. The purpose of testing, once again, is not so much to see how much individuals know when they come to your nursery to apply for a job but rather how well they can follow instructions and do the work.

After applicants are asked to begin the practical test, they are evaluated in terms of both the number of grafts in a 20- to 30-minute period as well as the quality of the grafting. Rather than giving a single half-hour grafting test, it is better to give two 15-minute tests so that a validity correlation coefficient may be carried out (showing how correct applicant's performance is) and a test reliability figure obtained (showing how consistent the applicant's performance is).

An example of a power test would be to set up four to six stations with equipment that does not work and ask applicants for a nursery mechanic position to fix these. Another power test may involve asking potential crew leaders or supervisors to role-play giving instructions, providing feedback, giving positive reviews, or listening to a complaint. A time limit is set for 20 or 30 minutes per station. Applicants are then evaluated as to what tools they used as well as the logic of their approach and whether they were able to have the equipment running.

Step 3: Assign a sequence to hurdles

The greenhouse and nursery manager can think of the selection process as a series of hurdles that applicants must clear in order to obtain the job. Each hurdle eliminates some applicants from contention. The sequence of these hurdles needs to be designed with care. Generally, the most expensive and time-consuming selection tools are used later in the selection process, unless there are very few applicants.

Step 4: Provide a realistic job preview

Applicants who have a clear understanding of what the job entails can make more informed decisions as to whether they want to apply. For instance, will the job meet their financial, emotional, and social needs? Selected applicants who have an accurate understanding of the job—of both its desirable and difficult aspects—are more likely to stay and succeed. If the selection process does not yield individuals who will excel on the job, it is best not to hire any of the applicants but start with the recruitment process one more time.

Performance Appraisal

After employee selection, performance appraisal is arguably the most important management tool a nursery manager has at his or her disposal. When employees know exactly what is expected of them and how they are performing in relation to those expectations, operations can become much more productive.

Employee Need for Feedback

Although employees vary in their desire for improvement, generally workers want to know how well they are performing. People need positive as well as course correction feedback and validation on a regular basis. Once an employee has been selected, few management actions can have as positive an effect on worker performance as encouraging affirmation. These are, in effect, goodwill deposits, without which withdrawals cannot be made. This does not mean you should gloss over areas needing improvement. When presented in a constructive fashion, workers will often be grateful for information on how to improve shortcomings.

In the traditional performance appraisal, the supervisor acts more as a judge of employee performance than as a coach. By so doing, unfortunately, the focus is on blame rather than on helping the employee assume responsibility for improvement. Next to employee discipline, performance appraisal interviews are probably the most dreaded management activity. Traditional performance appraisals put the supervisor in a position of being the expert on the employee's performance. The worker often reacts with passive resistance or noticeable defensiveness. No wonder supervisors are often hesitant to deliver bad news to workers. It is easier to ignore it and hope it goes away. In the next section, we introduce a new model of performance appraisal.

Negotiated Performance Appraisal

You can make the task easier by putting more responsibility on the worker for the performance appraisal. While there are many ways to achieve this, here is an approach that has worked well for me and has been tested in farms for many years. It is based on effective negotiation techniques. A nursery manager can ask the employee being appraised to bring three lists to the performance appraisal interview: (1) areas where the employee performs well (what he contributes to the container nursery operation); (2) areas where the worker has shown recent improvement (perhaps over the last year); and (3) areas where the worker feels weak, or thinks his supervisor would like to see improvement. It is appropriate to give the employee time to think through these lists, and so you may want to give employees a couple of weeks to complete the assignment.

Because as the supervisor you will also fill out the three lists, employees are more likely to bring candid responses to the table. The supervisor may say something like "I will fill out these three lists also" and then repeat the purpose of each list again: "That is, areas where you perform well and what you contribute to the organization from my perspective; areas where you have shown recent improvement; and weak areas where you still need to improve." The key point here is that employees will hear you say that there are performance areas, you value in their work, and just as importantly, that the employee can still improve in other aspects of the job.

This latter point is particularly critical from a psychological perspective. It is human nature not to want to bring up our faults, but it is also human nature to prefer to point out our own shortcomings rather than have someone else do it. This process allows the subordinate to think in terms of both her own performance expectations and perceived supervisor expectations.

There is a fourth list, filled out only by the subordinate, just as vital as the first three. While we will talk about this list later, it is important to include it as an assignment ahead of time so the employee has time to think about it and come prepared. The fourth list is the employee's response to the question "What can I do differently, as your supervisor, so you can be more effective (or thrive) in your job?" If a supervisor is not truly willing to listen to what the employee may have to say here, the negotiated performance appraisal will not work as it should, and a more traditional performance appraisal would work better.

Although the appraisal process can take place between supervisor and employee alone, the use of a qualified third-party facilitator can greatly improve the success of the approach. The message is thus clearly sent to all involved that this process is important to the nursery business. The third party is there mostly to listen to each individual in a separate meeting (or precaucus) and help the individual brainstorm and prepare for the joint meeting.

The performance appraisal joint meeting

When the time has arrived to sit and discuss the employee's performance, a relaxed, positive atmosphere should prevail before entering into the substance of the performance appraisal meeting. A location without distractions or interruptions is essential. All phones need to be turned off.

The worker is asked to read each list, beginning with the first. Managers should listen intently and take notes if needed, but they should not interrupt the employee except to ask questions that help clarify an issue. Interrupting to clarify one's understanding is almost always a good move. If the employee says something you find strange, troubling, or that you do not understand, it is good to ask the employee to amplify or explain a point. People seldom mind being interrupted when it means having the opportunity to offer clarification. Such questions should not put the employee on the defensive, nor should they be comments disguised as questions.

First list

The main purposes of the first list are to

- recognize employees' strong points and let them know these have not passed unobserved
- honestly build up employees so they can be more receptive to constructive criticism (an employee who is so concerned about his self-esteem, or about being attacked, will naturally become defensive and less receptive to suggestions for improvement)
- help avoid coloring all of an employee's behaviors with the same ink (e.g., thinking of her as a "difficult person" rather than as an employee who resorts to some "unproductive behaviors")

As the employee reads her first list out loud, do not hesitate to add anything you may have forgotten to include in yours. Acknowledge what is being said by careful listening. After the employee finishes reading his or her first list, read your list to the employee. Make sure to praise the worker's good points—even if the employee has already mentioned them.

The first list is the vital foundation to the performance appraisal process. Time spent developing and discussing what employees do well is never wasted. In the rush of everyday activities, supervisors often focus on what an employee is doing wrong. How often do we take time to stop and give carefully thought out compliments? Not infrequently, employees will visibly smile when honestly complimented by their supervisor. Employees who feel that they are performing well in

Photo: J. K. Clark.

at least one area of responsibility (and feel validated by their supervisors) are more likely to want to improve their performance in other areas, too. Remember, sincere compliments are goodwill deposits without which withdrawals cannot be made.

With over 15 years of experience with this technique, I have come to realize that the manager needs to spend at least 20 minutes focused on what the employee does well. Within this time period, the time spent by the subordinate reading the first list is also included. In order to be successful, the supervisor not only talks about each area that is being commended, but explains why it is important to the operation and provides specific examples that made a difference. For instance, a supervisor may say something like "I put a huge value on duty fulfilled. For me this means that an individual goes out of the way to get things done even when it may not be convenient. Paula, you are such a person! What this means to me is that I can focus more on my own work because I know I have another set of eyes in you. For example—and I thanked you for this a few months ago—I was staring out the office window at quitting time as several pickups left the nursery. I noticed you stopped the pickup and got out and got a shovel and some tools and then, after about 5 minutes, you drove off. I was so curious as to what had happened that I drove over there and noticed that you had repaired one of the irrigation pipes that was leaking. Everyone else had ignored this and was so focused on quitting time, but not you. Having that sort of a person at this nursery is invaluable to me. And just 2 weeks ago, that day when the storm hit us hard, you . . ."

It is the positive force, or momentum, that gives an employee the strength and determination to try harder in areas of weakness. Employees can quickly sense, however, when a compliment is not sincere. Furthermore, when supervisors are negative and find little to compliment in an employee, their subordinates are less likely to have the desire to make needed changes.

If the worker brings up, as one of his or her good points, a performance issue that you consider a weak point, attempt to understand the employee's perspective, and under no circumstance disagree with the employee at this point. While there may be disagreements between you and the person being appraised as to whether something is a positive trait, this is not the time to bring such issues up. The opportunity will present itself when discussing areas that the worker needs to improve. Nor should you cloud the positive issues by telling the employee now that this point belongs both under the positive employee contributions and the list of items that need improvement. Nevertheless, when discussing employees' weak points later on, it can be very beneficial to remind them of their positive traits.

Second list

The function of the second list is to permit employees to discuss weak areas they have been working on. Of course, when an employee says he has improved in an area it does not mean he or she has totally conquered the problem. As before, the supervisor listens and asks for clarification, without interrupting the employee. The supervisor acknowledges the comments of the employee and then reads her own list.

Third list

Everyone has areas in which he or she can improve. Just as we may color employees with negative strokes and not recognize the good in them, we can also neglect to help outstanding employees reach their full potential. This may be done by failing to acknowledge strengths or by ignoring weak areas, as insignificant as they may appear. In the process of sharing lists, areas of misunderstanding can be cleared up.

Again, allow the employee to go first. Permit the individual to read his complete list uninterrupted, except to ask for clarification when needed. When it comes time to read your third list, there is no need to repeat what the employee has said. Instead, bring up any issues that have not been raised. You may say, "Thanks so much for your list. I see you have put a lot of effort into it. Let me share a few items from my list." (And then only share those things they have not mentioned in theirs.) A key point to remember is that when employees acknowledge something as a weak point, they have taken ownership of that problem. The more concrete and specific the solutions, the greater the potential for success. This is an example where a facilitator may help a subordinate prepare a list of items that he or she wants to improve as well as specific ways to go about improving.

Fourth list

The fourth list, based on the question "What can I do differently as your supervisor so you can be more effective in your job?," is just as crucial as the first three. When workers are sincerely asked, and when they are given time to prepare a thoughtful answer, especially after the employee is put on notice that his own performance is being evaluated in such detail, this question can improve the performance appraisal process many fold. Also, because this question is asked last, I feel that employees are more likely to speak up, especially now that they know how serious the process is. The wording is such that it elicits genuine worker input. When the employee speaks in response to such a question, the supervisor needs to control the natural tendency to defend or explain past behaviors.

As soon as the employee realizes that the purpose of the performance appraisal is to solve problems rather than assign blame, difficulties are more likely to be raised and shared. A series of follow-up meetings to discuss where the employee has improved, as well as areas that need special attention, can often be profitable.

Compensation

Jobs that call for creativity, autonomy, analysis, and personal growth may provide the best motivator of all: intrinsic rewards. Such satisfaction originates from within the worker. An intrinsically motivated worker does not obtain his motivation from external stimulation provided by the employer. An overemphasis on external rewards may be responsible for elimination of internally originated ones. There are personal and organizational objectives that simply cannot be realized through pay.

On the downside, intrinsic motivators, as wonderful as they may appear, are not equally found among all of your nursery workers, nor do they always motivate the type of performance you may desire. Pay can be a powerful management tool and a compelling motivator. Employees often consider pay as a measure of individual achievement and social status. The importance of pay, then, ought to be neither over- or underrated.

To be effective, pay must be tied to performance. While incentives can yield the clearest link between performance and pay, they are not suitable to all jobs. We will first look at internal wage structures and then speak of incentives.

Internal Wage Structure

In a casual survey I conducted, workers said that they expected wages to do the following: cover basic living expenses, keep up with inflation, leave some money for savings or recreation, and increase over time.

Even if a grower devises a wage structure to satisfy these expectations, worker dissatisfaction may arise if either internal or external equity principles are violated. Simply put, internal equity refers to the relative fairness of wages received by other employees in the same organization. External equity means fairness relative to wages outside your nursery operation. Depending on the type of work and location, tests of external equity may involve comparisons with other nurseries or even nonagricultural corporations.

Decisions need to be made as to whether you wish to pay all workers the same if they are doing the same type of task or whether you wish to pay (reward) them differentially based on their performance, seniority, or both. Within each pay grade you can have steps. Generally, these steps (or raises) are smaller at the bottom of a grade, but given more frequently. Toward the end of an employee's career, the raises are much more substantial but require more effort to achieve. When pay increases are not given carefully, you may end up feeling that your experienced employees do little more than your new recruits yet get paid substantially more.

To the degree that wage structures reward employees for achieving specific knowledge or performance levels (in contrast to just paying them for surviving another couple of years at the nursery), your internal wage structure can promote worker productivity. For example, you may have several levels of pay for plant propagators: a basic or entry level for jobs that are easy to do and higher levels for those employees who have proved their ability to graft and understand propagation needs of multiple types of plants. Keep in mind that once raises have been given, these are mostly permanent—in contrast to incentive pay programs where employees only receive the added pay when they perform well. This is why salaries and the potential for salary increases are only motivating to employees when they are tied to merit, not just number of years worked. The clearer and more transparent the requirements are for attaining these raises, the more motivating these internal wage structures will be.

Cost of living adjustments (COLAs)

Inflation can have especially devastating effects on a worker's ability to make ends meet. If you find yourself hiring inexperienced workers at a higher step within their pay grade, it may be a reflection that the pay grade has not kept up with inflation.

Increases in the minimum wage can cause pay compression in agricultural enterprises paying at, or near, the legal minimum. Pay differences acknowledge variations in the needed preparation and skill required for any particular job. When the minimum wage raises the pay for the lower-end jobs, wage increases at the other levels are also needed in order not to obscure the reasons these pay differences were instituted to begin with. Once a pay raise is earned, it can seldom be taken away. That is the advantage of pay for performance or incentive pay, which we discuss next.

Incentive Pay

When properly designed and implemented to protect both management and nursery personnel, structured incentives work well. An effective structured incentive must

- be capable of fluctuating (variable pay) as performance changes
- be based on a specific accomplishment-reward connection understood by both management and workers

- yield positive results for both the nursery and the employees in the long run

Here are some basic steps to understand regarding incentives:

Step 1: Analyze the challenge and determine if incentives are appropriate

The purpose of an incentive program needs to be clear and specific. Just because a goal can be measured in clear and specific terms, however, does not mean incentives are called for. Incentives may not be appropriate to motivate employees who lack the resources or skills to perform.

Also, rewarding workers with incentives for bringing their poor work up to standard would be like paying twice for the same job: once for having the employee show up, the other for working. Instead, an incentive pay program can reward workers who continue to produce superior work or encourage those who already produce good work to excel.

Incentives are often needed to counteract the effect that crew dynamics has on performance. Hourly-paid workers tend to perform as fast as the slowest worker when they work in a crew. Workers paid by the hour tend to cling together, while those paid by piece tend to spread out, some working much faster than others.

Step 2: Link pay with performance

Some growers offer profit-sharing plans "because we did well this year." Lamentably, there are too many factors that affect nursery profits besides worker productivity. Workers do not always see a link between their efforts and profits. Another danger is a streak of ever-increasing profits followed by several years of deficits during a bad economy. While some workers may understand a reduced profit-sharing paycheck for a year, few will tolerate a longer drought without experiencing considerable dissatisfaction. One manager shared with me his excitement about a substantial profit-sharing bonus. As a result, he worked much harder the next year and felt defrauded when the bonus ended up substantially reduced compared with the first year. He soon left that enterprise.

Instead of being a motivator, profit sharing can discourage employees. Not only do profits depend on the efforts of the whole organization, but profits can be fickle. This is true for any organization, but it is especially true in the nursery industry where there may be a rash of good years followed by bad ones.

Any time employees are rewarded or punished for that which they cannot control, nursery managers are asking for a cynical or disillusioned workforce. Much better than profit sharing is breaking down all elements under the control of employees or management that affect profits and then rewarding personnel for achieving those results.

One way to categorize incentive pay is according to whether individuals, small groups, or all nursery personnel are covered. Individual incentive plans offer the clearest link between a worker's effort and the reward. Probably the best-known individual or small-group incentive pay plan in agriculture is piece rate. But some jobs at the nursery lend themselves better than others to piece-rate pay. A separate section on piece rate (whether individual or group) is provided below.

Step 3: Anticipate loopholes

Being so specific about a single result may cause workers to achieve it at the expense of all others. For instance, a manager increased yields but spent more on production than what the extra yields meant in profits. The number one loophole for quantity production incentives is often quality. Quality control, then, as well as rewarding quality, must be part of the process.

A systematic approach toward quality improvement involves a number of steps:

1. Test and calibrate the judgments of management and quality control personnel so that all are evaluating quality with the same eyes. Top personnel can be asked to independently make evaluative decisions (e.g., the quality of a graft), which are then compared against the decisions of others. This team discusses differences in opinions and further clarifies required quality standards. The process is repeated again until two or more persons are scoring a 92 percent agreement rate. Also, a trusted quality specialist may have to confirm the accuracy of the quality evaluations.
2. Train other quality control personnel and workers on quality standards and then evaluate them on their ability to make quality determinations.
3. Select applicants and employees on the basis of their skills in understanding quality determinations.
4. Make determinations on how to sample for quality control purposes using random inspections.
5. Develop a required range of standards for quality, in terms of what constitutes expected minimum quality, high quality, or below-minimum quality. More categories may be created as needed.
6. Require employees to prove through performance testing (step 3) that they can do the job. With new employees it may be necessary to have an hourly-paid crew for workers who are in the process of proving their knowledge and understanding of quality factors. Only workers who prove to the grower that they understand quality can move to piece-rate-paid or incentive-paid crews.
7. Design a feedback mechanism so workers know how they are doing and where they can improve. A card with a score broken down into areas where the employee is doing well or can improve is very helpful.
8. Require minimum quality through a combination of effective selection, training, and a disciplinary process with consequences for noncompliance.
9. Reward high quality through an additional bonus. Quality achievements are often obtained in incremental steps. Once employees have attained the

highest quality levels without fluctuation, it is possible to eliminate the steps that served as training wheels, or safety nets, to reach the high standard. This is not true if quality levels fluctuate widely.

10. Make sure that those who are evaluating quality are rewarded for good quality. When farm labor contractors, supervisors, or crew leaders are paid in proportion to worker earnings, farmers may inadvertently be encouraging less attention to quality—that is, unless worker earnings are already tied to quality.

The number of farm labor contractors (FLCs) used in agriculture is climbing. One key point is that when FLCs are rewarded as a percentage of employee earnings, FLCs may simply be focused on employees earning more and not necessarily on efficiency or quality. When jobs are paid on a piece-rate basis, furthermore, FLCs may not be as interested in your quality issues (unless quality is an integral portion of your compensation) as they are in the total units produced. FLCs, then, need to be part of the solution so that it is not to their advantage for you to have to hire more employees than you really need or to only look at total production while ignoring quality issues.

Step 4: Establish standards and determine pay

It is important to clarify exactly what is expected of the nursery employees and how their work will be judged. Since we are dealing with plants, variations have to be considered and accounted for. Work that requires more effort needs to be paid at a higher rate. When thinking of a piece rate, for instance, one considers the standard to be the easiest possible work (e.g., plants that lend themselves to the work and are not diseased). With increasing effort required, the percentage of pay increase goes along with this increased effort. So, say your base piece rate is 80 cents per unit and the work is twice as hard; you would pay $1.60 per unit.

Anytime we begin to measure productivity or efficiency per unit, such as per square foot or per plant, the possibility of a piece-rate-type incentive exists. The ideal in a piece-rate design is to reward employees for individual productivity. When this is not possible, remember that small-group incentives are more effective than those given for large groups. As stated above, the fastest crew worker performs at the same speed as the slowest one when paid by the hour. This can be easily verified when observing crews at work. All the workers seem to be working at the same pace. When workers are moving very fast or running, or when they are well spread out through the nursery, it usually means that the piece-rate pay has been well designed—at least from the worker perspective. If workers are moving faster than they would if paid by the hour but not as fast as if paid by a motivating piece rate, it often means that workers are paid on a group piece rate. In other words, it is more motivating than hourly work but not so much as an effectively designed, individual piece rate.

What does it mean when crew workers perform as if they were paid by the hour but are in reality paid by the piece? It signifies that the crew workers have not bought into the piece-rate plan. Four key reasons why nursery workers may act as if they were paid by the hour—even when paid by the piece—are the following:

- The piece rate is too low, and they are hoping the employer will raise it.
- The piece-rate design is faulty (such as an hourly base pay plus a piece-rate bonus or other types of incentives that do not reward effort).
- There is not enough work (workers realize that if they hurry, they will work themselves out of a job).
- They fear that the piece rate will be reduced (when the employer sees what workers are capable of).

Instead, the piece rate should begin with the first unit and continue to increase proportionately so that those who produce double, earn double—or produce triple, earn triple. Two of the reasons why piece rates often fail are (1) lack of quality control and (2) fear that employees may be earning too much. We already spoke about paying a quality-based bonus and not one for productivity alone. The solution to the second problem, a concern that nursery employees are making too much money, is to measure productivity in terms of cost per square foot or per plant. If labor costs are reduced, there is no need to be anxious that employees are earning too much. Instead, having employees who are making double the minimum wage (or more) actually means working with fewer but more productive employees and not having to worry about labor-supply issues.

Step 5: Protect workers from negative consequences

Employees have a number of reservations related to the use of incentives. These include such things as fear of job loss, unfair pay, and rate reductions. Many workers worry that when they show the manager exactly what they are capable of, more will be expected of them without an accompanying proportional pay raise. In the end, incentives can become a game in which everyone loses. Instead, incentives will work well in an operation if they are designed so that the more a nursery manager spends on them, the greater the nursery profits will be. Any other formula is not sustainable.

Step 6: Improve communications

Positive interpersonal relations between management and employees, as well as among employees, are a must before installing a successful incentive pay program. Incentives often add some tension and stress, especially at first, before results showing success are clear. Added demands for positive two-way communication, feedback, and teamwork will increase.

A simple program will help build trust. At minimum, all workers need to know what is expected of them and how their performance will translate into pay. It helps when the incentive plan is presented to workers for review and comments before implementation. If an expectation is set that employees can easily make the top incentive goal (e.g., for improving quality), the incentive may act as a demotivator. Instead, growers should encourage employees to try their best and begin by shooting for the lowest level. If the accomplishment exceeds the workers' expectations, all the better.

Step 7: Periodically review the program

Record keeping and statistical analysis are critical to determine the success of the incentive pay program. Good controls are crucial so incentive pay results can be isolated and correctly attributed to the pay system. If a grower introduces other changes simultaneously, he or she may never know the impact of the incentive program.

Negotiating through Conflict

Wherever choice exists, there is a potential for disagreement. Such differences, when handled properly, can result in richer, more effective, creative resolutions and interaction. But, alas, it is difficult to consistently turn conflicts into opportunities.

The very thought of interpersonal negotiation—or dealing with these differences in opinion among managers or with employees—is intimidating; yet we are all experienced negotiators. The process of taking turns in a conversation, or of deciding who says hello first, involves tacit negotiation. Some types of negotiation may be almost subconscious, such as holding a door open for another to pass through. It is one thing to negotiate, another to be a skilled negotiator.

As we put into practice effective interpersonal negotiation techniques, we gain confidence in our ability to find agreement and overcome challenges. This confidence can be contagious. Effective dialogue entails as much listening as talking. When disagreements emerge, it is easy to hear without listening. While effective two-way exchanges will happen naturally some of the time, for the most part they need to be carefully planned.

When disagreement is poorly dealt with, the outcome can be contention. Contention creates psychological distance between people through feelings of dislike, bitter antagonism, competition, alienation, and disregard.

When faced with problems, the human brain is capable of taking large amounts of data, quickly analyzing it, and coming up with the best solution. Unwanted options are discarded. This is fine when it comes to making quick decisions under time constraints. Unfortunately, we are often eager to accept the first solution that seems to work rather than the truly creative one. While some decisions may require careful consideration and even agony, we make others almost instinctively. Our best solution becomes our position or stance in the matter. Our needs, concerns, and fears play a part in the process of establishing our position.

Misunderstanding and dissent arise when our solutions are at odds with other people's positions. Several foes combine to create contention:

- Our first enemy is the natural desire to explain our side first. After all, we reason, if they understand our perspective, they will come to the same conclusions we did.
- Our second enemy is our ineffectiveness as listeners. Listening is much more than being quiet so we can have our turn. It involves a real effort to understand other perspectives.
- Our third enemy is fear. Fear that we will not get our way. Fear of losing something we cherish. Fear we will be made to look foolish or lose face. Fear of the truth—that we could be wrong.
- Our fourth enemy is the assumption that one of us has to lose if the other is going to win—that differences can only be resolved competitively.

It takes more skill, effort, and commitment—and in the short run, more stress—to face disagreement directly. Below are some keys to effective interpersonal negotiation skills.

Search for Interests

Some of the most powerful concepts are the simplest. One such principle was developed by the Harvard University negotiation team and is described in the book *Getting to Yes* (Fisher and Ury 1981). People in disagreement can benefit from focusing on their needs, fears, and interests rather than on their positions. Once we take the time to understand another person's needs and interests, we see that there are many solutions to challenges that seemed impossible.

In traditional negotiations we are inclined to focus exclusively on our own needs and assume it is the other party's responsibility to worry about having his or her needs met. Yet, by showing a sincere interest in the needs of others, we increase the chances of having our own needs met. While talking about our expectations and fears may have been considered selfish in traditional negotiation, creative negotiation entails the consideration of not only our needs and fears but also those of the other individual.

Seek to Understand

Steven Covey (1989) reinforced an important notion in his book *Seven Habits of Highly Effective People*: "Seek first to understand, then to be understood." If we encourage others to explain their side first, they will be more apt to listen to ours.

Problems are likely to increase, however, if we put all our needs aside to focus on another person's perspective. The other party may think we have no needs and be taken aback when we introduce them all of a sudden, almost as an afterthought. In order to avoid such unproductive shocks, I like the idea of establishing a psychological contract with the other individual in the conversation.

I might say "While I want to share my needs and views with you later, let me first focus on your thoughts, needs, and observations." At this point, I attempt to put my own needs aside and truly listen. I might say "So, help me understand your concerns regarding . . ."

That is the easy part. The difficulty comes in fulfilling the resolution to listen—to resist the tendency to interrupt with objections, no matter how unfounded the comments we hear may be. In trying to comprehend, we may need to express our understanding in the form of a tentative question and avoid being judgmental.

Control Emotions

Our emotions regularly get in the way of effective negotiations. Nothing kills creativity quicker than anger, pride, embarrassment, envy, greed, jealousy, or other strong negative emotions. Anger is often an expression of fear or lack of confidence in our ability to get what we think we want. Emotional outbursts tend to escalate rather than resolve a conflict.

If we can improve our ability to manage our emotions and respond without getting defensive, we have gone a long way toward creative negotiation. It is extremely difficult to hide our emotions, especially when we feel there is much in the balance. Our body language, particularly our facial gestures and voice qualities, often give us away. We are not emotionless robots. However, it is better to describe negative emotions (e.g., a feeling of disappointment) than to display them.

When it comes to managing emotions, I recommend the chapter "Master My Stories" in the book *Crucial Conversations* (Patterson et al. 2002). The authors contend that negative emotion is always preceded by telling ourselves a story. This may happen in milliseconds. The more critical the situation, or the more important our relationship with an individual, the more likely it is that we are vulnerable to such storytelling. When we presume to understand another's feelings or intentions, we may come up with the worst possible scenario in terms of future consequences.

Some years ago, I was asked to address a group of young adults. I noticed that as I spoke, a young man would lean toward the young lady beside him and whisper in her ear. I found this to be very distracting and annoying. I feel very strongly that only one person should speak at a time, so every time he began to talk, I stopped. When I stopped, he stopped, and so it went. I later learned he was interpreting for a visitor from Japan. The story I had constructed, however, was that this individual was flirting with the young lady and was being rude. As we give people the benefit of the doubt and consider alternative narratives that avoid the presumption of ill will, allowing for more honorable or even noble motives, we will succeed in managing our emotions.

Avoid the Presumption of Ill Will

One individual tended to think, anytime he saw two people conversing, that they were talking about him. This is called negative attribution. It is all too easy to incorrectly interpret another person's innocent behavior and assume the worst.

An effective practice, when we do not know how to interpret something, is to very briefly describe a situation, behavior, or apparent fault without offering an interpretation—and then permit the other person to explain. Such a description should avoid inferences as to why someone did something. We will often find out there was a good reason for what took place when we at least give others the opportunity to explain their perspective.

Break Down Big Issues into Smaller Ones

An effective negotiator is constantly looking for ways to break down challenges into smaller, more easily solvable issues. For instance, if a supervisor is resisting the introduction of new technology to track employee performance, it helps to talk it over and find out specific concerns. There may be some apprehension about the reliability of the system, the setup time, or even the challenge of staying on top of production data. Each of these concerns can be addressed separately.

Move Away from Blame

It is unfortunate that people often feed on faultfinding. If individuals are sufficiently introspective, they will often acknowledge that they had some blame in the matter. But as long as the situation is about blame, peaceful resolution will be unlikely.

Focus on the Problem, Not the Solution

The suggestion of concentrating on the problem rather than the solution may sound counterintuitive. Yet, for a number of reasons, it is one of the keys to effective negotiation. The more complex the situation, the greater the importance of this principle. When someone comes with the solution, even when that solution is a good one, it gives the other party the feeling of not having any control.

Reject Weak Solutions

As negotiators, it helps to learn about other people's preferences and to make our own clear. One grower explained that it was hard enough to understand his own needs and preferences, let alone be able to concentrate on someone else's. And perhaps that is one of the reasons we do not see interest-based negotiation used as frequently. It does take a certain amount of exertion, especially at first. With time, it can begin to feel more natural.

Sometimes a person will yield or pretend to yield—asserting, out of frustration, "That's fine; do it your way." By accepting another's yielding, individuals reduce their future negotiating power.

Maintain Integrity

At a time when many decisions were made on a handshake, my parents—grape growers in Chile's central valley—invited their children to a family conference.

"Earlier this year, we came to an agreement with the winery for a price," they explained. "Since then, many vineyards were affected by a terrible freeze—one that has meant a huge decline in supply. Had we waited a few more months, we could have gotten a much better deal."

My parents asked each of their five children for his or her opinion. The answer was a unanimous decision to honor the oral agreement. At the time, I was impressed that my parents would ask for our input. Since then, I have come to the conclusion that they knew the answer all along but wanted to teach us an important lesson about integrity.

Understand Time Pressures

Deadlines are often self-imposed. How often do we feel obligated to respond right away when facing a difficult situation? Why not solicit a little more time to study a matter or to accomplish a task? Do not be afraid to explain: "This is a tough one. It is now 8:15 and I'm tied up for the next 2 hours. If I call you between 11:00 and 11:30 this morning, will that work for you?" This type of detail takes only a few minutes longer to negotiate.

If we can build a little cushion for the unexpected, that is helpful. Most people do not mind waiting longer if they know what the real situation is. If a deadline seems hard to meet, ask to renegotiate an extension before the due date. An effective negotiator will ask the other party to suggest, or take a role in establishing, a deadline rather than arbitrarily impose one.

"I will call you back as soon as I can" or "I will call you right back," on the other hand, can leave much to be desired. The recipient of that message will wonder whether a call will come in the next half hour, 2 hours, or week. "Can I go to lunch," the person may question, "or do I need to sit here and wait?"

Lack of clarity can also come across as an avoidance tactic. To be credible, we need to be specific about time and about the nature of the task to be accomplished.

To do what we say we will do, in a timely fashion, builds trust. People who can be counted on to follow through with what they say they will do are considered invaluable.

Admit Error and Apologize

If the foundation is wrong, we may have to undo all our work and begin from scratch. Depending on how far into a project we are, this can be quite painful and expensive. We must first recognize that we have been wrong before we can make things right.

If we notice that the concrete foundation for the structure we are building is faulty, we can close our eyes and continue work at our own peril. As painful as it may seem, the sooner we recognize our mistake and make the necessary expenditures to break up and remove the concrete so we can start over, the better off we will be.

Sometimes, we become overinvested in an idea. It may be as hard to admit we are wrong as it is to break up that concrete. People who are willing to admit their mistakes are more likely to be considered trustworthy.

A proper apology is extremely convincing. So is sharing a new interpersonal skills goal. If we have been critical in the past, it helps to let people know we will be working to eliminate that negative trait.

To be genuine, an apology must not come across as a justification for what we have done wrong. A true apology is also accompanied, when possible, by an offer to make restitution. Furthermore, a sincere apology implies a willingness to make the appropriate changes commensurate with the wrong that has been done.

Value Others and Oneself

Everyone brings inputs (or contributions, such as a person's job, education, skills, or efforts) into a relationship. People put a value on each other's inputs. The best way of preserving the value of our own contributions is by valuing the contributions of others. The value placed on a person's time is a good proxy for power, which helps explain why quality time spent with people can be so meaningful (Brown 1986).

Conflict may arise when other people's assets are not valued. One young supervisor, a college graduate, may look at her formal education as an asset. A more seasoned supervisor at the nursery may look at her life experiences. Neither may value the other's assets. Both may compete for privileges or status based on their perceived contributions. Instead, they would be better off acknowledging each other's strengths.

For some people, it is very hard to say something kind about another. "I shouldn't have to say it," they reason, "because my actions should show my positive feelings." Others have trouble accepting the sincerity of affirming comments. Part of healthy interpersonal relationships is being able to both offer and accept positive comments: "Thanks. I appreciate your kind words. They made my day."

Separate Problems from Self-Worth

Without a doubt, it is a mistake to permit an issue to become intermixed with our self-worth. It is ineffective and manipulative, for instance, to imply that disagreement with our ideas is equivalent to a vote of no confidence in us. Such an approach will sooner or later make us feel rejected.

While we often have a great need to share feelings with those who can be supportive, we ought to choose such confidants with care. People who do not help us see from another's perspective beside our own may not be doing us any favors.

Keep Pretending

We tend to put our best foot forward when meeting people for the first time. During this period we make every effort to be the person we really would like to become someday. Sooner or later our human weaknesses show, and others see the real us. It is at this point that we may reason, "Why try and pretend anymore to be someone I am not?" The key, however, is to never stop trying to be the person we hope to someday become. To decide to continue to put our best foot forward despite our weaknesses. This requires a willingness to forgive others and ourselves for our imperfect condition.

Employee Discipline

At times workers do not meet expectations. As a first step, a supervisor will want to honestly consider if his or her own behavior is causing problems. Employees tend to perform better for supportive supervisors. Unfortunately, the opposite can be just as true. The first instinct of most supervisors is to tighten the reins and increase control over those who are perceived as having failed to meet their expectations. These apparent underperformers are quick to sense a lack of confidence in their work and in their decisions; they often become more defensive, refusing to make decisions they feel their bosses may overturn anyway, and they may withdraw mentally or physically.

Other common reasons for poor worker performance are lack of skill, knowledge, or ability. Lack of motivation or even purposeful misconduct may also be involved. Regardless of where the problems originated, a well carried out disciplinary process is yet another avenue to deal with performance challenges. When discipline is properly carried out, challenges are often resolved before they get out of hand. Much of the burden for improvement is placed, as it should be, back on the subordinate.

As a nursery manager, you get to make the rules and determine the consequences for their violation, as long as these rules are fair and defensible. For almost any misbehavior, there are many shades of wrongdoing.

Just as there are different degrees of fault, there are different consequences to deal with offenses. Tools to respond to infractions include communication of the standard, disapproval, verbal warning, written warning, suspension, and termination (Rosenberg 2002). Because it is so difficult for nursery crew leaders and supervisors to give written warnings, I prefer to substitute the verbal warning for an informal written warning. For example, this could be one in which the supervisor tells the worker that "this time the warning is between you and me as long as it does not happen again in so many months." I believe that supervisors are more likely to give such warnings, and they are much more powerful than a verbal warning. The supervisor also explains that if the employee does repeat the offense that led to the warning within the time period, both this first warning and the next one will be placed in the employee's folder.

If a rule is particularly important to you, the consequences for its violation may be more severe than those at the neighboring nursery. A useful guide in determining the fairness of consequences for disciplinary violations is to ask, for every rule and consequence, questions such as "What would I do if my best employee did not call in when he was sick? Came to work late? Got into a quarrel?" One may then be confident that the rule will not do more harm than good.

A progressive disciplinary approach combines the concept of stiffer penalties for more serious violations with that of increasingly more severe penalties for repeat offenses. The manager is thus forced to deal with less serious offenses before they become a major irritation. There will be no surprise terminations. When an employee's behavior is bad enough to require prompt action, even then the impending termination will not be a surprise.

With time, employees may be able to clear their record. For instance, an employee who was to be terminated the next time he was involved in horseplay most likely should receive a lesser penalty after several years of a perfect record.

The basic principles in this section follow a just-cause approach (Bureau of National Affairs 1992). That is, will your actions seem fair to your employees or to a judge if an employee sues the nursery?

Clearly Communicate Policies

Communication is the key link to a successful disciplinary process. Rules and consequences must be known by both those who apply them and those who are subject to them. It is not possible to conceive of every case of worker misbehavior, however.

When any infraction is repeated, the severity of the reprimand can progressively increase until a repeat offender is eventually terminated. No matter how bad the infraction, it should not result in immediate termination. An investigative suspension for a couple of days will permit you to contact a labor specialist and an attorney to make sure your case is defensible. The fact that California is an "at-will" state (an employee can quit or be fired at will, with some exceptions) will not keep a disgruntled employee from suing for wrongful termination.

Conduct a Fair Investigation

Listen to the accused employee's story first. A preliminary interview should be conducted with the employee before assigning penalties—from the least to the most serious infractions. This interview may be part of a more in-depth investigation. It is not uncommon to see a supervisor begin to lecture, nag, accuse, or scold an employee first, and then, almost as an afterthought, ask for the employee's perspective. By then, the damage has been done. The employee may have had a very good reason for her behavior. While some supervisors may apologize at this point (which, while nice, will not totally remove bad feelings nor prevent the erosion of trust and good morale), others are just as likely to continue to chastise the worker in an effort not to lose face before her.

A supervisor who truly gives the accused worker an opportunity to explain first, however, will often find that there is no need for discipline. The purpose of the investigation is, in part, to determine if there were any mitigating circumstances that could reduce, but not necessarily eliminate, disciplinary action. Could the employee's action have some justification?

Use Corrective—Not Punitive—Action

After the investigation, if the incident is worth documenting, it is serious enough to merit taking official disciplinary action. A consequence must be formulated keeping in mind the purpose of the disciplinary process. To obtain both maximum management and legal benefits from discipline, the response ought not be punitive in nature. The supervisor must act as if truly interested in helping the worker with the problem. To do so effectively, a friendly tone ought to be maintained throughout. It is a good idea, after listening to the employee and determining that the individual will be receiving formal discipline, to point out some of the qualities of the worker before getting into the disciplinary formalities. This will help set the right tone for a positive, nonconfrontational discussion. Just as important, after the process is completed, is to once again focus on some positive aspects of the employee.

The employee needs to feel the supervisor's concern for

her. These positive comments may be needed over the next few days, also. Large differences in status between supervisor and employee may cause workers to accept chastisement now but resent the supervisor later. Most important, the supervisor who makes it clear that this is not something personal against the employee, but just against a specific behavior, is more likely to succeed as a coach and mentor.

One of the most valuable lessons I have learned in this respect is to trust my feelings. If I feel uncomfortable jumping in to point out a fault, or discipline an employee, it is for a reason. It is critical to communicate and connect with the employee as a person before talking about behaviors that need correcting. If we skip this step now, we are likely to pay the price later in terms of increased stress and reduced interpersonal effectiveness and trust.

A formal disciplinary episode needs to include the following four elements (it helps if the employee can be involved in explaining what went wrong and contributing ideas toward improvement):

- Be specific about what the employee did wrong—without getting bogged down in minutiae. Explaining the reasons for the needed change may be appropriate at this time.
- Be clear about what the employee must do to improve (this is not always obvious).
- Advise the employee of the official nature of the discipline (that it will be documented and a copy will go to the employee and the other in his file).
- Inform the employee about future consequence(s) if there is no improvement. (Option: after telling the employee about the next consequence for nonimprovement, explain that if the problem continues, eventually it may lead to termination.)

A written document should be clear to someone who knows nothing about the situation. Do not finalize a written expression until someone you trust can read it over and give you constructive suggestions. This person should check for the following:

- Does the tone and substance of the warning show that you care about the worker?
- Are positive, sincere comments made about the employee? (These comments should be sincere and specific. "We really value your excellent welding skills" is better than "You are such a great worker.")
- Is the language so clear that anyone would understand it?
- Are all four of the basic disciplinary warning elements included?

References

Billikopf, G. 2003. Labor management in agriculture: Cultivating personnel productivity. Oakland: University of California Agriculture and Natural Resources Publication 3417. UC Agricultural Labor Management website, http://nature.berkeley.edu/ucce50/ag-labor/7labor/00.pdf.

———. 2008. Piece-rate pay design. University of California Agricultural Labor Management website, http://www.cnr.berkeley.edu/ucce50/ag-labor/7research/7calag06.htm.

———. 2009. Party-directed mediation: Helping others resolve differences. 2nd ed. University of California Agricultural Labor Management website, http://www.cnr.berkeley.edu/ucce50/ag-labor/7conflict/.

Brown, R. 1986. Social psychology. 2nd ed. New York: Free Press.

Bureau of National Affairs. 1992. Grievance guide. 8th ed. Washington, DC: BNA Books.

Covey, S. 1989. Seven habits of highly effective people. New York: Simon and Schuster.

Fisher, R., and W. Ury. 1981. Getting to yes: Negotiating agreement without giving in. New York: Penguin.

Patterson, K., J. Grenny, R. McMillan, and A. Switzler. 2002. Crucial conversations: Tools for talking when stakes are high. New York: McGraw-Hill.

Rosenberg, H. R., R. Carkner, and J. P. Hewlett. 2002. Ag help wanted: Guidelines for managing agricultural labor. Western Farm Management Extension Committee.

Taylor, F. W. 1947. Scientific management. New York: Harper and Row.

Glossary

abiotic. Related to nonliving factors. An abiotic disease is a physiological disorder caused by factors other than pathogens.

above ground system (AGS). A variation of the pot-in-pot system, this double-container system is for use aboveground. A production container nests in a double-walled holder container that serves to insulate the plant's roots from temperature extremes. A ballast can be added to prevent blow-over. The AGS allows easier harvesting than the inground, pot-in-pot system.

abscission. Falling-off of a plant part.

accounts payable. Found on the balance sheet, right side. Examples include the bill for ceramic containers or the telephone bill. They are bills due soon (usually within 30 days).

accounts receivable. Found on the balance sheet, left side. Examples would be in-store credit extended to retail customers to pay for plant purchases. Typically these bills are due within 30 to 90 days.

accrual record keeping. Accounting method in which income is recorded when the sale is completed (regardless of whether the payment has been received), and expenses are recorded at the time they are incurred (regardless of whether the expense has been paid). The IRS generally requires any firm with manufacturing, wholesale, or retail business activities to use the accrual method. However, an exemption exists for sod and nursery farms.

action, or treatment, threshold. The level at which plant injury or pest population size is sufficient to warrant some type of control action; below this level the presence of pests and amount of damage can be tolerated. By applying remedial action only if pest populations exceed this threshold, overall pesticide use and liabilities associated with it can be reduced. However, regulations such as quarantines and shipping protocols may impose threshold levels of zero pests.

adventitious organ. An organ that arises from an unusual, abnormal, or unexpected location (e.g., roots growing from leaves, petioles, or stems).

air-filled porosity. The percentage of a given volume of a potting mix that air occupies after irrigation and drainage.

annual plant. A plant that completes its life cycle in one growing season.

apical dominance. The suppression of lateral bud growth by the apical bud through the hormone auxin.

apical meristem. Undifferentiated tissue at the tip of a shoot or root formed of meristematic cells (see meristem) that elongates when growing.

apomixis. Seed (embryo) formation without sexual union and exchange of gametes.

asexual propagation. Reproduction of plants from vegetative tissues (e.g., stems, leaves) resulting in genetically identical offspring.

assets. Found on the balance sheet, all on the left side. These are things the business owns. Typically they are divided into two categories: current assets (cash or things that can be easily sold or converted to cash with 90 days) and fixed assets (things the business owns that cannot be easily converted or sold within 90 days—for example, it may take longer to sell irrigation equipment to raise cash). Assets include the accounts receivable, the plants and containers in inventory, the buildings, and vehicles and other equipment. The sum of the value of assets is always equal to liabilities plus owner's equity.

asset turnover (AT). This is a key financial ratio and one of the five calculated on the strategic profit model (SPM). An asset turnover is the rate or speed at which a business turns assets into sales. The equation for calculating asset turnover is AT = sales ÷ assets.

atmometer. An instrument for measuring the rate of evaporation from a wet surface to the atmosphere. When calibrated, it can be used to estimate plant evapotranspiration.

at-will state. A state that permits employers to fire individuals without having to show just cause. (However, employees may sue their employers if they are fired without just cause.)

augmentation. Biological control strategy that involves the periodic release or application of beneficials into the production system.

automation. Technology designed to perform a process by means of programmed commands combined with automatic feedback control.

available water. The percentage of a given volume of a potting mix that is occupied by water that the plant can extract.

bait station. A device used to house baits to attract target species of a pest.

balance sheet (BS). This is one of the two key financial statements (the other being the income statement). It has one date at the top, typically the last day of the calendar year or the fiscal year. It is a snapshot of the financial fitness of the business. It reflects financial fitness at one point in time.

balled and burlapped plant (B&B). Evergreen or deciduous woody plant that is dug with the soil mass attached; the root ball and soil are covered with heavy burlap to keep them together during shipment and sale.

bare-root plant. A plant that is sold with the roots exposed rather than in soil. Bare-root crops include roses, small deciduous shrubs and trees, deciduous berries, small ground covers, herbaceous perennials, and ornamental grass divisions.

base temperature. Temperature below which no plant growth occurs.

bedding plant. Any herbaceous plant that is produced and sold for planting in the landscape, garden, or patio.

benchmarking. A measure of performance that permits comparison, such as healthy plants cultivated per square foot.

bench-top heating system. A system that heats water that is then piped through tubes on benches below potted plants.

beneficials. Organisms that provide biological control of pests. Also called natural enemies or biological control agents.

best management practices (BMPs) for water pollution. Structural or managerial practices used to treat, prevent, or reduce water pollution.

biocontainer. A container that is not comprised of petroleum-based plastics, so it degrades rapidly when planted into the field or when placed in a composting operation.

biological control. The use or release of natural enemies to control pests.

biotic. Related to living organisms.

biotype. A group of organisms having an identical genetic constitution.

black sooty mold. Sooty molds are associated with sucking insect pests (aphids, soft scales, mealybugs, psyllids) that extract sap from the phloem tissue and secrete honeydew. Sooty molds lessen the aesthetic value of ornamental plants and lower the vigor of plants by blocking sunlight essential for photosynthesis.

book value. The original purchase cost of the asset minus the asset's accumulated depreciation. Owned capital in buildings, improvements, and equipment is usually assessed at book value.

broadcast baiting. In vertebrate pest control, a baiting method that typically involves the use of a mechanical seed spreader to thinly scatter rodenticide bait over a designated area.

broad-spectrum insecticide. A pesticide that kills a large number of unrelated species.

Bt *(Bacillus thuringiensis)*. Soil-dwelling bacterium that produces an endotoxin that may be extracted and used as a pesticide.

bulk density. A measure of the mass per unit volume of a dry soil or potting mix.

business plan. A formal statement listing a set of business goals and describing how they will be reached. A business plan provides an essential road map for business success; it outlines the route a company intends to take to produce revenues. Plan formats vary, but they generally include descriptions of the business, products and services, market, forecasts, management team, and financial analysis.

callus. An unorganized mass of cells usually formed by a plant in response to a wound.

capillary action. The movement of water through the pores of a medium due to the attraction of water to the solid particles of the medium.

capillary rise. The height above a free water surface to which water will rise by capillary action.

carbamate. A salt or ester of carbamic acid. Carbamates include synthetic organic insecticides that affect the nervous system by disrupting an enzyme that regulates acetylcholine, a neurotransmitter.

cash-basis accounting method. Record-keeping accounting method in which income is recorded when payments are received and expenses are recorded at the time of payment.

cast skin. The empty exoskeleton or shed skin of an insect.

CDFA pest rating policy. Pest ratings (Q, A, B, C, or D) represent the CDFA policy on specific pests so that the appropriate action, such as quarantine, eradication, control, hold, or inspection, is standardized.

centrifugal spreader. A dispenser of granular materials that uses a spinning disk with vanes to fling granules over a distance. These types of spreaders vary greatly in size, capacity, and throwing distance, and they may be hand-held or mounted on a cart, tractor, or other vehicle.

check valve. A valve that allows water to move in only one direction, preventing drainage.

chemigation. The application of a chemical, usually a pesticide, through the irrigation system.

chilling requirement. A period of cold temperature required by some plants before they will resume growth and flowering in spring.

chilling-sensitive plant. A plant that is prone to damage after exposure to low temperatures above freezing (50°F, for example).

chimera. Genetically distinct tissues growing adjacent to one another.

chlorination. A process that kills pathogens through the oxidation of cell membranes, proteins, and other organic biochemicals. Methods of incorporating chlorine into recycled irrigation water include the use of sodium hypochlorite, calcium hypochlorite, and chlorine gas.

Clean Water Act (CWA). A series of legislative acts that form the foundation for protection of U.S. water resources, including the Water Quality Act of 1965, the Federal Water Pollution Control Act of 1972, the Clean Water Act of 1977, and the Water Quality Act of 1987.

climate zone. A geographic area with a distinct climate defined by parameters such as latitude, elevation, length of growing season, minimum and maximum temperature, solar radiation, rainfall, and humidity.

clone. A population of plants derived asexually from a single individual and maintained using asexual means, thus making all members genetically identical. Also, any single member of the asexually propagated population.

Code of Federal Regulations (CFR). The codified rules of U.S. executive departments and agencies that have been published in the *Federal Register*.

cold frame. A structure that protects plants from precipitation and perhaps retains some ambient heat provided by the sun, usually with no temperature control technologies.

conditional waiver (ag waiver). A conditional waiver of the waste discharge requirements permit, issued by a California regional water board to owners and operators of irrigated agricultural land to protect surface water and groundwater.

conservation. Biological control strategy that involves protecting naturally occurring beneficials that keep pests in check.

constructed wetland. An artificial marsh or swamp for treating wastewater, controlling floodwater, and reducing erosion.

contact pesticide. Generally controls a pest as a result of direct contact. Insects are killed when the pesticide is sprayed directly on them or when the pests crawl across surfaces treated with a residual contact insecticide. Weed foliage is killed when enough surface area is covered with a contact herbicide. Contact herbicides affect only the part of the plant where they are applied and do not move to other parts of the plant.

container. The object in which plants are grown, such as a pot, basket, flat, or tray.

container capacity. The maximum amount of water that a soil or potting mix can hold in a given container. It is usually expressed as a percentage of the soil volume.

controlled-release fertilizer. A granular fertilizer that slowly releases nutrients through a porous coating.

convection tube. Plastic film tube installed in a greenhouse to distribute air throughout the production area to increase temperature uniformity.

conventional insecticide. As defined under the U.S. EPA Office of Pesticide Programs' Conventional Reduced Risk Pesticide Program, any pesticide other than biological pesticides and antimicrobial pesticides.

corion. The eggshell of an insect.

corporation ("C" corporation). An independent legal entity owned by shareholders, in which the corporation itself, not the shareholders that own it, is held legally liable for the actions and debts that the business incurs.

cost of goods sold (COGS). Found on the income statement, this is typically the second line. COGS shows, in dollars, everything that went directly into making products and services to sell. COGS does not include the salary for management or office staff. The other term for COGS is direct cost, since the costs go directly into making the product or service.

cull pile. Area where discarded plants are placed.

current assets. Found on the left side of the balance sheet. These are things the business owns that can be quickly converted to cash within 90 days or less. These include cash, accounts receivable (since these bills are typically paid within 30 to 90 days) and inventory (since the plants typically could be sold within 90 days to get cash).

current liabilities. Found on the right side of the balance sheet, this is something the business owes. It refers to bills that are paid typically within 30 to 90 days. These would include the payroll, insurance payments, and the telephone and utility bill. Also might include a short-term line of credit from the bank (or a credit card bill).

current ratio (CR). This is a key financial indicator and a calculation not found on the income statement or balance sheet. It shows the business's ability to pay bills. The equation for calculating current ratio is current assets ÷ current liabilities. The current ratio should be greater than 1.00. This means the business has $1 in current assets (cash or almost cash) for every $1 in current bills or liabilities. Less than 1.00 means the business doesn't have enough current assets to pay its bills.

dedifferentiation. The loss of specialized function of a cell resulting in a meristemlike state (see meristem).

depreciation. The scheduled reduction in the value of an asset. The federal government has a schedule according to which all assets are depreciated. It appears as an indirect cost, or expense, on the income statement. The value of assets declines every year because assets (with the exception usually of land) depreciate every year. A company needs to plan to have profits reinvested every year. If not, the owner's equity declines as the assets are depreciated.

destructive sampling. Removal of plant material for the purpose of pest sampling. Destructive sampling is necessary if a microscope is required to inspect samples, which is common when monitoring for parasitism in biological control programs.

deutonymph. The third stage of a mite.

diatomaceous earth. Finely ground fossilized shells of minuscule, hard-shelled algae called diatoms. Diatomaceous earth is used as a desiccant against insects.

differentiation. Development of many cells or tissues (from a single one) that acquire and maintain a specialized function.

direct spraying. Directing pesticide spray toward the part of the plant where the pest resides. Direct spraying can result in the overall reduction of pesticides.

discharge rate. The rate at which water leaves an emitter.

distribution uniformity (DU). A numerical measure of irrigation uniformity, with 100 percent being perfectly uniform and lower percentages being less uniform. It is determined by comparing the average amount of water applied to the area receiving the lowest 25 percent to that applied on the average over the entire irrigated area.

dormancy. A state when a plant stops growing in response to environmental signals and requires certain conditions to resume growth (e.g., a period of cold temperature to resume growth and flowering in spring).

ecdysis. To shed the exoskeleton.

electrical conductivity. A measure of salinity in terms of the ability of soil water to conduct electricity.

electrostatic sprayer. A type of ultra-low-volume sprayer that imposes a small electrical charge to the spray droplets. These charged droplets are attracted to plant surfaces, which results in a more uniform application of the spray material.

embryogenesis. The *de novo* (renewed) formation of an embryo from unorganized cells.

emission uniformity. A measure of irrigation uniformity used for microirrigation systems. It is determined following the same procedure as for distribution uniformity.

eradication. The elimination or removal of a pest or disease species conducted when pest or disease population growth is not rapid or when infestation or infection is not widespread. Eradication as a strategy is dependent on biological, ecological, social, and operational factors, including public support.

estivation. Dormancy of an animal during a prolonged hot or dry period.

evapotranspiration (ET). Evaporation from the soil surface combined with transpiration of the plant. Irrigation amounts should reflect the replacement of water used for plant evapotranspiration.

exhaust fan. A large fan that blows air out of a greenhouse at a rate that can result in several air exchanges per minute. The rate is determined by the size of the motor, fan blades, and the degree to which air can enter the greenhouse opposite the exhaust.

exoskeleton. An external skeleton or outside covering of the body.

exotic pest or disease. Pest or disease organisms (plants, animals, and microorganisms) that are not native to a particular region. The impact of exotic pests varies considerably, depending on the species and the area being invaded. Species that rapidly colonize an area are often called exotic invasives (see invasive species). Some exotic species have been intentionally introduced, while others are inadvertently introduced by humans, other species, or natural forces.

explant. The plant part used to initiate a tissue culture (e.g., shoot tip, leaf section, node, petiole).

external wage equity. Assessment of wage fairness by comparing earnings with those at similar nursery operations or even non-nursery operations.

fan-and-pad cooling. An evaporative cooling system consisting of exhaust fans at one end of the greenhouse and a pump circulating water through and over a porous pad installed at the opposite end of the greenhouse.

fan nozzle. A sprayer nozzle that produces an inverted "V" output pattern.

feeder bin (hopper). A container that dispenses large volumes of dry material, such as soil and soil components, at a fixed rate.

fiber container. A biodegradable container made of compressed wood fiber, peat moss, or other organic materials, such as rice hulls.

field stock. Nursery plants grown in outdoor settings from seed or transplants, then dug up by hand or machine and marketed as bare-root, balled and burlapped, or containerized plants.

filter strip. A band of vegetation that can be used between a runoff source and a water body to slow runoff from the production area and trap sediment, fertilizers, and pesticides before they reach the water body.

financial benchmark analysis. The use of key indicators to evaluate a company's operational and financial performance in comparison with industry standards or benchmark values.

fixed assets. These are found on the left side of the balance sheet. Fixed assets can't be sold or converted to cash easily, thus they are "fixed." For the nursery, these would include the irrigation equipment, for example, and the skid-steer loader used to unload trucks and move inventory.

flocculation. The removal of sediment from a fluid by using an agent that facilitates aggregation of the particulates. If clay and other small particles in a colloidal solution are flocculated (i.e., aggregated electrically and physically into larger particles) by flocculating agents such as polyacrylamides, alum, or ferric sulfate, these particles will settle much more easily.

flocculence. White, waxy or flaky covering on some insects, such as mealybugs and aphids.

flower induction. The process by which a bud is able to form a flower.

flushing. Using water in the line to remove accumulated material.

fog system. A system that produces a fine fog with small water particles that stay suspended in the air, used for creating a high-humidity environment or for greenhouse cooling.

forb. A broadleaf herbaceous plant that is not a grass.

forklift (lift truck). A powered industrial truck used to lift and transport materials.

front-end (or bucket) loader. A heavy equipment machine used in nurseries to move or load materials such as sand and wood chips onto another type of machinery such as a conveyor belt. Wheeled loaders are the most commonly used type due to their ability to lift large loads and their ease of operation. A farm tractor fitted with buckets can serve multiple purposes as a loader and an implement carrier.

frost hardy. Ability of a plant to withstand temperatures below freezing to a certain threshold.

frost pocket. A depression in the topography of the land where cold air can settle.

generally accepted accounting principles (GAAP). A standard framework of guidelines for financial accounting developed by the accounting profession and governed by the Financial Accounting Standards Board (FASB). These include the standards, conventions, and rules that accountants follow in recording, summarizing, and preparing financial statements. These standards ensure that the financial information has a consistent meaning, allowing valid comparisons over time or across different businesses. For publicly held companies, these standards ensure full disclosure of information that may be important to stockholders and potential investors.

germination room. Special room equipped with precise temperature, light, and humidity controls used to achieve uniform seed germination.

girdling. Severing and removing a ring of cambium around the plant stem or tree branch, which removes the phloem and causes plant death.

glazing material. The transparent cover that is part of a greenhouse. Common materials consist of glass, plastic film, or rigid plastic panels that can be either clear or translucent.

gravitropism. Growth away from or toward gravity.

gross margin, or gross profit margin (GM). This is a key financial indicator calculated from (not found on) the income statement. Margin means percentage and profit means dollars.

gross profit. The dollar amount left after cost of goods sold (COGS) is subtracted from sales. It appears on the income statement. It is typically the third line, or just below COGS. It shows how many dollars are left over after direct costs, or COGS, are taken from sales.

gross sales. Sales receipts from all sales, including items that were returned, plants that died in transit, or items for which a credit was issued.

growth chamber. Growth room equipped with heating, cooling, lighting, and humidity controls (and sometimes carbon dioxide injection) that create the specific environmental conditions desired for a plant.

hanging basket. A growing container with attached supports that can be hung from a hook or rack.

hardening off. Acclimating plants gradually to tougher conditions (e.g., to full sun conditions from growing in a protected environment such as a greenhouse or under shade).

hemimetabolous. Indicating an insect that undergoes incomplete metamorphosis, a form of simple metamorphosis. Immature insects (nymphs, often called naiads) are aquatic and breathe via gills. This pattern of development occurs in Ephemeroptera (mayflies), Odonata (dragonflies and damselflies), and Plecoptera (stoneflies).

herbaceous plant. A plant lacking a permanent woody stem.

herbicide. A chemical used to kill a plant or otherwise disrupt its growth process.

herbicide residual. The amount of active herbicide remaining for a given time after application.

herbivorous. Feeding on plants; an herbivore is an animal that feeds primarily on plants.

hermaphrodite. An organism that is normally equipped with both male and female reproductive organs.

high-density polyethylene (HDPE). A flexible plastic, usually black, that is typically used in blow-molded containers (up to 65 gallons) for outdoor nursery stock. Also extensively used for containing pesticides and herbicides, because it is thermally and chemically resistant.

high-impact polystyrene (HIPS). A plastic material that is inexpensive, extremely lightweight, and more rigid than polypropylene. Commonly used in containers for seedlings and small plants.

high-intensity discharge (HID) lamp. A lamp that produces visible light at high output levels. Various types of HID lamps are available, each with a specific spectral distribution. Those that produce a substantial amount of photosynthetically active radiation relative to the input energy may be used in greenhouse and nursery production.

holometabolous. Indicating an insect that undergoes complete metamorphosis. Holometabolous insects have a four-stage life cycle including egg, larva, pupa, and adult.

homoptera pest. Insect pests in the suborder Homoptera, such as leafhoppers, aphids, psyllids, whiteflies, and scale insects.

honeydew. An excretion from insects, such as aphids, mealybugs, whiteflies, and soft scales, consisting of modified plant sap.

hoophouse. A simple greenhouse constructed with bent pipe and a glazing material made from transparent plastic. Hoophouses generally have minimal environmental temperature control.

hopper rain. Liquid substance excreted by glassy-winged sharpshooters as they feed on xylem tissue.

horizontal airflow fan (HAF). A fan used to stir the air in a greenhouse to make the temperature more uniform and to push warmer air (from higher up in the greenhouse) down toward the plants. Horizontal airflow fans are necessary because heating and cooling result in temperature gradients that cause lack of uniformity in greenhouse production.

impact sprinkler. A rotating sprinkler with a weighted arm that drives the emitter to move in a circle.

impaired water body. A water body that is too polluted or otherwise degraded to support its designated and existing beneficial uses.

income statement (IS). This is one of the two key financial statements or forms (the other being the balance sheet). It has a time period at the top of the form, usually a calendar or fiscal year. You can have an income statement for any time period (quarter or month) if desired. While the balance sheet shows a snapshot of fiscal fitness, this is the movie — it shows fitness over time. It shows where the money went and what profit was earned. The "bottom line" refers to the bottom line of the income statement, which is net profit (or net profit before taxes).

indemnity. In insurance, a compensation paid for loss, damage, or liability (payment).

indirect costs. Found on the income statement, these are costs that indirectly create sales. These would include managerial salary (not any pay that goes into creating or delivering sales), office or support staff wages, telephone bills, advertising, payment to accountant or lawyer, depreciation of fixed assets, uniforms, and other items. Also might be called overhead expenses.

infrared (IR) film. A transparent plastic film that reduces the transmission of infrared light in comparison with regular glazing material. Infrared light is radiation with a wavelength longer than visible light (or photosynthetically active radiation, PAR).

infrared heating system. A heating system that creates radiant heat energy by firing combustible fuel directly into a steel tube. As a result, the steel tube glows and heats all plant material that is exposed to it. The aim is to heat only the plant without heating the rest of the greenhouse.

insect pheromone. A substance secreted by an animal, especially an insect, that causes a specific reaction in other individuals of the same species, such as a sex attractant, alarm substance, and so on.

instar. The immature insect between molts during development.

integrated pest management (IPM). An ecosystem-based strategy that focuses on long-term prevention of pests or their damage through a combination of pest control strategies including biological, cultural, mechanical, physical, and chemical control. These strategies are implemented according to established guidelines only after monitoring indicates that such treatment is appropriate. This involves knowledge of pest life cycles and an understanding of crop production. Pest control materials are selected and applied in a manner that minimizes risks to human health, beneficial and nontarget organisms, and the environment, with the goal of removing only the target organism.

internal wage equity. Assessment of wage fairness by comparing earnings with others in the same nursery operation.

invasive species. Nonindigenous species that are especially successful at establishing viable populations in their new location.

invertebrate. An organism that lacks a backbone or spinal column.

irrigation duration. The amount of time required for an irrigation event.

irrigation efficiency. A measure of how much of the applied water is beneficially used. The major beneficial water use is supplying the plant water needs.

irrigation frequency. How often irrigation occurs. The frequency of irrigation is based on factors such as the amount of available water in the plant root zone, the plant evapotranspiration, and the plant susceptibility to negative water stress impacts.

irrigation scheduling. Determining when and how much to irrigate.

irrigation uniformity. A measure of how evenly water is applied in an irrigation zone.

job sample test. A test that replicates an actual task, such as the grafting of plants.

larva (pl., larvae). The immature and wingless form that hatches from the egg of a holometabolous insect and eventually transforms into a pupa prior to reaching adulthood. Also, sometimes, the immature form (or nymph) of a hemimetabolous insect.

leaching. Removing accumulated salts from the root zone with water.

leaching fraction. The ratio of the volume of water leached (e.g., the water that runs out of the bottom of a pot) to the volume of water applied (e.g., the total amount of water applied to a pot).

leverage. One of five key financial ratios calculated on the strategic profitability model, it is an indication of how much money is borrowed. The equation for calculating leverage is (current + fixed liabilities + owner's equity) ÷ (owner's equity). Or, everything on the right side of the balance sheet ÷ owner's equity. It is always 1.0 or higher. Companies that are leveraged have used their money (owner's equity) to borrow money.

liability. Found on the right side of the balance sheet, a liability is something the company owes. This would be bills (insurance, payroll, or plants) or loans (short-term line of credit or longer mortgage payment). Liabilities also include the owner's equity, or net worth (the owner's investment in a company or the net worth of a company).

light-emitting diode (LED). Diodes are electronic devices (semiconductors) that can emit radiation, among other features. LEDs, or light-emitting diodes, are devices that emit light. They can do so in narrow wavelength ranges, enabling them to be monochromatic. LED technology is relevant in horticulture since it is possible to create lighting systems with arrays of LEDs of various colors to match the needs of the plants.

limited liability company (LLC). A hybrid type of legal structure that provides the limited liability features of a corporation and the tax efficiencies and operational flexibility of a partnership. The owners of an LLC are referred to as members. Depending on the state, the members can consist of a single individual (one owner), two or more individuals, a corporation, or other LLCs.

liquid asset. Also known as a current asset, found on the left side of the balance sheet. An asset is considered liquid if it is cash or can be easily converted to cash.

liquid feeding (fertigation). The application of fertilizers through an irrigation system.

loan covenant. A condition tied to a commercial loan or bond issue that requires the borrower to fulfill certain responsibilities and forbids the borrower from undertaking certain actions; it can possibly restrict certain activities to circumstances when other conditions are met.

long-term liability. Found on the right side of the balance sheet, this refers to bills that typically cannot be paid within 30 to 90 days. This would include a 30-year mortgage or a 3-year vehicle loan.

low- and ultra-low-volume (LV and ULV) sprayer. A sprayer that dispenses materials, such as pesticides, in a low volume of carrier. In comparison with high-volume sprayers, which use a large volume of water as the carrier, LV and ULV applicators may use a smaller volume of some other liquid as a carrier to dispense the material.

low-density polyethylene (LDPE). A relatively inexpensive film plastic, used as a covering for greenhouses and hoophouses.

macronutrient. An essential nutrient that is required by plants in relatively large amounts.

mandibles. The first of the paired mouth appendages in insects and other arthropods; usually jawlike (in chewing forms) or needlelike (in sucking forms).

mass merchandiser. Large retail store that offers a wide range of different kinds of products. Usually characterized by high-volume, fast-turnover selling of staple goods for less than conventional prices.

mechanization. The application of machines to do work. In the production of ornamental plants, industrial equipment such as forklifts are relatively common, but often the machines are designed for tasks specific to the nursery industry (e.g., potting machines).

medium. The substrate in which plants are grown; the plural of medium is media.

meristem. Undifferentiated tissue made up of cells that divide to form specialized tissues and organs.

metamorphosis. A series of marked and more or less abrupt changes in the form of a developing insect.

microirrigation. The application of water at low discharge rates by means of bubbler, drip, trickle, mist, or spray irrigation systems.

micropropagation. The aseptic process (free of pathogenic microorganisms) whereby asexual means are used to reproduce many whole plants from cells, tissues, or organs of a single plant on culture media in vitro (in a test tube or culture dish).

mist system. A system that produces a mist with water particles that fall rather than stay suspended in the air (as in a high-pressure fog system). Mist systems are used in horticulture to make plants wet when this is desired (propagation). They may also be used to help plants cool themselves.

mode of action (MoA) (mode of activity). The way a pesticide works. It refers to the action of a pesticide at its target site (the physical location within an organism where the pesticide acts). Pesticides can be classified based on the commonality of the active ingredient's mode of action, which is a useful categorization for resistance management.

moisture-release curve. A graph that expresses the relationship between the volumetric moisture content and the tension at which water is held in the medium.

molt. In arthropods (e.g., insects and crustaceans), the process of periodically shedding the exoskeleton in order to grow.

monitoring (scouting). The regular inspection of crops and growing areas for pests, including insects, other arthropods, vertebrates, pathogens, nematodes, and weeds. Monitoring is a basic component of IPM programs.

mosaic. Common virus symptom expressed in leaves with variable light and dark green areas, giving a mottled appearance.

mulch. Any material used to cover the soil or potting media surface, such as bark chips, plastic sheet, or fabric.

necrosis. The death of most or all of the cells in plant tissue due to disease, injury, or failure of the vascular system. Cells turn black or dark brown and decay.

neonicotinoid. A class of insecticides that act on the central nervous system in a similar way to nicotine. Neonicotinoids are systemic and move throughout the plant in the vascular system.

net profit. The dollar amount of profit or the bottom line of the income statement. It is important not to think of net profit as a residual or what is left over after cost of goods sold (COGS) and expenses are paid. Rather, think of it as a planned residual or profit.

net profit margin (NPM). The percentage of net profit dollars over sales. It is one of the key financial indicators that does not appear on the income statement (IS) or balance sheet (BS) but on the strategic profitability model (SPM).

net sales. Sales receipts after any returns or credits are issued to purchasers.

net worth (NW). Also called owner's equity, it is found at the bottom right of the balance sheet. This shows the owner's (or multiple owners') investment in the company. This is always equal to assets minus liabilities, since the balance sheet must balance.

nocturnal. Active at night.

nonpoint source pollution. Any source of water pollution that is not a "point source" as defined in Section 502(14) of the Clean Water Act. Nonpoint source (NPS) pollution comes from many diffuse sources, with agriculture as the single largest source. (See point source pollution.)

nonselective herbicide. An herbicide that has an effect on nearly all types of plants.

nonselective pesticide. A pesticide that kills a wide range of plants or animals; also called a broad-spectrum pesticide.

noxious weed. As defined by California Food and Agricultural Code, Section 5004, "any species of plant that is, or is liable to be, troublesome, aggressive, intrusive, detrimental, or destructive to agriculture, silviculture, or important native species and difficult to control or eradicate, which the director, by regulation, designates to be a noxious weed. In determining whether or not a species shall be designated a noxious weed for the purposes of protecting silviculture or important native plant species, the director shall not make that designation if the designation will be detrimental to agriculture."

ohmic heating (joule heating, resistive heating). A method of heating that uses an electrical current passing through a conductor to generate heat. Ohmic heating has been researched as a treatment for removing pathogens and weed seeds from nursery soils.

optivisor. A visor with a built-in magnifying lens, similar to a jeweler's magnifying visor, used for monitoring pests.

organogenesis. The *de novo* (renewed) formation of an organ from unorganized cells.

organophosphate. Any of several organic compounds containing phosphorus, some of which are used as fertilizers and pesticides.

origin inspection. Some nursery stock can enter California if the nursery stock has been inspected at the origin by the local authorities and if California is satisfied that the authorities are qualified to do an inspection and ensure that the stock is free from agricultural pests.

overspray. Applied irrigation water that is not applied to the target plants. Usually associated with sprinkler irrigation.

oviposition. The process of laying eggs.

ovipositor. An organ used by insects for depositing eggs in a place suitable for their development.

owner's equity (OE). Also called net worth, as defined above. This shows the owner's (or multiple owners') investment in the company.

ozonation. Process for infusing water with ozone to kill pathogens by disrupting cell membranes, proteins, and organic compounds.

pallet. A flat transport structure that supports goods in a stable fashion while being lifted by a forklift, pallet jack, front loader, work saver, or other jacking device.

palletization. The use of a pallet.

palps. Segmented appendage found near the mouth in invertebrate organisms such as mollusks, crustaceans, and insects.

parasitoid. Beneficial insect that is smaller than the host pest species and spends part or all of its life cycle with its host. Insect parasitoids that attack insect pests such as aphids, scales, and whiteflies are generally tiny wasps or flies that use their ovipositor or stinger to lay eggs on or in their insect hosts.

parthenogenesis. Reproduction without fertilization.

partnership. A single business where two or more people share ownership. Each partner contributes to all aspects of the business, including money, property, labor, or skill. In return, each partner shares in the profits and losses of the business.

passive cooling system. Form of cooling achieved by opening roof and side vents or roll-up walls to allow natural ventilation (compared with a forced ventilation system that depends on fans). Passive cooling relies on convection, whereby the hot humid air naturally rises out of a greenhouse while cooler air is drawn in through the lower openings.

pathogen. Disease-producing microorganism; causative agent of a disease, including a pathogenic virus, mycoplasma, bacterium, or fungus.

perched water table. A zone of saturation separated from permanent groundwater by a low, permeable soil layer. The soil layer that perches above water serves as a barrier, preventing contaminants from moving to the permanent groundwater supply, reducing risk.

perennial plant. An herbaceous or woody plant that lives at least 2 years (and sometimes much longer, as in the case of trees).

permeability. The rate at which water moves through soil. Permeability is controlled by the size and continuity of the soil pores, and it is influenced by soil texture, organic matter, and density.

pest. An organism in the crop environment that causes economic or aesthetic damage to crops. Numerous pests are found in nurseries due to the wide range of crops that are grown. Examples include insects and other arthropods, pathogens, nematodes, weeds, and vertebrates. Many pests are exotic species. In container nursery production, pests also include organisms that do not necessarily cause damage to the host plant but are capable of causing damage to other plants if disseminated during shipping. Specific legal definitions of "plant pest" and "pest" that are important in the shipping of nursery materials can be found, respectively, in the Code of Federal Regulations, Title 7, Section 330.100, and the California Food and Agricultural Code, Section 5006.

pesticide adjuvants. Products used to enhance the performance of pesticide applications. In general, they cause the pesticide to spread, stick to, or penetrate the plant material, resulting in more effective pest control. The use of adjuvants can allow for reduced pesticide use by providing more uniform coverage.

pesticide adsorption. How well a pesticide adheres to soil particles.

pesticide persistence. A measure of how long a pesticide resists degradation by soil microbial and chemical reactions and stays in its original form in the soil.

pesticide resistance. The genetically acquired ability of an organism to survive a pesticide application at doses that once killed most individuals of the same species.

pest management unit (PMU). A contiguous area used in pest monitoring for quantitative sampling. A PMU usually contains one crop, with all the plants at the same stage of growth. If a large area is separated into smaller PMUs, scouting can be done more efficiently, and pesticide use will be limited to the PMUs that need treatment.

pheromone mating disruption. Mass release of an insect pheromone to confuse the target insect pest so that mating cannot occur.

phloem. Tissue transporting carbohydrates from the leaves (where carbohydrates are synthesized) throughout the plant.

photosynthesis. The process by which green leaves produce carbohydrates (their own energy source) by using carbon dioxide from the air and water, as well as energy from sunlight.

photosynthetically active radiation (PAR). The part of the light spectrum from 400 to 700 nanometers. This range includes the wavelengths that stimulate the various systems in plants that drive photosynthesis. (It is also the range of wavelengths that humans can see.)

phototropism. Growth away from or toward a light source.

phytophagous. Feeding on plants, including shrubs and trees. Used especially in reference to certain insects.

phytotoxicity. A toxic effect on plant growth. Plant injury may occur when chemicals are used to protect plants from pests, fertilize plants, regulate plant growth, and so on. For example, phytotoxicity can occur when a pesticide is applied to the plant during adverse environmental conditions.

plant growth regulator. Synthetic substance that causes hormonelike responses in plants.

Plant Protection and Quarantine (PPQ). A plant protection organization within USDA-APHIS that is responsible for inspecting ships, planes and their cargo, passengers, and luggage arriving from foreign countries. Also responsible for domestic programs to manage or eradicate infestations.

plug tray. A growing tray, usually 10 by 20 inches, containing numerous pockets (cells) of growing medium used for the production of young plants that will later be transplanted into larger containers. The number of cells may range from 36 to 800 per tray.

point source pollution. Water pollution from a single, discrete place such as a wastewater treatment plant or an industrial waste treatment facility. The legal definition from Section 502(14) of the Clean Water Act is "discernible, confined and discrete conveyance, including but not limited to any pipe, ditch, channel, tunnel, conduit, well, discrete fissure, container, rolling stock, concentrated animal feeding operation, or vessel or other floating craft, from which pollutants are or may be discharged. This term does not include agricultural stormwater discharges and return flows from irrigated agriculture."

polyacrylamides (PAM). Polymer-based materials used as a flocculating water treatment additive to aid in the removal of suspended particles from runoff. They are also used to facilitate erosion control and decrease soil sealing by binding soil particles (especially clays) to a site.

polyembryony. The formation of more than one embryo in a single seed.

polypropylene (PP). The most commonly used type of plastic in the nursery industry, including use in decorative containers, injection-molded flats and plug trays, and hanging baskets. It is generally durable, lightweight, and stiff but flexible. Many PP products are shiny and tear resistant. Spunbonded polypropylene (or woven polypropylene) is used to make row covers for frost protection and to reduce or diffuse light or to retain humidity in the nursery.

postemergence herbicide. An herbicide that acts on the plant after it has emerged from the soil.

pot-in-pot system. A specialized type of container production system involving burying a holder pot, or moat pot, in the ground and placing a containerized plant inside it to reduce problems associated with conventional container production, such as root-zone temperature extremes and top-heavy trees.

power test. A test that requires an in-depth response, unlike a repetitive skill test.

preemergence herbicide. An herbicide that acts on the plant during the germination stage, prior to emerging from the soil.

pressure compensating. Design characteristic of drip emitters and microsprinklers that can discharge at a constant rate across a range of operating pressures.

propagule. A propagation unit, such as a seed or cutting, used to reproduce a whole, intact plant.

protonymph. Second instar of a mite.

pseudopupa. A larva in a quiescent, pupalike condition, one or two instars before the true pupal stage.

pulse irrigation. The application of water in small, frequent amounts to improve infiltration.

pyrethroid. Any of several synthetic compounds similar to pyrethrin, used as an insecticide.

QC 650. A type of master permit that authorizes compliance agreements to be written to allow shipment of nursery stock from Hawaii to California.

quarantine. Restraints because of health and safety concerns; refers either to the (1) prohibition of movement from a particularly defined region; (2) period of time during which plants or animals or products are held, observed, or tested prior to treatment, destruction, return, or release; or (3) law or regulation prohibiting entrance or allowing entrance contingent upon meeting requirements. Federal quarantines are either foreign, territorial, or domestic; state quarantines are exterior or interior. The Nursery Stock, Plant and Seed Quarantine 37 (7 CFR 319.37) are the U.S. quarantine regulations pertaining to entrance of nursery stock and other plant propagative materials from foreign countries.

quick ratio (QR). This is a calculation not found on the income statement or balance sheet. It shows the business's ability to pay bills in cash and accounts receivable. The equation for calculating this is cash + AR ÷ current liabilities. The current ratio includes inventory, so that is how the two differ. The QR should be greater than $0.75. When it gets below $0.75, the business doesn't have nearly enough cash on hand to pay current bills. It is a good measure of liquidity, or how much cash a company has.

rapid sand filtration (RSF). A physical filtration process that separates particulates suspended in water. Rapid sand filters do not remove pathogens, dissolved pollutants, or very fine particles.

reduced-risk pesticide. A pesticide that poses less risk to the environment and human health. Reduced-risk pesticides usually are targeted to specific pests and are IPM compatible.

respiration. The process of using the energy from carbohydrates that were produced by photosynthesis for growing or other plant processes.

restricted-entry interval (REI). The required number of hours or days after a pesticide application during which entry into treated areas is prohibited.

restricted-use pesticides. Restricted materials are pesticides deemed to have a higher potential than other pesticides to cause harm to public health, farmworkers, domestic animals, honey bees, the environment, wildlife, or other crops. With certain exceptions, restricted materials may be purchased and used only by or under the supervision of a certified commercial or private applicator. California requires permits for restricted materials so that the local county agricultural commissioner can assess, in advance, the potential effects of the proposed application on health and the environment.

retractable-roof greenhouse. A greenhouse structure with a glazing material that is mounted in a way that allows the roof covering to be pulled aside to expose plants to ambient outside conditions.

return on assets (ROA). This is a key financial ratio, not found on either the income statement or balance sheet but instead on the strategic profitability model. It means the return, or how much a company earned, for every dollar invested in assets. An ROA of 4.2 percent means that the business earns or returns $0.042 on every dollar invested in assets. It is calculated by multiplying net profit margin by asset turnover.

return on net worth (RONW). This is a key financial ratio, not found on either the income statement or balance sheet but on the strategic profitability model. It means the return, or how much a company earned, for every dollar it invested in owner's equity, or net worth. It is calculated by multiplying net profit margin by asset turnover by leverage or by multiplying return on assets (ROA) by leverage.

return on owner's equity (ROE). This is the same thing as RONW, but you will see it written both ways. The calculations and meaning are identical to RONW.

reverse osmosis (RO). Also known as hyperfiltration, this technology is used to remove a large majority of contaminants from recycled water by pushing the water under pressure through a semipermeable membrane. The pores are so small that the filters remove dissolved salts and most pathogens. Reverse osmosis is also used in nurseries that have water sources with high concentrations of dissolved salts.

rhizome. An underground horizontal stem that puts out lateral shoots and adventitious roots at intervals.

runoff. Water that moves off-site as a result of being applied in an amount that is in excess of soil uptake.

schmutzdecke. A layer of trapped organic matter at the surface of a slow sand filter in which a diverse and dynamic community of microorganisms such as algae, bacteria, diatoms, and zooplankton develops. These microorganisms within the biofilm are effective in removing pathogens and other pollutants.

scout. The person who monitors for pests, recording and summarizing information used in making pest management decisions. The scout may or may not be responsible for making pest control decisions. If the scout is not the decision maker, the scout presents summarized data for the decision-making process.

screenhouse. A plant growing structure made with insect screens to exclude various pests.

secondary pest. A pest that is normally not a problem until pesticide applications, applied for another pest or pests, kill its natural enemies.

Section 303(d). A section of the Clean Water Act that requires states to periodically identify waters that do not or are not expected to meet applicable water quality standards. These waters are identified on the Section 303(d) Impaired Waters List. A total maximum daily load (TMDL) must be developed for each water body on this list. If a listed water body has multiple impairments, a TMDL must be developed for each impairment.

sediment basin. An engineered structure that temporarily detains sediment-laden runoff under quiescent conditions, allowing sediment to settle out before the runoff is discharged.

seed. The result of fertilization in sexual propagation.

seeding machine. A device that distributes seed into a tray of soil for germination in order to produce seedlings. Variants use needles, drums, and optical or other methods to place a single seed into each cell of the tray.

select agent list. Developed by USDA-APHIS in response to the Public Health Security and Bioterrorism Preparedness and Response Act of 2002. Lists the plant pathogens considered as high risk for deliberate use against U.S. crops.

selective pesticide. A pesticide that kills only certain groups or species of plants or animals. For example, a selective herbicide is one that has an effect on one or more groups of plants but does not affect other groups of plants, such as the nontarget crop plants.

self-contained breathing apparatus (SCBA). A portable device that provides air for breathing when ambient air is limited or contaminated. For divers, this is commonly called SCUBA, where the "U" is for "underwater." A SCBA may be a necessary part of the pesticide safety equipment used in nurseries.

senescence. Aging of plant tissue such as leaves or flowers.

sessile. Attached or fastened, incapable of moving from place to place.

sexual propagation. Reproduction of plants involving the exchange of genetic material from a male and a female parent.

shadehouse. A plant growing structure designed to impose a significant, constant amount of shade to the plants grown inside.

shading compound. A paintlike substance used on transparent greenhouse glazing materials to reduce light entering the greenhouse during times when the ambient light level is too high and temperature control (cooling) is difficult.

sign. The visible presence of a pathogen or causal agent.

signal words. Words that are used to describe the acute (short-term) toxicity of the formulated pesticide product. The signal words are CAUTION, WARNING, and DANGER. The U.S. Environmental Protection Agency (EPA) requires a signal word on most pesticide product labels. The EPA also requires it to be printed on the front panel, in capital letters, to make it easy for users to find.

site production potential. The number of plants of a certain size that can be produced per unit of land area in a nursery.

skate-wheel conveyor. A type of wheel conveyor making use of a series of skate-wheels mounted on common shafts or axles or mounted on parallel-spaced bars on individual axles.

slow sand filtration. A biological treatment method that has been in use for centuries to produce drinking-quality water. The filter consists of a bed of sand through which the contaminated water slowly passes. A biofilm develops on the surface of the sand grains and is responsible for the removal of pathogens and other pollutants that may be in the water (see schmutzdecke).

soil adsorption potential. The binding of chemicals to particles of organic matter and clay in the soil. Adsorption retains pesticides in the soil, where they can be degraded.

solarization. A nonchemical method for controlling soilborne pests using high temperatures produced by capturing radiant energy from the sun.

sole proprietorship. The simplest and most common structure chosen to start a business. It is an unincorporated business owned and run by one individual, with no distinction between the business and the owner, who is entitled to all profits and is responsible for all the business's debts, losses, and liabilities.

somatic. Nonreproductive, vegetative tissue, such as leaves, stems, and roots, arising through mitosis.

speed test. A repetitive skill test that measures how many times an applicant can complete a task within a time frame, such as number of plants grafted per hour.

spot-baiting. In vertebrate pest control, a method of bait application that involves the scattering of bait around the target area, which is usually a burrow entrance or animal trail.

spot spraying. Focusing pesticide applications in areas where pests have been detected rather than spraying the entire crop. Spot spraying can significantly reduce overall pesticide use.

step controller. System that divides the action of the greenhouse heating and cooling equipment into steps, or stages, ranging from the opening of vents to more intensive control methods such as the use of fans and heaters. Within a particular greenhouse, as the control at one level becomes ineffective, more intensive control methods are implemented.

stippling. Small, light green or chlorotic specks; also called flecking.

stock-keeping unit (SKU). A unique identifier for each product sold. The SKU is the billable entity, which may be an individual item such as a plant or a bundle of items such as a tray of plants.

supplemental lighting. Light provided to plants in protected cultivation that supplements sunlight. It can be provided during the day at times of inclement weather or during the night to extend day length.

SWOT analysis. A structured planning method used to evaluate the strengths, weaknesses, opportunities, and threats involved in a business venture.

symphylans. An order of small wingless insects possessing an elongated body, three pairs of thoracic legs, and nine pairs of abdominal legs.

symptom. The visual manifestation of a disease or abiotic disorder.

systemic pesticide. A pesticide that is absorbed by plants and moves to untreated tissues. Systemic herbicides move through the plant's xylem or phloem from the point of application and affect plant growth and development away from the site of herbicide application. Systemic insecticides and fungicides move throughout treated plants and kill certain insects or fungi.

systems approach. As defined in the federal Plant Protection Act in 2000, "a defined set of phytosanitary procedures, at least two of which have an independent effect on mitigating pest risk associated with movement of commodities."

tank mix. Combining different pesticides to broaden the spectrum of activity of the application.

taper. The gradual decrease of stem circumference, starting from the base and moving up a tree stem.

tensiometer. An instrument used to measure soil moisture in terms of the tension of water held in the soil.

terracing. Level earthen steps that follow the contour of a hillside, breaking a long slope into a series of shorter segments and intercepting the flow of water. Terracing can be an effective treatment to control erosion on steep slopes.

test simulation. A test that uses a simulation, not a real task, to measure applicant skills.

tissue culture. The culture of plant tissues, often directed toward the micropropagation of plants. Media used in tissue culture usually contain mineral nutrients, sugars, vitamins, and plant growth regulators.

total maximum daily load (TMDL). The maximum amount of a pollutant that a water body can tolerate on a daily basis while meeting established water quality standards. TMDLs can be expressed in terms of mass per time, toxicity, or other appropriate measures that relate to a state's water quality standard.

total porosity. The percentage of a given volume of a potting mix that pores occupy (i.e., the space not occupied by solids).

totipotency. A demonstrated capability in plants where a single, isolated cell can develop into a whole plant.

toxicant. Poison or poisonous agent.

transpiration. The loss of water through very small pores (stomata) that open in the leaves, which occurs when plants perform photosynthesis.

trolley. In nurseries, a cart that is suspended from an overhead rail system for carrying materials.

ultraviolet light. UV light has a wavelength between 100 and 400 nanometers. UV light reduces the effective life of greenhouse plastics. Many insects are attracted to the wavelengths from 350 to 370 nanometers.

USDA plant hardiness map. A map of zones within the United States separated into average minimum winter temperatures in 10-degree increments.

UV stabilizer. Material used in plastics to reduce the effect that ultraviolet radiation has on degradation in plastics, thus prolonging the life of plastic glazing materials.

validity correlation coefficient. Statistical measure that is used to determine how well a test can predict on-the-job behavior.

vector. In biology, an agent capable of transmitting the causative agent of disease (e.g., Asian citrus psyllid and glassy-winged sharpshooter).

vegetated bioswale. A conveyance system that channels stormwater or irrigation runoff water and improves water quality by reducing flow velocity, increasing sedimentation, promoting soil infiltration, and filtering pollutants.

vegetated buffer. Area or strip of land maintained in permanent vegetation to prevent erosion and improve water quality by trapping and treating contaminants. Vegetated buffers help reduce runoff and encourage the settling of soil particles. In addition, plant roots take up water and dissolved pollutants from runoff water.

vegetative propagation. Method of reproducing plants utilizing vegetative tissues or organs such as stems, leaves, or roots.

vertebrate. An organism that possesses a backbone or spinal column.

waste discharge requirements (WDRs). Rules adopted by a California regional water board for a specific waste discharge (or type of discharge) to waters of the state. WDRs include limitations on the chemical or physical quality of waste discharges or set other conditions on waste discharges to ensure that discharges of wastes do not unreasonably affect surface and groundwater quality.

water body. A geographically defined portion of navigable waters, waters of the contiguous zone, and ocean waters under the jurisdiction of the United States, including segments of rivers, streams, lakes, wetlands, coastal waters, and ocean waters.

water-break (water bar). A dip built at an oblique angle across the road with a berm at the end to reduce flow velocity. Water-breaks are recommended on steep grades where runoff and erosion is anticipated.

water-holding capacity. The percentage of a given volume of a potting mix that water occupies after irrigation and drainage.

water impoundment. An engineered structure such as a retention basin, infiltration basin, collection basin, pond, or reservoir constructed to collect large runoff flow and retain (store) runoff water.

water quality. The biological, chemical, and physical conditions of a water body. It is a measure of a water body's ability to support beneficial uses.

weed spectrum. The common types of weeds found in a cropping system.

weed spectrum shift (weed shift). A change from the common weeds found in a cropping system to other weed species, generally due to repeated applications of the same herbicide (or other management practices), resulting over time in the number of susceptible weed species decreasing and tolerant weed species increasing.

windbreak. A physical barrier to reduce wind speed.

xylem. Tissue transporting water from the roots up into the plant.

Measurement Conversion Table

U.S. customary	Conversion factor for U.S. customary to metric	Conversion factor for metric to U.S. customary	Metric
inch (in)	2.54	0.394	centimeter (cm)
foot (ft)	0.3048	3.28	meter (m)
acre (ac)	0.4047	2.47	hectare (ha)
square inch (in2)	6.45	0.15	square centimeter (cm^2)
square foot (ft2)	0.0929	10.764	square meter (^{2})
ounce (oz)	28.35	0.035	gram (g)
quart, liquid (qt)	0.946	1.056	liter (l)
quart, dry (qt)	1.1	0.91	liter (l)
gallon (gal)	3.785	0.26	liter (l)
gallon per acre	9.36	0.106	liter per hectare (l/ha)
acre-inch (ac-in)	102.8	0.0097	cubic meter (m^3)
acre-foot (ac-ft)	1,233	0.000811	cubic meter (m^3)
cubic foot	28.317	0.353	liter (l)
cubic foot per minute per pound (cfm/lb)	0.0010405	961.11	cubic meter per second per kilogram ($m^3 \cdot s^{-1} \cdot kg^{-1}$)
pound (lb)	0.454	2.205	kilogram (kg)
pound per acre (lb/ac)	1.12	0.89	kilogram per hectare (kg/ha)
pounds per square inch (psi)	6.89	0.145	kilopascal (kPa)
BTU	1.055	0.947	kilojule (kJ)
Fahrenheit (°F)	°C = (°F – 32) ÷ 1.8	°F = (°C × 1.8) + 32	Celsius (°C)

Index

Tables are indicated with *t* and figures (illustrations and photographs) are indicated with *f*.

Q

R

S